A Systems Perspective on Financial Systems

T0303854

Communications in Cybernetics, Systems Science and Engineering

ISSN: 2164-9693

Book Series Editor:

Jeffrey Yi-Lin Forrest

International Institute for General Systems Studies, Grove City, USA
Slippery Rock University, Slippery Rock, USA

Volume 6

A Systems Perspective on Financial Systems

Jeffrey Yi-Lin Forrest

*College of Economics and Management, Nanjing University of
Aeronautics and Astronautics, Nanjing, China
Department of Mathematics, Slippery Rock University,
Slippery Rock, PA, USA*

CRC Press
Taylor & Francis Group
Boca Raton London New York

CRC Press is an imprint of the
Taylor & Francis Group, an **informa** business

A BALKEMA BOOK

CRC Press
Taylor & Francis Group
6000 Broken Sound Parkway NW, Suite 300
Boca Raton, FL 33487-2742

First issued in paperback 2019

© 2014 by Taylor & Francis Group, LLC
CRC Press is an imprint of Taylor & Francis Group, an Informa business

No claim to original U.S. Government works

ISBN-13: 978-1-138-02628-5 (hbk)
ISBN-13: 978-0-367-37879-0 (pbk)

Typeset by MPS Limited, Chennai, India

Library of Congress Cataloging-in-Publication Data

Lin, Yi, 1959–
    A systems perspective on financial systems / Jeffrey Yi-Lin Forrest.
        pages cm. — (Communications in cybernetics, systems science and engineering, ISSN 2164-9693; volume 6)
    Includes bibliographical references and index.
    ISBN 978-1-138-02628-5 (hardback : alk. paper)   1. Financial institutions.   2. Money.
3. System theory.   I. Title.
    HG173.L56   2014
    332.01'1—dc23
                                                                            2014000239

**Visit the Taylor & Francis Web site at
http://www.taylorandfrancis.com**

**and the CRC Press Web site at
http://www.crcpress.com**

Table of contents

PART 2
Domestic financial system: Seen as a closed system

Editorial board

Preface

Currency, banking, and financial markets have been an exciting area of research in the entire spectrum of economics. They not only interweave with the daily lives of the ordinary people from all walks of life, but also involve major flows of capital in the economy. These movements of funds affect the level of profits of individual enterprises, the production of goods and services, as well as the economic wellbeing of other countries. In other words, when each economy is seen as a system, the study of currency, banking and financial markets deals with internal structures of individual economic systems and their interactions, where money plays the role of "fluids" that flow across the boundaries of economies and facilitate the interaction of different economies with financial institutions and markets being the hubs through which these "fluids" flow into and out of each and every economic system.

Along with the rapid development of the financial markets and the introduction of new financial instruments with each passing day, the once well-organized and conservative banking industry now conducts its business very aggressively in order to keep pace with the increasing speed of flow of money from one business adventure to another, one geographic location to another, and from one nation to another. Because of the development of international trades and international financial markets, an integrated environment for a world economy has been created so that events that take place in the financial market of one nation could produce and have produced magnificent impacts on the operation of the financial markets of other nations. Speaking differently, as the movement speed of money increases, as more and more barricades that affect the smooth transactions of money are removed, once distant economic systems are now interacting with each other in an unprecedented fashion. Therefore, implementations of monetary policies of various scales have become the core of debate on monetary policies. The constant theoretical refinement and new understandings of money have forever sharpened our comprehension on the role of money in economy.

Through exploring the role of money in the economy, how financial institutions and financial markets function, and how international monetary capital is turned over and circulates by employing some of the most recent scientific breakthroughs in systems research, this book addresses many issues of concern related to money and finance, while reflecting some of the most exciting progresses in the recent investigations of money and banking. The purpose of this book is to establish a new unified analytical framework and logic of economic thinking, on the basis of systems thinking in general and the systemic yoyo model in particular, to study money, banking, and international financial markets. This provocative logic of thinking and new analytic framework will lead the reader's learning by using the basic economic principles while providing the help for the reader to comprehend the regional and international financial markets, management of banks, and the roles of money in the economy.

Based on the logical order and structural arrangement from macro- to micro-scales, and from the financial operations of money and banking of one nation to those of the international landscape, this book presents a rigorous yet complete thread of logical reasoning and theoretical system of the specific subject matter. This book adopts the method of model analysis, established on the intuition of the systemic yoyo structure, to carefully depict those stable and invariant economic variables, to clearly derive, and to rigorously analyze the models through step-by-step procedures. After establishing all these careful underlying theoretical foundations, various economic phenomena are explained on top of these established models. Such detailed, step-by-step model approach surely makes the learning of this relatively challenging subject matter easier and more systematic for the reader.

This book concludes with an empirical analysis on the development of China's monetary and financial industry. Although since the time it began its economic reform and opened itself up to the rest of the world over thirty years ago in the early 1980s, China's capital market has developed considerably, the development of its capital markets still lags behind severely and has been seriously hampering the improvement in the efficiency of macroeconomic monitoring and regulation. With China's entrance into the WTO, will Chinese financial sector, which will soon be fully liberated with foreign capital, have enough strength and hands-on experience to stand against "long-range precision strikes" of financial derivatives and other financial means? This concern makes people worry about and ponder over the self-defense capability of the Chinese financial system, and in turn about the economic future of China. Even if for the time being one puts aside all issues regarding the exchange rate of RMB and the foreign exchange reserve of over three trillion dollars, one still has to concern about China's situation and position in the game played along with the political hot money, which flows across national borders and out of the sight of normal financial operations. How to speed up the development of China's capital and financial markets has been an important theoretical and practical problem that has concerned with the efficiency of China's monetary policies. This book, considering its unconventional theoretical foundation, is expected to provide a scientifically sound methodology that can be potentially employed to resolve these and other relevant problems.

It is our hope that you, the reader, will benefit from reading this book and referencing this book time and again in your professional endeavors. If you have any comments or suggestions, I can be reached at either Jeffrey.forrest@sru.edu or Jeffrey.forrest@iigss.net.

Jeffrey Yi-Lin Forrest

About the author

Dr. Yi Lin, also known as Jeffrey Yi-Lin Forrest, holds all his educational degrees (BS, MS, and PhD) in pure mathematics respectively from Northwestern University (China) and Auburn University (USA) and had one year of postdoctoral experience in statistics at Carnegie Mellon University (USA). Currently, he is a guest or specially appointed professor in economics, finance, systems science, and mathematics at several major universities in China, including Huazhong University of Science and Technology, National University of Defense Technology, Nanjing University of Aeronautics and Astronautics, and a tenured professor of mathematics at the Pennsylvania State System of Higher Education (Slippery Rock campus). Since 1993, he has been serving as the president of the International Institute for General Systems Studies, Inc. Along with various professional endeavors he organized, Dr. Lin has had the honor to mobilize scholars from over 80 countries representing more than 50 different scientific disciplines.

Over the years, Professor Lin has served on the editorial boards of 11 professional journals, including Kybernetes: The International Journal of Systems, Cybernetics and Management Science, Journal of Systems Science and Complexity, International Journal of General Systems, and Advances in Systems Science and Applications. And, he is the editor-in-chief of two book series, one of which is entitled "Systems Evaluation, Prediction and Decision-Making," published by CRC Press (New York, U.S.A.), an imprint of Taylor and Francis, since 2008, and the other one entitled "Communications in Cybernetics, Systems Science and Engineering," published by CRC Press (Balkema, The Netherlands), an imprint of Taylor and Francis, since 2011.

Some of Dr. Lin's research was funded by the United Nations, the State of Pennsylvania, the National Science Foundation of China, and the German National Research Center for Information Architecture and Software Technology.

Professor Yi Lin's professional career started in 1984 when his first paper was published. His research interests are mainly in the area of systems research and applications in a wide-ranging number of disciplines of the traditional science, such as mathematical modeling, foundations of mathematics, data analysis, theory and methods of predictions of disastrous natural events, economics and finance, management science, philosophy of science, etc. As of the end of 2013, he had published nearly 300 research papers and about 40 monographs and edited special topic volumes by such prestigious publishers as Springer, Wiley, World Scientific, Kluwer Academic (now part of Springer), Academic Press (now part of Springer), and others. Throughout his career, Dr. Yi Lin's scientific achievements have been recognized by various professional organizations and academic publishers. In 2001, he was inducted into the honorary fellowship of the World Organization of Systems and Cybernetics.

Acknowledgements

This book contains many research results previously published in various sources, and I am grateful to the copyright owners for permitting me to use the material. They include the International Association for Cybernetics (Namur, Belgium), Gordon and Breach Science Publishers (Yverdon, Switzerland, and New York), Hemisphere (New York), International Federation for Systems Research (Vienna, Austria), International Institute for General Systems Studies, Inc. (Grove City, Pennsylvania), Kluwer Academic and Plenum Publishers (Dordrecht, Netherlands, and New York), MCB University Press (Bradford, UK), Pergamon Journals, Ltd. (Oxford), Springer, Taylor and Francis, Ltd. (London), World Scientific Press (Singapore and New Jersey), and Wroclaw Technical University Press (Wroclaw, Poland).

I like to use this opportunity to express my sincere appreciation to many individuals who have helped to shape my life, career, and profession. Because there are so many of these wonderful people from all over the world, I will just mention a few. Even though Dr. Ben Fitzpatrick, my PhD degree supervisor, has left this material world, he will forever live in my professional works. His teaching and academic influence will continue to guide me for the rest of my professional life. My heartfelt thanks go to Shutang Wang, my MS degree supervisor. Because of him, I always feel obligated to push myself further and work harder to climb high up the mountain of knowledge and to swim far into the ocean of learning. To George Klir – from him I acquired my initial sense of academic inspiration and found the direction in my career. To Mihajlo D. Mesarovic and Yasuhiko Takaraha – from them I was affirmed my chosen endeavor in my academic career. To Shoucheng OuYang and colleagues in their joint research group, named Blown-Up Studies, based on their joint works, Yong Wu and I came up with the systemic yoyo model, which eventually led to completion of the earlier books *Systemic Yoyos: Some Impacts of the Second Dimension* (published by Auerbach Publications, an imprint of Taylor and Francis in 2008), and *Systemic Structure behind Human Organizations: From Civilizations to Individuals*, (jointly with Bailey Forrest, published by CRC Press, an imprint of Taylor and Francis in 2011). To Zhenqiu Ren – with him I established the law of conservation of informational infrastructure. To Gary Becker, a Nobel laureate in economics – his rotten kid theorem has initially brought me deeply into economics, finance, and corporate governance.

Part 1

Preparation

Chapter 1

Overview

As the introduction, this chapter looks at the basics of money, such as the concept, functions, and classifications of money, why a systemic approach is important in terms of knowledge exploration, and what kinds of problems systems science could address. At the end, the organization and main results of this book are outlined.

1.1 THE BASICS OF MONEY

To prepare the reader with the answer to the question of why we need another book on money and financial institutions, in this section we look at the theoretical meanings, practical functions, and classifications of money. So, after learning relevant aspects of systems science in the following sections, we will establish the need for a different approach to the investigation of money and its movement.

1.1.1 The meaning and function of money

Money, for economists, has only one specific meaning. It stands for and can be anything that is commonly accepted to exchange for goods and services or as payments of debts. Currency (that is, the bank notes and coins) obviously satisfies the definition of money given by economists. So, currency is a kind of money. However, if money is merely defined as currency, then for economists the meaning is too narrow. Because checks are also accepted as payments for purchases, checking account deposits are also seen as money. As for traveler's checks, savings deposits, and others alike, if they can be quickly and conveniently converted to currency or checking deposits, they can also play the role of money and of paying bills. Hence, there is a need to have a more general definition of money.

Economists distinguish various forms of money, which can be used to make purchases, such as currency, checking account deposits, etc., from wealth, which represents the totality of different kinds of assets that can be kept due to their values. Wealth includes not only money, but also bonds, stocks, artworks, lands, furniture, cars, houses, etc. Additionally, money and income are two different concepts. In fact, income stands for the inflow of earnings of a unit time, while the word "money" is a concept regarding accumulation. That is, money stands for a definite sum at a certain particular time moment.

In each economic society, money, no matter whether it consists of gold, silver, shells, stones or pieces of paper, has three functions: media of trade, standard of value, and store of value. Among these three functions, both media of trade and standard of value are more fundamental, and represent the basic conditions for an object to become money, whereas the function as a store of value is merely an extension of the other two functions. Hence, the function of store of value is secondary. It is the function that money can play the role of media of trades that money is distinguished from other assets, such as stocks, bonds, houses, etc.

In all market transactions of the modern economic society, money is indispensable. By using money as the media of trades, large amounts of time, otherwise necessary for the exchange of goods and services, are saved so that economic efficiency is greatly improved. For a good to effectively play the role of money, it has to satisfy certain requirements:

1   It can be easily standardized so that people can conveniently recognize its value;
2   It has to be widely accepted;
3   It can be divided so that changes can be made easily;
4   It is convenient to array; and
5   Its quality will not deteriorate quickly.

The second function of money is its standard of value. In other words, in each economic society people use money to calculate various goods and labor. By relying on this function, the number of values that have to be considered is greatly reduced so that the cost of trades is lowered. When the complexity of the economy increases day by day, the benefit of the function of money as a standard of value becomes forever more obvious. Additionally, it is because money is employed as a unified measure of value that accounting becomes possible. It is exactly because money is used as the unified standard of value that all kinds of incomes and payments can be compared, leading to the development of the concept of accounts in modern society. When the practice of accounting is expanded to a nation, there then appear various national accounts. That is to say, only with money as the standard of value, can various macroeconomic indicators be introduced so that the consequent statistics of these indicators can be collected and computed. That leads to the development of various systems of macroeconomic indicators. On the basis of these systems of macroeconomic indicators, economists are able to investigate the macroeconomic situations of a region or a nation.

Money also plays the role of storing value. That is an extension of the function that money is the media of trades. When money is used as the media of trades, the original direct "object-to-object" exchanges of goods and services become the indirect "object-to-money-object" exchanges. This function is very useful, because most people do not like to instantly spend all the money at the moment when they receive their incomes. Instead, they wait until their needs arise and then they decide on how to spend their money. Using money to store value is not the only method of preserving value and is not necessarily the best method for such purposes, either. Each asset, no matter whether it is money, stocks, lands, houses, artworks, or jewelry, can be employed as a way to store value; and for the purpose of preserving value, many of these kinds of assets are actually better than money. They can often provide better interest benefits than money can. They can either increase in value or provide services, such as housing, etc. Then a natural question arises: If these different kinds of assets can do better than money to preserve value, then why would people still hold money?

To address this question, we have to involve an important economic concept: liquidity. This concept stands for how easy and how fast an asset can be converted to a media of trades. The liquidity of an asset is very necessary for people's daily needs; the more liquid an asset is, the more people like to have the asset, because if it can be quickly and easily converted to a media of trades then people can use it to purchase goods or services to satisfy their immediate needs. Money has the highest liquidity among all kinds of assets, because it represents the media of trades and does not need to be converted to anything else before it is used to make purchases. When any other kind of asset is converted to the media of trades, frictions appear, leading to transaction costs.

The pros and cons of using money to preserve value depend on the price level of goods and services. It is because the value of money is determined by the price level of goods and services. For example, if the prices of all goods and services are doubled, it then means that the value of money has been reduced by 50%. Conversely, if the prices of all goods and services are discounted by 50%, it then means that the value of money has been doubled. During times of inflation, prices of goods and services rise quickly so that money also loses its value quickly. In such times, if people use money to preserve value, they will not be able to accomplish what they want. The situations are much worse during times of hyperinflation, which are when the monthly rate of inflation is higher than 50%.

1.1.2 Different kinds of money

The payment system is the method of trades in economic societies. For centuries, the payment system has been constantly evolving. Accordingly, the form of money has also been changing. For an article to play the role of money, it has to be willingly acceptable by all people as evidence of payments for goods and services. So, anything that is obviously valuable to each person can hopefully become money. Hence, people naturally selected gold, silver, and other kinds of precise metals or articles of high values as money. These kinds of money are known as commodity money.

The next stage of development of the payment system is the introduction of paper notes. The so-called paper notes are pieces of paper that have the functionality of trade media. At the beginning, the introduction of paper notes contained such promise: They could be exchanged for coins or precise metals. Later on, the paper notes in most countries evolved into the fiat money, which the governments authorized as the legal money, that is, people had to accept it as means of payment for debts or goods, without the need to exchange into coins or precise metals. The advantage of paper notes is that it is convenient to carry; it is much lighter than coins and precise metals. However, only when the authorities who issue the paper notes possess certain level of trust, and the printing technology is so advanced that counterfeiting is made extremely difficult, can paper notes be accepted as the media of trades. Because the issue of paper notes is a legal arrangement, the amount of paper notes to be issued can be adjusted according to the national demand.

The main weakness of paper notes and coins is that they can be easily stolen. And when the amount of transaction is large, it becomes expensive to transport the huge volume of the money. To overcome this problem, along with the development of modern banks, the payment system evolved further by introducing checks. A check represents a debenture that is accepted as money, and helps to complete trades without the need to carry a large sum of money. The introduction of checks represents a major innovation that makes the payment system much more efficient. Payments often go back and forth, cancelling each other. If checks were not employed, payments would cause a movement of large amounts of currency. With checks in use, mutually cancelling payments can be cleared by cancelling the relevant checks without the need to transport the necessary currency. That is how the usage of checks has reduced the transaction cost of the payment system, while enhancing the economic efficiency. Another advantage of checks is that they can be written in any amount against the available funds in the bank account so that major business deals are made convenient. In the area of reducing theft losses, checks also have their advantage. Additionally, checks can be used as convenient receipts for purchases.

However, the check-based payment system also suffers from two problems. Firstly, it takes time to deliver checks. If the payee has to have the money quickly, then check payments cannot satisfy his needs. Additionally, it takes several bank-business days for the money to be deposited into the payee's account after he deposited the check. Hence, if the depositor is desperately in need of cash, the form of check payments will make him very frustrated. Secondly, there are a lot of expenses involved in dealing with checks. For example, the United States spent over $5 billion in the clearing of issued checks.

The development of modern communication and computer technology has brought forward a new development in the payment system. People have been conducting business transactions by making use of various advanced methods. For example, payments are done through electronic mails, which help to eliminate all the relevant paper work of the past. That is the so-called electronic funds transfer system (EFTS), a new development in the direction of an electronic payment system.

Although the EFTS system is more effective than the paper-based payment system, many specific circumstances still warrant the existence of a paper-note payment system, which makes the development toward a checkless society much slower than expected earlier. Paper notes have their advantage: They provide receipts and traces, making stealing more difficult. On the other hand, the electronic payment system does not have such an advantage. Another problem with the electronic payment system is that there are currently still many unsolved legal issues involved. For example, when a person gets hold of the password of a client and illegally transfers all the money from the client's account, who is legally responsible for the loss of money? Can the client stop the payment on the electronic payment system just as he could when using checks?

In terms of its form, money can be roughly divided into several categories as physical money, metal money, paper money (or notes), and deposit money. In particular, by physical money is meant those particular articles that were once used as the media of trades before metal money appeared. For example, such articles as rice, cloth, lumber, shells, domestic animals, horns, hunting tools, etc., had played the role of trade media during different times in history. When these articles were used as money, they were basically maintained in their original, natural form. The main weaknesses of these articles are that their volumes are large and that qualities vary; they cannot be subdivided into small units. Due to their relatively small values per unit, large sums are difficult to carry, while their qualities easily deteriorate over time.

The so-called metal money stands for such money that is made up of metal materials. The initial kinds of metal money looked like sticks, and were known as weighing money. More recent metal money was molded into different shapes to distinguish different colors and weights, which has been known as mint. Gold, silver, copper, iron, and other kinds of metals were once used as the physical materials of money. These materials had steady supplies, could be cut and processed to serve the purpose of money with uniform quality, standardized value, lasting durability, and convenience for carrying. Because this kind of money can surely play the roles of trade media, standard of value, and preservation of value, as of the present day, metal money is still in circulation.

Paper notes (or money), as the term suggests, are made of paper and shaped differently with clearly labeled face values. There are two kinds of paper notes: convertible and inconvertible notes. Here, convertible notes can be used at any time to exchange with the issuing bank or the government for minted money or gold/silver bullions. Their usefulness is completely the same as that of metal money, whereas they can be easily carried while avoiding the possibility of being damaged. On the other hand, inconvertible notes cannot be exchanged into metal money or bullions. They possess only money value without any implied material value. Currently, almost all paper notes that are in circulation in different countries are inconvertible.

By deposit money, it is meant to be demand deposits. Against a demand deposit account, one can write a check at any time to pay for goods or service. That is, the check is used directly as a media of trade and a tool of payment. Deposit money occupies an important position in all developed commercial and industrial countries. Most of the business dealings are conducted by using this kind of money as the media of trade.

In terms of the relationship between face values and the values of the money materials, money can be classified into three groups: commodity money, representative money, and credit money. In particular, commodity money is also known as physical money or real money. It is the money that equates commodity values with money values. Hence, the face value of this kind of money is equal to the value of the money material, which is made up of such high valued materials as precise metals (as gold, silver, etc.). Early commodity money was made of cows, sheep, cloth, and other commodities; later, the commodity money consisted of gold coins, silver coins, etc.

Representative money is such a kind of money that represents the real money to circulate in the market place. It generally is made of paper, issued by the government or a bank, representing the metal money. Although paper notes are circulating in the market place, as the media of trade, they are backed by sufficient gold and silver money or equivalent bullions. The holder of paper notes has the right to exchange his notes with the government or bank for gold/silver money or bullions. Therefore, even though representative money's own value is less than its face value, the public holding representative money is equivalent to having the right to demand for the real money. The advantage of representative money includes the low cost of production, convenience to carry, and practical saving on gold and silver so that these precise metals can be used for other purposes. A further development in form of representative money leads to the appearance of credit money. That is a form of money currently all countries are using.

The appearance of credit money has its historical roots, and is a direct consequence of the collapse of the metal currency system. The worldwide financial crisis of the 1930s forced each and every major country to get out of the gold and silver standards so that the currencies issued by these countries could no longer be converted into metal money. As a consequence, credit money (currency) appears. Not only is its value of credit money lower than its money value, but also is it different from representative money. It does not represent any precise metal. Credit money consists of three main forms: subsidiary coins, paper currency, and bank deposits, where the main function of subsidiary coins is to play the role of media for micro- and sporadic trades. These coins are generally minted with cheap metals, such as copper, nickel, aluminum, and others, where the government has the exclusive right to mint these coins. The main function of paper currency is to serve as the tool for people to purchase their daily necessities. Again, the government owns the exclusive distribution right of paper currency, while the issuing authority differs from one country to another. For many countries, it is their central bank, or department of treasury, or some specifically established money authority that issues the paper currency.

By bank deposits, it means the various forms of bank deposits that can be used as means of trade media. They mainly include bank demand deposits. The reason why such deposits can be used as a means of trade media is because they have the following advantages: (i) they can avoid the risk of loss and damage; (ii) they can be transported conveniently with very low cost; (iii) they can be used to pay any specific amount without the need to make changes; and (iv) after the payee receives the payment, the check can be circulated within a certain range.

According to the classification of money based on the value relationship and that based on forms of money, it can be seen that credit money contains both commodity and metal money, representative money is convertible. That is, credit money includes inconvertible money, deposit money, and metal subsidiary coins.

Based on the land territory within which currencies play their roles, currencies can be seen as domestic currency and international currency. The currency that can only be used within a country is a domestic currency; those that can be used beyond the border of one country are known as international currencies. In international exchanges and international payments, gold is a commonly accepted media of trade and method of payment. So, gold is an international currency. The currencies of certain countries, such as U.S. dollar, British pound, Japanese yen, and others, can be exchanged into the currencies of other countries without any restriction. So, they can be used within many different countries, making them international currencies. There are some other forms of international payments, such as special drawing right (SDR), euro, etc. They are also considered international currencies.

Whether or not the currency of a particular country can become an international currency, the key is whether or not it can be converted. That is, it is determined by whether or not the specific currency can be exchanged into that of another country without any restriction. Only the national currency that can be freely exchanged can become international. International currencies can play the roles of media, standard of value, and method of payment for international trades and international movements of capital. Certain international currencies whose values are relatively stable can also be used as the international reserve of a nation. The portion of the international reserve that is made up of foreign currencies is known as foreign reserve or reserve currencies, which are mainly used for settling the balance of international payments and intervening the foreign exchange market in order to maintain the stability of exchange rates. The most important reserve currency is known as the dominant currency. Here, the dominant currency is definitely a reserve currency, while a reserve currency is not necessarily the dominant currency.

1.2 WHY IS A SYSTEMIC APPROACH IMPORTANT?

Since 1924 when von Bertalanffy pointed out that the fundamental character of living things is its organization, the customary investigation of individual parts and processes cannot provide a complete explanation of the phenomenon of life, this holistic view of nature and social events has spread over all corners of science and technology (Lin and Forrest, 2011). Accompanying this realization of the holistic nature, in the past 80 some years, studies in systems science and systems thinking have brought forward brand new understandings and discoveries to some of the major unsettled problems in the conventional science (Lin, 1999; Klir, 1985). Due to these studies of wholes, parts, and their relationships, a forest of interdisciplinary explorations has appeared, revealing the overall development trend in modern science and technology of synthesizing all areas of knowledge into a few major blocks, and the boundaries of conventional disciplines have become blurred ("Mathematical Sciences," 1985). Underlying this trend, we can see the united effort of studying similar problems in different scientific fields on the basis of wholeness and parts, and of understanding the world in which we live by employing the point of view of interconnectedness. As tested in the past 80 plus years, the concept of systems has been widely accepted by the entire spectrum of science and technology (Blauberg, et al., 1977; Klir, 2001).

1.2.1 A two-dimensional spectrum of knowledge

Similar to how numbers are theoretically abstracted, systems can also be proposed out of any and every object, event, and process. For instance, behind collections of objects, say, apples, there is a set of numbers such as 0 (apples), 1 (apple), 2 (apples), 3 (apples), . . .; and behind each organization, such a regional economy, there is an abstract, theoretical system within which the relevant whole, component parts, and the related interconnectedness are emphasized. And, it is because these interconnected whole and parts, the totality is known as an economy. In other words, when internal structures can be ignored, numbers can be very useful; otherwise the world consists of dominantly systems (or structures or organizations).

Historically speaking, on top of numbers and quantities has traditional science been developed; and along with systemhood comes the systems science. That jointly gives rise of a 2-dimensional spectrum of knowledge, where the classical science, which is classified by the thinghood it studies, constitutes the first dimension, and the systems science, which investigates structures and organizations, forms the genuine second dimension (Klir, 2001). That is, systems thinking focuses on those properties of systems and associated problems that emanate from the general notion of structures and organizations, while the division of the classical science has been done largely on properties of particular objects. Therefore, systems research naturally transcends all the disciplines of the classical science and becomes a force making the existing disciplinary boundaries totally irrelevant and superficial.

The importance of this supplementary second dimension of knowledge cannot be in any way over-emphasized. For example, when studying dynamics in an *n*-dimensional space, there are difficulties that cannot be resolved within the given space without getting help from a higher-dimensional space. In particular, when a one-dimensional flow is stopped by a blockage located over a fixed interval, the movement of the flow has to cease. However, if the flow is located in a two-dimensional space, instead of being completely stopped, the 1-dimensional blockage would only create a local (minor) irregularity in the otherwise linear movement of the flow (that is how nonlinearity appears (Lin, 2008)). Additionally, if one desires to peek into the internal structure of the 1-dimensional blockage, he can simply take advantage of the second dimension by looking into the blockage from either above or below the blockage. That is, when an extra dimension is available, science will gain additional strength in terms of solving more problems that have been challenging the very survival of the mankind.

1.2.2 The systemic yoyo model

Even though systems research holds such a strong promise as what is described in the previous subsection, the systems movement has suffered a great deal in the past 80 some years of development due to the reason that this new science does not have its own particular speaking language and thinking logic. Conclusions of systems research, produced in this period of time, are drawn either on ordinary language discussions or by utilizing the conventional mathematical methods, making many believe that systems-thinking is nothing but a clever way of rearranging conventional ideas. In other words, due to the lack of an adequate tool for reasoning and an adequate language for speaking, systems research has been treated with less significance than they were thought initially since the 1970s when several publications criticized how systems enthusiasts derived their results without sufficient rigorous means (Berlinski, 1976; Lilienfeld, 1978), even though most of the results turned out to be correct with 20-20 hindsight.

Considering the importance of the Cartesian coordinate system in modern science (Kline, 1972), (Wu and Lin, 2002) realizes that the concepts of (sizeless and volumeless) points and numbers are bridged beautifully together within the Cartesian coordinate system so that this system plays the role of intuition and playground for modern science to evolve; and within this system, important concepts and results of modern mathematics and science are established. Recognizing the lack of such an intuition and playground for systems science, on the basis of the blown-up theory (Wu and Lin, 2002), the yoyo model in Figure 1.1 is formally introduced by (Lin, 2007) in order to establish the badly needed intuition and playground for systems science.

In particular, on the basis of the blown-up theory and the discussion on whether or not the world can be seen from the viewpoint of systems (Lin, 1988; Lin, Ma and Port, 1990), the concepts of black holes, big bangs, and converging and diverging eddy motions are coined together in the model shown in Figure 1.1. This model was established in (Wu and Lin, 2002) for each object and every system imaginable. In other words, each system or object considered in a study is a multi-dimensional entity

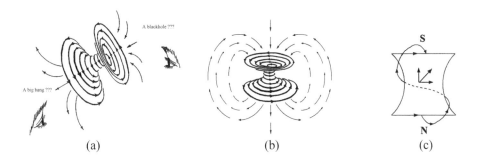

Figure 1.1 The eddy motion model of the general system.

that spins about its either visible or invisible axis. If we fathom such a spinning entity in our 3-dimensional space, we will have a structure as shown in Figure 1.1(a). The side of black hole sucks in all things, such as materials, information, energy, profit, etc. After funneling through the short narrow neck, all things are spit out in the form of a big bang. Some of the materials, spit out from the end of big bang, never return to the other side and some will (Figure 1.1(b)). For the sake of convenience of communication, such a structure as shown in Figure 1.1(a), is referred to as a (Chinese) yoyo due to its general shape. More specifically, what this systemic model says is that each physical or intellectual entity in the universe, be it a tangible or intangible object, a living being, an organization, a culture, a civilization, etc., can all be seen as a kind of realization of a certain multi-dimensional spinning yoyo with either an invisible or visible spin field around it. It stays in a constant spinning motion as depicted in Figure 1.1(a). If it does stop its spinning, it will no longer exist as an identifiable system. What Figure 1.1(c) shows is that due to the interactions between the eddy field, which spins perpendicularly to the axis of spin, of the model, and the meridian field, which rotates parallel to axis of spin, all the materials that actually return to the black-hole side travel along a spiral trajectory.

To show this yoyo model can indeed, as expected, play the role of intuition and playground for systems researchers, (Lin, 2008; Lin and Forrest 2011) have successfully applied it to investigate Newtonian physics of motion, the concept of energy, economics, finance, history, foundations of mathematics, small-probability disastrous weather forecasting, civilization, business organizations, the mind, among others.

At this junction, as an example, let us look at how a workplace can be investigated theoretically as such a spinning structure. In fact, each social entity is an objectively existing system that is made up of objects, such as people and other physical elements, and some specific relations between the objects. It is these relations that make the objects emerge as an organic whole and a social system. For example, let us look at a university of higher education. Without the specific setup of the organizational whole (relationships), the people, the buildings, the equipment, etc., will not emerge as a university (system). Now, what the systemic yoyo model says is that each imaginable system, which is defined as the totality of some objects and some relationships between the objects (Lin, 1999), possesses the systemic yoyo's rotational structure so that each chosen social system, as a specific system involving people, has its own specific multi-dimensional systemic yoyo structure with a rotational field.

To this end, there are many different ways for us to see why each social entity spins about an invisible axis. In particular, let us imagine an organization, say a business entity. As it is well known in management science, each firm has its own particular organizational culture. Differences in organizational cultures lead to varied levels of productivity. Now, the basic components of an organizational culture change over time. These changes constitute the evolution of the firm and are caused by inventing and importing ideas from other organizations and consequently modifying or eliminating some of the existing ones. The concept of spin beneath the systemic yoyo structure of the firm comes from what ideas to invent, which external ideas to import, and which existing ones to eliminate. If idea A will likely make the firm more prosperous with higher level of productivity, while idea B will likely make the firm stay as it has been, then these ideas will form a spin in the organizational culture. Specifically, some members of the firm might like additional productivity so that their personal goals can be materialized in the process of creating the extra productivity, while some other members might like to keep things as they have been so that what they have occupied, such as income, prestige, social status, etc., will not be adversely affected. These two groups will fight against each other to push for their agendas so that theoretically, ideas A and B are actually "spin" around each other. For one moment, A is ahead; for the next moment B is leading. And at yet another moment no side is ahead when the power struggle might have very well reached an equilibrium state. In this particular incidence, the abstract axis of spin is invisible, because no one is willing to openly admit his underlying purpose for pushing for a specific idea (either A or B or other ones).

As for the concept of black hole in a social organization, it can be seen relatively clearly, because each social organization is an input-output system, no matter whether the organization is seen materially, holistically, or spiritually. The input mechanism will be naturally the "black hole," while outputs of the organization the "big bang". Again, when the organization is seen from different angles, the meanings of "black hole" and "big bang" are different. But, together these different "black holes" and "big bangs" make the organization alive. Without the totality of "black holes" and that of "big bangs", no organization can be physically and holistically standing. Other than intuition, to this end the existing literature on civilizations, business entities, and individual humans does readily testify.

From this example, a careful reader might have sensed the fact that in this systemic yoyo model, we look at each system, be it a human organization, a physical entity, or an abstract intellectual being, as a whole that is made up of the 'physical' body, its internal structure, and its interactions with the environment. This whole, according to the systemic yoyo model, is a high dimensional spin field. Considering the fact that the body is the carrier of all other (such as cultural, philosophical, spiritual, psychological, etc.) aspects of the system, in theory the body of the system is a pool of fluid realized through the researcher's sensing organs in the three-dimensional space. The word "fluid" here is an abstract term totalling the flows of energy, information, materials, money, etc., circulating within the inside of, going into, and giving off from the body. And in all published references we have searched these flows are studied widely in natural and social sciences using continuous functions, which in physics and mathematics mean flows of fluids and are widely known as flow functions. On the other hand, as it has been shown and concluded in (Lin, 2008; Lin and Forrest, 2011) that the universe is a huge ocean of eddies, which changes and evolves constantly. That is, the totality of the physically existing world can be legitimately studied as fluids, their mechanism of internal working, and their interactions with each other.

To make this introductory chapter complete for the reader, let us look briefly at the justification of this yoyo model of systems.

In theory, the justification for such a model of general systems is the blown-up theory (Wu and Lin, 2002). It can also be seen as a practical background for the law of conservation of informational infrastructures. More specifically, based on empirical data, the following law of conservation is proposed (Ren, Lin and OuYang, 1998): For each and every given system, there must be a positive number a such that

$$AT \times BS \times CM \times DE = a \tag{1.1}$$

where A, B, C, and D are some constants determined by the structure and attributes of the system of concern, and T stands for the time as measured in the system, S the space occupied by the system, M and E the total mass and energy contained in the system.

Because M (mass) and E (energy) can exchange into each other and the total of them is conserved, if the system is a closed one, equ. (1.1) implies that when time T evolves to a certain (large) value, space S has to be very small. That is, in a limited space, the density of mass and energy becomes extremely high. So, an explosion (a big bang) is expected. Following the explosion, space S starts to expand. That is, time T starts to travel backward or to shrink. This end gives rise of the well-known model for the universe as derived from Einstein's relativity theory (Einstein, 1983; Zhu, 1985). In terms of systems, what this law of conservation implies is that: Each system goes through such cycles as: ... → expanding → shrinking → expanding → shrinking → ... Now, the geometry of this model of universe established from Einstein's relativity theory is given in Figure 1.1.

Empirically, the multi-dimensional yoyo model in Figure 1.1 is manifested in different areas of life. For example, each human being, as we now see it, is a 3-dimensional realization of such a spinning yoyo structure of a higher dimension (for more details please consult Part 3 in (Lin and Forrest, 2011)). To this end, consider two simple and easy-to-repeat experiences. For the first one, imagine we go to a swim meet. As soon as we enter the pool area, we immediately fall into a boiling pot of screaming and jumping spectators, cheering for their favorite swimmers competing in the pool. Now, let us pick a person standing or walking on the pool deck for whatever reason, either for her beauty or for his strange look or body posture. Magically enough, before long, in fact, almost instantly, the person from quite a good distance will feel our stare and she/he will be able to locate us in a very brief moment out of the reasonably sized and boiling audience. The reason for the existence of such a miracle and silent communication is because each side is a high dimensional spinning yoyo. Even though we are separated by space and possibly by informational noise, the stare of one side on the other has directed that side's spin field of the yoyo structure into the spin field of the yoyo structure of the other side. That is the underlying mechanism for the silent communication to be established instantly over space and informational noise.

As the second example, let us look at the situation of human relationship. When an individual A has a good impression about another individual B, magically, individual B also has a similar and almost identical impression about A. When A does not like B and describes B as a dishonest person with various undesirable traits, it has been clinically proven in psychology that what A describes about B is exactly who A is himself (Hendrix, 2001). Once again, the underlying mechanism for such a quiet and unspoken evaluation of each other is because each human being stands for a spinning yoyo and its rotational field. Our feelings about other people are formed through the interactions of our invisible yoyo structures and their spin fields.

For more in-depth discussion of both theoretical and empirical justifications of the systemic yoyo model, please go to Chapter 2 of this book.

1.3 PROBLEMS INVESTIGATED USING SYSTEMS THINKING AND METHODOLOGY

Because of the availability of the yoyo model, there is now an intuition and playground that is commonly available for systems theorists and practitioners to house their abstract reasoning and thinking, just as what people are accustomed to do with the Cartesian coordinate system when they think about how to resolve a problem in the classical science. Now, one of the most important aspects of systems research is what kinds of problems systems science could and should attack and attempt to resolve. Historically speaking, the importance of this question is that only if systems science can provide new and powerful means to resolve at least some of the age-old whys which have challenged the mankind since the dawn of the recorded history, systems science will have a chance to become firmly recognized as a legitimate branch and the second dimension of knowledge. The never-fading effort of man invested in studying these forever important whys fundamentally signals the relevance of these endeavours to the very survival of the human race.

1.3.1 Problems systems science could address

Here is a common belief in the community of systems researchers about what kinds of problems systems science could and should attempt to address (Armson, 2011): Systems science resolves problems that are related to systemhood instead of thinghood. In other words, systems science is good at addressing such a problem that when one tries to start looking at one aspect, he realizes that several other aspects of the issue should be first addressed. That is, the issue seems to be messy with neither any beginning nor an ending and is surely not representable by using linear causalities. Many factors influence the outcome, while the outcome simultaneously affects the influencing factors.

Although such a belief is in line with how systems science is perceived, it has been quite misleading. In fact, based on this belief, it should be readily recognized that the limitation of modern science is its ignorance of all structure related aspects of issues it addresses. For instance, for the very basic arithmetic fact that $1 + 1 = 2$ to hold true, modern science has purposefully give away many structural characteristics of the issue of concern. In particular, if $1 + 1 = 2$ stands for the fact that when one object is placed together with another object, there will be two objects totally, then the internal structures of the objects have been ignored. It is because when the internal structures of the objects are concerned with, the togetherness of the objects can be zero, or one, or two, or any other possible natural number. To illustrate this fact, let us assume that $1 + 1 = 2$ represent placing two systems together. (That is, the objects we put together now have their individual internal structures.) Does that mean consequently we will have two systems totally? The answer is: Not necessarily. In particular, let us imagine each system as a spin field, as what we have seen in the previous section. When we place two spin fields together, what do we have? Do we really have two spin fields?

1.3.2 Is 1 + 1 = 2 universally true?

To answer this question, let us consider all possibilities when two spin fields N and M are placed alongside of each other, Figure 1.2. Considering their directions of spin and their divergence and convergence of the fields N and M, Figure 1.2(a) will produce the outcome of two systems. The spin fields N and M in Figure 1.2(b) will remain separate while creating a joint rotational field, leading to three spin field with two nested inside a large field. Because of their convergence, the fields N and M in Figure 1.2(c) will combine to become one spin field that is greater than the original fields individually. If we look back at Figure 1.2(b), then they are simply the "big-bang" sides of the

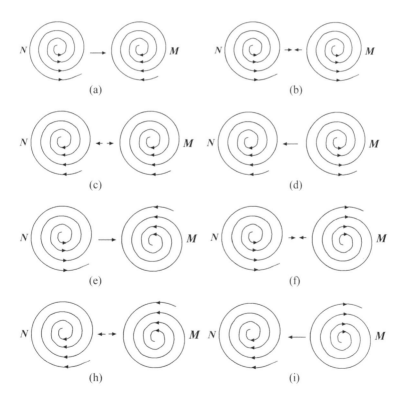

Figure 1.2 The internal structure of a two-body system.

convergent fields in Figure 1.2(c). So, if the fields N and M do combine with each other, then there will be only one rotational field resulted. If these fields do not combine into a greater field, then while they stay separate from each other, they also create many smaller fields in the areas between themselves. That is, in this case, $1 + 1 = ?$ that is definitely more than 2. Similar to the situation in Figure 1.2(a), the fields N and M in Figure 1.2(d) will stay separate without creating any small field. This represents another scenario where $1 + 1 = 2$.

Similar analysis indicates that the fields N and M in Figure 1.2(e, i) will either destroy each other so that no more rotational field is resulted (that is $1 + 1 = 0$), or if they stay separate, they will create many smaller fields in the zones between N and M (that is $1 + 1 > 2$). For the fields N and M in Figure 1.2(f, h), they either destroy each other ($1 + 1 = 0$) or simply stay separate ($1 + 1 = 2$).

In short, $1 + 1 = 2$ is a very special case among all the eight possibilities as depicted in Figure 1.2. As a matter of fact, we can conclude the same fact that $1 + 1 = 2$ is a very special case as follows: When one positron is placed together with an electron, the outcome is nothing; that is $1 + 1 = 0$. When a woman and man combine into a family, the outcome can be (in theory) any number of people. In comparison, of course, the previous spin-fields analysis is systemically more scientifically significant than this short version of modelling.

1.3.3 How modern science resolves problems?

Now, let us look at how modern science resolves problems by analysing the following basic algebraic scenario: Suppose that John and Ed together can finish a job in 2 hours, John and Paul together can do the same job in 3 hours, while Ed and Paul together in 4 hours. The question that needs to be resolved is: If John, Ed, and Paul work together, how long do they need to finish the job?

The standard method of solving this algebraic problem at the high school level of mathematics is first assume that John can do the job alone in x hours, Ed alone in y hours, and Paul alone in z hours. Then, the following system of equations is established to describe the relationship among the quantities x, y, and z.

$$\begin{cases} \dfrac{1}{x} + \dfrac{1}{y} = \dfrac{1}{2} \\[2mm] \dfrac{1}{x} + \dfrac{1}{z} = \dfrac{1}{3} \\[2mm] \dfrac{1}{y} + \dfrac{1}{z} = \dfrac{1}{4} \end{cases} \tag{1.2}$$

Next step is to solve this system mathematically, producing the answer that in $24/13 \approx 1.85$ hours John, Ed, and Paul can complete the job jointly.

To confirm the answer is correct, one is required to go back to the problem to check his answer. Here is one way to do just that: Because John and Ed can finish the job in 2 hours together, with Paul added, the additional manpower should surely make completing the job quicker. So, 24/13 is potentially the correct answer.

Now, if we look at the resolution of the previous problem from the angle of systems thinking, we can see at least two problems: One is that the job might have an internal structure so that extra hand may not help at all. The other is that if each of the three workers is seen righteously as a living system, then when they are put together, they will surely interact with each other so that instead of speeding up the work progress, the interaction may also very well slow down the progress. Although a greatly simplified version of the first problem has been addressed in operations research, the second problem represents an extremely difficult issue for modern science, known as the three-body problem. For more detailed discussion about this end, please consult (Lin, 2008).

What these two examples are intended for is that by simply holding onto the thinking of systemhood, almost all, if not all, basic concepts and methods of modern science can be either generalized or improved or both. For example, the following are some of the success stories among many others of thinking and doing along this line:

1   By applying the systemic yoyo model to the forecasting of nearly zero probability disastrous weather conditions, the current prediction accuracy that is commercially available has been greatly improved (Lin, 2008; Lin and OuYang, 2010).
2   By employing the rotation structure of the systemic yoyo model and the observational facts of the dishpan experiment, we are led to the discovery of the fourth crisis in the foundations of mathematics (Lin (guest editor), 2008).
3   By relying on the systemic yoyo model as a road map, a sufficient and necessary condition is established for when Becker's rotten kid theorem can hold true, where the theorem is widely used in the research of economics (Lin and Forrest, 2008).
4   By employing the systemic yoyo model as the theoretical basis and intuition for logical reason, the research of civilizations, business organizations, and the mind is brought up to a much higher level with many new conclusions produced (Lin and Forrest, 2011).

In short, the discussion in this section simply implies that by making use of the systemic yoyo model and the thinking logic of systems research, most of the traditional studies of science, which of course includes the studies of money and financial institutions, can be enriched while new discoveries and understandings can be made and achieved accordingly. Such fruitful efforts will in turn truly make systems science the second dimension of knowledge.

1.4 MAIN RESULTS OF THIS BOOK

Combining what are discussed in Sections 1.1–1.3 above, the reader can see clearly why this book is written: When the flow of money is seen analogous to the movement of fluids, such as water and air, then the thinking logic and methodology of systems science in general and the systemic yoyo model in particular can naturally come into play. Based on the already published works along this line of reasoning, it can be righteously expected that by doing so, brand new discoveries and theoretical breakthroughs could and will be forthcoming. That is indeed what will be shown throughout this book.

This book contains a total of three parts and twelve chapters. Part 1 contains two chapters and focuses on the relevant foundations of the systemic yoyo model. In this part, other than this introductory chapter, Chapter 2 is a preparation for the rest of this book. In this chapter, the structure of meridian fields, which helps to hold the dynamic spin field of the yoyo model together, and some of the elementary properties of this model will be studied. And for the completeness of this book, the theoretical and empirical justifications on why each and every system can be seen as a rotational yoyo are provided briefly. Additionally, the fundamental laws on the state of motion of materials are introduced, and how figurative analysis is valid as a tool for scientific research is addressed.

Part 2 treats each domestic financial system as a closed system and contains Chapters 3–7. In particular, after Chapter 2 lays down the foundation that the universe generally can be seen as an ocean of systems, which in turn can be thought of as abstract rotational yoyo fields of different scales that compete with each other, then when one focuses on the finance of a nation as a system, the particular financial system of the nation is made up of fluids, such as money, information, knowledge, etc. Other than studying information and knowledge as fluids by making use of continuous and differentiable functions, money could also be treated as a fluid that circulates around all corners of the nation and fuels the entire economy. If in a specific geographic area the money supply is low, then the regional economic development will suffer and the local residents would have to look for employment and other opportunities somewhere else. If this trend continues, the specific region would have to be abandoned eventually. At the same time, money should also be seen as a carrier of other relevant fluids, such as information and knowledge, which dictate where the opportunities are so that money as a fluid flows to where it could find the maximal returns. In this part of the book, we look at the financial system of a general industrialized nation from the viewpoint of fluids and fluid dynamics, while assuming that the nation is in isolation so that no (major) external effect exists to influence its financial stability, its social structure, and its political operation. Similar to what is done in natural science, we first look at closed systems. After we have gained sufficient understanding of such systems, we then investigate the interaction between such relatively closed systems.

Chapter 3 contains 8 sections. Section 3.1 looks at the composition of the national financial system of a typical industrialized nation. Section 3.2 focuses on the monetary standard of currencies. Section 3.3 investigates the concept of currency. Section 3.4 studies how credit can help to boost the economy. In Section 3.5 we look into more details of the financial market, while in Section 3.6 we do the same with the banking system. After the basics of the financial infrastructure of a nation are laid out systematically, in Section 3.7 we look at why innovation is the key for the national economy to stay stable and prosperous by using the observational facts of the dishpan experiments and by using the Theorem of Never-Perfect Value Systems of the microeconomics. At the end, Section 3.8 concludes the presentation of this chapter.

The supply and demand of money constitute the corner stone of monetary theory, where money supply reveals various factors that affect the amount of money available in the economy. It is the basic starting point of analyzing how monetary policies affect the economy. Correspondingly, the demand of money represents another corner stone for analyzing the effect of money on the economy. Here, changes in money supply are related to the healthy operation of the entire economy. Money plays the role of economic blood, which flows constantly through all parts of the economy, stimulates the economic development, helps accumulate capital, compartmentalize production, and promotes effective use of natural resources.

In Chapter 4, we will look at such important problems as: How much money is needed for an economy to operate effectively? Who is really in control of the money supply? What are the factors that change the money supply? Through analyzing the mechanism underneath the money supply, this chapter will address these questions. In terms of the demand of money, it stands for the hottest debated part of the monetary theory, where formed are various different theories, some of which will be illustrated in this chapter, including the classical theory of money demand developed before the 1930s, the Keynesian (John Maynard Keynes) theory developed since 1936, and the modern quantity theory developed since 1956 by a group of economists, as represented by Milton Friedman. At the end, we will also discuss another important problem of the monetary theory: the effect of interest rates on the demand of money.

Chapter 4 contains 4 sections. In Section 4.1, we will investigate the concept of supply and demand and see how supply and demand in fact constitutes two interacting economic forces. In Section 4.2, we focus on the supply of money by looking at how money is created, the concept of money multipliers, and various theoretical models that describe the supply of money. Section 4.3 emphasizes on the other side of the coin: the demand of money, by looking closely at the more difficult problem of where and why money is needed. It is in this section that we will study the classical quantity theory of money, Keynesian demand for money, following developments of Keynesian theory, modern quantity theory of money, and related empirical tests. Then, Section 4.4 concludes this chapter with some relevant comments.

In Chapter 4, interest rates are considered as a factor that influences the demand and supply of money. That in fact treats interest rates as an exogenous variable. So, in Chapter 5, we treat interest rates as an endogenous variable of the economic system. By considering its interaction with the demand and supply of money, we investigate how interest rates are determined and how they fluctuate. At the same time, we study the problem of how interest rates and the total outputs are determined, while expanding our discussion to the analysis of why there are differences among the interest rates of different bonds. At the end, we look at how the marketplace reacts respectively to safe and risky investment opportunities.

Chapter 5 contains 7 sections. In Section 5.1, we look at loanable funds. In Section 5.2, we study the theory of liquidity preference. In Section 5.3, the relationship between interest rate and total production output is considered. In Section 5.4, the risk structure of interest rates is analyzed, in Section 5.5, the term structure of interest rates is investigated. Different of the existing literature, in Section 5.6, we look at the market behaviors and response to different-scale investment opportunities. This chapter ends with a few concluding words in Section 5.7.

In the development of market economy, to avoid certain undesirable economic difficulties, national macro-controls have become gradually visible and have been playing ever-increasing important roles, where the focus of the macro-controls is the financial regulation. The so-called financial regulation means that through introducing monetary policies and implementing these policies by the central bank, the national government materializes its adjustment and control of the economic activities of the entire society. The effect and influence of financial regulation are not about a specific area of the economic activities of the society; instead, they are about the overall situation of the entire national economy. Through the particular position and the monetary policy tools in the control of the central bank, the national government actualizes its management of the country and intervention of the economy of the society by adjusting the variables that are related to the scale of credit and the amount of money in circulation, and by influencing the activities of the entire national economy through changes in the supply of money. For measuring the goal and effect of financial regulation are there multiple criteria, which can be as comprehensive as those including price stability, economic growth, sufficient employment, international trade balance, etc. To realize these goals, it is often necessary to employ monetary and fiscal policies jointly, where the former represent the central contents of financial regulation. With the emphasis on monetary policies, Chapter 6 discusses the relevant problems. In particular, in Section 6.1, we will look at the tasks, contents, and goals of monetary policies. In Section, we focus on the medium-term goals of monetary policies. Then the available tools necessary for carrying out the established monetary policies are studied in Section 6.3. The particular mechanism of how monetary policies affect the economy is discussed in Section 6.4. Relevant empirical evidence on the importance of money is provided in Section 6.5. In Section 6.6, we will look at how the IS-LM model can be employed to the understanding of money and fiscal policies, their effects on the total output and interest rates, and the crowding-out effect. Section 6.7 considers how inflation is caused, how inflation is related to the prices of goods, and how inflation can be dealt with. At the end, Section 6.8 concludes this chapter by exploring how systemic structure actually underlies monetary policies.

As the last chapter of Part 2 on the domestic financial system, seen as a closed system, Chapter 7 looks at the studies on the portfolio of assets. In other words, in this chapter we will learn as an individual, such as a person, a fund manager, an investment firm, how he could select financial assets to invest his available money in order to maximize his return. The theory of asset demand is the theoretical foundation of all financial investments. It focuses on how a person, a fund manager, or an investment firm could increase his return in order to maximize his utility (satisfaction) by selecting appropriate financial assets from a huge pool of various available choices. This chapter is organized as follows. Section 7.1 addresses the problem of why investment opportunities always exist from the angle of the systemic yoyo model. Section 7.2 presents the concepts of interest rates and yields. Different factors that affect the investment portfolio returns are listed in Section 7.3. In terms of balancing the risk and yields, Section 7.4 investigates the need of diversity of investment, the concept of mean variance utility, and matters related to the determination of the portfolio proportion. Then this chapter is concluded by Section 7.5, where we study the expected yields of portfolios, demands of assets, and the advantages of diversity.

Part 3 contains Chapters 8–12 and looks at the international financial system as an ocean of interacting semi-closed regional systems. Specifically, the International Monetary System (IMS) stands for the institutional arrangement that emerges out of the most fundamental problems of the international monetary and financial relations, including the international exchange rate system, international payment and settlement system, international payment coordination, international solvency supply, and international financial institutions, etc. The international monetary system represents an important aspect of the modern international economic relation. It not only plays a huge role on the exchange rate system, foreign exchange policy, and the management of reserve assets of each country, but also creates long lasting effects on the trade patterns and economic developments throughout the world. Chronologically, the development of the international monetary system consists of three periods: the period of international gold standard, the period of Bretton Woods system, and the current period of Jamaican system. After introducing the basic concepts and main contents of the international monetary system, Chapter 8 then presents the historical backgrounds, missions, and business activities of major international financial institutions, and analyzes the evolutionary process of the system, explores the reforms of the system, and the current state of the European monetary system by walking along the thread of history. At the end, we investigate the systemic three-ringed structure of viable, independent economic systems. This chapter is organized as follows. In Section 8.1, we learn the composites, classification and functions of the international monetary system. In Section 8.2, we look at the development history of the international monetary system by going through the international gold standard, the Bretton Woods and the Jamaica system. Section 8.3 introduces the main suggestions about how to potentially improve the international monetary system and an expected future of the system. Section 8.4 studies the European monetary system by going over the history, the European union of money, and the impacts of the Euro. International organizations of finance, such as the International Monetary Fund and the World Bank, are introduced in Section 8.5, where the proposal of bankruptcy of nations is discussed. After learning the basic composites of the international monetary system, in Section 8.6 we consider the systemic structure that underlies each stable national economic system. Section 8.7 concludes this chapter with a few final words.

In the names of international reserves and capital flows across national borders, Chapter 9 studies how national economies affect and influence each other. Here, international reserves have been one of the central problems involved in the reforms of the international monetary system since World War II. On the one hand, it is related to each country's capability to regulate its balance of payments and to stabilize its exchange rate. On the other hand, international reserves also impact majorly on the price levels of the world and the development of international trades. Along with the deepening international economic interdependencies, the payment behavior of international economic transactions becomes increasingly important. The amounts, the structure, and the management of international reserves can directly affect many aspects of a nation, such as the nation's regulation on the balance of international payments, the stability of exchange rates, the prevention of currency crises, etc. At the same time, the flow of international capital represents the worldwide movement of capital that transcends national borders. Along with the development of globalization of the world economy, flows of international capital are increasingly becoming the most active factor of the world economy. However, at the same time when the flows of international capital help to grow the world economy, they also provide the grounds for wide-spread international financial crises to occur. Since the 1990s, along with the frequent occurrence of international financial crises of increasing severity, heightened attention has been given to the flows of these capitals. This chapter is organized as follows. Section 9.1 looks at the characteristics and components of international reserves and borrowed international reserves. Section 9.2 presents how international reserves are managed in terms of magnitudes and structures. As case studies, we will closely exam the reserves of the USA, Europe, Japan, and Great Britain. Section 9.3 considers such potential problems regarding foreign currency reserves as capital flight, credibility, and currency substitution. Capital flows across national borders are presented in Section 9.4, where what's given includes types of international capital flows, causes, present state and expected future of capital flows, impacts of international capital flows, and the relationship between international capital flows and individual national financial systems. The final Section 9.5 looks at the interaction of individual national economic systems from the angle of the systemic yoyo model by showing that small economies do not have much choice in dealing with large economies; if conditions are right, these small economies could potentially chain up together to form a heavy weight economic player in the international arena.

Because of the advances in communication technology, the world economy has shown the tendency of globalization; and the magnitude of the international flow of capitals has gradually surpassed the scale of international trades, forming a huge international financial market. This international financial market, which exists beyond geographic and spatial constraints, makes it possible for the supply and demand sides of money located in different countries to trade quickly and conveniently with each other. That enhances both the depth and breadth of business transactions, and helps to bring forward an unprecedented development to the world economy. However, at the same time, high-speed flows of huge amounts of capital also create both directly and indirectly far-reaching effects on each nation's international payments and domestic economy through capital and financial accounts, causing monetary and financial crises of multiple nations. Chapter 10 studies the structure of the international financial market, causes and effects of international capital flow, impacts of financial globalization, and why financial attacks are always possible. In particular, Section 10.1 will look at the international currency market, capital market, and foreign exchanges. Section 10.2 investigates the reasons, characteristics, and effects of international capital flows. Section 10.3 considers debt and bank crises caused by international capital flows. In Section 10.4, such topics as the concept and causes currency crises, the Asian currency crisis of the 1990s and its resolution are discussed. The final Section 10.5 considers theoretically the question of whether or not financial attacks are always possible by looking at how environmental conditions determine the natures of individual economies, why there will always be financial centers under different sets of rules and regulations, and why emerging economies will always exist, providing easy targets for financial attacks.

As the reader should have been accustomed to by this point in this book, when seen from systems research, each economy can be righteously treated as a pool of rotational fluids of information, knowledge, money, etc. That is, the international economy is an ocean of spinning fields, some of which exist peacefully alongside of each other, while others fight against each other, destroy each other, and takeover each other. Analogous to the phenomenon of tsunamis that appear in the earthly ocean, there should be crises in the international financial world. That is exactly what we are going to study in Chapter 11. Specifically, this chapter is organized as follows. Section 11.1 introduces the concepts of financial and currency crises and glances over several representative theories. Section 11.2 focuses on speculative attacks across national borders and relevant currency crises, their characteristics, and then analyzes the success story of Chilean capital account liberation. In Section 11.3, we investigate currency wars and a possible strategy of self-defense against currency wars based on recent advances in the research of feedback systems.

To help make this book as self-contained as possible, Chapter 11 contains an appendix, where the necessary systems research on feedback systems is detailed.

As a new, heavy weight player in the international arena, it is very important for other members of this arena to understand China with reasonable expectations in order to bring about mutually beneficial economic consequences. To achieve this goal, Chapter 12 is devoted to the discussion of modern China by looking at its current state of economic and financial operations in particular and by looking at China at the height of civilization in general. By doing so, it is expected that all the players of the international economy and the international financial system could understand how decisions are made in China and what could be expected from Chinese leadership. This chapter is organized as follows. In Section 12.1, after providing a brief development history of the central banks of China, it introduces the current hierarchy of Chinese central bank, the business coverage of the People's Bank of China, and how money is supplied in China. Section 12.2 is devoted to the study of Chinese financial system by looking at Chinese banks, different kinds of financial organizations, and the practice of lending and investing. Section 12.3 focuses on the financial reform China has embarked on. In Section 12.4 the international reserve of China is the topic of discussion. It looks at how fast the foreign reserves in China have been increasing, how these foreign reserves are managed, and how the risk associated with the large foreign reserves could potentially be avoided. Section 12.5 attempts to understand China at the height of civilization in order to understand how China might behave differently from the norm of the Western nations in the international economy and the international financial system. The concluding Section 12.6 looks at the question: Will economic prosperity visit China soon?

In terms of literature, other than the references cited directly within the context of this book, Chapters 3–7 have been heavily based on (Campbell, Lo and McKinley, 1997; Hayek, 1966; Huang and Litzenberger, 1988; Li, 2005; Mishkin, 1995; Wu, 2006; Ye and Lin, 1998). And, Chapters 8–10 are based on (Jiang, 2008; Wang and Hu, 2005; Eiteman, Stonehill and Moffett, 2001; O'Hara, 1998; Krugman and Obstfeld, 1997; Obstfeld and Rogoff, 1996).

Chapter 2

The yoyo mode: Properties and justifications

To prepare the theoretical foundation for the rest of this book, in this chapter, we will study based on the basic attributes of the yoyo model the structure of meridian fields that helps to hold the dynamic spin field of the yoyo model together and some of the elementary properties of this model. Then, for the completeness of this book, we will glance through the theoretical and empirical justifications on why each and every system can be seen as a rotational yoyo.

2.1 PROPERTIES OF SYSTEMIC YOYOS

2.1.1 The field structure

Because each yoyo spins as in Figure 2.1, other than the eddy field, which is perpendicular to the axis of rotation as shown in Figure 2.1, there also exists a meridian field accompanying each yoyo (see Figure 2.2). The invisible meridians go into the center of the black hole, through the narrow neck, and then out the big bang. They travel through the space and return to the center of the black hole. Somehow we can imagine that these meridians help to hold different layers of the eddy field of the yoyo structure together. For the convenience of our communication, for any given yoyo structure, the black hole side will be referred to as the south pole of the structure and the big bang side the north pole.

In this model, the word "spin" is used to capture the concept of angular momentum or the presence of angular momentum intrinsic to a body as opposed to orbital angular momentum of angular momentum that is the movement of the object about an external point. For example, the spin of the earth stands for the earth's daily rotation about its polar axis. The orbital angular momentum of the earth is about the earth's annual movement around the sun. In general, a two-dimensional object spins around a center (or a point), while a three-dimensional object rotates around a line called an axis.

Mathematically, a spin of a rigid body around a point or axis is followed by a second spin around the same point (respectively axis), a third spin results. The inverse of a spin is also a spin. Thus, all possible spins around a point (respectively axis) form a group of mathematics. However, a spin around a point or axis and a spin around a different point (respectively axis) may result in something other than a rotation, such as a translation. In astronomy, spin (or rotation) is a commonly observed phenomenon. Stars, planets, galaxies all spin around on their axes. In social science areas, spin

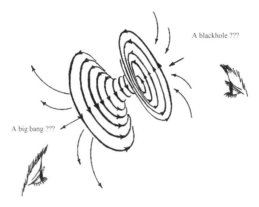

Figure 2.1 Eddy motion model of a general system.

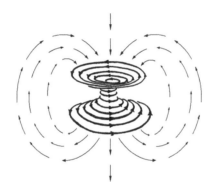

Figure 2.2 The distribution of a meridian field.

appears in the study of many topics. An old saying goes as things always "go around and come around." The theory and practice of public relations heavily involve the concept of spin, where a person, such as a politician, or an organization, such as a publicly traded company, signifies his often biased favor of an event or situation. In quantum mechanics, spin is particularly important for systems at atomic length scales, such as individual atoms, protons, or electrons (Griffiths, 2004).

Because systems are of various kinds and scales, the universe can be seen as an ocean of eddy pools of different sizes, where each pool spins about its visible or invisible center or axis. At this junction, one good example in our 3-dimensional physical space is the spinning field of air in a tornado. In the solenoidal structure, at the same time when the air within the tornado spins about the eye in the center, the systemic yoyo structure continuously sucks in and spits out air. In the spinning solenoidal field, the tornado takes in air and other materials, such as water or water vapor on the bottom, lifts up everything it took in into the sky, and then it continuously spays out the air and water from the top of the spinning field. At the same time, the tornado also breathes in and out with air in all horizontal directions and elevations. If the amounts of air

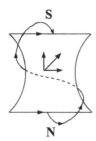

Figure 2.3 Slanted meridians of a yoyo structure.

and water taken in by the tornado are greater than those given out, then the tornado will grow larger with increasing effect on everything along its path. That is the initial stage of formation of the tornado. If the opposite holds true, then the tornado is in its process of dying out. If the amounts of air and water that are taken in and given out reach an equilibrium, then the tornado can last for at least a while. In general, each tornado (or a systemic yoyo) experiences a period of stable existence after its initial formation and before its disappearance. Similarly, for the general systemic yoyo model, it also constantly takes in and spits out materials. For the convenience of our discussion in this book, we assume that the spinning of the yoyo structures follows the left hand rule 1 below:

Left-Hand Rule 1:   When holding our left hand, the four fingers represents the spinning direction of the eddy plane and the thumb points to the north pole direction along which the yoyo structure sucks in and spits out materials at its center (the narrow neck). (Note: It can be seen that in the physical world, systemic yoyos do not have to comply with this left hand rule.)

As influenced by the eddy spin, the meridian directional movement of materials in the yoyo structure is actually slanted instead of being perfectly vertical. In Figure 2.3, the horizontal vector stands for the direction of spin on the yoyo surface toward the reader and the vertical vector the direction of the meridian field, which is opposite of that in which the yoyo structure sucks in and spits out materials. Other than breathing in and out materials from the black hole (the south pole) and big bang (the north pole) sides, the yoyo structure also takes in and gives out materials in all horizontal directions and elevations, just as in the case of tornadoes discussed earlier.

2.1.2 The quark structure of systemic yoyos

The so-called three-jet event of the particle physics is an event with many particles in a final state that appear to be clustered in three jets, each of which consists of particles that travel in roughly the same direction. One can draw three cones from the interaction point, corresponding to the jets, Figure 2.4, and most particles created in the reaction appear to belong to one of these cones. These three-jet events are currently the most direct available evidence for the existence of gluons, the elementary particles that cause quarks to interact and are indirectly responsible for the binding of protons and

Figure 2.4 A "snapshot" a three-jet event.

neutrons together in atomic nuclei (Brandelik, et al., 1979). Because jets are ordinarily produced when quarks hadronize, the process of the formation of hadrons out of quarks and gluons, and quarks are produced only in pairs, an additional particle is required to explain such events as the three-jets that contain an odd number of jets. Quantum chromodynamics indicates that this needed particle of the three-jet events is a particularly energetic gluon, radiated by one of the quarks, which hadronizes much as a quark does. What is particularly interesting about these events is their consistency with the Lund string model. And, what is predicted out of this model is precisely what is observed.

Now, let us make use of this laboratory observation to study the structure of systemic yoyos. To this end, let us borrow the term "quark structure" from (Chen, 2007). Because out of the several hundreds of different microscopic particles, other than protons, neutrons, electrons, and several others, most only exist momentarily, it is a common phenomenon for general systemic yoyos to be created and to disappear constantly in the physical microscopic world. All microscopic systemic yoyos can be classified (Chen, 2007, p. 41) into two classes using the number of quarks involved. One class contains 2-quark yoyos, such as electrons, π-, κ-, η-mesons, and others; and the other class 3-quark yoyos, including protons, neutrons, Λ-, Σ-, Ω-, Ξ (Xi) baryons, etc. Here, electrons are commonly seen as without any quark. However, Chen (2007) showed that yes, they have 2 quarks. Currently, no laboratory experiment has produced 0-quark or n-quark particles, for natural number $n \geq 4$.

Following (Chen, 2007), each spinning yoyo, as shown in Figure 2.1, is seen as a 2-quark structure, where we imagine the yoyo is cut through its waist horizontally in the middle, then the top half is defined as an absorbing quark and the bottom half a spurting quark. Now, let us study 3-quark yoyos by looking at a proton P and a neutron N. At this junction, the three jet events are employed as the evidence for the structure of 3-quark yoyos, where there are two absorbing and one spurting quarks in the eddy field. The proton P has two absorbing u-quarks and one spurting d-quark (Figure 2.5), while the neutron N has two spurting d-quarks and one absorbing u-quark (Figure 2.6). In these figures, the graphs (b) are the simplified flow charts with the line segments indicating the imaginary axes of rotation of each local spinning column. Here, in Figure 2.5, the absorbing u-quarks stand for local spinning pools while together they also travel along in the larger eddy field in which they are part of. Similarly in Figure 2.6, the spurting d-quarks are regional spinning pools. At the same time when they spin individually, they also travel along in the large yoyo structure of the neutron N. In all these cases, the spinning directions of these u- and d-quarks are the same except that each u-quark spins convergently (inwardly) and each d-quark divergently (outwardly).

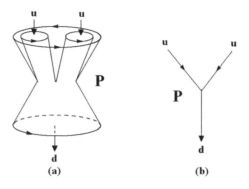

Figure 2.5 The quark structure of a proton P.

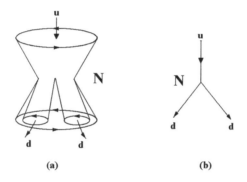

Figure 2.6 The quark structure of a neutron N.

Different yoyo structures have different numbers of absorbing u-quarks and d-quarks. And, the u-quarks and d-quarks in different yoyos are different due to variations in their mass, size, spinning speed and direction, and the speed of absorbing and spurting materials. This end is well supported by the discovery of quarks of various flavors, two spin states (up and down), positive and negative charges, and colors. That is, the existence of a great variety of quarks has been firmly established.

Hide (1953) used two concentric cylinders with a liquid placed in the ring-shaped region between the cylinders. He placed the container on a rotating turntable subjected to heating near the periphery and cooling at the center. The turntable generally rotated counter clockwise, as does the earth when viewed from above the North Pole. Even though everything in the experiment was arranged with perfect symmetry about the axis of rotation, such as no impurities added in the liquid, the bottom of the container is flat, Hide observed the flow patters as shown in Figure 2.7. Briefly, with fixed heating, a transition from circular symmetry (Figure 2.7(a)) to asymmetries (Figure 2.7(b)), then Figure 2.7(c) would take place as the rotation increased past a critical value. With sufficiently rapid but fixed rate of rotation, a similar transition would occur when the heating reached a critical strength, while another transition back to symmetry would occur when the heating reached a still higher critical strength. Also, in stage

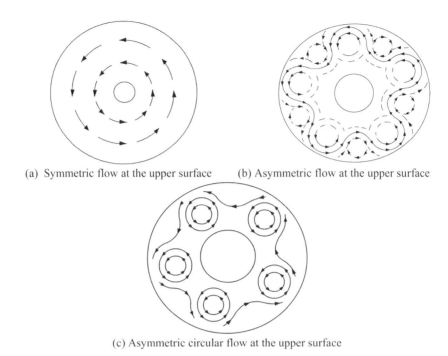

(a) Symmetric flow at the upper surface   (b) Asymmetric flow at the upper surface

(c) Asymmetric circular flow at the upper surface

Figure 2.7 Patterns observed in Hide's dishpan experiment.

Figure 2.7(c), a chain of identical eddy motions would appear. As they travel along, they would alter their shapes in unison in a regular periodic fashion, and after many rotations of the turntable, they would regain their original shape and then repeat the cycle. Fultz and his colleagues (1959) conducted another dishpan experiment without the inner cylinder, producing similar results.

Now, if we fit Fultz's dishpan experiment to the discussion above by imagining both the top and the bottom of each yoyo as a spinning dish of fluids, then the patterns as observed in the dishpan experiment suggest that in theory, there could exist such a yoyo structure that it has n u-quarks and m d-quarks, where $n \geq 1$ and $m \geq 1$ are arbitrary natural numbers, and each of these quarks spins individually and along with each other in the overall spinning pool of the yoyo structure.

From (Lin, 2007; Lin, 2008), it can be seen that due to uneven distribution of forces, either internal or external to the yoyo structure, the quark structure of the spinning yoyo changes, leading to different states in the development of the yoyo. This end can be well seen theoretically and has been well supported by laboratory experiments, where, for example, protons and neutrons can be transformed into each other. When a yoyo undergoes changes and becomes a new yoyo, the attributes of the original yoyo in general will be altered. For example, when a two-quark yoyo is split into two new yoyos under an external force, the total mass of the new yoyos might be greater or smaller than that of the original yoyo. And, in one spinning yoyo, no matter how many u-quarks (or d-quarks) exist, although these quarks spin individually, they also spin in the same direction and at the same angular speed. Here, the angular speeds of u-quarks and d-quarks do not have to be the same, which is different of what is observed in the dishpan experiment, because in this experiment everything is arranged with perfect symmetry, such as the flat bottom of the dish and perfectly round periphery.

2.2 LAWS ON STATE OF MOTION

Based on the discovery (Wu and Lin, 2002) that spins are the fundamental evolutionary feature and characteristic of materials, in this section, we will study the figurative analysis method of the systemic yoyo model and how to apply it to establish laws on state of motion by generalizing Newton's laws of motion. More specifically, after introducing the new figurative analysis method, we will have a chance to generalize all the three laws of motion so that external forces are no longer required for these laws to work. As what's known, these laws are one of the reasons why physics is an "exact" science. And, in the rest of this book, we will show that these generalized forms of the original laws of mechanics will be equally applicable to social sciences and humanity areas as their classic forms in natural science. The presentation in this section is based on (Lin, 2007).

2.2.1 The first law on state of motion

Newton's first law says: An object will continue in its state of motion unless compelled to change by a force impressed upon it. This property of objects, their natural resistance to changes in their state of motion, is called inertia. Based on the theory of blown-ups, one has to address two questions not settled by Newton in his first law: If a force truly impresses on the object, the force must be from the outside of the object. Then, where can such a force be from? How can such natural resistance of objects to changes be considered natural?

It is because uneven densities of materials create twisting forces that fields of spinning currents are naturally formed. This end provides an answer to the first question. Based on the yoyo model (Figure 2.1), the said external force comes from the spin field of the yoyo structure of another object, which is another level higher than the object of our concern. These forces from this new spin field push the object of concern away from its original spin field into a new spin field. Because if there is not such a forced traveling, the said object will continue its original movement in its original spin field. That is why Newton called its tendency to stay in its course of movement as its resistance to changes in its state of motion and as natural. Based on this discussion and the yoyo model (Figure 2.1) developed for each and every object and system in the universe, Newton's first law of mechanics can be rewritten in a general term as follows:

The First Law on State of Motion: *Each imaginable and existing entity in the universe is a spinning yoyo of a certain dimension. Located on the outskirt of the yoyo is a spin field. Without being affected by another yoyo structure, each particle in the said entity's yoyo structure continues its movement in its orbital state of motion.*

Because for Newton's first law to hold true, one needs an external force, when people asked Newton where such an initial force could be from, he answered (jokingly?): "It was from God. He waved a bat and provided the very first blow to all things he created (Kline, 1972)." If such an initial blow is called the first push, then the yoyo model in Figure 2.1 and the stirring forces naturally existing in each "yoyo" created by uneven densities of materials' structures will be called the second stir.

2.2.2 The second law on state of motion

Newton's second law of motion says that when a force does act on an object, the object's velocity will change and the object will accelerate. More precisely, what is claimed is that its acceleration $\vec{a}$ will be directly proportional to the magnitude of the total (or net) force $\vec{F}_{net}$ and inversely proportional to the object's mass m. In symbols, the second law is written:

$$\vec{F}_{net} = m\vec{a} = m\frac{d\vec{v}}{dt} \tag{2.1}$$

Even though equ. (2.1) has been one of the most important equations in mechanics, when one ponders over this equation long enough, he has to ask the following questions: What is a force? Where are forces from and how do forces act on other objects?

To answer these questions, let us apply Einstein's concept of "uneven time and space" of materials' evolution (Einstein, 1997). So, we can assume

$$\vec{F} = -\nabla S(t, x, y, z), \tag{2.2}$$

where $S = S(t, x, y, z)$ stands for the time-space distribution of the external acting object (a yoyo structure). Let $\rho = \rho(t, x, y, z)$ be the density of the object being acted upon. Then, equ. (2.1) can be rewritten as follows for a unit mass of the object being acted upon:

$$\frac{d\vec{v}}{dt} = -\frac{1}{\rho(t, x, y, z)}\nabla S(t, x, y, z). \tag{2.3}$$

If $S(t, x, y, z)$ is not a constant, or if the structure of the acting object is not even, equ. (2.3) can be rewritten as

$$\frac{d(\nabla x \times \vec{v})}{dt} = -\nabla x \times \left[\frac{1}{\rho}\nabla S\right] \neq 0 \tag{2.4}$$

and it represents an eddy motion due to the nonlinearity involved. That is, when the concept of uneven structures is employed, Newton's second law actually indicates that a force, acting on an object, is in fact the attraction or gravitation from the acting object. It is created within the acting object by the unevenness of its internal structure.

By combining this new understanding of Newton's second law with the yoyo model, we get the models on how an object m is acted upon by another object M (see Figures 2.8 and 2.9).

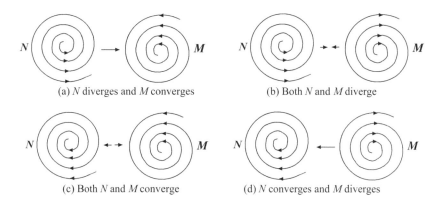

(a) N diverges and M converges   (b) Both N and M diverge

(c) Both N and M converge   (d) N converges and M diverges

Figure 2.11 Same scale acting and reacting spinning yoyos of inharmonic patterns.

involved, the divergent spin field (N) exerts more action on the convergent field (M) than M's reaction peacefully in the case of Figure 2.10(a) and violently in the case of Figure 2.11 (a).

(2) *For the cases (b) in Figures 2.10–2.11, there are permanent equal, but opposite, actions and reactions with the interaction more violent in the case of Figure 2.10(b) than in the case of Figure 2.11(b).*

(3) *For the cases in (c) of Figures 2.10–2.11, there is a permanent mutual attraction. However, for the former case, the violent attraction may pull the two spin fields together and have the tendency to become one spin field. For the later case, the peaceful attraction is balanced off by their opposite spinning directions. And, the spin fields will coexist permanently.*

That is to say, Newton's third law holds true temporarily for cases (a), permanently for cases (b) and partially for cases (c) in Figures 2.10–2.11.

If we look at Newton's third law from the second angle: One spinning yoyo m is acted upon by an eddy flow M of a higher level and scale. If we assume m is a particle in a higher-level eddy flow N before it is acted upon on by M, then we are looking at situations as depicted in Figures 2.8 and 2.9. Jointly, we have what is shown in Figure 2.12, where the sub-eddies created in Figure 2.12(a) are both converging, since the spin fields of N and M are suppliers for them and sources of forces for their spins. Sub-eddies in Figure 2.12(b) are only spinning currents. They serve as middle stop before supplying to the spin field of M. Sub-eddies in Figure 2.12(c) are diverging. And, sub-eddies in Figure 2.12(d) are only spinning currents similar to those in Figure 2.12(b).

That is, based on our analysis on the scenario that one object m, situated in a spin field N, is acted upon by an eddy flow M of a higher level and scale, we can generalize Newton's third law to the following form:

The Fourth Law on State of Motion: *When the spin field M acts on an object m, rotating in the spin field N, the object m experiences equal, but opposite, action and reaction, if it is either thrown out of the spin field N and not accepted by that of M*

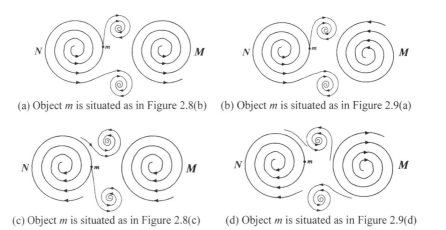

(a) Object *m* is situated as in Figure 2.8(b)   (b) Object *m* is situated as in Figure 2.9(a)

(c) Object *m* is situated as in Figure 2.8(c)   (d) Object *m* is situated as in Figure 2.9(d)

Figure 2.12 Object *m* might be thrown into a sub-eddy created by the spin fields of *N* and *M* jointly.

(Figure 2.8(a), (d), Figure 2.9(b) and (c)) or trapped in a sub-eddy motion created jointly by the spin fields of N and M (Figure 2.8(b), (c), Figure 2.9(a) and (d)). In all other possibilities, the object m does not experience equal and opposite action and reaction from N and M.

2.2.4 Validity of figurative analysis?

In the previous three subsections, we have heavily relied on the analysis of shapes and dynamic graphs. To any scientific mind produced out of the current formal education system, he/she will very well question the validity of such a method of reasoning naturally. To address this concern, let us start with the concept of equal quantitative effects. For detailed and thorough study of this concept, please consult (Wu and Lin, 2002; Lin, 1998).

By equal quantitative effects, it means the eddy effects with non-uniform vortical vectorities existing naturally in systems of equal quantitative movements due to the unevenness of materials. Here, by equal quantitative movements, it means such movements that quasi-equal acting and reacting objects are involved or two or more quasi-equal mutual constraints are concerned with. What's significant about equal quantitative effects is that they can easily throw calculations of equations into computational uncertainty. For example, if two quantities x and y are roughly equal, then $x - y$ becomes a computational uncertainty involving large quantities with infinitesimal increments. This end is closely related to the second crisis in the foundations of mathematics.

Based on recent studies in chaos (Lorenz, 1993), it is known that for nonlinear equation systems, which always represent equal quantitative movements (Wu and Lin, 2002), minor changes in their initial values lead to dramatic changes in their solutions. Such extreme volatility existing in the solutions can be easily caused by changes of a digit many places after the decimal point. Such a digit place far away from the decimal point in general is no longer practically meaningful. That is, when equal quantitative effects are involved, we face with either the situation where no equation can be reasonably established or the situation that the established equation cannot be solved with valid and meaningful solution.

That is, the concept of equal quantitative effects has computationally declared that equations are not eternal and that there does not exist any equation under equal quantitative effects. That is why OuYang (Lin, 1998) introduced the methodological method of "abstracting numbers (quantities) back into shapes (figurative structures)." Of course, the idea of abstracting numbers back to shapes is mainly about how to describe and make use of the formality of eddy irregularities. These irregularities are very different of all the regularized mathematical quantifications of structures.

Because the currently available variable mathematics is entirely about regularized computational schemes, there must be the problem of disagreement between the variable mathematics and irregularities of objective materials' evolutions and the problem that distinct physical properties are quantified and abstracted into indistinguishable numbers. Such incapability of modern mathematics has been shown time and time again in areas of practical forecastings and predictions. For example, since theoretical studies cannot yield any meaningful and effective method to foretell drastic weather changes, especially about small or zero probability disastrous weather systems, (in fact, the study of chaos theory indicates that weather patterns are chaotic and unpredictable. A little butterfly fluttering its tiny wings in Australia can drastically change the weather patterns in North America (Gleick, 1987), the technique of live report has been widely employed. However, in the area of financial market predictions, it has not been so lucky that the technique of live report can possibly applied as effectively. Due to equal quantitative effects, the movements of prices in the financial market place have been truly chaotic when viewed from the contemporary scientific point of view.

That is, the introduction of the concept of equal quantitative effects has made the epistemology of natural sciences gone from solids to fluids and completed the unification of natural and social sciences. More specifically, after we have generalized Newton's laws of motion, which have been the foundations on which physics is made into an "exact" science, in the previous three subsections, these new laws can be readily employed to study social systems, such as military conflicts, political struggles, economic competitions, etc.

Since we have briefly discussed about the concept of equal quantitative effects and inevitable failures of current variable mathematics under the influence of such effects, then how about figurative analysis?

As for the usage of graphs in our daily lives, it goes back as far as our recorded history can go. For example, any written language consists of an array of graphic figures. In terms of figurative analysis, one early work is the book, named *Yi Ching* (or the *Book of Changes*, as known in English (Wilhalm and Baynes, 1967)). For now, no one knows exactly when this book was written and who wrote it. All known is that the book has been around since about three thousand years ago. In that book, the concept of Yin and Yang was first introduced and graphic figures are used to describe supposedly all matters and events in the world. When Leibniz (a founder of calculus) had a hand on that book, he introduced the binary number system and base p number system in modern mathematics (Kline, 1972). Later on, Bool

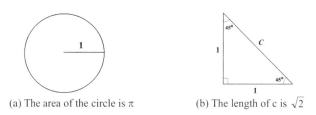

(a) The area of the circle is π   (b) The length of c is $\sqrt{2}$

Figure 2.13 Representing π and $\sqrt{2}$ figuratively and precisely.

furthered this work and laid down the foundation for the modern computer technology to appear.

In our modern days, figures and figurative analysis are readily seen in many aspects of our lives. One good example is the number π. Since we cannot write out this number in the traditional fashion (in either the decimal form or the fraction form), we simply use a figure π to indicate it. The same idea is employed to write all irrational numbers. In the area of weather forecasting, figurative analysis is used each and every day in terms of weather maps. In terms of studies of financial markets, a big part of the technical analysis is developed on graphs. So, this part of technical analysis can also be seen as an example of figurative analysis.

From the recognition of equal quantitative effects and the realization of the importance of figurative analysis, OuYang invented and materialized a practical way to "abstract numbers back into shapes" so that the forecasting of many disastrous small or zero probability weather systems becomes possible. For detailed discussion about this end, please consult Appendix D in (Wu and Lin, 2002) and (Lin, 1998). To simplify the matter, let's see how to abstract numbers π and $\sqrt{2}$ back into shapes with their inherent structures kept. In Figure 2.13(a), the exactly value π is represented using the area of a circle of radius 1. And, the precise value of $\sqrt{2}$ is given in Figure 2.13(b) by employing the special right triangle. By applying these simply graphs, the meaning, the precise values and their inherent structures of π and $\sqrt{2}$ are presented once for all.

2.3 THEORETICAL JUSTIFICATIONS

This section looks at the theoretical foundation on why such an intuition as the yoyo model of general systems holds for each and every system that is either tangible or imaginable.

2.3.1 Blown-ups: Moments of transition in evolutions

When we study nature and treat everything we see as a system (Klir, 2001), then one fact we can easily seen is that many systems in nature evolve in concert. When one thing changes, many other seemingly unrelated things alter their states of existence accordingly. That is why we propose (OuYang, Chen, and Lin, 2009) to look at the evolution of a system or event of pour concern as a whole. That is, when developments and changes naturally existing in the natural environment are seen as a whole, we have the concept of whole evolutions. And, in whole evolutions, other than continuities, as well studied in modern mathematics and science, what seems to be more important and more common is discontinuity, with which transitional changes (or blown-ups) occur. These blown-ups reflect not only the singular transitional characteristics of the whole evolutions of nonlinear equations, but also the changes of old structures being replaced by new structures. By borrowing the form of calculus, we can write the concept of blown-ups as follows: For a given (mathematical or symbolic) model, that truthfully describes the physical situation of our concern, if its solution $u = u(t; t_0, u_0)$, where t stands for time and u_0 the initial state of the system, satisfies

$$\lim_{t \to t_0} |u| = +\infty, \tag{2.7}$$

and at the same time moment when $t \to t_0$, the underlying physical system also goes through a transitional change, then the solution $u = u(t; t_0, u_0)$ is called a blown-up solution and the relevant physical movement expresses a blown-up. For nonlinear models in independent variables of time (t) and space $(x, x$ and y, or x, y, and $z)$, the concept of blown-ups are defined similarly, where blow-ups in the model and the underlying physical system can appear in time or in space or in both.

2.3.2 Mathematical properties of blown-ups

To help us understand the mathematical characteristics of blown-ups, let us look at the following constant-coefficient equation:

$$\dot{u} = a_0 + a_1 u + \cdots + a_{n-1} u^{n-1} + u^n = F, \tag{2.8}$$

where u is the state variable, and $a_0, a_1, \ldots, a_{n-1}$ are constants. Based on the fundamental theorem of algebra, let us assume that equ. (2.8) can be written as

$$\dot{u} = F = (u - u_1)^{p_1} \ldots (u - u_r)^{p_r} \left(u^2 + b_1 u + c_1\right)^{q_1} \ldots \left(u^2 + b_m u + c_m\right)^{q_m}, \tag{2.9}$$

where p_i and q_j, $i = 1, 2, \ldots, r$ and $j = 1, 2, \ldots, m$, are positive whole numbers, and $n = \sum_{i=1}^{r} p_i + 2\sum_{j=1}^{m} q_j$, $\Delta = b_j^2 - 4c_j < 0$, $j = 1, 2, \ldots, m$. Without loss of generality, assume that $u_1 \geq u_2 \geq \cdots \geq u_r$, then the blown-up properties of the solution of equ. (2.9) are given in the following theorem.

Theorem 2.1. The condition under which the solution of an initial value problem of equ. (2.8) contains blown-ups is given by

1. When u_i, $i = 1, 2, \ldots, r$, does not exist, that is, $F = 0$ does not have any real solution; and
2. If $F = 0$ does have real solutions u_i, $i = 1, 2, \ldots, r$, satisfying $u_1 \geq u_2 \geq \ldots \geq u_r$,

   (a) When n is an even number, if $u > u_1$, then u contains blow-up(s);
   (b) When n is an odd number, no matter whether $u > u_1$ or $u < u_r$, there always exist blown-ups.

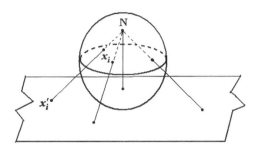

Figure 2.14 The Riemann ball – relationship between planar infinity and three-dimensional North Pole.

A detailed proof of this theorem can be found in (Wu and Lin, 2002, pp. 65–66) and is omitted here. And for higher order nonlinear evolution systems, please consult (Lin, 2008).

2.3.3 The problem of quantitative infinity

One of the features of blown-ups is the quantitative infinity ∞, which stands for indeterminacy mathematically. So, a natural question is how to comprehend this mathematical symbol ∞, which in applications causes instabilities and calculation spills that have stopped each and every working computer.

To address the previous question, let us look at the mapping relation of the Riemann ball, which is well studied in complex functions (Figure 2.14). This so-called Riemann ball, a curved or curvature space, illustrates the relationship between the infinity on the plane and the North Pole N of the ball. Such a mapping relation connects $-\infty$ and $+\infty$ through a blown-up. Or in other words, when a dynamic point x_i travels through the North Pole N on the sphere, the corresponding image x_i' on the plane of the point x_i shows up as a reversal change from $-\infty$ to $+\infty$ through a blown-up. So, treating the planar points $\pm\infty$ as indeterminacy can only be a product of the thinking logic of a narrow or low dimensional observ-control, since, speaking generally, these points stand implicitly for direction changes of one dynamic point on the sphere at the polar point N. Or speaking differently, the phenomenon of directionless, as shown by blown-ups of a lower dimensional space, represents exactly a direction change of movement in a higher dimensional curvature space. Therefore, the concept of blown-ups can specifically represent implicit transformations of spatial dynamics. That is, through blown-ups, problems of indeterminacy of a narrow observ-control in a distorted space are transformed into determinant situations of a more general observ-control system in a curvature space. This discussion shows that the traditional view of singularities as meaningless indeterminacies has not only revealed the obstacles of the thinking logic of the narrow observ-control (in this case, the Euclidean space), but also the careless omissions of spatial properties of dynamic implicit transformations (bridging the Euclidean space to a general curvature space).

Summarizing what has been discussed above in this section, we can see that nonlinearity, speaking mathematically, stands (mostly) for singularities in Euclidean spaces, the imaginary plane discussed above. In terms of physics, nonlinearity represents eddy motions, the movements on curvature spaces, the Riemann ball above. Such motions are a problem about structural evolutions, which are a natural consequence of uneven evolutions of materials. So, nonlinearity accidentally describes discontinuous singular evolutionary characteristics of eddy motions (in curvature spaces) from the angle of a special, narrow observ-control system, the Euclidean spaces.

2.3.4   Equal quantitative effects

Another important concept studied in the blown-up theory is that of equal quantitative effects. Even though this concept was initially proposed in the study of fluid motions, it essentially represents the fundamental and universal characteristics of all movements of materials. What's more important is that this concept reveals the fact that nonlinearity is originated from the figurative structures of materials instead of non-structural quantities of the materials.

The so-called equal quantitative effects stand for the eddy effects with non-uniform vortical vectorities existing naturally in systems of equal quantitative movements due to the unevenness of materials. And, by equal quantitative movements, it is meant to be the movements with quasi-equal acting and reacting objects or under two or more quasi-equal mutual constraints. For example, the relative movements of two or more planets of approximately equal masses are considered equal quantitative movements. In the microcosmic world, an often seen equal quantitative movement is the mutual interference between the particles to be measured and the equipment used to make the measurement. Many phenomena in daily lives can also be considered equal quantitative effects, including such events as wars, politics, economies, chess games, races, plays, etc.

Comparing to the concept of equal quantitative effects, the Aristotelian and Newtonian framework of separate objects and forces is about unequal quantitative movements established on the assumption of particles. On the other hand, equal quantitative movements are mainly characterized by the kind of measurement uncertainty that when I observe an object, the object is constrained by me. When an object is observed by another object, the two objects cannot really be separated apart. At this junction, it can be seen that the Su-Shi Principle of Xuemou Wu's panrelativity theory (1990), Bohr (N. Bohr, 1885–1962) principle and the relativity principle about microcosmic motions, von Neumann's Principle of Program Storage, etc., all fall into the uncertainty model of equal quantitative movements with separate objects and forces.

What's practically important and theoretically significant is that eddy motions are confirmed not only by daily observations of surrounding natural phenomena, but also by laboratory studies from as small as atomic structures to as huge as nebular structures of the universe. At the same time, eddy motions show up in mathematics as nonlinear evolutions. The corresponding linear models can only describe straight-line-like spraying currents and wave motions of the morphological changes of reciprocating currents. What is interesting here is that wave motions and spraying currents are local characteristics of eddy movements. This fact is very well shown by the fact that linearities are special cases of nonlinearities. Please note that we do not mean that linearities are approximations of nonlinearities.

The birth-death exchanges and the non-uniformity of vortical vectorities of eddy evolutions naturally explain where and how quantitative irregularities, complexities and multiplicities of materials' evolutions, when seen from the current narrow observ-control system, come from. Evidently, if the irregularity of eddies comes from the unevenness of materials' internal structures, and if the world is seen at the height of structural evolutions of materials, then the world is simple. And, it is so simple that there are only two forms of motions. One is clockwise rotation, and the other counter clockwise rotation. The vortical vectority in the structures of materials has very intelligently resolved the Tao of Yin and Yang of the "*Book of Changes*" of the eastern mystery (Wilhalm and Baynes, 1967), and has been very practically implemented in the common form of motion of all materials in the universe. That is when the concept of invisible organizations of the blown-up system comes from.

The concept of equal quantitative effects not only possesses a wide range of applications, but also represents an important omission of modern science, developed in the past 300 plus years. Evidently, not only are equal quantitative effects more general than the mechanic system of particles with further reaching significance, but also have they directly pointed to some of the fundamental problems existing in modern science.

In order for us to intuitively see why equal quantitative effects are so difficult for modern science to handle by using the theories established in the past 300 plus years, let us first look at why all materials in the universe are in rotational movements. According to Einstein's uneven space and time, we can assume that all materials have uneven structures. Out of these uneven structures, there naturally exist gradients. With gradients, there will appear forces. Combined with uneven arms of forces, the carrying materials will have to rotate in the form of moments of forces. That is exactly what the ancient Chinese Lao Tzu (English and Feng, 1972) said: "Under the heaven, there is nothing more than the Tao of images," instead of Newtonian doctrine of particles (under the heaven, there is such a Tao that is not about images but sizeless and volumeless particles). The former stands for an evolution problem of rotational movements under stirring forces. Since structural unevenness is an innate character of materials, that is why it is named second stir, considering that the phrase of first push was used first in history (OuYang, et al., 2000). What needs to be noted is that the phrases of first push and second stir do not mean that the first push is prior to the second stir.

Now, we can imagine that the natural world and/or the universe be composed of entirely with eddy currents, where eddies exist in different sizes and scales and interact with each other. That is, the universe is a huge ocean of eddies, which change and evolve constantly. One of the most important characteristics of spinning fluids, including spinning solids, is the difference between the structural properties of inwardly and outwardly spinning pools and the discontinuity between these pools. Due to the stirs in the form of moments of forces, in the discontinuous zones, there exist sub-eddies and sub-sub-eddies (Figure 2.15, where sub-eddies are created naturally by the large eddies M and N). Their twist-ups (the sub-eddies) contain highly condensed amounts of materials and energies. Or in other words, the traditional frontal lines and surfaces (in meteorology) are not simply expansions of particles without any structure. Instead, they represent twist-up zones concentrated with irregularly structured materials and energies (this is where the so-called small probability events appear and small-probability information is observed and collected so such information (event) should also be called irregular information (and event)). In terms of basic energies, these twist-up zones cannot be formed by only the pushes of external forces and cannot

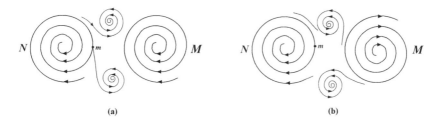

Figure 2.15 Appearance of sub-eddies.

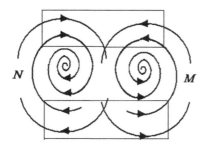

Figure 2.16 Structural representation of equal quantitative effects.

be adequately described by using mathematical forms of separate objects and forces. Since evolution is about changes in materials' structures, it cannot be simply represented by different speeds of movements. Instead, it is mainly about transformations of rotations in the form of moments of forces ignited by irregularities. The enclosed areas in Figure 2.16 stand for the potential places for equal quantitative effects to appear, where the combined pushing or pulling is small in absolute terms. However, it is generally difficult to predict what will come out of the power struggles. In general, what comes out of the power struggle tends to be drastic and unpredictable by using the theories and methodologies of modern science.

2.4 EMPIRICAL JUSTIFICATIONS

Continuing on what was done in the previous sections, we will in this section study several empirical evidences and observations that underline the existence of the yoyo structure behind each and every system, which either tangibly exists or is intellectually imaginable.

2.4.1 Bjerknes' circulation theorem

Based on the previous discussions, it is found that nonlinearity accidentally describes discontinuous singular evolutionary characteristics of eddy motions (in curvature spaces) from the angle of a special, narrow observ-control system, the Euclidean spaces

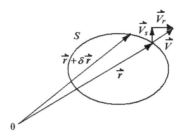

Figure 2.17 The definition of a closed circulation.

(the imaginary plane). To support this end, let us now look at the Bjerknes' Circulation Theorem (1898).

At the end of the 19th century, Bjerkes (1898) (Hess, 1959) discovered the eddy effects due to changes in the density of the media in the movements of the atmosphere and ocean. He consequently established the well-know circulation theorem, which was later named after him. Let us look at this theorem briefly.

By a circulation, it is meant to be a closed contour in a fluid. Mathematically, each circulation Γ is defined as the line integral about the contour of the component of the velocity vector locally tangent to the contour. In symbols, if $\vec{V}$ stands for the speed of a moving fluid, S an arbitrary closed curve, $\delta\vec{r}$ the vector difference of two neighboring points of the curve S (Figure 2.17), then a circulation Γ is defined as follows:

$$\Gamma = \oint_S \vec{V}\delta\vec{r}. \tag{2.10}$$

Through some very clever manipulations, we can produce the following well-known Bjerknes' Circulation Theorem:

$$\frac{d\vec{V}}{dt} = \iint_\sigma \nabla\left(\frac{1}{\rho}\right) \times (-\nabla p) \cdot \delta\sigma - 2\Omega\frac{d\sigma}{dt}, \tag{2.11}$$

where $\vec{V}$ stands for the velocity of the circulation, σ the projection area on the equator plane of the area enclosed by the closed curve S, p the atmospheric pressure, ρ the density of the atmosphere, and Ω the earth's rotational angular speed.

The left hand side of equ. (2.11) represents the acceleration of the moving fluid, which according to Newton's second law of motion is equivalent to the force acting on the fluid. On the right hand side, the first term is called a solenoid term in meteorology. It is originated from the interaction of the p- and ρ-planes due to uneven density ρ so that a twisting force is created. Consequently, materials' movements must be rotations with the rotating direction determined by the equal p- and ρ-plane distributions (Figure 2.18). The second term in equ. (2.11) comes from the rotation of the earth. In short, when a force is acting on a fluid, a rotation is created.

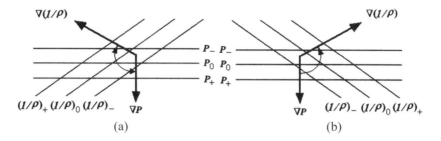

Figure 2.18 A diagram for solenoid circulations.

2.4.2 Conservation of informational infrastructure

Some branches of modern science were made "exact" by introducing various laws of conservation, even though, at the times when they were proposed, there might not have been any "theoretical or mathematical" foundations for these laws. Walking along the similar lines, Lin (1995) developed a theoretical foundation for some laws of conservation, such as the laws of conservation of matter-energy, of fundamental particles, etc., on the basis of general systems theory. By addressing some problems related to the discussions in (Lin, 1995), Lin and Fan (1997) systematically showed that human understanding of nature can be very much limited by our sensing organs, even though our constant attempts do help us get closer to the true state of the nature.

To forms the heuristic foundation of the conservation law of informational infrastructure, let us first look at the intuitive understanding of the concept of general systems. From a practical point of view, a system is what is distinguished as a system (Klir, 1985). From a mathematical point of view, a system is defined as follows (Lin, 1987): S is a (general) system, provided that S is an ordered pair (M, R) of sets, where M is the set of objects of the system S and R a set of some relations on the set M. The sets M and R are called the object set and the relation set of the system S, respectively. (For those readers who are sophisticated enough in mathematics, for each relation r in R, it implies that there exists an ordinal number $n = n(r)$, a function of r, such that r is a subset of the Cartesian product of n copies of the set M.) The idea of using an ordered pair of sets to define the general system is to create the convenience of comparing systems. In particular, when two systems S_1 and S_2 are given, by writing each of these systems as an ordered pair (M_i, R_i), $i = 1, 2$, we can make use of the known facts of mathematics to show that $S_1 = S_2$, if, and only if $M_1 = M_2$ and $R_1 = R_2$. When two systems S_1 and S_2 are not equal, then with their ordered pair structures, we can readily investigate their comparisons, such as how and when they are similar, congruent, or one is structurally less than the other, and other properties between systems. For more details about this end, please consult (Lin, 1999).

By combining these two understandings of general systems, we can intuitively see the following: Each thing that can be imagined in human minds is a system according to Klir's definition so that this thing would look the same as that of an ordered pair (M, R) according to Lin's definition. Furthermore, relations in R can be about some information of the system, its spatial structure, its theoretical structure, etc. That is, there should exist a law of conservation that reflects the uniformity of all tangible and imaginable things with respect to:

1 The content of information;
2 Spatial structures;
3 Various forms of movements, etc.

Based on this intuition of (general) systems and by looking at the available data from particle physics, field theory, astronomy, celestial mechanics, geo-physics, and meteorology, it is shown that between the macrocosm and the microcosm, between the electromagnetic interactions of atomic scale and the strong interactions of Quark's scale, between the central celestial body and the circling celestial bodies of celestial systems, between and among the large, medium, small, and micro-scales of the earth atmosphere, there are exist laws of conservation of products of spatial physical quantities. In particular, it can be conjectured that there might be a more general law of conservation in terms of structure, in which the informational infrastructure, including time, space, mass, energy, etc., is approximately equal to a constant. In symbols, this conjecture can be written as follows:

$$AT \times BS \times CM \times DE = a \tag{2.12}$$

or more generally,

$$AT^{\alpha} \times BS^{\beta} \times CM^{\gamma} \times DE^{\varepsilon} = a, \tag{2.13}$$

where α, β, γ, ε, and a are constants, T, S, M, E and A, B, C, D are respectively time, space, mass, energy, and their coefficients. These two formulas can be applied to various conservative systems of the universal, macroscopic, and microscopic levels. The constants α, β, γ, ε and a are determined by the initial state and properties of the natural system of interest.

In equ. (2.12), when two (or one) terms of choice are fixed, the other two (or three) terms will vary inversely. For example, under the conditions of low speed and the macrocosm, all the coefficients A, B, C, and D equal 1. In this case, when two terms are fixed, the other two terms will be inversely proportional. This end satisfies all principles and various laws of conservation in the classical mechanics, including the laws of conservation of mass, momentum, energy, moment of momentum, etc. So, the varieties of mass and energy in this case are reflected mainly in changes in mass density and energy density. In the classical mechanics, when time and mass are fixed, the effect of a force of a fixed magnitude becomes the effect of an awl when the cross section of the force is getting smaller. When the space and mass are kept unchanged, the same force of a fixed magnitude can have an impulsive effect, since the shorter the time the force acts the greater density the energy release will be. When time and energy are kept the same, the size of working space and the mass density are inversely proportional. When the mass is kept fixed, shrinking acting time and working space at the same time can cause the released energy density reaching a very high level.

Under the conditions of relativity theory, that is, under the conditions of high speeds and great amounts of masses, the coefficients in equ. (2.12) are no longer equal to 1, and equ. (2.13) becomes more appropriate, and the constants A, B, C, D, and a

and the exponents α, β, γ, and ε satisfy relevant equations in relativity theory. When time and space are fixed, the mass and energy can be transformed back and forth according to the well-known mass-energy relation:

$$E = mc^2.$$

When traveling at a speed close to that of light, the length of a pole will shrink when the pole is traveling in the direction of the pole and any clock in motion will become slower. When the mass is sufficiently great, light and gravitation deflection can be caused. When a celestial system evolves to its old age, gravitation collapse will appear and a black hole will be formed. We can imagine based on equ. (2.12) that when our earth evolves sufficiently long, say a billion or trillion years, the relativity effects would also appear. More specifically speaking, in such a great time measurement, the creep deformation of rocks could increase and solids and fluids would have almost no difference so that solids could be treated as fluids. When a universe shrinks to a single point with the mass density infinitely high, a universe explosion of extremely high energy density could appear in a very short time period. So, a new universe is created!

If the previous law of conservation of informational infrastructure holds true, (all the empirical data seem to suggest so), its theoretical and practical significance is obvious. The hypothesis of the law contains the following facts:

1   Multiplications of relevant physical quantities in either the universal scale or the microscopic scale approximately equal a fixed constant; and
2   Multiplications of either electromagnetic interactions or strong interactions approximately equal a fixed constant.

In the widest domain of human knowledge of our modern time, this law of conservation deeply reveals the structural unification of different layers of the universe so that it might provide a clue for the unification of the four basic forces in physics. This law of conservation can be a new evidence for the big bang theory. It supports the model for the infinite universe with border and the oscillation model for the evolution of the universe, where the universe evolves as follows:

$$\ldots \rightarrow \textit{explosion} \rightarrow \textit{shrinking} \rightarrow \textit{explosion} \rightarrow \textit{shrinking} \rightarrow \ldots$$

It also supports the hypothesis that there exist universes "outside of our universe". The truthfulness of this proposed law of conservation is limited to the range of "our universe", with its conservation constant being determined by the structural states of the initial moment of "our universe".

All specific examples analyzed in establishing this law show that to a certain degree, the proposed law of conservation indeed holds true. That is, there indeed exists some kind of uniformity in terms of time, space, mass, and energy among different natural systems of various scales under either macroscopic or microscopic conditions or relativity conditions. Therefore, there might be a need to reconsider some classical theoretical systems so that our understanding about nature can be deepened. For example, under the time and space conditions of the earth's atmosphere, the traditional view in atmospheric dynamics is that since the vertical velocity of each atmospheric huge scale system is much smaller than its horizontal velocity, the vertical velocity is ignored. As a matter of fact (Ren and Nio, 1994), since the atmospheric density difference in the vertical direction is far greater than that in the horizontal direction, and since the gradient force of atmospheric pressure to move the atmospheric system 10 m vertically is equivalent to that of moving the system 200 km horizontally, the vertical velocity should not be ignored. The law of conservation of informational infrastructure, which holds true for all scales used in the earth's atmosphere, might provide conditions for a unified atmospheric dynamics applicable to all atmospheric systems of various scales. As a second example, in the situation of our earth where time and mass do not change, in terms of geological time measurements (sufficiently long time), can we imagine the force, which causes the earth's crust movements? Does it have to be as great as what is believed currently?

As for applications of science and technology, tremendous successes have been made in the macroscopic and microscopic levels, such as shrinking working spatial sectors, shortening the time length for energy releasing, and sacrificing partial masses (say, the usage of nuclear energy). However, the law of conservation of informational infrastructure might very well further the width and depth of applications of science and technology. For example, this law of conservation can provide a theory and thinking logic for us to study the movement evolution of the earth's structure, the source of forces or structural information which leads to successful predictions of major earthquakes, and to find the mechanisms for the formation of torrential rains and for the arrival of earthquakes (Ren, 1996).

Philosophically speaking, the law of conservation of informational infrastructure indicates that in the same way as mass, energy is also a characteristic of physical entities. Time and space can be seen as the forms of existence of physical entities with motion as their basic characteristics. This law of conservation connects time, space, mass, and motion closely into an inseparable whole. So, time, space, mass, and energy can be seen as attributes of physical entities. With this understanding, the concept of mass is generalized and the wholeness of the world is further proved and the thoughts of never diminishing mass and never diminishing universes are evidenced.

2.4.3 Silent human communications

In this subsection, we will look at how the systemic yoyo model is manifested in different areas of life by briefly visiting relevant experimental and clinical evidences. All the omitted details can be found in the relevant references.

Based on the systemic yoyo model, each human being is a 3-dimensional realization of a spinning yoyo structure of a certain dimension higher than three. To illustrate this end, let us consider two simple and easy-to-repeat experiments.

Experiment #1: Feel the Vibe

Let us imagine we go to a sport event, say a swim meet. Assume that the area of competition contains a pool of the Olympic size and along one long side of the pool there are about 200 seats available for spectators to watch the swim meet. The pool area is enclosed with a roof and walls all around the space.

Now, let us physically enter the pool area. What we find is that as soon as we enter the enclosed area of competition, we immediately fall into an excited crowd of screaming and jumping spectators, cheering for their favorite swimmers competing in the pool. Now, let us pick a seat a distance away from the pool deck anywhere in the seating area. After we settle down in our seat, let us purposelessly pick a voluntary helper standing or walking on the pool deck for whatever reason, either for her beauty or for his strange look or body posture, and stare at him intensively. Here is what will happen next.

Magically enough, before long, our stare will be felt by the person from quite a good distance; she/he will turn around and locate us in no time out of the reasonably sized and excited audience.

By using the systemic yoyo model, we can provide one explanation for why this happens and how the silent communication takes place. In particular, each side, the person being stared at and us, is a high dimensional spinning yoyo. Even though we are separated by space and possibly by informational noise, the stare of one party on the other has directed that party's spin field of the yoyo structure into the spin field of the yoyo structure of the other party. Even though the later party initially did not know the forthcoming stare, when her/his spin field is interrupted by the sudden intrusion of another unexpected spin field, the person surely senses the exact direction and location where the intrusion is from. That is the underlying mechanism for the silent communication to be established.

When this experiment is done in a large auditorium where the person being stared at is on the stage, the afore-described phenomenon does not occur. It is because when many spin fields interferes the field of a same person, these interfering fields actually destroy their originally organized flows of materials and energy so that the person who is being stared at can only feel the overwhelming pressure from the entire audience instead of from individual persons.

This easily repeatable experiment in fact has been numerously conducted by some of the high school students in our region. When these students eat out in a restaurant and after they run out of topics to gossip about, they play the game they call "feel the vibe." What they do is to stare as a group at a randomly chosen guest of the restaurant to see how long it takes the guest to feel their stares. As described in the situation of swim meet earlier, the chosen guest can almost always feel the stares immediately and can locate the intruders in no time.

Experiment #2: She Does Not Like Me!

In this case, let us look at the situation of human relationship. When an individual A has a good impression about another individual B, magically, individual B also has a similar and almost identical impression about A. When A does not like B and describes B as a dishonest person with various undesirable traits, it has been clinically proven in psychology that what A describes about B is exactly who A is himself (Hendrix, 2001).

Once again, the underlying mechanism for such a quiet and unspoken evaluation of each other is based on the fact that each human being stands for a high dimensional spinning yoyo and its rotational field. Our feelings toward each other are formed through the interactions of our invisible yoyo structures and their spin fields. So, when person A feels good about another person B, it generally means that their underlying yoyo structures possess the same or very similar attributes, such as spinning in the same direction, both being either divergent or convergent at the same time, both having similar intensities of spin, etc. When person A does not like B and lists many undesirable traits B possesses, it fundamentally stands for the situation that the underlying yoyo fields of A and B are fighting against each other in terms of either opposite spinning directions, or different degrees of convergence, or in terms of other attributes. For a more in depth analysis along a line similar to this one, please consult Part 4: The Systemic Yoyo Model: Its Applications in Human Mind, of this book.

Such quiet and unspoken evaluations of one another can be seen in any working environment. For instance, let us consider a work situation where quality is not and cannot be quantitatively measured, such as a teaching institution in the USA. When one teacher does not perform well in his line of work, he generally uses the concept of quality loudly in day to day settings in order to cover up his own deficiency in quality. When one does not have honesty, he tends to use the term honesty all the time. It is exactly as what Lao Tzu (exact time unknown, Chapter 1) said over 2,000 years ago: "The one who speaks of integrity all the time does not have integrity."

When we tried to repeat this experiment with local high school students, what is found is that when two students A and B, who used to be very good friends, turned away from each other, we ask A why she does not like B anymore. The answer is exactly what we expect: "Because she does not like me anymore!"

Part 2

Domestic financial system: Seen as a closed system

Chapter 3

The financial infrastructure

As what has been shown in Chapter 2 earlier, the universe generally can be seen as an ocean of systems, which in turn can be thought of as abstract rotational yoyo fields of different scales that compete with each other. When one focuses on the finance of a nation as a system, the particular financial system of the nation is made up of fluids, such as money, information, knowledge, etc., where other than studying information and knowledge as fluids by making use of continuous and differentiable functions, money should also be treated as a fluid that circulates around all corners of the nation and fuels the entire economy. If in a specific geographic area the money supply is low, then the regional economic development will suffer and the local residents would have to look for employment and other opportunities somewhere else. If this trend continues, the specific region would have to be abandoned eventually. At the same time, money should also be seen as a carrier of other relevant fluids, such as the relevant information and knowledge, which dictate where the opportunities are so that money as a fluid flows to where it could find the maximal returns.

In this part of the book, we look at the financial system of a general industrialized nation from the viewpoint of fluids and fluid dynamics, while assuming that the nation is in isolation so that no (major) external effect exists to influence its financial stability, its social structure, and its political operation. Similar to what is done in natural science, we first look at closed systems. After we have gained sufficient understanding of such systems, we then investigate the interaction between such relatively closed systems.

This chapter is organized as follows. Section 3.1 looks at the composition of the national financial system of a typical industrialized nation. Section 3.2 focuses on the monetary standard of currencies. Section 3.3 investigates the concept of currency. Section 3.4 studies how credit can help to boost the economy. In Section 3.5 we look into more details of the financial market, while in Section 3.6 we do the same with the banking system. After the basics of the financial infrastructure of a nation are laid out systematically, in Section 3.8 we look at why innovation is the key for the national economy to stay stable and prosperous by using the observational facts of the dishpan experiments and by using the Theorem of Never-Perfect Value Systems of the microeconomics. At the end, Section 3.8 concludes the presentation of this chapter.

3.1 COMPOSITION OF THE FINANCIAL SYSTEM

When looking at the abstract concept of nation that resembles one of the currently existing nations from around the world, we in fact assume the existence of a relatively stable political, social entity within which different parts are connected by common beliefs, philosophical values, and bonds (Lin and Forrest, 2010a; b). For such a nation to operate smoothly, stably, and effectively, such as an industrialized country in general, the nation has to have a mature and perfecting financial infrastructure, which consists mainly of the banking system and the financial market. The purpose of such an establishment is to effectively direct the movement of such fluids as money and related matters so that the nation can sustain itself as an existing social and political organization (Lin and Forrest, 2010c). In such a financial infrastructure, the central bank is the core of the entire system, and the financial market is such a place through which capitals move from those who have surplus into the hands of those who are in need of funds. The financial market is made up of bond market, stock market, and foreign exchange market, where both bond and stock markets play the important role of transferring capitals from the hands of people who cannot put the money into reproduction into the hands of those who can. So, the economic efficiency is improved.

The exchange market represents such a place of trading activities where demanders and suppliers buy and sell their needed and idling foreign currencies through intermediaries. It is an important component of the international financial market. Jointly, the activities of the financial market are directly connected to our wealth, and directly affect the behaviors of the industrial and commercial enterprises. In particular, in the bond market, a security stands for a contingent claim on the future incomes and assets of the issuer. A bond represents a debt certificate which promises that within certain timeframe timely repayments will be made. The bond market is especially important for economic actors. It helps either business firms or governments to raise money by borrowing.

In the stock market, a stock stands for a contingent claim on the corporate earnings. The reason why the stock market is the place that has been attracting the most attention is multi-faceted. Although many people look at the stock market as a place where they could become rich overnight, the real underneath reason is that the government has introduced various laws that stimulate the activities in the stock market for the reason that the magnitude of such activities directly reflects how creatively the nation explores new opportunities (Lin and Forrest, 2011). On the other hand, the volatility of stock prices constantly changes the magnitudes of wealth people owe so that this market dictates the spending inclinations of people. In terms of business firms, stock prices determine how much funds companies could raise for the purpose of investing in their business ventures from selling their stocks.

If one thinks of interactions between nations in terms of monetary movements, the exchange market comes into play. Here exchange is the abbreviation of foreign exchange, the general meaning of which is the method of international payment using foreign currencies. That is the method of how foreign debts can be directly paid in order to materialize the purchasing power of a nation in the international market of trades. The so-called method of international payments means foreign currencies, bank deposits, and various kinds of bills and securities that can be used to make international payments. Because the value one nation's currency represents is mostly likely different from that of another nation and the currency names and units are also different from one country to another, the legal tender of one nation in general cannot be circulated in another nation. So, international remittances have to be accommodated to clear the large amounts of national claims and debt relations created through international economic transactions. In other words, a currency is exchanged into another currency in order to materialize the transfer of capital from one nation to another to pay off the claims and debts between nations. Such activities between nations represent the dynamic concept of foreign exchange. The exchange market is composed of the demanders, suppliers, and intermediaries that provide the service needed for the activities of purchasing and selling foreign currencies. It is an important component of the international financial market and plays an important role in the areas of international credit and debt settlements, supply of credits, elimination of exchange rate risk, etc. The so-called foreign exchange rate, or simply exchange rate, stands for the value of the currency of one nation in terms of another nation's currency. Exchange rate fluctuates constantly, affecting the consumers directly, because it reflects the costs of foreign goods. Therefore, it can be seen that the fluctuation in the exchange market can greatly affect the economy of one nation.

Other than flowing into production through the financial market, capitals can also indirectly finance industrial and commercial productions through financial institutions. And for business firms, the indirect financing is more important than the direct participation. To guarantee the effective and stable operation of the economic and financial systems, banks play an extremely important role. The modern bank system consists of two layers: the central bank as the top layer, and commercial banks and other financial institutions as the second layer, where the central bank occupies the core position. In particular, the central bank possesses its particular position and functionality. Its main responsibilities include:

1   The issue and circulation of the nation's currency. Hence, the central bank is also the nation's issuing bank;
2   The strong backing of commercial banks, the supervision and administration of the nation's financial industry. Hence, it is also the bank of all banks;
3   The exchequer agent that develops and implements the nation's monetary policies, and represents the national government to manage its international financial affairs. Hence, it is also the bank of the government.

On the other hand, commercial banks (also known as deposit banks) occupy the dominant position in the modern financial system of each nation. They represent the framework of a nation's financial system. Their main business activities consist of attracting the deposits from the public and giving out loans to the public. That is also the major difference between commercial banks and other financial institutions. Specialized banks have their respectively focused ranges of business, representing institutions that provide specific financial services. Their main characteristic is that they do not accept demand deposits (deposits in checking accounts). The services provided by specialized banks are created based on the need of the economic development so that the business activities of these banks are irreplaceable. Specialized banks include:

Investment banks, which are specialized for the investment in stocks and bonds of industrial and commercial companies, issue and/or underwrite securities and provide long-term credits to business entities.
Savings banks, which are specialized for opening passbook savings accounts by attracting savings deposits from residents, issues loans and mortgages that are secured with collaterals.

Development banks, which are specialized for issuing investment loans for the purpose of economic developments, can be classified into three groups: international, regional, and national. The mission of national development banks is to support the construction of the nation's infrastructure through long-term financing. So, such development banks are also known as construction banks. The capital of development banks is mainly from the government and by issuing bonds.

Agricultural banks, as the name suggests, are specialized for providing preferential credits to agricultural agencies and farmers.

Secured banks are specialized for providing secured loans for lands, houses, and other real estate properties. Their capitals come mainly from issuing securities backed by mortgages. They can also raise funds through discounting short-term bills and issuing bonds.

Export-import banks are specialized for foreign trades. Instead of making profits, their goal is to promote the outflows and inflows of one nation's goods. Therefore, their capitals are mainly from the investments of governments, borrowings from the governments, issuing bonds, etc.

Non-banking financial institutions include:

Insurance companies, which are established for insurance business. The surplus of insurance premium income minus insurance expenses represents the steady capital funds of the insurance companies. These funds are then used for investment activities and loan business. That makes insurance companies important financial enterprises.

Pension funds, which are the financial institutions that provide steady incomes in terms of annuities for retired senior citizens who participate in pension plans. The capitals in pension funds come from two sources: one is from deductions out of employees' wage payments and the corresponding contributions of the employers, and the other the investment income for the accumulated capitals. Because both the employees and employers each month pay into the pensions funds far more than the amounts the pension funds need to pay out to the annuitants, large amounts of additional money can be used for steady investments.

Credit unions, which are a cooperative form of financial organizations commonly existing in industrialized nations. They are generally established within specific industries or particular geographic regions, such as rural credit cooperatives, urban credit cooperatives, credit union of light industries, credit union of contractors, etc. The general scale of credit unions is quite limited. Their source of funding comes mainly from members' share capitals and the deposits. The main investments of their funds are production and consumer loans made to the members.

Finance corporations (also known as financial companies). They acquire their capitals from issuing commercial papers, bonds, and stocks. The acquired capitals are mainly used for the financial enterprises that make specific consumer loans and industrial and commercial enterprise loans. These corporations do not accept deposits; their methods of financing consist of making small loans against borrowed large funds. They mainly make installment loans for such purchases of durable goods as automobiles.

Trust and investment companies, or briefly known as trust companies, or investment companies, or fund companies, etc. They raise funds generally through issuing stocks, bonds, and investment income certificates. Then they invest their acquired funds in other companies' stocks, bonds, and various kinds of business projects. Trust companies can also use their purchased securities as collaterals to issue their new trust and investment securities to raise additional funds. However, they do not give out loans to industrial and commercial firms. Trust and investment companies collect capitals from investors of medium to small sizes and invest the funds diversely in different countries, regions, and in different types of securities of various economic sectors. By doing so, they expect to reduce the investment risk, while maintaining stable investment returns. Because of this reason, these investment companies have attracted the major attention of small to medium size investors, and have been powerful competitors of commercial banks for deposits from these investors.

In short, the financial infrastructure of a nation is established and further enriched for the purpose of encouraging economic activities in such a way that most of the idling money can be effectively mobilized for either the purpose of production or other reasons. This fact explains why in our modern time, prosperous industrialized nations all have similar financial systems that direct the movement of money purposefully.

3.2 MONETARY STANDARD OF CURRENCIES

According to the definition, within any chosen country money is any object or record that is commonly and generally accepted as payment for goods and services and repayment of debts or socio-economic context (Mishkin, 2007, pp. 8). That is, money plays the roles of medium of exchange, unit of account, and standard of deferred payment (Mankiw, 2007, pp. 22–32). Any kind of object or record that plays these roles can serve as money. In order for any chosen object or record to play these roles in business and trade, there must be a kind of criterion, known as the monetary standard, behind the object and record. In fact, the monetary standard of a nation is the basic unit of the money and the standard of value assumed within the nation's monetary system. The so-called standard of value was once expressed with a certain amount of precise metal, such as gold or silver, during the era of metal currencies. In that case, the name of the monetary standard is given after the metal used for the standard of value. For example, when a certain amount of gold is used to represent the basic unit of money and the standard of value, we then have a gold standard; when a certain amount of silver is employed to stand for the basic unit of money and the standard of value, we then have a silver standard; when both gold and silver are used as the basic unit of money and the standard of value, we then have a bimetal standard. When no amount of metal is used as the basic unit of money and the standard of value, we then have a currency of paper standard. In order to establish its monetary system, a nation first needs to clearly determine the currency standard. The currency introduced on the basis of this standard is known as the standard money. This standard money represents the basis for all other kinds of money used in this nation; all other kinds of money can be exchanged into this standard money without any constraint and are maintained in a fixed price relationship with the standard money.

Commonly seen monetary standards include metal standards and paper standards, where the metal standards include gold standard, silver standard, and various combinations of both. Specifically, the gold standard is such a monetary system that applies a certain amount of gold to represent and to calculate the unit price of the currency. Corresponding to the way a currency can be exchanged into gold, the gold standard is further divided respectively into the gold coin standard (also known as the gold specie standard system), gold bullion standard, and gold exchange standard. When the gold coin standard is used, the currency that circulates in the market consists of gold coins, where the unit currency is determined to be a certain weight and fineness of gold. The public is free to apply for the permission to use gold to cast coins and is allowed to melt gold coins. Other permitted coins of the currency system can be freely exchanged into gold coins of the same value. In terms of the prices of goods, when the price level floated higher, the amount of goods the unit currency can purchase reduces, and the amount of gold the unit currency can buy also decreases. That means that when the value of the currency falls, the price of gold rises. So, people would melt gold coins and sell the resultant gold so that the total amount of gold coins in circulation would correspondingly decrease and the price of goods would consequently fall, leading to a rebound of the currency value to realign with the price of gold. Conversely, when the price of goods slides downward, the value of the currency rises so that the public would cast gold back into coins, increasing the number of gold coins in the circulation. That causes the price of goods to increase and the value of the currency to fall, which eventually helps the gold coins to align with the price of gold.

The gold bullion standard was a product of World War I and was mainly caused by a shortage in gold supply. In order to maintain the gold standard, this method of effectively using gold appeared. In the gold bullion system, although the standard currency was still measured by using a certain amount of gold as the value standard of the unit currency, the gold was stored in a centralized location by the government; instead of having gold coins circulate in the market, people used paper notes issued by the government or the central bank. These paper notes could not be freely traded with gold. For the purpose of meeting the need of output, gold bullions could be obtained with at least a certain minimum amount of paper notes. Because gold coins were no longer circulated, gold was stored centrally by the government, and exchanges could be done only with a certain minimum amount of paper notes, the outflow of gold into foreign countries was greatly reduced. That helped protect the nation's ability to make international payments, and strengthened the ability for the monetary authorities to manage their currencies.

The gold exchange standard in fact is a virtual gold standard. Although it requires gold to be the standard currency, no gold coins are casted and circulated. Instead, what is issued and in circulation is paper notes, which are kept in a fixed exchange rate with the gold standard currency of another country. The issuing of paper notes needs to be backed up by the available amount of gold and the foreign reserves stored in the national central bank or the central bank of a foreign country. So, when needed, either the gold or the foreign reserves can be used in international payments. When exchange is needed, what is obtained is either gold or a foreign currency, which the public does not have any choice. Generally speaking, the domestic currency cannot be directly exchanged into gold without first going through a foreign currency. The adoption of the gold exchange system saves even more gold than the system of gold bullion standard. However, the automatic regulation effect of the gold exchange system on the economy is relatively minor. In order to achieve an balance between international payments and the domestic monetary supply, a large amount of human intervention is needed.

For the system of silver standard, it employs a certain amount of silver to represent and to calculate the unit value of the currency. Based on how the currency can be exchanged into silver, the system of silver standard can also be classified as silver coin standard, silver bullion standard and silver exchange standard. Similar to the gold coin standard, in the system of silver coin standard, what is issued and in circulation as the currency in the market is silver coins, where the unit currency is determined by using a certain amount and certain fineness of silver. The public can freely apply for using silver to cast silver coins, and can also freely melt silver coins. All other forms of currency allowed in the system can be freely traded equivalently with silver coins. Silver and silver coins can be freely converted into each other, while the silver coins are the unlimited legal tender. The existence of these freedoms represents the spirit of the silver standard. It is also because of these freedoms that the system of silver standard can play the role of automatically adjusting the economy. In the development history of currency systems, the silver standard once occupied an important position. However, with the economic development, the amount of trades has been growing increasingly so that the silver coins of large volumes and small values could no longer satisfy the needs of business transactions. So, at the beginning of the 19th century, major European countries, such as France, Italy, Belgium, Switzerland, etc., started to expand their silver standard system into a multiple-standard system. And the silver standard system became almost completely extinct since after the 1930s.

The so-called two-standard system stands for that within which both gold and silver are used as the material of the currency, consisting of gold coins and silver coins, both circulating in the market. With time, the system of this kind standard evolved through three different stages: the parallel standard system, the multiple standard system, and the limping standard system. Here, the characteristics of the parallel standard system include:

1 Both gold coins and silver coins are the standard currency;
2 Both of the coins are unlimited legal tenders;
3 They can be freely casted and melted; and
4 The exchange rate between gold coins and silver coins is completely determined by the market price of the gold coins. It is adjusted through economic forces instead of any human factor.

The characteristics of the so-called system of multiple standards include:

1 Both gold coins and silver coins are the standard currency;
2 Both of the coins are unlimited legal tenders;
3 They can be freely casted and melted; and
4 The exchange rate between gold coins and silver coins is determined through the law.

So, this system is such a currency system that combines the systems of gold standard and silver standard, where both the gold and silver coins are allowed and can be traded using the exchange rate given by the law. The government clearly describes the amount of gold contained in the unit gold coin and the amount of silver contained in the unit of silver coin.

The multiple-standard system can maintain the market exchange rate of gold and silver the same as the legal exchange rate of gold and silver. Hence, the value standard of the unit currency is stabilized. For example, assume that the legal exchange rate between gold and silver is 1:10, while the market rate is 1:12. Then it means that the market price of gold is higher than the price allowed by the law and the market price of the silver is lower than that allowed by the law. In this case, the public would be more willing to melt gold coins into gold and trade the gold for silver using the market price ratio of 1:12. And then the public would cast the purchased silver into silver coins and exchange these silver coins for gold coins according to the legal ratio of 1:10. Through such transactions, one unit of gold coin becomes 1.2 units, earning the extra 0.2 units of gold coins. Such an opportunity of money making continuously drives the public to melt gold coins into gold and to cast silver into silver coins, causing a growing reduction of gold coins available in the circulation while a growing amount of gold traded in the market place. At the same time, the number of silver coins in circulation grows increasingly, while the amount of silver traded in the market place reduces continuously. The increasing amount of gold available for the market trading reduces the market price of gold, and the reducing amount of silver makes the market price of silver rise. If the price changes make gold traded at a price lower than that allowed by the law, then the price of silver will be higher than that allowed by the law. So, a situation opposite of what is described above will occur: The public would melt silver coins into silver and trade the silver for gold; then the purchased gold would be casted into gold coins that are in turn used to exchange for silver coins. Consequently, the amount of silver increases in the market place so that the price of silver will fall. At the same time, because the amount of gold in the market place decreases, the price of gold will rise. In short, as long as the price of gold is higher than the legal price, accordingly the price of silver is lower than the legal price, driven by the opportunity of profit making, the market price of gold will fall and the price of silver will rise; as long as the market price of gold is lower than the legal price, correspondingly the price of silver is higher than the legal price, the market price of gold will rise while the price of silver will fall. That is, when the market prices of gold and silver are not identical to the legal prices, the market will push the prices toward the legal prices until the market prices equal to the legal prices. That is how multiple-standard system corrects itself when needed.

The so-called limping standard system means that

1   Both gold coins and silver coins are the standard currency;
2   Both of the coins are unlimited legal tenders;
3   The gold coins can be freely casted and melted, while silver coins are not allowed; and
4   The exchange rate between gold coins and silver coins is determined by the government through the law.

The gold and silver coins in the multiple-standard system are like two legs. However, for this new system, the freedom of casting and melting silver coins is eliminated, which is like the situation that one leg is now missing. That is how the original multiple-standard system becomes a limping standard system. The appearance of the limping standard system is mainly caused by the drastic fall in the silver price in the 1870s. To maintain the status of the silver standard and its legal exchange rate with gold, France and the United States decided to disallow the public to freely cast silver coins, making the original multiple-standard system a limping standard system.

In terms of the paper standard system, it has been a monetary standard adopted gradually by all countries from around the world since the 1930s. The characteristics of the paper standard system include that the standard currency is issued either by the government or the central bank, and that the standard currency does not maintain any equivalence relationship with any metal. Because the issuance of paper notes (currency) can be altered easily without being constrained by other conditions of the country, it can also been seen as a standard-free system. Because the paper currency, as the standard currency, does not maintain any equivalence relationship with any metal, the issuance of paper notes is not backed by gold. So, with the changes of the market, the supply of money can be artificially adjusted in a timely fashion, by using policies on the basis of practical need and carefully and thoughtfully considerations and planning. By doing so, both inflation and deflation could be eliminated, and steady economic growth could be maintained. At the same time, to maintain the stable valuation of the paper currency against foreign trades and the balance of international payments, the exchange rates need to be closely managed and a foreign reserve fund established. Therefore, in terms of economic development it is more advantageous than the system of gold-coin standard if a nation implements a managed system of paper-money standard. That is the theoretical reasoning behind the development of managed system of paper-money standard. At the same time when the managed system of paper-money standard possesses greater number of advantages over the system of gold-coin standard, it also suffers from certain problems. First of all, because the issuance of paper currency is not limited by any reserve, it makes the elasticity of the money supply very big, which often leads to the occurrence of hyperinflations. Secondly, the exchange rate is not determined by how much gold is concerned with. Instead, it can be freely adjusted by the nations involved, making the value of a currency change constantly. That directly affects the development of international trades and the flow of capitals from one country to another. Lastly, the success and failure of a managed system of paper-money standard are heavily dependent on human factors. So, whether or not the managers have the relevant knowledge of money and the capability of decision making becomes a very important question.

Although the modern managed paper money is fiat, to prevent excessive issuance, to strengthen the trust of the public on the paper money, and to maintain the stability of the paper money and of the exchange rate to guarantee the balance of international payments, some nations still establish their preparation system for paper-money issuance by applying gold, silver, foreign reserves, etc., as their backing. The issuance reserves are divided into cash reserves and warranty provisions. The cash reserves are also known as goods ready, generally including gold, silver, real money, foreign reserves, etc. On the other hand, short-term notes that can be easily converted into cash, the marketable securities of the national treasury, reliable stocks and company bonds generally act as warranty provisions. When seeing from the angle of cash security and the angle of maintaining the stability of exchange rates, cash reserves are better than warranty provisions. It is because the cash reserves can be used to directly honor the paper money and to export the cash, while the warranty provisions cannot achieve any of these goals. However, from the angle of profit making for the banks, warranty provisions are better than cash reserves, because the warranty provisions interest-bearing assets, which can create incomes for the banks, while cash reserves cannot. Considering the scaling capability on the amount of money in circulation, it can be seen that the scaling capability on the amount of money in circulation is small for the system of cash reserves (for example, if the amount of gold and silver or foreign reserves is insufficient, then it will be difficult to issue additional paper currency), while the scaling capability is greater for the system of warranty provisions (for example, when the amount of circulating currency is insufficient, additional paper notes can be issued promptly; and additional money supply can also be injected into the economic system through purchasing marketable securities and certificates). Thus, warranty provisions are better than cash reserves.

Based on the previous discussion on the advantages and disadvantages of the system of paper-money standard, which has been the norm of the modern world, it can be readily expected that human factors will play an ever increasing role in the issuance of paper money, in the making of monetary policies, regulation of exchange rates, and other related matters. So, by looking at all possible ways rotational yoyo fields could potentially interact with each other, and considering the fact that each human being is also a high-dimensional yoyo field, it can be reasonably concluded that under particular circumstances, the decision makers would have to bend to various external pressures. And when this happens, the normal, rational expansionary and lessening adjustment of money would be interrupted, leading to hyperinflations; exchange rate management would be thrown into chaos, causing imbalances in international payments. That in turn will adversely affect the development of international trades and flows of capitals between nations. More importantly, because human factors are heavily involved in the management of the system of paper-money standard, the success and failure of the management becomes a function of the knowledge of money, judgment, and the ability of decision making. However, the knowledge of money changes with time and circumstances. In other words, past experience might not be applicable in the future, while the judgment and the ability of decision making vary to a great extend with the magnitude of the circumstantial pressures. In short, the common acceptance of the system of paper-money standard has prepared the material ground for the arrival of currency wars between geographic regions within a nation and between nations.

3.3 DEFINITION OF MONEY

There are two ways to define what is considered as currency. One is from the theoretical angle; the other from the practical standard point of view. Theoretically, currency is applied as the media of trades. Only those assets that can play the role of media of trades without any ambiguity can be considered as part of the category of money. Because currencies, the deposits in checking accounts, and traveler's checks can all be applied to pay for goods and labors, they undoubtedly function as the media of trades. Therefore, they are considered as money. Additionally, theoretical studies only treat these assets as the categorical parts of money. So, when theoretical methods are used to define money or supply of money, it implies these three classes of assets: currencies, the deposits in checking accounts, and traveler's checks. Although this theoretical definition of money seems to be quite clear, in practice the concept of money is not as clear as expected. Other than those monetary assets as clearly defined in theory, there are still other assets that also function to a certain extend as money. Although these assets do not have the same level of liquidity as currencies and the deposits in checking accounts, they do possess a degree of liquidity. For example, the customers of brokerage firms can write and sign checks against the values of bonds kept at the brokerage firms; the savings deposits at banks can be quickly converted to cash without incurring any cost. Are these assets, which although not completely liquid do possess some degree of liquidity, considered money? By using theoretical means, this question cannot be addressed satisfactorily. So, there are ambiguities in the definition of money if only theoretical methods are employed. This fact indicates that there is need to further consider how to define money.

The category of money as outlined by using the theoretical means is known as the narrow money, denoted by M_1. That is to say, M_1 is an indicator of money supply defined by using the narrowest definition of money. Symbolically, we have

$$M_1 = \text{currency} + \text{deposits in checking accounts} + \text{traveler's checks}$$

If checks can be written against only the deposits in checking accounts, then both such checks and traveler's checks are commonly known as demand deposits, denoted by D. Let us use C to denote the amount of money that is beyond banks. Because currency is the asset with the strongest liquidity, it is used to represent the totality of all currencies outside banks. In this way, we have the following formula for calculating the narrow money

$$M_1 = C + D = M_0 + D$$

From the angle of the government, the narrow money M_1 is the nation's most commonly accepted and strictly controlled indicator of money

From a practical standard point of view, because there is ambiguity in the theoretical definition of what assets belong to the category of money, it is suggested to define the concept of money from a relatively empirical angle. When deciding on which asset should be treated as money, the economic variable that needs to be explained and predicted has to be predetermined. The monetary indicator that can best predict the changes of the particular economic variable describes the boundary of the monetary category of this variable. However, unfortunately, the empirical reality is extremely complicated. The monetary indicator that can well predict the economic activities of concern in one time period might not work at all in another time period for the same set of economic activities.

In short, no matter whether it is using theoretical methods or employing empirical means to define the concept of money, the result is not ideal. And it is difficult to guarantee that a relatively useful monetary indicator of the past will continue to function as expected in the future. So, to truly and correctly know what money is is an impossible task. Even so, economists have still developed several different measurable monetary indicators for various purposes based on empirical experiences, which are widely known as the general money M_2, M_3, and L, defined as follows:

$$M_2 = M_1 + T = M_0 + D + T = C + D + T$$

where M_1 stands for the narrow money, T those assets that are closely related to nominal incomes, although they are not very liquid, such as small time deposits, savings deposits, money market deposit accounts, money market mutual funds held by the public, overnight repurchase agreements, etc. Even though these assets cannot become the media of payments immediately, they can be easily converted into such media within short periods of time. And, M_3 is the total of M_2 plus all other assets with small levels of liquidity, such as the time deposits of large-denominations, money market mutual funds held by institutions, long-term repurchase agreements, etc. The most general measurable monetary indicator is L. It is defined as the sum of M_3 and relatively liquid bonds and receipts, such as short-term treasury bills, commercial papers, savings bonds, bank promissory notes, etc. This indicator L covers all such assets that are completely liquid to those receipts and papers that are relatively liquid. Therefore, L is really not a statistical monetary indicator; instead it is an indicator that conveys some aspect of the information of all those assets that are relatively liquid.

Because each asset is liquid to a degree, any chosen asset possesses the features of money to a certain level. So, we can say that any chosen asset plays the some aspects of the role of money. For instance, small time deposits plays 60% of the role of money; large time deposits can be seen as 40% of money; stocks can be seen 50% the same as money. This end leads to the possibility of applying the liquidity of assets to design an overall weighted monetary indicator. In particular, convert each chosen asset into its equivalent monetary value based on its liquidity; then compute the total monetary value of all converted assets. To this sum is the weighted total M_3' of money obtained by adding M_1. Assume that an economic society has n kinds of assets that are not considered in M_1, the total of the ith kind of asset is A_i, whose level of liquidity is α_i, a value between 0 and 1. Then we have

$$M_3' = M_1 + \sum_{i=1}^{n} \alpha_i A_i$$

In short, because of the difficulty of actually defining what money is and what can be used as currency either in theory or in practice, what is discussed in the previous section implies that under pressure while facing crises, the decision makers could always find a way out in making the humanly desirable decisions regarding money supply, exchange rates, etc. These decisions might look ok in front of the particular circumstances; however, most likely they bring forward with them unexpected and often disastrous consequences.

3.4 CREDIT

Credit represents a promise to pay in the future for the goods purchased or the debt owed. It is an agreement between claims and debts. It stands for the two sides of the activity of borrowing and lending. In lending activities, one party is known as the creditor, while the other party the debtor. The creditor lends out goods or money, known as providing credit; the debtor receives the goods or money, known as fiduciary or receiving credit. If the debtor keeps his promise and returns the goods or money on time, he is then seen as trustworthy. In lending activities, the obligation for the debtor to return the goods or debt on a future date is known as the debt. Within any timeframe, the total of debts is always equal to the total of credits. After money enters the economy, it helps quantify and develop credit money as an extension of money. As a matter of fact, to a large extent, the development of credits has substituted the role of money and become the foundation of money supply. By the liquidity of credit, it means that a lending activity has occurred with a written instrument. That instrument can be traded in the marketplace so that the underlying capital can be used for other purposes.

Accompanying the concept of credit, various financial instruments are introduced to testify the existence of debts and relevant conditions. As short-term instruments of credit for the time durance of within one year, there are promissory notes, drafts (or drafts of exchange), checks, and credit cards, where:

A promissory note is such a bill that promises to pay the holder without any condition a certain amount of cash at a specified time. That is, it is a written, unconditional guarantee that the debtor promises to repay the creditor on a predetermined date. The holder can transfer the ownership of the note using the method of endorsement. According to the difference of issuers, promissory notes are divided into commercial papers and cashier's checks. The former in general are issued by large scale enterprises with good reputation to raise short-term capital. These papers are normally guaranteed by financial institutions. Cashier's checks are certificates that unconditionally pay the holders pre-determined amounts of cash. They are mainly used to substitute for cash. Corresponding to differences in payment terms, promissory notes are classified into spot notes and future notes, where the former stand for notes which demand payments immediately while the latter do not have to be paid until the agreed dates.

Draft or draft of exchange represents such an issued certificate that demands unconditional payment from the payer to the receiver (the holder of the note) according to the specified date. The main difference between drafts and promissory notes is that each draft involves three parties: the issuer, the payer, and the receiver, while each promissory note only involves the issuer and the receiver, where the issuer is the payer.

Checks stand for such notes that are issued by the owners of demand deposits accounts. They require the banks where the accounts are located to pay the holders of the papers the pre-determined amounts. Here, each check also involves three parties: the issuer, the payer, and the fund receiver. After endorsements, checks can be circulated freely. However, checks are different of drafts in two main aspects: (1) the issuer of a check has to be the owner of a demand deposit account at the bank and the payer the bank, while drafts do not have such limitations; and (2) each check is a spot note that demands payment upon the receipt of the check without any issue with promise, while drafts are different.

Credit cards are a form of consumption credit with the characteristic of consuming first and paying next. As a payment tool, they bring huge profits to the issuing banks and merchants, while providing convenience for the consumers.

Long-term instruments of credit with over one year time durance consist mainly of various certificates, such as stocks and bonds, where stocks are certificates showing the ownership, profit sharing rights, and management rights of a stock company. Other than the proportional ownership of the company, they also represent the proportional risk sharing of the company. The holders of stocks cannot retire from their ownerships of the company, but can transfer their stocks, or use the stocks as collaterals. Bonds are certificates of indebtedness; the issuers promise to pay the holders of the bonds interests according to some pre-determined rates and to repay all the principals upon the maturity of the certificates. Same as stocks, bonds are also a tool for issuers to raise funds and for investors to invest their money. They can also be transferred so that both stocks and bonds have their market prices. However, bonds are clearly different from stocks: (1) What is raised through stocks is owned capital, while what is collected from bonds is additional capital; (2) Holders of stocks are considered owners of the company, while the holders of bonds the creditors of the company; (3) Stocks are risky, do not pay fixed dividends, and the original capital does not need to be repaid, while bonds are less riskier, bear fixed interest incomes, and recoup the original investments on time.

The use of credit makes capitals flow from one department to another. As a tool of analyzing this flow and impacts of capital, the table of capital flows is employed. For any economic entity, its wealth stock is known as asset. The physical assets with over one-year lengths of use are known as real assets; all the assets that exist in the form of credit instruments, such as currencies, deposits slips, bonds, stocks, etc., are known as financial assets. Because all financial assets can be traded and transferred, they are also known as financial products. Therefore, financial assets, financial products, financial instruments, and credit instruments all represent the same things. When an economic organization purchases real assets, it is said to have made real investment, or investment for short. When it purchases financial assets, it is said to have made financial investment.

For an individual economic organization, its capital consists mainly of four aspects: (1) recurrent payments or expenditures, which are used to purchase the products and services for daily consumption; (2) real investments, which are used to increase real assets; (3) financial investments, which are employed to increase financial assets; and (4) holding and deposit of cash, which are utilized to increase the balance of money. The main source of funds used for all these aspects of expenditure comes from recurrent revenue and loans. For example, the recurrent income of a family consists of wages and bonuses; the recurrent income of a business firm comes from the revenue of selling its products; and the recurrent income of a government is made up of the tax revenue. Taking out loans generally increases the debt of the organization.

When recurrent expenditures are deducted from the recurrent incomes, what is obtained is the net savings, also known as owned funds, of the company, which can be treated as the source of funds of the company. Other than borrowing money, another source of capital, each firm has also to repay debts, another expense. When debts and repayments of debts are combined, the result is the net indebtedness of the firm. Similarly, for real investments, financial investments, and the holdings and deposits of cash, the concepts of net amounts are also employed. That is, by real investments, it means the net result after combining the purchase and sale of real assets; by financial investments, it means the net result of combining the purchase and sale of financial assets; and by the holdings and deposits of cash, it means the net result of combining the amount of increase and decrease of cash holdings. With these explanations in place, one can obtain such a table that reflects the sources and applications of funds of a firm, Table 3.1, which represents the changes of assets and debts of the firm between two pre-determined time moments.

In general, the savings S of a firm is not the same as the investment I. When this happens, there must appear a difference between the sources and applications of funds. In particular, if there is a surplus due to the reason that the investments are smaller than the savings, then the funds in the surplus have to be dealt with through either financial investments (giving out loans), or repayment of debts, or holdings and deposits of cash. Similarly, if there is a deficit due to the reason that investments are greater than savings, then the deficit has to be handled by either taking out loans, or selling financial

Table 3.1   Sources and applications of funds (the capital account) of a firm.

Applications of funds	Sources of funds
Increase in real assets (real investment): $\Delta RA(I)$ Increase in financial assets (financial investment): $\Delta FA(F)$ Increase in cash holdings and deposits: $\Delta M(H)$	Net increase in incomes (savings): $\Delta NW(S)$ Net increase in debts (loans): $\Delta L(B)$
Total: $\Delta RA + \Delta FA + \Delta M$	Total: $\Delta NW + \Delta L$

Table 3.2   Capital flows of the entire economy.

Business category	Sector 1 Application	Sector 1 Source	Sector 2 Application	Sector 2 Source	Sector 3 Application	Sector 3 Source	Entire economy Application	Entire economy Source
Savings		S_1		S_2		S_3		S
Real assets	I_1		I_2		I_3		I	
Loans		B_1		B_2		B_3		B
Financial Investment	F_1		F_2		F_3		F	
Holdings of cash	H_1		H_2		H_3		H	

assets, or decreasing the holdings and deposits of cash. That explains why in Table 3.1 the total of fund applications is always equal to the sum of fund sources.

When all economic sectors are seen collectively, each payment of one sector becomes an income of another sector. That is how money starts to flow from one economic sector into another. In order to record such capital interactions between various economic sectors, one can assemble all individual firms' or economic sectors' tables of sources and applications of funds together to construct an overall table of capital flows for the entire economy, Table 3.2, where it is assumed that there are a total of three economic sectors, and the symbols S_k, I_k, B_k, F_k, H_k are respectively used to represent the savings, the real investments, borrowings, financial investments, and cash holdings and deposits of the kth economic sector, $k = 1, 2, 3$. Because for each economic sector, the total of fund applications is equal to the total of fund sources, that is, $I_k + F_k + H_k = S_k + B_k$, for the entire economy it must be that the total of fund applications $(I + F + H)$ is the same as the total of fund sources $(S + B)$. Symbolically, we have

$$I + F + H = \sum_{k=1}^{3}(I_k + F_k + H_k) = \sum_{k=1}^{3}(S_k + B_k) = S + B$$

The table of capital flows reflects not only the business trades of real assets between different economic sectors, but also, more importantly, the trades of financial assets. Through tables of capital flows, one can analyze the forms and effects of the financial trades between the individual economic sectors; he can explain how each economic sector internally deals with and makes up the differences between savings and investments. Therefore, the financial structure of the entire economy, along with the distribution of capital and the demand for money, can be understood, providing the necessary bases for the government to introduce and implement fiscal and monetary policies.

To summarize, the introduction of credit helps to speed up the movement of money, circulating from consumers to providers. That, of course, consequently stimulates the further development of the economy. This end coincides with the fact that when extreme spatial unevenness exists in the distribution of materials, (gradient) forces will naturally appear. In terms of the current context, when the money is not homogeneously distributed over space and time, credit appears naturally to play the role of promise of future payments for consumptions of goods and services and for the repayments of monetary loans, and the role of promise between claims and debts.

3.5 FINANCIAL MARKETS

In this section, we look at the financial markets in more details. As they have evolved over time, the financial markets collectively have become a huge scale and complicated market place. It provides a tunnel to promote the circulation of capitals from those people who do not employ them in any way of production to those who do. The fundamental characteristic of the financial markets is that though trading financial instruments the flow of capitals from non-producers to producers is facilitated. This characteristic determines that the basic elements of the financial markets include suppliers and demanders, financial instruments, and financial intermediaries.

Figure 3.1 illustrates the processes of how capitals flow from savers to investors, where the suppliers of capitals include individual people and families, industrial and business firms, financial institutions, and government entities. However, as suppliers and demanders, these four parties carry different weights. When seeing from the angle of supply of capitals, financial institutions, specifically commercial banks, are the largest suppliers. They are not only organizations, to which various forms of deposits

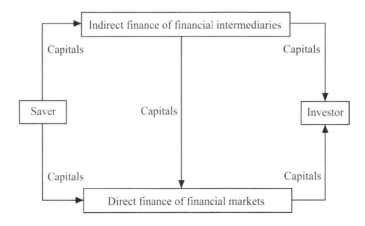

Figure 3.1 Capital flows in a financial system.

are made, but the commercial banks can also create money multiple times through their demand deposits. In particular, the demand deposits strengthen the credit capability of the commercial banks. In the system of fractional reserves, commercial banks give out loans in the form of crediting the demand deposits of the borrowers. As long as these deposits are not completely withdrawn, they become a new source of funds for the banks. They can then give out additional loans against these available funds. If the borrower takes his borrowed amount by transferring it into his account in a different bank, that transferred amount becomes a source of capital of the other bank and this bank could give out loans against this newly available funds. When this process continues indefinitely, the deposits of the banking system are actually multiplied. That is the so-called capability of credit creation. Because of this credit creation, commercial banks can strengthen their credit capability by using demand deposits so that the total money supply of the society is influenced. It is because of this reason that commercial banks occupy an important position in the financial system.

What are inline next are the industrial and commercial enterprises. Due to the time differences between incomes and payments, idling capitals are formed, which becomes an important source of money supply. Next are the balance funds of individual families. These funds are generally deposited in bank accounts or used to purchase priced securities. Lastly, because government agencies spend their fiscal allocations after first having received the funds, there appear idling funds. That also constitutes a source of money supply.

When seeing from the angle of demand, industrial and commercial enterprises are the largest demanders of money. They need large amounts of money for such purposes as capital formation, expansion of production, circulation, etc. Next inline are government organizations. They raise money through issuing bonds in order to fund various public needs, to make up the deficit, etc. The third class of demanders includes families and individuals. They need money to purchase homes, automobiles, and other durable goods, or to satisfy their urgent necessities. Lastly, the demanders of money are financial institutions. When they are short of funds, they would borrow from other financial institutions or from the central bank, or sell their financial assets through the financial markets.

The trading objects of the financial markets are known as financial instruments. Without well-designed financial instruments, there would not be any smooth flow of capital. These financial instruments represent financial assets; they are the objects traded on the market place. That is why financial instruments are also known as financial products, which include bonds, treasuries, stocks, corporate bonds, loan contracts, mortgage deeds, as well as the financial bonds issued by financial institutions, transferrable deposit slips, drafts, commercial papers, bank's acceptance bills, and other various kinds of notes.

Most of the trades of the financial market are completed through financial intermediaries. These appropriate organizations greatly facilitate the intermediation of funds. They include banks, insurance companies, investment firms, and other financial institutions, as well as such individuals as brokers, dealers, jobbers, etc. Brokers are the middlemen of the financial market who, while charging a service fee, play the role of media between the buyers and sellers; dealers, also known as traders, are those speculators who buy and sell securities or drafts for the purpose of making profits from price differences for their own accounts; jobbers of the financial markets are those who play the double roles of brokers and dealers who consign and sell securities for particular issues.

There are different types of financial markets. Based on the terms of the traded objects, financial markets can be divided into two large classes: money markets and capital markets. Here money markets represent those financial markets where traded are short-term credit instruments, including the short-term bond market (the most important money market), deposit slip market (short-term lending market), discounted bills market (such as promissory notes, drafts, etc.), interbank market, where banks adjust their individual money demands and supplies, intermediate short-term funds. Capital markets are those financial markets where the objects of trading are long-term credit instruments, including security markets and long-term lending markets. Security markets can in turn be finely divided into stock markets and (long-term) bond markets.

Based on how traded objects are classified, the financial market can be divided into local currency markets, foreign currency markets, gold markets, stock markets, spot markets, and futures markets. Considering the nature and method of trading, the financial market can be divided into lending markets, rental markets, stock markets, open-outcry markets, and bargaining markets. The financial market is also classified by levels: the primary markets and secondary markets. The primary markets are the financial markets where either companies or governments sell their newly issued securities, either bonds or stocks, to the initial buyers; the secondary markets are those markets where initially purchased securities are further traded.

Because the financial market provides a way to stimulate the circulation of money so that money is transferred from people who do not use the money for production purposes to those who will, the quality of a nation's financial markets can be reasonably used as an indicator of the state of health of the nation in general and how the nation is engaged in the exploration of new opportunities in particular.

3.6   THE BANKING SYSTEM

Other than through the financial market capitals are flown to purposes of production, there is still another relatively less risky path. That is through the indirect financing of financial institutions. For industrial and commercial enterprises, such indirect financing is more important than direct financing. And banks, among all financial institutions, play an extremely important role in terms of guaranteeing the efficiency and stability of the operation of the economic and financial systems.

3.6.1   The modern banking system

In the modern, industrialized nations, there commonly exists such a financial system that is made up of two-tiered banking system. The first tier consists of the central bank as the core of the entire system; and the second tier includes commercial banks and other financial institutions, which have shown the tendency of becoming more diversified over time.

The central bank is the core of a nation's financial system. It occupies a specific position and plays specific roles. Its main functions include: (1) the responsibility for the issuance and circulation of the nation's money so that the central bank is also the issuing bank; (2) the strong backing for commercial banks. It is responsible to monitor the financial industry of the nation so that it is also known as the bank of all banks; and (3) the fiscal agent of the nation. It introduces and reinforces the monetary policies of the nation, and handles the international financial business transactions on behalf of the national government. That is why it is also known as the bank of the national government.

The second tier of the banking system includes commercial banks, specialized banks, and other non-banking financial institutions. In particular, commercial banks play a dominating role in the financial system of each nation, and are the backbone of the financial system. They are mainly engaged in attracting deposits from the public and giving out loans back to the public. That is the main signature separating commercial banks from other financial institutions. Specialized banks have their particularly defined ranges of business and provide much focused financial services. Their most noticeable characteristic is that they do not accept demand deposits (checking account deposits). The ranges of service of the specialized banks are established based on the needs of the economic developments. That makes these banks irreplaceable. Although there are many kinds of specialized banks, the following are a few of the major types.

Investment banks: They are specifically designed to deal with the stocks and bonds of industrial and commercial enterprises. They issue and underwrite securities, and provide long-term credits for business firms.

Savings banks: They open passbook savings accounts to satisfy residents' need of savings deposits, and then use these funds as the source of capital to give out mortgages. These banks do not involve in the business of checking accounts deposits and in that of providing loans to industrial and commercial firms. However, they do actively participate in the investment activities of bonds issued by governments and companies. In recent years, the business activities of these banks have also touched on such non-traditional areas as commercial credits, consumer credits, etc.

Development banks: They are specifically designed to provide investment loans for the purpose of economic developments. There are three kinds of development banks: International, regional, and domestic. The World Bank, whose full name is the International Bank for Reconstruction and Development (IBRD), is the most well-known development bank of the international nature. The funds of international and regional development banks are mainly from the fees paid by the membership nations and the bonds issued in the international capital market. The mission of a nation's development bank is to support the construction of the nation's infrastructure through long-term financing. That is why these banks are also known as construction banks. The funds of development banks are mainly from governments' appropriations and issuance of bonds.

Agricultural banks: They mainly provide preferential credits to agricultural agencies and farmers. Because, in areas of agriculture, credit terms tend to be long, the interest rates low, loans riskier, while it is difficult to deal with the collaterals, commercial banks are not willing to handle credit issues and loans related to agriculture.

Mortgage banks: They specialize in the business of mortgage loans using real estates, such as lands, houses, etc., as the collaterals. That is why they are also known as real estate mortgage banks. The funds of mortgage banks are mainly from the issuance of mortgage backed securities. These banks can also raise funds by discounting short-term notes and issuing bonds.

Export-import banks: They are specially established to deal with the nation's foreign trades. They are of official or semi-official nature. Their goal is to promote the import and export of commercial goods instead of producing profits. Therefore, their funds are mainly from governments' investments, loans from the governments, issuance of bonds, etc.

As for the other non-banking financial institutions, they include

Insurance companies: Although the business of insurance does not belong to the category of financial activities, the premium income of the insurance companies is much greater than the insurance payouts. This surplus constitutes the steady monetary funds of the insurance companies, and is applied to make various investments, such as giving out loans. That explains why insurance companies are considered important financial firms.

Pension funds: They are financial organizations that provide annual retirement incomes to retirees who participate in pension plans. The funds of pension funds are from two sources: One is from the proportional deductions of the employees' wages and the corresponding contributions of the employers, and the other the investment profits of the accumulated capital. Because the amounts employees and employers pay into pension funds each month are much greater than those the pension funds have to pay out, a large surplus of funds can be steadily applied to make investments.

Credit unions: They represent a commonly existing financial organization of the cooperative form in industrialized nations. Credit unions generally develop with specified industries or regions, such as rural credit unions, urban credit unions, handcrafts credit unions, credit unions of construction workers, etc. The sizes of credit unions are limited. Their funds are mainly from the share capitals paid by the members and the attracted deposits. The collected funds are mainly used to give out productive or consumer loans to the members.

Financial firms: They acquire funds through issuing commercial papers, bonds, and stocks, while they apply the raised funds to mainly give out particular consumer loans and loans for industrial and commercial enterprises. They do not accept deposits. Their form of financing is borrowing in large sums of money while giving out loans in small amounts. Their loans are mainly amortized and designed for the purchases of durable consumer goods, such as automobiles.

Trust and investment companies: They are also known as trust companies, investment companies, fund companies, etc. These companies acquire their funds mainly from issuing stocks, bonds, and investment income certificates. Then, they invest the collected funds in the stocks and bonds of other companies and in various industrial projects. Trust companies can also use their purchased securities as collaterals to issue new trust and investment notes to raise additional funds. However, these companies do not participate in the business of industrial and commercial loans. By collecting funds from medium to small investors, trust and investment companies diversify investment risk while acquire steady investment incomes through investing in various kinds of securities of different countries, different regions, and different types of industries. Because of this reason, trust and investment companies are very attractive to medium and small investors and constitute powerful competitors to commercial banks in terms of attracting deposits.

3.6.2 The banking business

First, let us look at the asset-debt structure of a bank. Each bank raises funds mainly through issuing debts and earns profits by applying the raised funds to purchase assets. All these activities of the bank are recorded on its balance sheet, also known as asset-debt table. This sheet reflects how the bank acquires his funds and how the bank applies the funds. That is, it shows the asset-debt structure of the bank. The so-called asset-debt structure of the bank stands for the totality of the indicators of all claims against external entities and the indicators of all debts owed at specific time moments.

For a commercial bank, the part of asset that exceeds the debt is known as net asset or the bank's capital. That represents the ownership of the shareholders of the bank. That is,

Net asset = totality of various assets – totality of all debts
Total asset = total debts + net assets

The bank's ownership of the shareholders can also be seen as debts of the bank. So, the net asset is often listed in the category of the bank's debts. Therefore, on the asset and debt table of a commercial bank, the total of assets is equal to the total of debts, making the table known as the bank's balance sheet. Table 3.3 is a typical asset-debt table of a commercial bank. The left-hand side of the table records the categories of assets and their percentage values, while the right-hand side the categories of debts and their corresponding percentages. The bank's net asset is placed in the table's categories of debts.

The bank uses the capital acquired from issuing debts to purchase assets. So, the asset categories of the bank are essentially about how the bank applies its funds. Through applying its funds, the bank earns its income and makes its profits. In

Table 3.3 The asset and debt table of a commercial bank.

	Asset category (applications of capital) (% of total)		Debt category (sources of capitals) (% of total)		
Cash assets	Cash reserves	2			
	Cash receivables	2	Check deposits		23
	Interbank deposits	2			
Bonds	Central bank bonds	19	Non-trading	Savings deposits	20
	Other bonds	5	deposits	Small face-value CDs	16
Loans	Commercial loans	16		Large face-value deposits	9
	Real estate mortgages	25	Debts		24
	Consumers loans	10	Bank assets		8
	Interbank loans	4			
	Other loans	8			
Other assets		7			
Total		100	Total		100

Table 3.3, all the asset categories are ordered based on their liquidities as cash assets, bonds, loans, and other assets, where cash assets have the most liquidity among all the bank's assets. However, the cash assets do not directly create income for the bank. Cash assets include cash reserves, cash receivables, and interbank deposits.

Cash reserves stand for the sum of the reserves of commercial bank deposited with the central bank and the cash on hand, which the commercial bank holds overnight in its treasury. The cash reserves do not bear any interest for the bank. But as required by the law, the bank must have them. The law dictates that for each unit of demand deposits (checking account deposits) the bank must maintain a proportion as deposit reserves kept with the central bank. This portion of the deposits is known as the statutory reserves. Other than the statutory reserves, the bank still needs to maintain a certain amount of cash on hand to cope with withdrawals. This portion of the maintained cash is known as the bank's excess reserve. The total cash reserve of the bank is the sum of the statutory reserve and the excess reserve.

Cash receivables represent the receipts, which the bank holds, that are waiting to be repaid by cash. These receipts will become cash inflow for the bank within a few days. For example, the bank just received a check payment against a checking account from another bank. This check is then treated as the receiving bank's cash receivable. It represents a contingent claim of the receiving banking on the paying bank. Within five business days, the receiving bank will be able to obtain the cash amount of the check from the relevant account in the paying bank.

Interbank deposits stand for the deposits of one bank with other commercial banks. Many small banks deposit their funds with large banks in order exchange for such services of the large banks as check receivables, foreign currency trading, purchase of securities, etc. That is a part of the system of bank agents or exchanges.

The bonds held banks represent an important class of income producing assets. For any commercial bank, it can only hold bonds, because the law does not allow it to own stocks. In the asset-debt Table 3.3, the commercial bank holds two kinds of bonds: those issued by the central bank and those issued by others, where the latter include bonds issued by local governments and business firms. Among all kinds of bonds, those issued by the central bank have the most liquidity, can be traded most easily, and can be converted into cash with the least costs. Because of their high levels of liquidity, the short-term bonds issued by the central bank are also known as the secondary reserve of the holding commercial bank. Other bonds, issued by either local governments or business firms, have lower levels of liquidity and are difficult to trade so that they bear higher levels of risk. The risk is mainly stemmed from the potential of default or breach of contract, where the bond issuer is unable to pay for the promised interests or to return for the principal on maturity. There two reasons for why commercial banks buy these riskier bonds: One is that the expected (after tax) return is higher than that of the bonds issued by the central bank, and the other is that the bonds issued by local governments generally enjoy some preferred tax benefits, and the local governments are more willing to do business with the banks that hold their bonds.

Banks earn their profits mainly through giving out loans, where the revenue produced by outstanding loans generally amounts to more than 50% of the banks' total income. It is exactly because loans bring in revenue that loans are considered assets of banks. However, loans do not have much liquidity and suffer from relatively high degrees of default risk. That makes loans most profitable part of banks' business. For commercial banks, the main loans are targeted at industrial and commercial firms and real estate mortgages. Other than these two kinds of loans, commercial banks also give out consumer loans, interbank loans (loans giving to other banks), and other kinds of loans. However, these loans are not the business focus of commercial banks. The most important differences between the asset-debt tables of various depository institutions are shown in their differences in the kinds of loans they give out. For example, the savings and loan association and mutual savings banks mainly partici-pate in the business activities of residential mortgages, while credit unions mainly issue consumer loans.

Other assets include physical real estate properties, such as the office buildings of the banks, computers, and other equipment.

Banks acquire their capital through issuing (selling) debts. That is often known as sources of capital, the right hand side of Table 3.3. By using the acquired capital, banks purchase assets; and the difference between banks' assets and debts constitutes the banks' net assets, which stand for the ownership of the banks' shareholders. Because banks' net assets are also one of the sources of capital and can be seen as funds borrowed by the shareholders, the net assets are included in the categories of debts.

Check deposits stand for deposits to be used for transaction purposes, and include deposits into all kinds of bank accounts against which checks can be written, such as non-interest checking accounts, interest-bearing accounts of transferrable withdrawal notices, and money market deposit accounts. Check deposits represent the capital of the depositors and a debt of the banks. They generally stand for a source of banks' funds of the lowest costs. The reason why the depositors are willing to give up some of the interest income by opening up check deposit accounts in banks is because such deposits are liquid assets that can be used to make various kinds of purchases, which greatly enhances the convenience of people's daily transactions of money. Check deposits used to be an important source of funds for banks. However, along with financial innovations, the attraction of check deposits has been continuously decreased; and the proportions of check deposits in the banks' overall amounts of debts have been accordingly shrinking.

Non-trading deposits stand for those deposits that are not for transaction purposes. Currently, non-trading deposits have become the main source of capital for banks. Although the owners of non-trading deposits cannot write out checks against their deposits, the interest rates of their non-trading deposits are generally higher than those of check deposits. There are two basic types of non-trading deposits: savings deposits and time deposits. Savings deposits are the most common form of non-trading deposits. The funds in these accounts can be added or withdrawn at any time. Additions, with-drawals, and interest payments are recorded on either the monthly statements or the deposit books. Technically speaking, such deposits are not payable on demand; banks can make the payments within 30 days. Hence, in the computation of the total money supply, savings deposits are not considered within M_1; instead, they are included in M_2. However as a consequence of the competition between banks, banks allow the owners of savings accounts to withdraw money without delay. Time deposits are also known as certificate deposits (CDs). They carry pre-determined terms, which can be as such short terms as several months, or as such long terms as several years. Fixed-term deposits are less liquid than savings deposits; but they carry higher interest rates than savings deposits. In terms of banks, time deposits stand for a relatively high cost source of capital. Additionally, time deposits are also divided into small- and large-value time deposits. In the United States of America, the time deposits in the amount of less than $100,000.00 are considered small valued, while those in the amount of more than $100,000.00 large valued. Small-value time deposits are mainly used by the ordinary residents; large value time deposits are designed to attract the capital of business firms and organizations. Large-value deposits are transferrable. They can be traded before their maturity on the secondary markets just as bonds. It is because of this reason that companies, mutual funds of the capital markets, and other financial institutions use these transferrable deposits as substitutes of treasury bills and short-term bonds. Since the year 1961 when transferrable deposits were initiated, they have become one of the important sources of capital for commercial banks.

Commercial banks can borrow funds from the central bank, other commercial banks, and business firms. When they borrow from the central bank, the borrowed funds are known as discounted loans. Commercial banks can borrow from other banks and financial institutions. Particularly, they can borrow from their parent companies or other companies. They can also borrow from foreign sources, such as introducing U.S. dollars from Europe. With time, borrowed funds gradually become a relatively important source of capital for banks.

Banks' net assets are equal to the differences of the banks' total assets minus the total debts. These assets are formed through either selling shares or retaining profits. They represent the ownerships of the shareholders of the banks. The net assets play the role of buffering potential falls in the values of the banks' assets. If the total asset value falls below the total debt value, then the bank becomes insolvent. To deal with potential falls in the value of assets and to prevent the potential of insolvency, one of the methods is for the bank to maintain a net asset, in particular, a reserve for loan losses. Some of the banks' loan projects might become "bad debts", from which the banks are unable to collect the interests and the principals. So, on some future dates, these debts will have to be dropped from the banks' balance sheets, indicating that the banks have suffered from losses. Reserves of loan losses are the preparations banks use to avoid adversarial effects of such losses. Before bad debts actually appear, banks have already had their reserves of loan losses listed on their balance sheets within the banks' assets. Such advanced preparation of course is better than no preparation when bad debts appear. At the same time, such reserves reduce the reported income of the banks so that they do not need to pay income tax on the reserves. Additionally, doing so provides a way for the banks to report future losses to their shareholders, depositors, and administrators.

For the central bank, as the core of the financial system, the structure of its assets and debts is different from that of a commercial bank. Any alteration in this structure of assets and debts can cause changes in the nation's supply of money so that the national economy is consequently affected. Table 3.4 is a representative assets and debts table of a central bank.

Under the heading of asserts, there are such items as

Securities: The purpose for the central bank to be involved in the operation of securities is not for profit; instead it is for the purpose of adjusting the money supply. Under the normal conditions, the securities the central bank trades are mainly the government bonds. When necessary, the central bank will also trade a small amount of the stocks

Table 3.4 The assets and debts table of a central bank in the unit of US$100 millions.

Assets (applications of capital)		Debts (sources of capital)	
Securities: governments bonds and banks' promissory notes	3445	Amount of currency issued: money in circulation	3439
Discount loans: for commercial banks	9	Bank deposits: reserves required by law	350
Cash assets	4	Treasury deposits	148
Cash receivables	65	Foreign and other deposits	8
Gold/SDR certificate accounts	191	Cash items to be paid	55
Other reserve assets	382	Other debts and capital projects	98
Total	4098	Total	4098

of business companies, commercial papers, banks' promissory notes, etc. The bonds held by the central bank stand for the major part of the assets of the central bank;

Discount loans: On the open market, the central bank has a discount window. When commercial banks or other financial institutions experience difficulties with their cash flows, they can take out loans from the central bank through the discount window. These loans then become parts of the assets of the central bank;

Cash assets: They represent the currency, mainly coins, held by the central bank in order to meet its daily needs;

Cash receivables: The central bank receives many checks in the process of the national clearance. Before the check capitals are received, they become the cash receivables of the central bank;

Gold and SDR (special drawing right) certificate accounts: Each national central bank maintains a certain amount of gold to support the credit of its national currency and to settle international trades. Special drawing rights represent the drawing certificates issued by the international monetary fund to each national government. In international financial transactions, these SDRs have replaced the role of gold and can be used to clear debts between nations; and

Other reserve assets: include foreign currency deposits and bonds, computers, office facilities, buildings, and other physical assets, held by the central bank.

Under the heading of debts, there are such items as

Amount of currency issued: The central bank issues currency, which is equivalent to the public's deposits of their assets at the central bank. So, the amount of currency issued stands for a debt of the central bank borrowed from the public. However, in terms of a holder of currency, he does not necessarily recognize that he owns claims on the central bank. Instead he might believe that he possess some social wealth. Hence, the debt, incurred from issuing currency, in fact represents a long-term debt the central bank does not need to repay;

Bank deposits: represent the reserves individual commercial banks maintain at the central bank, as required by the law;

Treasury deposits: The central bank functions as the national treasury. So, all treasury deposits (fiscal budget deposits) have to be kept at the central bank, which becomes the debt of the central bank;

Foreign and other deposits: They include the deposits in the domestic central bank of foreign governments, foreign central banks, international organizations, such as the World Bank and the United Nations, and the deposits of domestic governments and some other organizations, such as the National Insurance Company, National Mortgage Banks, etc.;

Cash items to be paid: In the national clearance system, the central bank needs to transfer money into the accounts of some checks. Before these amounts are transferred, they constitute the cash items for the central bank to pay; and

Other debts and capital projects: This line includes all debts that are not included in the previous lines. For example, the shares of the central bank, purchased by member banks, which constitute the capital of the central bank, are listed in this line.

Secondly, let us look at the business of commercial banks. Along with the constant financial innovation, the types of business activities commercial banks pursue have grown in numbers. Even so, from the attraction of funds to the application of the funds, the business of commercial banks can be roughly classified into three groups: liabilities, assets, and intermediaries.

In terms of liabilities, it means the business for commercial banks to attract funds. First of all, commercial banks increase their capitals through issuing new shares, which are also seen as liabilities. Other liability business includes:

Demand deposits: They stand for such a form of deposits that the depositors can withdraw funds against the balances from the accounts at any time. And when depositors withdraw funds, they have to use particular checks as specified by the banks. That is the reason why demand deposits are also known as checking deposits. The demand deposits have always represented a major business of commercial banks. These banks have to maintain reserves as required by law at the central bank for demand deposits; and other than the reserves, these banks also have to maintain certain amounts of excess reserves on hand. The cost of pursuing the business of demand deposits includes the payment of interests and all other expenses related to the relevant services: the processing and possession of used checks, the editing and delivery of monthly statements, the availability of cashiers and relevant equipment, the design and maintenance of the banks' images (business locations, halls of operations, branch offices, etc.), advertisement, marketing, etc. Because the liquidity of demand deposits is very high, involving various kinds of services related to frequent deposits and withdrawals, commercial banks pay relatively low interest rates to demand deposits. The 1933 banking law of the United States also prohibited commercial banks to even pay any interest to demand deposits. Although the later U.S. laws on the control of interest rates have been relaxed, this particular law that prohibits the payment of interest on deposits has still been in effect.

Time deposits: they represent the kinds of deposits that carry pre-fixed terms, where the pre-fixed terms could be 3 months, 6 months, 1 year, 2 years, 3 years, 5 years, or even longer. When the depositor withdraws his funds, he must present the certificates of deposits as specified by his bank. That is why time deposits are also known as certificate deposits (CDs). Although the interest rates vary with the lengths of the terms, they are all fixed and higher than those of demand deposits. The funds in a time deposit can also be withdrawn ahead of the pre-determined time. When that is done, a high rate of penalty is imposed. Because time deposits carry relatively longer terms, they have been a steady source of capital for commercial banks to make various kinds of investments.

Savings deposits: They are generally deposit accounts established in commercial banks by individuals and non-profit organizations for the purpose of accumulating funds while earning some interest income. Savings deposits do not carry any term; the deposit book needs to be used for any addition and/or withdrawal of funds. That is, the account owner can deposit additional money into his account by using his deposit book at any time; and he can also take money out of his account, but not overdraw the account, by using his deposit book at any time. Because savings deposits do not involve checks or certificates, and the procedure of taking money out and adding money into the account is relatively simply, they have been widely accepted by the working class. Deposit books do not have any liquidity, cannot be transferred by trading. When compared to demand deposits and time deposits, the interest rates of savings deposits are in between the two. That is, the interest rates of savings deposits are higher than those of demand deposits and lower than those of time deposits.

Borrowings: The borrowing business of commercial banks is generally made up of four areas: issuance of bonds, interbank lending, borrowing from the central bank, and borrowing from abroad. The bonds issued by commercial banks are such debt certificates that pay periodic interests and repay the principals on maturity. They possess the characteristics of promissory notes with varying terms from 10 to 30 years, where the majority of the bonds carry terms from 20 to 25 years. The so-called interbank lending means loans obtained from other commercial banks. Generally, they are overnight loans and do not need to have any collateral. When a commercial bank is short of capital, it can also take out a loan from the central bank; or it can take some of the unmature commercial papers it holds to the discount window of the central bank to rediscount them. Other than borrowing funds from domestic money markets, it is also common for commercial banks to reach out to the international money markets to borrow funds to make up their capital shortages.

In terms of asset business, it means the business for commercial banks to apply their capitals. It is the most important business from which commercial banks earn their incomes. In the following, let us look at two most important asset business activities of commercial banks: lending and investing.

Lending is the business focus of commercial banks. It is the main way for banks to apply their capitals to create profits. Each commercial bank must formulate its clear lending guidelines so that the movement of its capital can be determined and the scale and structure of its lending can be under controlled. The lending guidelines detail specified geographic regions, types of loans to give out, kinds of acceptable collaterals, loan amounts, lending terms, loan commitments, size of annual loan business, the repayment of loans, and other major contents. When a potential borrower applies for a loan from a commercial bank, he needs to first submit a written application, within which he clearly describes such main information as the purpose of the loan, how the funds will be used, how much is applied for, the intended term, how the loan will be repaid, how the loan will be guaranteed, etc. Upon receiving the written application, the commercial bank investigates the potential borrower in terms of his credit conditions, including the person's operational capability, management quality, profitability, development future, liabilities, credit history, etc. At the end, a loan contract is signed. That is, both the lender and the borrower sign through their respective legal representatives the legal documents that spell out the rights and obligations of both parties and that specify such details of the loan as the amount, purpose, interest rate, term limits, collateral requirements, method of repayment, handling of default, debt settlement when the borrower declares bankruptcy, etc.

The investment business of commercial banks means their activities of purchasing securities, including the bonds of governments, various governmental agencies, local governments, and commercial firms. That is the business of investing in securities. This is also an important asset business and represents a major source of income of commercial banks.

Both commercial banks and investment banks participate in the investment activities of securities. In order to strengthen the management of these activities, through legislation the developed countries strictly distinguish the security business of commercial banks from those of investment banks. For the purpose of retaining a safe and stable financial system, the relevant laws stipulate that:

1   No commercial bank is allowed to issue, subscribe, underwrite, broker securities, and other activities that belong to the business area of investment banks;
2   No official of any commercial bank can also hold a position in an investment bank; and no boards of directors that chain both a commercial and investment bank are allowed;
3   No commercial bank is allowed to establish branches and subsidiaries that engage in securities investment business.

Other than undertaking the conventional lending and investment business, commercial banks also play the role of intermediaries. That is, without applying their own funds, these banks provide payment and other entrusted services for their customers for fees. These intermediary services mainly include the following several kinds:

Agiotage: Through this service, a customer delivers a certain amount of money to a commercial bank; then the bank on behalf of the customer remits the funds to a particular person located in a different location. By undertaking this service, commercial banks collect service fees on the one hand, and occupy part of the customer's funds on the other hand through the time difference between when the customer delivers the money to the bank and when the receiver at another location actually collects the funds.

Credit operation: In international trades, based on the application of a buyer, a commercial bank provides the seller a letter of credit to guarantee payment. This service resolves the problem of mistrust between the buyer and seller living in different places. When providing letters of credit, commercial banks collect from their customers not only the deposits or collaterals, but also handling fees. These fees constitute the most important source of income for commercial banks from the business of providing letters of credit.

Collection service: In the payment process of international trades, what is commonly practiced is that the seller writes up the invoice and authorizes a commercial bank to collect the payment from the buyer. The income the commercial bank generates in the collection service is from the service fees collected from the sellers.

Bank credit cards: Credit cards are a tool for commercial banks to give out consumer credits. The card-issuing banks provide consumers with the convenience of first consumption and second payment and allow for a certain degree of well-intentioned overdrafts. The two largest credit card organizations of the world are the visa group, consisting of the Bank of America and other banks from over thirty countries, that issues the visa cards, and the master group, consisting of 25,000 financial institutions, that issues the master cards.

Thirdly, let us look at the business of the central bank. The central bank plays the three roles of the issuing bank, the bank of all banks, and the bank of the government. Through these three functions, the central bank macro-monitors and macro-adjusts the national finance. In terms of its debts, the business of the central bank includes:

Currency issuance: It is the most important liability business of the central bank. Currently, the issuance of currency in all nations is monopolized by their central banks. The paper notes of the central bank are introduced into circulation to satisfy the monetary need of economic development through discount, loans, purchase of securities, acquisition of gold, silver, and foreign currencies, and other methods.

Agent of national treasury and maintenance of fiscal deposits: As the bank of the government, the central bank plays the role of agency for the national treasury and accepts fiscal deposits. It provides conditions for government financing and provides no interest on the treasury deposits.

Centralized management of deposit reserves: The central bank centrally manages the statutory deposit reserves of all commercial banks without paying any interest on these funds. It uses these funds as loans to those commercial banks that experience difficulties in their cash flows. That saves individual commercial banks from maintaining their own deposit reserves and makes sufficient use of the capital. The central bank is obligated to determine the rates of deposit reserves for all commercial banks, and to urge commercial banks to hand over their required reserves.

Handling nationwide clearance: The credit and debt relationship between business firms is generally handled by banks so that the business firms' credit and debt relationship becomes that between banks. Through the accounts of individual commercial banks, the central bank clears all interbank creditor-debtor relations that exist throughout the nation so that the trouble of transporting cash between different locations is avoided. This service makes the movement of capital more convenient and speeds up the flow of commercial goods.

Other businesses: Other than the four kinds of debt businesses as mentioned previously, the central bank also has liability business with international financial organizations, the business of honoring treasury bonds, etc.

In terms of its assets, the business of the central bank includes:

Discounted loans: The central bank holds days of discount window, rediscounts unmature discounted bills held by commercial banks, and gives out loans to commercial banks. In some countries, the central banks also provide loans to the governments;

Gold and foreign reserves: In order to centralize reserves, mediate funds, adjust the speed of circulation of money, and stabilize the foreign exchange rates and the financial markets, the central bank participates in trading gold, silver, foreign currencies, etc., in both the domestic and international financial markets. This activity represents an important asset business of the central bank;

Investment in securities: To mediate the tightness of and to control the supply of money, the central bank also engages in the operation of securities trading but not for the purpose of making profits. The securities the central bank trades consist of mainly the government bonds. When necessary, the central bank also trades a small volume of company stocks.

3.7   INNOVATION: LIVELIHOOD OF NATIONAL FINANCIAL INFRASTRUCTURE

In this section we will look at why innovation and openness to different ideas are the essence for a relatively self-sustained national financial infrastructure to stay healthy and viable. To this end, let us first introduce the dishpan experiment initially designed and conducted in 1953 by Hide of Cambridge University, Great Britain, and then modified in 1959 by Fultz of the University of Chicago, USA.

In particular, Hide (1953) used two concentric cylinders with a liquid placed in the ring-shaped region between the cylinders. He placed the container on a rotating turntable subjected to heating near the periphery and cooling at the center. The turntable generally rotated counter clockwise, as does the earth when viewed from above the North Pole. Even though everything in the experiment was arranged with perfect symmetry about the axis of rotation, such as no impurities added in the liquid, the bottom of the container is flat, Hide observed the flow patterns as shown in Figure 2.7. Briefly, with fixed heating, a transition from circular symmetry (Figure 2.7(a)) to asymmetries (Figure 2.7(b)), then Figure 2.7(c) would take place as the rotation increased past a critical value. With sufficiently rapid but fixed rate of rotation, a similar transition would occur when the heating reached a critical strength, while another transition back to symmetry would occur when the heating reached a still higher critical strength. Also, in stage Figure 2.7(c), a chain of identical eddy motions would appear. As they travel along, they would alter their shapes in unison in a regular periodic fashion, and after many rotations of the turntable, they would regain their original shape and then repeat the cycle.

Referencing to what Hide did in 1953, Fultz and his colleagues (1959) conducted another version of the dishpan experiment without the inner cylinder, producing similar but different results. Specifically, in the late 1950s, Dave Fultz (Fultz et al., 1959) constructed the following experiment: He partially filled a cylindrical vessel with water, placed it on a rotating turntable, and subjected it to heating near the periphery and cooling near the center. The bottom of the container is intended to simulate one hemisphere of the Earth's surface; the water, the air above this hemisphere; the rotation of the turntable, the Earth's rotation; and the heating and cooling, the excess external heating of the atmosphere in low latitudes and the excess cooling in high latitudes.

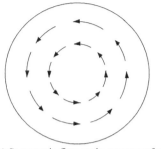

(a) Symmetric flow at the upper surface   (b) Asymmetric flow at the upper surface

Figure 3.2 Patterns observed in Fultz's dishpan experiment.

To observe the pattern of flows at the upper surface of the water, which was intended to simulate atmospheric motion at high elevations, Fultz sprinkled some aluminum powder. A special camera that effectively rotated with the turntable took time exposures so that a moving aluminum particle would appear as a streak, and sometimes each exposure ended with a flash, which could add an arrowhead to the forward and end of each streak. The turntable generally rotated counterclockwise, as does the Earth when viewed from above the North Pole.

Even though everything in the experiment was arranged with perfect symmetry about the axis of rotation, such as no impurities added in the water, and the bottom of the container was flat, Fultz and his colleagues observed more than they bargained for. First, both expected flow patterns appeared, as shown in Figure 3.2, and the choice depended on the speed of the turntable's rotation and the intensity of the heating. Briefly, with fixed heating, a transition from circular symmetry (Figure 3.2(a)) would take place as the rotation increased past a critical value. With the sufficiently rapid but fixed rate of rotation, a similar transition would occur when the heating reached a critical strength, while another transition back to symmetry would occur when the heating reached a still higher critical strength. Here, the number of eddy leaves, as shown in Figure 3.2(b) and Figure 2.7(b–c), id determined by the temperature difference between the polar circle and the periphery. When this temperature difference decreases, the number of eddy leaves decreases; if the temperature difference disappears totally, the fluid in the spinning dish flow around the polar circle as a large whirlpool. However, as the temperature difference increases, the number of eddy leaves also increases until eventually the leaves become so fine that they seem to become flows moving toward the center of the spinning dish.

No matter which version of the dishpan experiment we look at, the fundamental principle behind the appearance of chaotic patterns is the existence of uneven distribution of forces. That is similar to the very reason why each and every system in the universe rotates: As long as the distribution of materials in the system is nonhomogeneous, gradient forces appear, leading to rotation of the relevant materials. From this fact it follows that the more uneven the distribution of materials is, the more intense the system rotates.

Now, we can naturally imagine the economy of our ideal, isolated nation as the Fultz's version of the dishpan experiment, where the currency in circulation is modeled by the spinning fluid, connected through information, knowledge, and other relevant matters that have something to do with money and its direction of movement. Then, at the beginning of time (of the nation), the circulation of money looks loosely like that pictured in Figure 3.2(a). As the economic activities increase in terms of their fluidity and magnitude, the assumption that the economy is closed from the outside world implies that along with the appearance of nonhomogeneous gradient economic forces within the economy, the circulation of money within the nation would soon become like that in Figure 3.2(b). However, due to the fact that in reality, the environmental conditions around the nation and the distribution of natural resources in the nation would not be homogeneous, the eddy leaves in the flow pattern of money as pictured in Figure 3.2(b) would not be as perfectly shaped and distributed as in this drawing. In terms of political economies, that means that the economic developments in some geographical regions would be more advanced than other areas from around the nation, creating additional non-homogeneity, which consequently leads to more powerful gradient economic forces so that the yoyo field behind the nation's economy rotates at a much faster speed than before.

What happens with the constantly increasing speed of rotation is that migration and relocation of people start to occur massively. That of course would bring forward with major social inequalities and political upheavals, making the nation politically and economically instable. To stabilize the situation, a coordinated effort, if successful, would be initiated to balance the economic development over the entire nation so that political upheavals could consequently be soothed.

As suggested by the relatively calm center in the flow pattern in Figure 3.2(b), it can be seen that one of the possible consequences of the coordinated effort will be the establishment of a central bank whose endowed tasks would include those described earlier and more if necessary. However, when such a central bank is established with its endowed authorities, theoretically, the nation's economy has now been transformed into one whose high-dimensional yoyo field looks like the fluid and dish in Hide's version of the dishpan experiment. In other words, a hard core in the 'middle' of the economy becomes visible. But, what is different of this new economy from the situation of Hide's dishpan experiment is that the 'speed of spin' of this core in general is not synchronized with that of the 'spin' of the economy itself. In particular, no matter what is forthcoming in the realistic economy, the 'core', the central bank, tends to guess ahead of time what might be happening. As limited by the present capability of science in terms of making effective and accurate predictions about the future, such guesses in most situations are either incorrect or making the economy worse off than it would be. Such non-synchronization between the actual economy and the 'core' surely makes the originally periodical and predictable pattern changes of the fluid in the Hide dishpan experiment more chaotic and unpredictable. That is the reason why some scholars believe that economic cycles cannot be effectively predicted (Friedman and Friedman, 1998, pp. 50; Friedman and Schwartz, 1993, pp. 678).

Next in this section, let us apply the following result of microeconomics to support claim that if an economy is closed from the outside world, then along with the appearance of nonhomogeneous gradient economic forces within the economy, the

Figure 3.3 Evolution of the head's benevolence and selfish members' behaviors.

circulation of money within the nation would transform from the state depicted in Figure 3.2(a) to that in Figure 3.2(b).

Theorem 3.1. (The Theorem of Never-Perfect Value Systems (Lin and Forrest, D., 2008)) *In a family of at least two members, one member h is the head who is benevolent and altruistic toward all other members. The head establishes a value system for all members of the family to follow so that in his eyes, every selfish member will be better off both for now and for the future. If a selfish member k measures up well to the value system, he will then be positively rewarded by the head h. Unfortunately, the more effort member k puts in to measure up to the value system, the more he will be punished by the reward system.*

As depicted in Figure 3.3 and as in practical situations, what is described in this theorem is that the benevolent head transfers his resources to other family members in a periodic fashion over time without an end in sight, where although the word of "benevolent" is used, based on the proof of this theorem it only means that he is a person responsible to distribute wealth around, also the phrase of "his resources" does not necessarily mean that he is the person who has made all the wealth. The idea of value systems implies that the head tells other members at some moment in time along the time line that starting at a certain pay period, each member's behavior will affect the amount he/she will receive from the head at the end of that period. And, the head will design his response to each member's behaviors in such a way as to maximize his own utility. That is, Theorem 3.1 explains a sequential behavioral change and reflection over time between the head and his family members. In Figure 3.3, the scale marks on the time line represent the moments of individual asset distributions and are given for reference purposes without much practical implications. Mark 0 can be located anywhere on the line as the beginning of our discussion or focus. Negative scale marks represent the moments of distribution of the past and the positive marks the future moments of distribution.

Proof.   Assume that Y_k is an index, satisfying $0 < Y_k < 1$, established to check how well member k measures up to the value system predetermined by the family head h. This index satisfies that the greater Y_k is, the better member k measures up to the value system. Because k has to put in extra effort to increase the value of Y_k, his utility function $U_k = U_k(X_k, Y_k)$ satisfies

$$\frac{\partial U_k}{\partial X_k} > 0 \quad \text{and} \quad \frac{\partial U_k}{\partial Y_k} < 0$$

where X_k stands for member k's total consumption of numeraire good. The reason why $\frac{\partial U_k}{\partial Y_k} < 0$ is because all people are lazy to a certain degree.

Let the head h's utility function be

$$U_h = U_h\left(X_h, U_1, \ldots, U_n\right)$$

where it is assumed that other than the head h, the family contains n other members who are all selfish, X_h is the head's own consumption, and U_i the utility function of family member i, $i = 1, 2, \ldots, n$.

Assume that the reward from the head h to member k is determined by

$$b_k = b_k(Y_k)$$

such that $\frac{\partial b_k}{\partial Y_k} > 0$. Now, the total consumption of member k is given by

$$X_k = I_k + b_k = I_k + b_k(Y_k)$$

where I_k is member k's own income unrelated to Y_k. To reflect the fact that the index Y_k represents k's performance on the head's value system, which is assumed to make member k better off, we did not explicitly list Y_k as an independent variable in the head's utility function. Instead, Y_k is shown in the total consumption $X_k = I_k + b_k(Y_k)$ of k. That is, an increase in the value of Y_k has a complex effect on member k, both positive and negative. Being positive is in the short term because the head h will help to bring k's consumption to a higher level, and it could be negative because member k has to put in more effort to raise the Y_k-value. So, to make our model more plausible, we can add the assumption that

$$\left|\frac{\partial U_k}{\partial X_k} \cdot \frac{\partial X_k}{\partial Y_k}\right| > \left|\frac{\partial U_k}{\partial Y_k}\right|$$

That is, the increased utility of k, brought forward by the higher-level consumption $X_k = I_k + b_k(Y_k)$, is more than enough to offset the decreased utility of k when he has to put in additional effort to incrementally raise the Y_k-value.

To the head h, his distribution plan of his own income to all other members has to maximize his utility function subject to the following budgetary constraint:

$$X_h + \sum_{j=1}^{n} X_j = X_h + \sum_{j=1}^{n}(I_j + b_j) = I_h + \sum_{j=1}^{n} I_j$$

If we ignore all members except the head h and member k, then the first-order condition for the head's optimization problem is given by

$$\begin{bmatrix} \dfrac{\partial U_h}{\partial X_h} \\[2ex] \dfrac{\partial U_h}{\partial X_k} \\[2ex] \dfrac{\partial U_h}{\partial Y_k} \end{bmatrix} = \begin{bmatrix} \dfrac{\partial U_h}{\partial X_h} \\[2ex] \dfrac{\partial U_h}{\partial U_k} \cdot \dfrac{\partial U_k}{\partial X_k} \\[2ex] \dfrac{\partial U_h}{\partial U_k} \cdot \dfrac{\partial U_k}{\partial Y_k} \end{bmatrix} = \lambda \begin{bmatrix} 1 \\[2ex] 1 \\[2ex] \dfrac{\partial b_k}{\partial Y_k} \end{bmatrix}$$

And member k chooses such a Y_k-value Y_k^* to satisfy his first-order condition

$$\frac{\partial U_k}{\partial Y_k} = \frac{\partial U_k}{\partial Y_k} + \frac{\partial U_k}{\partial X_k} \frac{\partial h_k(Y_k)}{\partial Y_k} = 0$$

so that he maximizes his utility.

Now, the third equation in the head's first-order condition implies that when the head's expected Y_k-value is greater than Y_k^*, his reward $h_k(Y_k)$ for member k would have a negative rate of increase. That is, we have shown that the more effort member k puts in to measure up to the value system, the more he will be punished by the reward system. QED.

Example 3.1. To see how the result in Theorem 3.1 materially acts out, let us assume that a family consists of two members: the benevolent and altruistic head h and a selfish member k, and Y_k the index outlined in the proof of Theorem 3.1. Let member k's utility function be given as follows:

$$U_k = U_k(X_k, Y_k) = X_k(1 - Y_k), \tag{3.1}$$

where X_k is member k's total goods consumption and $(1 - Y_k)$ stands for his degree of laziness, and the head's utility function be

$$U_h = U_h(X_h, U_k) = X_h + \sqrt{U_k} = X_h + \sqrt{X_k(1 - Y_k)}. \tag{3.2}$$

Assume that the reward from the head h to member k is determined by

$$h_k = wY_k, \tag{3.3}$$

where w is a fixed constant > 0. Then the total consumption of k is

$$X_k = I_k + h_k = I_k + wY_k,$$

where I_k is member k's own income unrelated to Y_k.

To the family head, he needs to maximize his utility function subject to the budgetary constraint:

$$X_h + X_k = X_h + (I_k + h_k) = I_h + I_k. \tag{3.4}$$

The first-order condition for this optimization problem is given by

$$\begin{bmatrix} \dfrac{\partial U_h}{\partial X_h} \\[2mm] \dfrac{\partial U_h}{\partial X_k} \\[2mm] \dfrac{\partial U_h}{\partial Y_k} \end{bmatrix} = \begin{bmatrix} 1 \\[2mm] \dfrac{1 - Y_k}{2\sqrt{X_k(1 - Y_k)}} \\[2mm] \dfrac{w(1 - Y_k) - X_k}{2\sqrt{X_k(1 - Y_k)}} \end{bmatrix} = \lambda \begin{bmatrix} 1 \\[2mm] 1 \\[2mm] -w \end{bmatrix}, \tag{3.5}$$

where $\lambda = 1$ the Lagrange multiplier. So, from equ. (3.5), it follows that

$$\sqrt{\frac{1 - Y_k}{X_k}} = 2 \quad \text{and} \quad \frac{X_k - w\,(1 - Y_k)}{\sqrt{X_k\,(1 - Y_k)}} = 2w. \tag{3.6}$$

So, $w = \frac{1}{8}$ and the selfish member k chooses $Y_k^* = \frac{1}{2} - 4I_k$ to maximize his utility, and when $Y_k \neq Y_k^*$, we have

$$X_k = \frac{1}{4}(1 - Y_k). \tag{3.7}$$

Substituting equ. (3.7) into equ. (3.4) and solving for h_k leads to

$$h_k = \frac{1}{4}(1 - Y_k) - I_k.$$

This equation implies that member k's reward h_k from the family head h is a decreasing function of Y_k, an index which measures how well k is doing in terms of the established value system. Q.E.D.

What the Theorem of Never-Perfect Value Systems implies is that in each economic system, which functions and evolves on a fundamental set of laws and regulations, has to go through moral decline over time, because the reward system of the economy in the long run punishes those who actually obey the laws and who act within the given boundaries. This fact very well illustrates why if an economy is closed from the outside world, while its set of rules is fixed, then along with the appearance of nonhomogeneous gradient economic forces within the economy, the circulation of money within the nation would transform from the state depicted in Figure 3.1(a) to that in Figure 3.1(b). It is because the overall moral of the participants of the economy continues to drop until something new is brought into the operational system of the economy.

3.8   SOME FINAL REMARKS

This chapter outlines how the financial infrastructure of a typical industrialized nation is constructed for the purpose of stabilizing and invigorating the society, the political system, and the economy, while creating opportunities for innovative thinking, new product design and manufacturing, and the establishment of prosperity. By introducing financial products, idling money is gathered for productive uses in the financial markets. By establishing various general-purpose and specialty banks, the banking system implicitly direct the flow of money along such paths that are most beneficial to the well-being of the society.

On the basis of the established fact that the movement of money can be conceptually investigated as fluids, and the fact that each system is a rotational field of fluids, it is then shown that innovation and constantly introducing new products and new organizational formats can potentially prevent a relatively stable economy from falling into chaos and self-destruction.

Chapter 4

Supply and demand of money

Money supply and demand is the cornerstone of the theory of money. The study of money supply reveals the various economic factors that affect the quantity of money in circulation. It provides a basic starting point for analyzing how monetary policies actually influence the development of the economy. Correspondingly, the study of money demand stands for another foundation for the analysis of how money affects the economic development. Here, any change in money supply affects the healthy development of the entire economy. Money plays the role of economic blood, which constantly flows through all major economic sectors, stimulating economic activities, capital accumulation, production specialization and division of labor, and effect use of natural resources.

In this chapter, we will look at such important problems as: In order to guarantee the effective operation of the economic system, how much money should be supplied? Who is controlling the money supply? What factors cause the money supply to change? By analyzing the mechanism of money supply, this chapter will address these problems. In terms of money demand, it stands for the most debated part of the theory of money, where various theories are formed and different hypotheses are introduced. This chapter will glance through these different theories on the demand of money, including the classical theory of money demand established before the 1930s, the Keynesian, demand for money theory, initiated by John Maynard Keynes and developed since 1936, and the modern quantity theory of money collectively developed since 1956 with Milton Friedman as a representative of this school of thought. At the end, we will also discuss another important question addressed in the theory of money: the effect of interest rates on the demand of money.

This chapter is organized as follows. In Section 4.1, we will investigate the concept of supply and demand and see how supply and demand in fact constitutes two interacting economic forces. In Section 4.2, we focus on the supply of money by looking at how money is created, the concept of money multipliers, and various theoretical models that describe the supply of money. Section 4.3 emphasizes on the other side of the coin: the demand of money, by looking closely at the more difficult problem of where and why money is needed. It is in this section that we will study the classical quantity theory of money, Keynesian demand for money, following developments of Keynesian theory, modern quantity theory of money, and related empirical tests. Then, Section 4.4 concludes this chapter with some relevant comments.

4.1   THE CONCEPT OF SUPPLY AND DEMAND: TWO INTERACTING ECONOMIC FORCES

In an ideal market where all economic activities are controlled completely by free competitions, the price P of a consumer good is closely related to and determined by the demand D and the supply S. Assume that change in P is directly proportional to the difference of the demand and supply:

$$\frac{dP}{dt} = k(D - S), \quad k > 0, \tag{4.1}$$

where t stands for time and k a constant. With all other variables fixed, assume that both the demand D and the supply S are functions of the price P:

$$D = D(P) \quad \text{and} \quad S = S(P). \tag{4.2}$$

The relationship between the demand D and the price P is generally linear and can generally be written as follows:

$$D(P) = -\lambda P + \beta, \quad \lambda, \beta > 0, \tag{4.3}$$

where λ is the rate of change of the demand with respect to the price P and β the saturation constant of the demand.

As for the relationship between the supply S and the price P, it is generally nonlinear, because when the price of a consumer good lowers, the number of buyers will increase and the demand consequently increases. The increased demand stimulates the production of the good so that the supply is increased. If the price gradually increases, the demand will accordingly decrease so that the supply will consequently be lowered. Since the demand and the supply are not correlated directly, when the price reaches certain height, even though the demand continues to drop, the supply might be increased because the increased price can stimulate the production (Figure 4.1). Therefore, the relation of the supply with respect to the price is generally nonlinear. This relationship can be written symbolically as (at least by means of local approximations):

$$S(P) = \delta + \alpha P + \gamma P^2, \tag{4.4}$$

where $\delta > 0$ is a constant, α and γ are respectively the linear and nonlinear intensities of the supply, satisfying $\alpha > 0$ and $\gamma < 0$.

Substituting equs. (4.3) and (4.4) into equ. (4.1) produces

$$\frac{dP}{dt} = AP^2 + BP + C, \tag{4.5}$$

where $A = -k\gamma > 0$, $B = -k(\lambda + \alpha) < 0$ and $C = k(\beta - \delta) < 0$.

For equ. (4.1) or (4.5), the majority of the literature focuses on the study of the price stability at the demand-supply equilibrium, while ignoring the whole evolutionary characteristics of the price. Since the price evolution model (equ. (4.5)) is quadratic, the

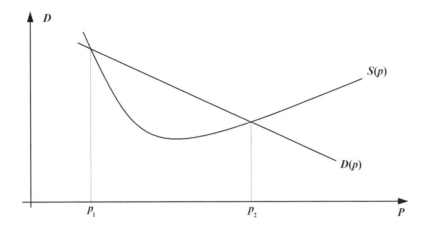

Figure 4.1 The functional relationships of the demand and supply on the price.

price in general changes discontinuously and has the characteristic of singular reversal transitions. In particular, let the discriminant of equ. (4.5) be

$$\Delta = B^2 - 4AC = k^2\{(\lambda + \alpha)^2 + 4\gamma(\beta - \delta)\},$$

then we have three possibilities: $\Delta = 0$, $\Delta > 0$, and $\Delta < 0$. In the following, let us analyze the evolution model in each of these cases.

Case 1. When $\Delta = 0$, at the equilibrium there is only one equilibrium price $P_1 = -\frac{1}{2}\frac{B}{A} > 0$. The characteristics of the whole evolution of the price are described by

$$P = -\frac{1}{2}\frac{B}{A} - \frac{1}{P_0 + At}, \tag{4.6}$$

where P_0 is the integration constant determined by the given initial condition. If $P_0 > 0$, the price P decreases continuously with time and approaches the equilibrium state $\left(-\frac{1}{2}\frac{B}{A}\right)$. If $P_0 < 0$, then when $t = t_b = -\frac{P_0}{A}$, a discontinuity occurs to the price change. When $t < t_b$, the greater the demand the higher the price. When $t = t_b$, the price falls abruptly, indicating the fact that either the demand reaches its level of saturation or the supply increases drastically, causing the abrupt drop in the price. When $t > t_b$, the price starts to rebound continuously with time and eventually approaches the demand-supply equilibrium P_1. This process indicates that through the market adjustment, the price is no longer growing as blindly as during the first period of time. Instead, by taking the market situation into consideration, to keep a reasonable equilibrium between the supply and demand, the price eventually stabilizes within a certain range.

This whole evolution analysis of the price is not only more complete than the stability analysis at the equilibrium, but also more practically realistic than studies of continuous evolutions when compared to the objective situations existing in the marketplace of free competitions.

Case 2. When $\Delta > 0$, solving equ. (4.5) leads to

$$\left| \frac{P + \frac{1}{2}\frac{B}{A} - \frac{1}{2}\sqrt{\frac{B^2}{A^2} - \frac{4C}{A}}}{P + \frac{1}{2}\frac{B}{A} + \frac{1}{2}\sqrt{\frac{B^2}{A^2} - \frac{4C}{A}}} \right| = \exp\left(A\sqrt{\frac{B^2}{A^2} - \frac{4C}{A}}\, t + P_{20} \right), \qquad (4.7)$$

where P_{20} is the integration constant. If $\left|P + \frac{1}{2}\frac{B}{A}\right| < \frac{1}{2}\sqrt{\frac{B^2}{A^2} - \frac{4C}{A}}$, then equ. (4.7) indicates that changes in price are continuous. Otherwise, equ. (4.7) indicates that the price evolution contains blown-ups.

Case 3. When $\Delta < 0$, solving equ. (4.5) produces

$$P = \frac{1}{2}\sqrt{\frac{4C}{A} - \frac{B^2}{A^2}}\, \tan\left(\frac{A}{2}\sqrt{\frac{4C}{A} - \frac{B^2}{A^2}}\, t + \frac{1}{2}P_{30} \right) - \frac{1}{2}\frac{B}{A}, \qquad (4.8)$$

where P_{30} is the integration constant. This equation implies that at $t = t_b = \dfrac{2}{\sqrt{\frac{4C}{A} - \frac{B^2}{A^2}}}$ $\left(\frac{\pi}{2} + n\pi - P_{30}\right)$, $n = 0, \pm 1, \pm 2, \ldots$, periodic blown-ups in the price occur. That is, the price of the consumer good shows the behavior of periodic rise and fall.

From the physical and mathematical characteristics of nonlinearity, as studied in Chapter 2 of this book, this simple analysis above shows that the concept of demand and supply is in fact about mutual restrictions or mutual reactions of different forces under equal quantitative effects (The concept of equal quantitative effects was initially introduced by OuYang (1994) in his study of fluid motions. And later, it is used to represent the fundamental and universal characteristics of all materials' movements (Lin, 1998). By equal quantitative effects, it means the eddy effects with non-uniform vortical vectorities existing naturally in systems of equal quantitative movements, due to the unevenness existing in the structures of materials. In this definition, by equal quantitative movements, it means such movements that quasi-equal acting and reacting objects are involved or two or more quasi-equal mutual constraints are concerned with. For example, in a principal-agent problem, the principal and the agent can be seen as quasi-equal parties. Each interaction between them can be seen as an equal quantitative movement. The effect of such a movement can be seen as an equal quantitative effect.). Lin (2007) shows that if Einstein's concept of "uneven time and space" of materials' evolution (Einstein, 1997) is employed, Newton's second law of motion actually indicates that a force, acting on an object, is in fact the attraction or gravitation from the acting object. It is created within the acting object by the unevenness of its internal structure. So, as soon as calculus-based mathematics is employed to the study of such a relationship as that between demand and supply, one has to face the problem of nonlinearity, which is an unsettled problem of mathematics. With the introduction of blown-up theory (Wu and Lin, 2002), it has been shown that mathematical nonlinearity is about structures of the entities involved. So, to understand nonlinearity, the quantitative, formal analysis needs to be modified accordingly. For instance, to overcome the difficulty encountered by the traditional mathematical methods in the area of weather forecasting, OuYang (Wu and Lin, 2002) proposed the idea of blown-ups and a practical procedure based on the vorticity of materials to resolve evo-lution problems. In particular, his method is developed on the basis of classifications of materials' structures according to their vectorities and has been proven effective in forecasting (nearly-) zero-probability, disastrous weather conditions. For more details to this end, please consult Part 5 in (Lin, 2008).

For our current purpose, let us see how such a structure of spinning yoyo fields can be employed to model and analyze the evolution of and interaction between the supply and demand of money.

4.2 THE SUPPLY OF MONEY

As the title suggests, this section studies the supply of money with its emphasis placed on the role various banks play on the creation of money. Then, the concept of money multipliers is introduced and various money supply models are established.

4.2.1 Creation of currency

The supply of money is a complex process that involves the behaviors of all economic entities. In this process, large amounts of money are created through the currency that is supplied by the central bank. In the creation of money, the bank system plays an important role. The most of the money supply is created by the bank system. In particular, the central bank holds the right to issue money, both the paper notes and coins, for the country. Additionally, the central bank also controls the deposit reserves of commercial banks. The asset-and-debt table of the central bank includes both the money issued and the deposits of banks. These two liabilities (or debts) constitute the foundation of the nation's money supply. Notice that the so-called money issued used in the current content means the total amount of the currency the central bank places into the circulation. It is also known as the currency in circulation. That is the total amount of currency (both the paper notes and coins) held by the public. The so-called deposits of banks stand for the deposit total of the commercial banks at the central bank, which represents the overall reserve of the entire banking system.

The currency in circulation is an important part of the money supply. It is equal to the amount of money issued as listed in the asset-and-debt table of the central bank. Through its asset business, such as giving out loans or investing in securities, the central bank places the created currency into circulation. When the central bank's holding of bonds increases, it means that the money supply is increasing, where symbol C is commonly used to represent the total amount of currency in circulation. The reserve of the banking system stands for the capital that might be necessary for dealing with the withdrawals of account owners of commercial banks. That is, it is the deposit reserve. It is divided into two parts. One is the statutory reserve, as required by the law, which is retained at the central bank. Although the central bank does not pay any interest on these deposits, commercial banks are allowed to withdraw their reserves based on their specific needs. The other part of the deposit reserve is known as excess reserve. It is kept in the treasury of individual commercial banks. That is why it also known as cash on hand. Because these reserves kept on the hands of commercial banks do not earn any interest income, these banks are not willing to hold too much excess reserves. Instead, they lend out the excess reserves as loans or use these funds to invest in securities. That is to say, as long as a bank's reserve increases, the total amount of loans and investments of the bank will increase. That is equivalent to say that the reserve has increased. Here, the symbol R is used to represent the total reserve of the banking system. The monetary base is defined to be the sum of the amount of currency in circulation and the reserve of the banking system, denoted by MB:

$$MB = C + R = \text{currency} + \text{banks' reserve}$$
$$= \text{currency in circulation} + \text{deposit reserve} \qquad (4.9)$$

The monetary base is relatively stable. That is, no matter whether deposits are converted into currency or currency is converted into deposits, the monetary base stays the same. That is because when deposits are converted into currency, banks need to pay these conversions by using their available reserves so that their reserves decrease accordingly. Since the decreased amount of reserves is exactly the same as the increase of currency, it guarantees the monetary base stays the same. Conversely, when a portion of the currency in circulation is converted into deposits, if the banks do not give out loans by using up the increased deposits, then the banks' reserves will increase; the amount of increase is equal to that of decrease of currency available for circulation. That once again explains why the monetary base stays the same. If the banks do lend out the increased deposits as loans, let us assume the amount of increase in deposits, that is the amount of decrease in currency available for circulation, is ΔD, then according to the law on the rate r of statutory reserves, the banks will need to maintain $r\Delta D$ as their reserves. The rest amount $(1 - r)\Delta D$ is loaned out. The borrowers of this amount deposit this money into their bank accounts. That causes the banking system to have an additional deposits in the amount of $(1 - r)\Delta D$. After storing away $r(1 - r)\Delta D$ as reserves, the rest $(1 - r)^2\Delta D$ is once again loaned out by the banks. When this process is repeated indefinitely, the increase in the banks' reserves is given as follows:

$$r\Delta D + r(1 - r)\Delta D + r(1 - r)^2\Delta D + \cdots = \sum_{n=0}^{\infty} r(1 - r)^n\Delta D = \Delta D \qquad (4.10)$$

So, it can be seen that the increased amount of bank reserves is still the same as the amount of currency decrease. In short, when currency is converted into deposits, no matter whether the banks lend the deposits out as loans or not, the increased amount of bank reserves is always equal to the decreased amount of currency so that the monetary base does not vary.

Both currency in circulation and deposits in the banking system have a great deal to do with the wills of the public. They can change wildly and randomly, making them difficult to control and to predict. However, when the two are summed up together, the monetary base, it stays fixed. It is exactly because of this stability that the central bank's focus of its money supply is placed on the control of the monetary base.

Through its business operations on the open market, the central bank can change the quantity of the monetary base. The central bank conduct two kinds of businesses on the open market: one is trading securities and the other providing loans to commercial banks through the discount window.

When the central bank purchases government bonds on the open market, it is equivalent to that it pumps new currency into the circulation or the banking system. If these bonds are purchased from industrial and commercial firms or the public, there exist two scenarios: One is that the sellers of the bonds do not deposit their cash obtained from selling their bonds into their bank accounts. In this case, the total amount of currency in circulation increases. Another scenario is that these sellers do deposit their cash into their bank accounts. In this case, the deposits become new reserves of the banks. If the bonds are purchased from commercial banks, there are also two possibilities: One is that the central bank pays the commercial banks using checks. In this case, the commercial banks deposit these checks into their accounts with the central bank so that the reserves of these commercial banks at the central bank increase. The other case is that the central bank applies cash to pay the commercial banks that sell bonds, or applies checks to pay the commercial banks, where these commercial banks cash the checks. In this case, the cash these commercial banks obtained from selling their bonds is maintained in the treasuries of the banks so that these banks increase their cash on hands. In short, when the central bank purchases bonds, it means that the amount of currency in circulation increases or the reserve of the banking system increases so that the monetary base goes up quantitatively. On the contrary, if the central bank sells bonds on the open market, it means that the monetary base decreases quantitatively. In summary, by trading bonds on the open market, the central bank can alter the supply quantity of the monetary base.

When the central bank gives out a discounted loan to a commercial bank, the proceeds of the loan will be credited into the account of the commercial bank at the central bank. That causes changes to the asset-and-debt table of the central bank. At the same time when a discounted loan appears in the list of assets, there also appears a deposit of the same amount in the list of debts. That implies that the reserve of the banking system has been increased, where the increment is equal to the proceeds of the discounted loan. Therefore, the monetary base also enlarges accordingly. So, it can be seen that through giving out discounted loans to commercial banks, the central bank can also vary the quantity of the monetary base, causing the monetary base enlarge.

By the statement that the central bank issues currency, it essentially means that the central bank injects new currency into circulation and into the economy or repossesses some of the circulating currency through purchasing securities or providing discounted loans or both. Hence, the monetary base can be divided into two portions: the increased currency and banks' reserves, known as the non-borrowed portion of the monetary base and denoted by NB, as caused by the central bank's selling of securities, and the increased banks' reserves, known as borrowed monetary base or discounted loans and denoted by DL, as caused by the central bank's providing of discounted loans, So, the monetary base MB is equal to the sum of NB and DL: $MB = NB + DL$.

The right of trading securities on the open market is in the hands of the central bank. Any time when it desires, it can buy and sell securities. However, although the amount of discounted loans is more or less related to the discount rate the central bank determines, it is not completely within the control of the central bank. It is also closely related to the decision making of commercial banks. Only when commercial banks are willing to borrow from the central bank, discounted loans can be given out. So, to a certain degree, the amount of discounted loans is mainly determined by commercial banks, where the central bank is in a passive position. From this discussion, it follows that the portion of the monetary base the central bank has the total control is the non-borrowed monetary base. The borrowed portion of the monetary base, the discounted loans, is not completely within the control of the central bank. That clearly describes the range of practical control of the central bank over the monetary base.

Now, let us turn our attention to commercial banks.

On the one hand, banks attract deposits; on the other hand, they give out loans using the attracted deposits. After going through various market activities, the loans re-enter the banking system in the form of deposits. Then, these deposits once again are giving out by the banks as loans. Such repeated entering of loans into the banking systems makes the number of deposits in banks expand constantly so that the so-called deposit currency is created out of the banking system. If the monetary base or the base currency is the currency of the central bank, then the deposit currency (that is the demand deposits) is the currency of commercial banks. That is, deposit currency is introduced by commercial banks. To this end, by manipulations in the open market, the central bank can make the amount of deposits in the banking system either multiple or shrink.

As for the process of multiplication of deposits, let us assume that the central bank purchased the bonds of worth ΔR dollars held earlier by a certain commercial bank. So, the reserve of this commercial bank (deposit reserve) increases as much as ΔR dollars. Because keeping excess reserve brings in no interest income, the commercial bank would give out loans against this new reserve. Note that because the amount of deposits did not increase, the bank does not need to prepare a deposit reserve for the newly acquired cash so that these ΔR dollars can be entirely loaned out. The general method for the bank to give out loans is to establish accounts for the borrowers and then transfer into these accounts the money, which becomes the deposits of the borrowers. From doing so, the bank creates the first around of deposits in the amount of ΔR. However, this amount would not be idling in the accounts within this bank too long. Instead, the borrowers would use the funds to purchase goods or services from other companies or individuals. When the borrowers sign their checks to pay their purchases, the funds, as represented by the checks, would flow from the borrowers' accounts into the bank accounts of the sellers of the merchandise. That makes the deposits in other banks increase ΔR dollars. For the sake of convenience for our analysis, assume that the bank within which the borrower opens his account is B_0 and then this ΔR dollars created enters into bank B_1. Similarly, bank B_1 does not like to hold excessive reserves, either. So, after retaining the statutory reserve $r\Delta R$ at the rate r, the rest $(1-r)\Delta R$ is loaned out. At this juncture, bank B_1 also establishes an account for its borrow to park the borrowed money and transfers the loan amount into the account. So, bank B_1 creates a deposit of $(1-r)\Delta R$ dollars. Similarly, this deposit will not stay with bank B_1 very long before it is moved to another bank B_2 and becomes a deposit of bank B_2. Bank B_2 also retains $r(1-r)\Delta R$ dollars as its statutory reserve at the rate r and loans out the rest. So, Bank B_2 creates a deposit in the amount of $(1-r)^2\Delta R$ dollars. When this process of deposit creation continues indefinitely, the increased reserve of bank B_0 brings forward a while series of created deposits, where bank B_n creates a deposit of the amount $(1-r)^n \Delta R$ dollars, $n = 0, 1, 2, 3, \ldots (n = 0, 1, 2, 3, \ldots)$. So the total increased amount of deposits for the entire banking system is

$$\Delta D = \sum_{n=0}^{\infty} (1-r)^n \, \Delta R = \frac{1}{r} \Delta R \qquad (4.11)$$

Because the rate r of the statutory reserve is a positive number less than 1, and mostly likely less than 50%, say, around 10%, the reciprocal $1/r$ of this rate is greater than 1, and generally is greater than 2, say, around 10. That illustrates that any increase in a bank's reserve causes the deposits in banks rise multiple times. Especially when the rate of statutory reserve is 10%, the money pumped into circulation by the central bank through purchasing bonds can make the banks' deposits increase 10 times the amount of the money pumped into the economy. That is the expansionary effect of bank deposits. What is described above is the expansionary process of demand deposits.

As for the contraction process of deposits, it means the effect that is opposite to the creation of multiple deposits. That is, if the central bank causes the bank reserve to decrease through selling bonds, then under the condition of requiring statutory reserves, the scale of bank deposits will decrease multiple times. For example, bank B_0 spends ΔR dollars to purchase bonds from the sale of the central bank. Because it does not maintain excessive reserve, it makes the bank's reserve lower than the required amount of statutory reserve. To make up the shortage, bank B_0 sells bonds of the value ΔR dollars or withdraws loans of the amount of ΔR dollars. The purchaser of the bonds from bank B_0 or the borrower who repaid his loans to bank B_0 has to withdraw ΔR dollars from his bank account to pay bank B_0. That leads to a deposit reduction of ΔR dollars from the banking system. For the sake of convenience for our communication, let us assume that the deposit reduction occurs at bank B_1. That is, the total amount of deposits at bank B_1 reduces ΔR dollars. That implies that the reserve of bank B_1 at the central bank is dropped by ΔR dollars, because it does not maintain the appropriate excess reserve. Because of the reduction in deposits, the statutory reserve of bank B_1 can be decreased by $r\Delta R$ dollars. Now, there is still a reserve short in the amount of $(1-r)\Delta R$ dollars, which needs to be covered. So, bank B_1 has to sell bonds in the value of $(1-r)\Delta R$ dollars or withdraw loans in the amount of $(1-r)\Delta R$ dollars. This action of bank B_1 leads to a further deposit reduction of $(1-r)\Delta R$ dollars in the banking system. Assume that this deposit reduction occurs at bank B_2. When this kind of deposit reduction continues indefinitely, bank B_2 causes a deposit reduction of $(1-r)^2\Delta R$ dollars to the banking system, bank B_3 causes a deposit reduction of $(1-r)^3\Delta R$ dollars to the banking system, $\ldots$, bank B_n causes a deposit reduction of $(1-r)^n\Delta R$ dollars to the banking system, $n = 0, 1, 2, 3, \ldots$ Consequently, the total amount of bank deposit reduction is

$$\Delta D = \sum_{n=0}^{\infty} (1-r)^n \Delta R = \frac{1}{r} \Delta R \qquad (4.12)$$

Because $0 < r < 1$, it means that the total amount of bank deposits decreases multiple times.

The process of deposit expansion tells us that when the reserve of the banking system increases, bank deposits will multiply many times over. This expansion multiple is known as deposit multiplier, which is equal to the reciprocal of the rate of the statutory reserve. This principle on deposit expansion is known as deposit multiplier principle. Note that for this principle to work one has to assume that banks do not maintain excess reserves; and when a borrower obtains his loan, he leaves his entire borrowed funds with the bank. Additionally, it is not just the central bank that can make the banks' reserve increase, causing multiple deposit expansions. The public's preference over holding currency can also affect the banks' reserve. If the public decides to put a portion of its hand-held cash into banks, that will surely make the banks hold additional cash, leading to increased bank reserves. According to the same principle, the banks' deposits will expand multiple times. In fact, as long as banks have excess reserves, the process for the banks to create deposits will not halt. Only when banks do not hold any excess reserves, that is when the banks' reserves are exactly the same as their statutory reserves the creation of deposits can be stopped. That is when the deposit and loan activities at banks will be in equilibrium. So, different levels of reserves support different and corresponding levels of demand deposits.

The afore-mentioned models of multiple deposits seem to suggest that as long as the central bank establishes the levels of statutory reserves and bank reserves, it can materialize the total control of the level of demand deposits. However, the reality is not so. The afore-mentioned models of multiple deposits are only some elementary, idealized, mechanical models. In real life, the creation of deposits is much more complicated process than what is suggested by the models, which have ignored many fine details, while these details make the ultimate level of demand deposits not as high as that described by these models. For example, when the deposit at bank B_2 increases, the bank can use the additional funds in three different ways. It can apply part of the funds as the statutory reserve; it can keep part of the funds as cash; and it can loan out the rest. Because the scenario of keeping additional cash on hands is missed in the previous models, the result of deposit expansion is not as ideal as that described by the models. For example again, if after bank B_2 lends a loan to an individual, instead of leaving the borrowed money in a bank account that person keeps the money all in cash, then the imagined process of deposit expansion is terminated right there and right then. Evidently, the created deposit total in this case will not be as large as what the models depict.

The deposit creation models assume that banks do not hold any excess reserve. However, in reality, each bank maintains a certain amount of excess reserve. The existence of excess reserves is an important omission of the deposit expansion model. It makes the deposit multiplier less than the reciprocal of the rate of the statutory reserve. After learning these shortcomings of the multiple deposit creation models, the significance of the models become clear. They mean that the amount of bank deposits that is supported by the reserves required by the law is the maximum possible amount of deposits instead of the realistic amount of deposits.

Additionally, other than the behavior of the central bank that affects the level of deposits, the decisions of individual borrowers to hold cash and the decisions for commercial banks to maintain amounts of cash on hands also influence the level of deposits. They can all cause the money supply to change. It can be said that the behaviors of all four capital providers of the financial market, financial institutions, industrial enterprises and commercial firms, government agencies, and families and individuals, can all impose effects on the level of deposits. Hence, in the study of money supply, these relevant aspects should not be ignored.

4.2.2 Monetary multiplier

The so-called monetary multiplier stands for the ratio of the money supply M and the monetary base MB, denoted by m. That is, symbolically, we have

$$m = \text{monetary multiplier} = \frac{\text{money supply}}{\text{monetary base}} = \frac{M}{MB} \qquad (4.13)$$

This expression indicates that the money supply is a multiple of the monetary base. That multiple is the money multiplier. Because of this reason, the base currency is also known as high powered money.

When the money supply M in the previous equation is understood as the narrow money M_1, the corresponding monetary multiplier is known the narrow monetary multiplier, denoted as m_1. When the money supply M is understood as a general money, the corresponding monetary multiplier is known as a general monetary multiplier. For example, when $M = M_2$, we have the monetary multiplier m_2 of M_2; when $M = M_3$, we have the monetary multiplier m_3 of M_3; and when $M = L$, we have the monetary multiplier ℓ of L; and when $M = M'_3$, we have the monetary multiplier m'_3 of M'_3, which is the weighted monetary multiplier.

In this section, we mainly study the narrow monetary multiplier. In the following, without specific explanation, all monetary multiplier m will stand for the narrow monetary multiplier $m = m_1$. That is, all the discussion on m_1 also applies to any of the general monetary multipliers.

It has been established that the monetary base is equal to the sum of the currency in circulation and the bank reserves; and the bank reserves consist of two parts: One is the statutory reserves of the banking system, which are required by the law and are kept at the central bank without bearing any interest; other is the excess reserves of individual banks, which are cash on the hands of banks and are maintained in the treasuries of individual banks. If symbol C is used to represent the currency in circulation, R the bank reserves, RR the statutory reserves, ER the excess reserves, and D the bank deposits (that is, checking account deposits), then the currency ratio r_c, the statutory reserve ratio r_d, and the excess reserve ratio r_e are defined as follows:

$$r_c = \frac{C}{D}, \quad r_d = \frac{RR}{D}, \quad \text{and} \quad r_e = \frac{ER}{D} \qquad (4.14)$$

Assume that these ratios are fixed constants. Then from $MB = C + R$ and $R = RR + ER$, it follows that $MB = r_c \times D + r_d \times D + r_e \times D = (r_c + r_d + r_e) \times D$ and that

$$D = \frac{MB}{r_c + r_d + r_e} \qquad (4.15)$$

From the definition of narrow money, the money supply M is equal to the sum of the currency C in circulation and the checking account deposits D so that the following holds true:

$$M = M_1 = C + D = r_c \times D + D = (1 + r_c)D = \frac{1 + r_c}{r_c + r_d + r_e} MB$$

This expression provides a computational formula for the monetary multiplier m:

$$m = m_1 = \frac{1 + r_c}{r_c + r_d + r_e} \tag{4.16}$$

Deposit multiplier is defined to be the reciprocal of the statutory reserve ratio so that monetary multiplier is less than the deposit multiplier. To help understand this point, let us take a look at the process of deposit expansion on the basis of $M = m \times MB$. It can be seen that when the base currency increases, the money supply also increases accordingly and the amount of increase is m times of the increase of the base currency. If the public does not hold any cash on hands, instead it deposits all its cash in bank accounts, and banks do not maintain any excess reserve, then both r_c and r_e are zero. So, when the monetary base increases, the entire amount of increased monetary base enters into the expansion process of the banks' deposits. The amount of increase of the money supply is equal to the increase in the banks' deposits. Hence, the monetary multiplier is the same as the deposit multiplier (that implies that the money supply is equal to the demand deposits, while the amount of currency in circulation is zero. What needs to be noted here is that the money supply does not include banks' reserves.)

However, the situation in real life is not the same as what is just described. Part of the increased currency will be held in the public's hands, that is, $r_c > 0$, and the rest will be deposited into bank accounts. Banks will also keep a portion of the increased inflow of funds as excess reserves, that is, $r_d > 0$, and lend out the rest as loans. Those amounts of the base currency that are either held by the public or stored in banks' treasuries do not have any effect on creation of bank deposits. They do not lead to the multiplication of bank deposits. Therefore, among the increased monetary base only a portion instead of the entire amount contributes to deposit expansions, while the rest only sits at where they are without producing any effect on the expansion. That is, the amount of increase in the money supply caused by an enlargement in the monetary base will surely be less than the amount of increase of money as computed by assuming that the entire enlargement in the monetary base participate in the deposit expansion. That implies that the monetary multiplier is less than the deposit multiplier, because the latter is equal to the multiple of expansion of the money supply when the entire enlargement of the monetary base helps to expand the bank deposits.

The fact that monetary multiplier is less than the deposit multiplier can be shown directly by using algebraic method. In fact, from $r_c > 0, r_e > 0, 1 > r_d > 0$ and the formulas of deposit and monetary multipliers, it follows immediately that

$$\frac{1}{r_d} - m = \frac{1}{r_d} - \frac{1 + r_c}{r_c + r_d + r_e} = \frac{r_c(1 - r_d) + r_e}{r_d(r_c + r_d + r_e)} > 0$$

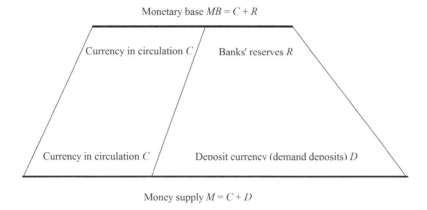

Figure 4.2 The relationship between money supply and monetary base.

Therefore, $m < \frac{1}{r_d}$. That is, the monetary multiplier is less than the deposit multiplier. By comparing this algebraic argument to the earlier economic reasoning, the latter is surely more straightforward.

Based on this new understanding of the fact that the monetary multiplier is less than the deposit multiplier, the relationship between the money supply and the monetary base can be depicted in Figure 4.2. In this figure, the part of currency of the monetary base does not affect the deposit expansion so that the currency part of the money supply is equal to that of the monetary base.

From the formula of monetary multiplier in equ. (4.16), it follows that the factors that affect the magnitude of the monetary multiplier are the currency ratio, the statutory reserve ratio, and the excess reserve ratio. In order to gain additional understanding of the monetary multiplier, let us look at how changes in these factors affect the monetary multiplier.

First, let us look at how changes in the currency ratio r_c affect the monetary multiplier m. To this end, assume that the currency ratio r_c changes while all other factors remain constant. If the currency ratio increases, that implies that depositors have converted some of their checking account deposits into currency (because the totality of the monetary base stays the same). Therefore, the checking account deposits will shrink multiple times while the portion of currency in the money supply only increases the same amount. That is to say, an increase in the currency ratio inevitably causes a decrease in the money supply. Because the monetary base remains stable, the consequence of increasing the currency ratio is that the monetary ratio decreases. In fact, this conclusion can be shown by using mathematical rigor. In fact, by computing the partial derivative of the monetary multiplier with respect to the currency ratio, we obtain the following:

$$\frac{\partial m}{\partial r_c} = -\frac{1 - r_d - r_e}{(r_c + r_d + r_e)^2} < 0, \quad \text{because } r_d + r_e < 1.$$

which means that as the currency ratio r_c increases, the monetary multiplier m decreases. Therefore, the monetary multiplier and the currency ratio are negatively correlated.

Secondly, let us look at how changes in the ratio r_d of statutory reserves affect the monetary multiplier m. To this end, assume that the statutory reserve ratio r_d changes while all other factors stay constant. If the statutory reserve rate increases, it means that banks have to maintain an increasing amount of reserves so that initially a shortage in banks' reserves would appear. To make up the difference, banks have to reduce the amount of loans that are giving out. This behavior would cause the demand deposits of the banking system shrink multiple times, leading to a decreased money supply. So, an increase in the rate of statutory reserves would cause the monetary multiplier decrease. This conclusion can also be shown mathematically. For example, by differentiating the monetary multiplier with respect to the rate of statutory reserves, we obtain

$$\frac{\partial m}{\partial r_d} = -\frac{1 + r_c}{(r_c + r_d + r_e)^2} < 0$$

That implies that as the rate of statutory reserves increases, the monetary multiplier m decreases. That is to say, the monetary multiplier and the rate of statutory reserves are negatively correlated.

Thirdly, let us look at how changes in the rate of excess reserves affect the monetary multiplier m. When only the rate r_e of excess reserves changes while all other factors remain constant, an increase in the rate of excess reserves implies that banks increased their excess reserves. However, because the monetary base stays the same and the currency ratio also stays invariant (that is, current currency is not converted into deposits), it means that the increase in the excess reserves causes the statutory reserves to decrease. In order to make up the shortage in the statutory reserves, banks have to reduce their amounts of loans being given out. That would cause the amounts of demand deposit in the banking system to decrease multiple times. So, it can be seen that an increase in the rate of excess reserves could also lead to currency ratio to drop. Mathematically, differentiating the currency multiplier with respect to the rate of excess reserves produces

$$\frac{\partial m}{\partial r_e} = -\frac{1 + r_c}{(r_c + r_d + r_e)^2} < 0$$

That means that as the rate r_e of excess reserves increases, the currency multiplier decreases. Therefore, the currency multiplier and the rate of excess reserves are negatively correlated.

Besides, from the fact that the partial derivative of the currency multiplier with respect to the rate of statutory reserves is equal to that with respect to the rate of excess reserves, it follows that the effect on the monetary multiplier from increasing the rate of excess reserves is identical to that from increasing the rate of statutory reserves. Their effects are of the same magnitude.

4.2.3 Money supply models

Money supply can be seen differently with narrow and general classifications and varied layers. In order to develop money supply models on each layer, the key is to find out the determining factors that affect the money supply. To this end, let us start with an analysis of the determining factors at the monetary base level. Through explaining the behaviors of depositors and banks, we will look for the factors that determine the money supply so that money supply models on different levels can be established gradually.

The base currency is the currency of the central bank and consists of two parts: One is the non-borrowed money NB, whose quantitative changes the central bank can control through trading securities on the open market. The other part of the base currency is the borrowed currency DL. That is the discounted loans, although whose quantity more or less is related to the discount rate (that is the rate used to discount loans), it is more likely determined by the decisions of commercial banks. That is, the quantity of discount loans is not completely within the control of the central bank. Because the central bank cannot totally control the amount of discount loans, discount loans DL become an important factor that influences the monetary base.

Because the central bank can completely control the non-borrowed monetary base NB, let us find out the factors that determine the magnitude of NB. As mentioned earlier, the central bank influences the magnitude NB through trading securities. Therefore, the amount of securities the central bank holds becomes an important factor that affect the magnitude of NB. Other than that, through trading gold, SDR (special drawing right), foreign currency deposits, and any other kinds of assets, the central bank can affect the quantity of the non-borrowed monetary base as effectively as through trading securities. So, any increase (or decrease) in these assets in the asset-debt table of the central bank causes equal amount of increase (or decrease) in the monetary base. Additionally, there are also two items on the asset-and-debt table of the central bank: One is the cash receivables in the list of assets, and the other is the cash payables in the list of debts. The difference of the cash receivables and the cash payables is known as the funds in transit.

Let us use an example to illustrate how the funds in transit affect the monetary base. Assume that the central bank currently has zero amount of funds in transit. That is, the amount of cash receivables is the same as that of cash payables. Now, the central bank receives from Bank A a check in the amount of $1,000 that is payable to Bank B. When the central bank receives this check, because temporarily there are no specific behaviors of receiving and paying funds, this check implies that the central bank owes Bank B $1,000 (that is, the debt of the central bank increases $1,000). At the same time, Bank A owes the central bank $1,000 (that is, the asset of the central bank increases $1,000). So, both the cash receivable and the cash payable of the central bank have increased $1,000. However, the corresponding funds in transit are zero. What the central bank does in general is that when it receives the check, the central bank first credits the account of the receiving Bank B at the central bank the amount of the check. Then after one or two days, the central bank debits this amount the account of the paying Bank A at the central bank. That is, by giving the check back to the paying Bank A, Bank A is demanded for payment. The credit cancels the debt of the central bank; that is, the cash payable is cleared, while the debit drops the asset, which clears the cash receivable. Between the credit and debit there is a time gap, especially weather changes could also prolong this time gap, where credit is ahead of debit. When a credit is given while a debit is yet to be assigned, funds in transit appear. That is, the cash payable has already disappeared, while the cash receivable still appear on the list of assets. That creates a non-zero difference between cash receivables and cash payables. After credit is given, the reserve of Bank B is increased by $1,000; after debit is assigned, the reserve of Bank A is reduced by $1,000. So, during the time after credit is given and before debit is assigned, the total bank reserves are increased by an amount of $1,000. That is exactly the amount of funds in transit.

From the discussion above, it can be seen that an increase in the funds in transit causes the monetary base enlarge in an equal amount. However, because funds in transit only exists temporarily, the fluctuation in the monetary base, as caused by the funds in transit, is also short lived. That is to say, funds in transit represent one of the main reasons for the monetary base to fluctuate temporarily, but not a main reason for long term fluctuations.

Based on the discussion above, it can be seen that all such factors as the amount of discount loans giving out by the central bank, the amounts of securities, gold, SDR (special drawing rights), foreign currency deposits, and other assets held by the central bank, and the amount of funds in transit are important factors affecting the monetary base, which is positively correlated with each of these factors: As these factor increase, the amount of the base currency also increase.

Now, from $M = m \times MB$ and $MB = NB + DL$, it leads to the following narrow money supply model:

$$M_1 = m_1 \times MB = \frac{1 + r_c}{r_c + r_d + r_e}(NB + DL) \tag{4.17}$$

By combining this model and the previous discussion on monetary multipliers, it follows immediately that the narrow money M_1 is negatively correlated to the currency rate r_c, the rate r_d of statutory reserves, and the rate r_e of excess reserves, and is positively correlated to the non-borrowed monetary base and discounted loans.

In order to better understand this model, there is a need to analyze why these five factors affect the money supply M_1. That is, there is a need to explain the behaviors of depositors and banks so that meaningful factors, which affect the money supply, can be located.

First, let us look at the currency ratio. This ratio stands for that of hand-held currency and demand deposits. It reflects the choice of the public between holding either cash or demand deposits. As for whether people like to hold cash or demand deposits, it is determined by the level of wealth people possess and the expected rate of return on the demand deposits. Additionally, the tax imposed on wealth affects the amount of wealth people are willing to hold so that the tax also influences the currency ratio and money supply. To this end, when the level of people's wealth changes, how will the currency ratio vary accordingly? To address this problem, we only need to notice such a fact in life: Cash is often widely used by low-income people or people who do not possess much wealth. So, comparing to demand deposits, cash becomes a necessity. When seeing as a necessity, the demand for cash has very small elasticity. So, when people's wealth (income) increases, the demand for the necessity decreases. That is, the speed of increasing need for cash is less than the speed of increasing of wealth. In other words, as the level of wealth increases, the magnitude of increase in demand deposits is greater than that in cash. That implies that the currency ratio decreases with the increase in wealth. That is how wealth affects the currency ratio and consequently imposes effect on the money supply: Increasing levels of wealth causes the currency ratio to decrease and money supply to increase.

The second factor that affects the currency ratio is the level of expected return of the checking account deposits in comparison with cash and other assets. The higher the expected return of checking account deposits is, the more willing people are to hold such deposits so that the lower the currency ratio. That is to say, the currency ratio is negatively correlated to the expected rate of return of checking account deposits. Therefore, money supply is positively correlated to the expected rate of return of checking account deposits. There are mainly three factors that affect the expected return of return on checking account deposits: the interest rate of checking deposits, bank panics, and illegal activities. In legal terms, checking account deposits do not bear any interest. However, financial innovations helped banks to get around this legal constraint and enabled banks to pay interest on checking account deposits. So, in real life checking account deposits do carry interest rates, which become an important factor affecting the return of checking account deposits. Under the condition that the probability for uncertain events to occur stays the same, the higher the interest rate on checking account deposits, the higher the expected rate of return on those deposits. So, increasing interest rate on checking account deposits causes the expected rate of return to rise, the currency ratio to drop, and the money supply to increase. If a bank panic occurs, then the probability that the depositors' interest might be harmed increases dramatically. So, instead of leaving their money in banks, people will be more willing to keep their cash under mattresses. Hence, bank panics cause the expected rate of return on checking account to decrease, the currency ratio to rise, and the money supply to fall. The expected rate of return on checking account deposits is also affected by the amount of illegal activities. If checks are used to pursue illegal transactions, a paper trail will be left behind, which will potentially cause the transactions profitless. So, using checks in this sense leads to extremely low rate of expected return. That is, using cash to do illegal transactions will leave behind no trace and the activities will be more easily covered up. For example, drug trafficking, illegal gambling, the trading of stolen goods, etc., are all done using cash as the medium of underground payment. Because such underground payment helps the illegal transactions to escape the detection of the authorities, the expected rate of return involved in these illegal activities is much higher than that of checking account deposits. If the quantity of illegal activities increases, it will clearly make the expected return of the checking account deposits very low. So, increasing amount of illegal activities causes the expected rate of return on checking account deposits to decrease, the currency ratio rise, and the money supply drop.

Another non-ignorable factor that affects the currency ratio and money supply is taxation. On the one hand, any taxation on wealth forces people to own less wealth so that the currency ratio increases and the money supply decreases. On the other hand, rising rate of taxation motivates people to evade taxes. Because when checks are used each transaction is well recorded on banks' statements, which makes it different to evade taxes, it encourages people to apply cash to conduct their business in order to hide their incomes from the tax authorities. Hence, when the rate of tax increases, the increasing pressure of taxation greatly motivates people to conclude business deals by using cash in order to dodge taxes. That surely causes people to hold more cash so that the currency ratio increases. In short, taxation and increasing rate of tax all make the currency rate rise and the money supply fall.

Secondly, let us look at the behavior of banks. Here we analyze what factors affect the money supply through the rate of excess reserves and the amount of discounted loans. In order to comprehend why a bank would accept this rate of excess reserve instead of other rates, we must consider the cost and benefit of holding excess reserves. A common reason is that if the cost of holding excess reserve increases, then the bank would decrease how much excess reserve it holds; if the benefit of holding excess reserve goes up, then the bank would increase the amount of excess reserve it holds. Now, the question becomes: What factors could affect the cost and benefit for a bank to hold excess reserve? Such factors include mainly the following two: The market interest rate and expected outflow of deposits. Here, the cost for the bank to hold excess reserve is the opportunity cost, because holding excess reserve means giving up the returns that could be potentially made through lending out the money or investing the money in securities. The scale of the opportunity cost is determined by the interest rates of the lending market and the securities market. If the interest rate i of the lending market and the securities market increases (or decreases), then the opportunity cost of holding excess reserve increase (respectively decreases). Therefore, banks would reduce (respectively increase) the amounts of excess reserves, making the rates of excess reserves fall (respectively rise), loans and investments increase (respectively decrease). That explains that the rate of excess reserve is negatively correlated to the interest rate of the market, and the money supply positively correlated to the interest rate of the market. The reason why banks hold excess reserves is because they want to reduce the loss caused by outflows of deposits, where the reduced loss is a gain of the banks. When banks expect the outflow of deposits would increase, they would expect an increase in the consequent loss caused by the outflow of deposits so that they would increase the expected return on the excess reserves. That is when banks would decide to hold additional excess reserves. This end explains that the rate of excess reserve is positively correlated to the expectation of increasing outflows of deposits; and the expected increasing outflow of deposits leads to an increased rate of excess reserve.

From the angle of analyzing the cost and benefit of taking discounted loans, it can be seen that the factors that affect a commercial bank to take out discounted loans from the central bank are the discount rate and the market interest rate, where the former is a factor that affects the cost, while the market interest rate is a factor that affects the return. Here, the discount rate i_d means the rate at which the central bank gives out its discounted loans. The higher the discount rate is the higher the cost will be for commercial banks to borrow from the central bank so that the less likely commercial banks are willing to take discounted loans and the fewer the total amount of discounted loans will be. So, the number of discounted loans is negatively correlated to the discount rate, making the money supply negatively correlated to the discount rate. The interest rate i of the loans and securities markets can help cancel some of the negative effects of the discount rate on discounted loans. If the market interest rate i is higher than the discount rate i_d, then banks are more willing to take discounted loans and then lend out the borrowed funds as loans or invest in securities in order to earn a return at the rate $(i - i_d)$. That explains that the higher the market interest rate i is, especially the greater the difference $(i - i_d)$ of the market interest rate and the discount rate is, the greater the number of discounted loans is, and the greater the money supply.

The previous paragraphs analyzed the various factors of the money supply model, explained why these factors could vary, and how changes in these factors could affect the monetary multiplier and the base currency so that the money supply is eventually affected. Let us now summarize what has been obtained in order for us to learn the mutual influence and effect of these factors on each other. First of all, the following 10 factors affect the monetary multiplier and the monetary base so that in turn the money supply is affected, where the first 7 factors affect the monetary multiplier, the last 2 factors affect the monetary base, while the 8th factor affects both the monetary multiplier and the monetary base.

1   When the rate of statutory reserve rises, the monetary multiplier becomes smaller and money supply decreases;
2   When the level of wealth rise, the currency ratio falls, the monetary multiplier gets greater and the money supply increases;
3   When the interest rate of checking account deposits rises, the currency ratio falls, the monetary multiplier becomes greater, and the money supply goes up;
4   When the probability for a bank panic to occur goes up, the currency ratio rises, the monetary multiplier falls, and the money supply reduces;
5   When the number of illegal activities increases, the currency ratio goes up, the monetary multiplier becomes smaller, and the money supply falls;
6   When the rate of income taxes gets higher, the currency ratio increases, the monetary multiplier becomes smaller, and the money supply reduces;
7   When the outflow of deposits is expected to increase, the rate of excess reserve goes up, the monetary multiplier becomes smaller, and the money supply falls;
8   When the market interest rate goes up, the rate of excess reserve falls down, the monetary multiplier becomes greater, the number of discounted loans increases, and the money supply increases;
9   When the discount rate rises, the number of discounted loans drops, and the money supply goes down; and
10 When the non-borrowed monetary base enlarges, the monetary base gets larger and the money supply goes up.

As for the mutual influences of these factors, changes in the level of wealth, the interest rate of checking account deposits, the number of illegal activities, and the rate of income taxes affect the decision of depositors in terms of how much cash to hold on hands. That in turn causes the currency ratio to vary, and changes in the currency ratio affect banks' decisions on their rates of excess reserves. The reaction of banks to changes in the currency ratio is to adjust both their rates of excess reserves and their interest rates on checking account deposits, which reciprocally affect the depositors' decisions on how much cash they like to keep on hands, which are the depositors' decisions on the currency ratio. So, in the process of supplying money, depositors and banks constantly influence each other and adjust their own decisions based on the decisions of the other side. The 10 factors interact continuously, making the supply of money situated in non-stopping changes. Only when these laws of changes are sufficiently comprehended, one can effectively control the supply of money. The complete money supply model should be able to fully reflect all these relevant changes.

In the following, we will employ the previously establish money supply model to explain the consequences caused by bank panics. When the probability for a bank panic to occur increases, due to the fear of potential economic harms if banks fall into bankruptcy, depositors rush to their banks to withdraw their deposits in order to hold cash by converting their deposits. That causes the currency ratio rise drastically, the money supply shrink majorly. Facing such situations, in order to protect themselves from the disastrous outflows of deposits, banks raise their rates of excess reserves. That in turn makes the monetary base, which supports the deposit expansion, reduce its magnitude, causing the amount of deposits shrink multiple times. That further makes the money supply drop. Therefore, under the mutual influence of the currency ratio and the rate of excess reserve, bank panics push the money supply to dry up drastically. That was exactly what actually happened during the Great Depression from 1930 to 1933. During that time period, although the monetary base rose 20%, the money supply slipped 25%. The reason is that the enlargement of the monetary base did not support deposit expansion; instead, the base currency that supported deposit expansion actually shrunk (both r_d and r_e became greater, making the monetary multiplier fell drastically).

In the rest of this section, let us study the general money supply model. The idea and method of developing the general money supply model are similar to what we did in the previous paragraphs when we established the narrow money supply model. The key is to derive the general monetary multiplier, which makes the general and narrow money supply models different from each other. From knowing this fact, we will focus our attention mainly on the deduction of the general monetary multiplier while we briefly introduce the general money supply model.

For the money M_2 supply, according to the definition of M_2, it includes the currency C in circulation, the checking account deposits D, the savings and small-valued time deposits S, and the total F that includes the money markets accounts, money market mutual funds, overnight repurchase agreements, and the overnight European US dollars:

$$M_2 = M_1 + S + F = C + D + S + F \tag{4.18}$$

We further assume that C, S, and F all grow with D at an identical rate. That is, we assume that r_c, r_d, r_e, $r_s = S/D$, and $r_f = F/D$ are all constant. Then from $D = MB/(r_c + r_d + r_e)$, we obtain the following computational formula for the general monetary multiplier m_2 and the supply model of the general money M_2:

$$m_2 = \frac{M_2}{MB} = \frac{1 + r_c + r_s + r_f}{r_c + r_d + r_e} = m_1 + \frac{r_s + r_f}{r_c + r_d + r_e} \tag{4.19}$$

$$M_2 = m_2 \times MB = \frac{1 + r_c + r_s + r_f}{r_c + r_d + r_e}(NB + DL)$$

$$= M_1 + \frac{r_s + r_f}{r_c + r_d + r_e}(NB + DL) \tag{4.20}$$

where the general monetary multiplier m_2 is negatively correlated to r_c, r_d, and r_e, and positively correlated to both r_s and r_f.

Notice that in this discussion the problem for banks to maintain reserves for S and F is not illustrated. In fact, when funds of these two categories flow into banks, through the loan activities between borrowers and banks these funds would be immediately converted into checking account deposits. Since D stands for the total of all checking account deposits, it in essence has already included all the checking account deposits that are converted from both S and F. Hence, both r_d and r_e have already contained the reserves prepared for S and F. If the reserves for these two items are considered again, the problem of repeated computation will have to emerge. In the following when we construct general money supply models of higher levels, similar convention will be assumed.

For the supply model of money M_3, other than containing M_2, M_3 also include large valued time deposits T, and money market mutual funds held by organizations, long-term repurchase agreements, and long-term European US dollars, which are collectively denoted by MF. That is, we have

$$M_3 = M_2 + T + MF = C + D + S + F + T + MF \tag{4.21}$$

Similarly, we assume that C, S, F, T, and MF grow with D at an identical rate. That is, we assume that r_c, r_d, r_e, r_s, r_f, $r_t = T/D$, and $r_m = MF/D$ are all constant. From $D = MB/(r_c + r_d + r_e)$, we obtain the following computational formula for the general monetary multiplier m_3 and the supply model of the general money M_3:

$$m_3 = \frac{M_3}{MB} = \frac{1 + r_c + r_s + r_f + r_t + r_m}{r_c + r_d + r_e} = m_2 + \frac{r_t + r_m}{r_c + r_d + r_e} \tag{4.22}$$

$$M_3 = m_3 \times MB = \frac{1 + r_c + r_s + r_f + r_t + r_m}{r_c + r_d + r_e}(NB + DL)$$

$$= M_2 + \frac{r_t + r_m}{r_c + r_d + r_e}(NB + DL) \tag{4.23}$$

where the general monetary multiplier m_3 is negatively correlated to r_c, r_d, and r_e, and is positively correlated to r_s, r_f, r_t, r_m.

For the supply of money L, L is equal to the sum of M_3 and other assets of relatively high levels of liquidity: $L = M_3 + A$, where A stands for the total of all those assets of relatively high levels of liquidity that are not included in M_3. That is, A includes commercial papers, savings bonds, banks' promissory notes, short-term treasury bills, etc. Other than what is assumed above, we further assume that $r_a = A/D$ is a constant. That is, A changes with D at a fixed rate. By using the same method of reasoning as above, we can obtain the general monetary multiplier ℓ and the supply model of the general money L as follows:

$$\ell = \frac{L}{MB} = \frac{1 + r_c + r_s + r_f + r_t + r_m + r_a}{r_c + r_d + r_e} = m_3 + \frac{r_a}{r_c + r_d + r_e} \tag{4.24}$$

$$L = \ell \times MB = \frac{1 + r_c + r_s + r_f + r_t + r_m + r_a}{r_c + r_d + r_e}(NB + DL)$$

$$= M_3 + \frac{r_a}{r_c + r_d + r_e}(NB + DL) \tag{4.25}$$

From these formulas, it follows that the general monetary multiplier ℓ is negatively correlated to r_c, r_d, and r_e, and is positively correlated to r_s, r_f, r_t, r_m, r_a.

For the weighted money supply, assume that other than the currency in circulation and checking account deposits, there are also other n kinds of assets of certain levels of liquidity: asset A_1, asset A_2, ..., asset A_n, where the liquidity of asset A_k is α_k, satisfying $0 < \alpha_k < 1$, $k = 1, 2, \ldots, n$. According to the definition of weighted money M_3', we have

$$M_3' = C + D + \sum_{k=1}^{n} \alpha_k A_k \tag{4.26}$$

Assume that the currency C and the various kinds of assets A_k, $k = 1, 2, \ldots, n$, all change with bank deposits at the same rate. That is, assume that r_c and $r_k = A_k/D$, $k = 1, 2, \ldots, n$, are all constant. Then the weighted monetary multiplier m_3' and the supply model of the weighted money are given as follows:

$$m_3' = \frac{M_3'}{MB} = \frac{1 + r_c + \sum_{k=1}^{n} \alpha_k r_k}{r_c + r_d + r_e} = m_1 + \frac{\sum_{k=1}^{n} \alpha_k r_k}{r_c + r_d + r_e} \tag{4.27}$$

$$M_3' = m_3' \times MB = \frac{1 + r_c + \sum_{k=1}^{n} \alpha_k r_k}{r_c + r_d + r_e}(NB + DL)$$

$$= M_1 + \frac{\sum_{k=1}^{n} \alpha_k r_k}{r_c + r_d + r_e}(NB + DL) \tag{4.28}$$

Therefore, the weighted monetary multiplier m_3' is negatively correlated to r_c, r_d, and r_e, and is positively correlated to each r_k, $k = 1, 2, \ldots, n$.

4.3 THE DEMAND OF MONEY

The quantity theory of money investigates the relationship between the prices of commercial goods and the quantity of money. Its basic assumptions include: the quantity of money determines the monetary value and the price level of goods; the monetary value is inversely proportional to the quantity of money; and the price level of goods is directly proportional to the quantity of money. The development of the quantity theory of money can be traced to the classical quantity theory of money. It is a theory about how to determine the nominal value of the total revenue. The classical quantity theory of money reveals the necessary quantity of money when the total revenue is fixed. Therefore, it is a theory on the demand of money.

This section consists of five subsections. The first subsection looks at the classical quantity theory of money; the second subsection introduces the Keynesian demand theory. Further developments of the Keynesian theory are studied in the third subsection; and the modern quantity theory is touched upon in the fourth subsection. In the fifth subsection, the importance of empirical tests of different assumptions and conclusions of the money demand theories is looked at.

4.3.1 Classical quantity theory of money

The classical quantity theory of money was initially introduced in the 18th century, rapidly developed in the 19th century, and perfected in the 1920s. This theory investigated the money demand within the system of gold standard, where paper notes were only an alternative currency. The main representatives of the theory include David Home, David Ricardo, Karl Marx, Irving Fisher, Alfred Marshall, Arthur Cecil Pigou, John Maynard Keynes, Joseph Alois Schumpeter, and others.

4.3.1.1 The theory of cash transactions

This theory is a quantity theory of money developed by classical economists during the end of the 19th century and the start of the 20th century. It leads to a classical theory of money demand. The most important characteristic of the theory is the belief that the demand for money has nothing to do with interest rate. That is, interest rate does not bear any impact on the demand for money. In his book, "Purchasing Power of Money," published initially in 1911, Irving Fisher (Fisher and Brown, 2010) presented this theory well. From the angles of the money supply M, which can be today's M_1 or M_2, and the speed V of circulation of money, which is defined to be the average times a unit of money is used to purchase goods within a year, this theory looks at the demand for money. It treats the monetary function of trading medium as the only function of money, believing that money is commonly accepted in the social and economic lives for the reason that it is a tool for making business transactions or trades. So, the total amount Y of trades of the entire society (that is the total amount of goods in circulation) and the total amount of payments should be the same as the total nominal income from selling all the goods. At the same time, money possesses a speed of circulation. That is, each unit of money is used multiple times in the trades of commercial goods. The total amount of trades and total amount of payments of the entire society should be equal to the multiplication of the average times a unit of money is used in business trades and the quantity of money in circulation. That leads to the following transaction equation, also known as the Fisher's equation:

$$M \times V = P \times Y \tag{4.29}$$

where P is price level (price index). So, $P \times Y$ stands for the total nominal income. This transaction equation relates the price level and the quantity of money, constituting the most basic proposition of the classical quantity theory of money.

As for the factors that affect the circulation speed of money, Irving Fisher believes that the speed is determined by the economic system and technological conditions that affect how individuals make their business transactions. For example, when credit cards are used in business transactions, less quantity of money will be needed than when no credit card is applied. The circulation speed V of money is surely faster in the former situation than in the latter scenario. However, in the ages of technological backwardness, the development of new technology is quite slow; each technological innovation takes many years to introduce. For example, credit cards initially appeared before World War II, and did not become a bank business until in the 1960s. Additionally, the evolution of economic systems also takes a gradual and slow process. Especially, when the society is relatively stable, its economic system generally stays invariant. Hence, the classical economists have all the reason to believe that within short periods of time, both systems and technologies do not change so that the circulation speed of money is constant. Therefore, equ. (4.29) implies that the nominal income is completely determined by the quantity of money; it is directly proportional to the quantity of money, and when the quantity of money doubles, so does the nominal income. In his book "Purchasing Power of Money", Irving Fisher points out that the transaction equation reveals the following three relationships:

1   When both V and Y stay constant, the price level P is directly proportional to the money demand M;
2   When M and Y are kept constant, the price level P is directly proportional to the circulation speed of money V; and
3   When M and V are invariant, the price level P is inversely proportional to the total output Y (the total amount of goods in circulation).

Among these relationships, Irving Fisher believes that the first one is most important; it constitutes the basic framework of the quantity theory of money. The classical economists believe that during the time of normal economic development, the quantity Y of goods in circulation is maintained at the level of sufficient employment so that within any short period of time, the variable Y in the transaction equation stays invariant. So, within a short period of time, both the circulation speed V of goods and the total output Y are constant. That is why the first kind of relationship $P = VM/Y$, as revealed by the transaction equation, is most important: The price level is completely determined by the quantity of money and is directly proportional to the quantity of money; when the quantity of money doubles, the price level also doubles. That is the theory of how price level is determined in the theory of cash transactions.

From the transaction equation, Irving Fisher obtains the following computational formula for the demand of money:

$$M = \frac{P \times Y}{V} = \frac{1}{V}PY \qquad (4.30)$$

based which Irving Fisher points out that because the circulation speed V of money is invariant, the demand M of money is completely determined by the level of nominal income PY ($= P \times Y$); that is, M is equal to the nominal income divided by the circulation speed of money. Therefore, the demand of money is only a function of income without anything to do with interest rate. That is, interest rate imposes on effect on the demand of money. This demand theory of money reveals the quantity of money needed for a fixed level of income.

Evidently, this kind of classical theory of money demand suffers from major weaknesses. First of all, it one-sidedly treats money only as the medium of trades and considers trades the only reason why money is demanded, while ignoring all other functions of money and the connection between money and assets. The fact is that these ignored aspects affect the demand of money enormously. Secondly, this theory boils down the factors that affect the circulation speed of money only to the trading system and technology without considering the effect of the monetary function of wealth storage on the circulation speed. It believes that without any exception money goes entirely into the circulation so that the conclusion of invariant circulation speed is derived. This end does not agree with the reality, where the circulation speed of money varies constantly due to the effects of wealth storage and production. Thirdly, in the relationship between the movement of goods and the movement of money, the theory of cash transactions places the movement of money ahead of that of goods by emphasizing on the dominating and promoting effects of the movement of money. It employs changes in the quantity of money to explain the cyclical evolutions of economies, leading to the conclusion that the demand for money is only determined by the nominal income. Evidently, that confuses what is primary and what is secondary. In an actual economic activity, what is primary is the movement of goods. Without the movement of goods, there would not be any movement of money. This relationship should not be reversed.

4.3.1.2 The cash balance theory

As represented by such scholars of Cambridge University, Great Britain as Alfred Marshall, Arthur Cecil Pigou, John Maynard Keynes, the quantity of money is analyzed directly from the angles of money demand and the amount of money on hands, as opposite to what Irving Fisher did, where the demand for money is seen from the angles of money supplies and circulations. These scholars propose the cash balance theory and provide a new explanation for the quantity of money. In particular, Pigou (1917), on the basis of the theory of his professor, Alfred Marshall, introduced the so-called Cambridge equation in 1917.

The Cambridge School believes that people apply their wealth in three ways: (1) Wealth is used in investment in order to earn interest; (2) wealth is employed for consumption for personal enjoyments; and (3) wealth is held in hands for convenience and safety. In the form of money, people hold their wealth in hands. That is, wealth is used in the third way as mentioned above, which constitutes cash balances. That is to say, when the functions of money as storage of wealth and delayed payments are utilized, people can temporarily store their wealth in the form of money so that when needed, it can be used at any time so that convenience and safety are provided. Because the portion of wealth that is used for investment and consumption is quickly exhausted, there is no need to keep that portion of wealth in the form of money. People need money only for the purpose of holding cash balance. Therefore, the demand for money should be equal to the cash balance.

Among the afore-mentioned three ways of applying wealth, by comparing the corresponding returns people decide what proportions they use to divide their wealth for each of these uses. Hence, the cash balance represents only a part of the entire wealth, the amount of wealth people retain in the form of money. Let us use Y to stand for the totality of wealth, P the price level, M the cash balance, which is the demand for money, k the proportion of the cash balance against the entire amount of wealth (that is, the proportion of the wealth used in the third way compared to the entire quantity of wealth). Then PY ($= P \times Y$) represents the total amount of wealth income written in the form of money. Based on this understanding, Pigou obtained the following equation

$$M = kPY \qquad\qquad (4.31)$$

This equation is known as the Cambridge equation or Pigou's equation. It expresses the quantity of money people hold in hands and are ready to use to purchase goods. That is the true demand for money. This equation also presents to us the determining factors of money demand: The demand for money is affected by the proportion k of how much wealth people hold in hands, the price level P, and the total level of wealth Y. This equation emphasizes on the relationship between money demand and price level.

When the total amount of wealth is seen as the totality of goods in circulation, the transaction equation and the Cambridge equation take the same form, where $k = V^{-1}$. Even so, the significances of these equations are different in the following aspects. Firstly, the theory of cash transactions believes that although the circulation speed V of money is determined by the system of trade and the available technology, it is invariant within short period of time. However, in reality, V is also affected by many other factors, and represents an extremely difficult quantity to be specified. On the other hand, the proportion k of the part of wealth kept in the form of moneu against the entire level of wealth is different. Although quantitatively k is equal to the reciprocal of V, k is determined by the attempt and desire people choose to hold their wealth in the form of money. It is much easier to be determined and controlled. Secondly, although the coefficient k in the Cambridge equation is equal to the reciprocal of the circulation speed V of money in the transaction equation, these quantities possess completely different meanings. The circulation speed of money only reflects the function of money as the medium of trades, while the coefficient k in the Cambridge equation embodies the motivation for people to hold money. So, the cash balance theory connects the demand for money with the motivation of economic entities. The reason for people to hold money is not merely for satisfying their current need of transactions, but also for preparing for the unexpected future and for emergent needs in order to have peace of mind. Thirdly, the circulation speed V of money does not reflect any aspect of the benefit for people to hold money, while the coefficient k in the Cambridge equation expresses the magnitude of benefit for people to hold wealth in the form of cash. That implies that the theory of cash transactions does not treat money as assets, while the cash balance theory considers money as assets, where the demand for money is a consequence of assets selections. That might be the biggest difference between the cash balance theory and the theory of cash transactions.

Treating money as assets is not really a new discovery of the cash balance theory. In fact, the mercantilism, the classical school, and Karl Marx have all recognized this fact in the 19th century (Marx, 2008), where Marx also analyzes the conversion of money into assets. Even so, Marx did propose that each individual person has the need to hold money as assets. Consequently, in analyzing the problem of monetary circulations, he does not treat money as assets.

Lastly, when seen from the angle of research methods, the theory of cash transactions analyzes the quantity of money macroscopically with emphasis placed on the system's structure; that is why it concludes that interest rate does not bear any influence on the demand for money. On the other hand, the cash balance theory possesses the characteristics and flavor of microeconomic analysis. It focuses on the convenience of money, while emphasizing the demand for money as a kind of asset, a need for storing value without simply limiting to the necessity of transactions. Therefore, this theory does not exclude the potential influence of interest rate on the demand of money. And, in their theories, both Marshall and Pigou implicitly hint the effect of interest rate on the demand for money. Keynes emphasizes the effect of interest rate, and further develops the theories of Marshall and Pigou.

4.3.1.3 The income quantity theory

This theory is an improvement of the theory of cash transactions. A group of scholars, as represented by Joseph Alois Schumpeter, believes that many goods included in the quantity Y of the transaction equation do not carry much significance. Symbol Y contains all goods in circulation. However, many goods exist only for the purpose of resale. That is, they are intermediate products. They do not create actual revenue. Additionally, intermediate stages of transaction are difficult to measure. So, Joseph Alois Schumpeter proposes to use the final product and labor of the economy as a substitute for the symbol T so that the following equation is obtained:

$$M \times V_y = P_y \times T_y \tag{4.32}$$

where M stands for the quantity of money, V_y the circulation speed of monetary income, which is defined as the average times a unit money is used to purchase final products and service within a year, P_y the price level of the final products and services within a year, and T_y the totality of final products and services within a year. This expression is known as income equation, the right hand side $P_y \times T_y$ of which is exactly the gross national products (GNP).

The demand for money, as obtained from the income equation, excludes the intermediate transactions that do not create value. That might make the outcome more realistic; and the values of P_y and T_y are more readily available. However, this formula still suffers from several weaknesses, especially V_y is more difficult to control than V. It is extremely difficult to estimate and determine on the average how many times one dollar is used within a year to purchase final products and labor. In practice, the estimate of V_y is often done by using the method of backward deduction: Based on known GNP (that is $P_y \times T_y$) and the money supply M, by dividing GNP by M produces the circulation speed of monetary income V_y:

$$V_y = \frac{P_y \times T_y}{M} = \frac{GNP}{M} \tag{4.33}$$

Additionally, between V_y and V there is not any fixed proportional relationship. However, empirical tests indicate that over long terms the directions of change of both V_y and V are the same. To this end, please consult (Mishkin, 1995) for more details.

4.3.2 Keynesian demand for money

Due to the Great Depression of the 1930s and the publication of the book, entitled "The General Theory of Employment, Interest, and Money", by John Maynard Keynes, in 1936, scholars replaced the classical quantity theory of money with the theory of income and expenditure, as initiated by Keynes in his theory of money. On the basis of the cash balance theory, Keynes modified the classical quantity theory of money by proposing the concept of liquidity preferences, known as Keynes' theory of the demand for money.

The conclusion that the nominal income is determined by the supply of money, as obtained by Irving Fisher, is established on the basis of invariant circulation speed of money $V = PY/V$, believing that within any short period of time, this speed is stable so that it can be seen as constant. Is this assumption reasonable? As a matter of fact, from the data collected from 1915 to 1993 (Mishkin, 1995) about how the circulation speed of money varies from one year to another, it follows that even within a short period of time, this speed changes quite drastically. So, it cannot be considered as a constant. In particular, due to the reason that after 1982 V_1 fluctuates drastically, while the circulation speed V_2 of money M_2 has been more stable when compared to V_1, the American Federal Reserve System decided in 1987 to give up M_1 and instead to utilize M_2 as the indicator of the total money supply. However, due to the instability in the circulation speed of M_2 in the early 1990s, Federal Reserve declared in July 1993 that it would no longer recognize the total of M_2 as a reliable indicator for monetary policies.

How didn't the classical economists discover the fact that the circulation speed of money could be variant? It is because at that time, which is before World War II, there were no accurate data on GDP (GNP) and money supply. These classical economists could not have recognized the mistake in their convention that the circulation speed of money is assumed to be constant. Therefore, after the Great Depression of 1930–1933 in the United States, a group of economists, led by Keynes, started to investigate other factors that influence the demand of money. That effort led to the Keynesian demand for money theory, which is the liquidity preference theory. In fact, Keynes is still a quantity theorist of money; his theory does not completely depart from and overthrow the classical quantity theory of money. Instead, his theory provides a more thorough modification of the classical theory than what the quantity theory of income does. The following represents a main difference between Keynes' theory of money and the classical quantity theory of money: The classical school of thought attempts to address the problem of whether or not changes in the quantity of money can affect the price level, while Keynes describes how alterations in the quantity of money affect the price level. The modifications Keynes introduces to the classical theory of money are mainly reflected in the following three aspects.

The traditional theory believes that supply can automatically create demand, and the supply has to be the same as the demand. The supply of money is no exception to this rule. Money supply can automatically create money demand; no matter how much money is supplied, there will be corresponding demand for money so that the economy is always situated in the state of full use of resources and sufficient employment, leading to fixed economic total output. Therefore, increasing money supply would surely cause equal-proportional rise in the price level. However, the Great Depression of the 1930s forces Keynes to reconsider this basic convention of the traditional economics. Keynes believes that the cause of the Great Depression is insufficient effective demand and insufficient consumption and investment, all of which together make resources idle and insufficient works, leading to increased unemployment. Keynes points out that when there is unemployment, the amount of employment varies in the same proportion with the amount of money so that the price level stays invariant; as long as sufficient employment is achieved, the price level starts to change in the same proportion with the quantity of money. Hence, he proposes the theory of "liquidity trap", believing that when the interest rate falls (the price of bonds rises) to a certain degree, the public would think it impossible for the current interest rate to fall further, and for the price of bonds to rise so that people dump bonds and prefer to hold cash. With a huge amount of cash in hands, the public utterly refuses to purchase any bonds in order to avoid potential losses from the expected falling prices of the bonds. In such scenarios, no matter how much cash the public holds in hands, people are still willing to maintain the cash. This phenomenon is referred to by Keynes as the "liquidity trap". When a liquidity trap appears, the demand for money becomes infinite. Even if the bank system increases the money supply, it cannot make the interest rate fall. That is, the demand for bonds will not increase so that the prices of bonds will not go up. So, the expansion of effective demand for money and any increase of price level are hindered. That is, the convention that supply automatically creates demand becomes a fairy tale.

The traditional theory divides prices into two kinds: individual prices and the general price. Each individual price is determined by such factors as production cost, the relationship between supply and demand, etc., while the general price is determined by the quantity of money and the circulation speed of money under the assumptions of sufficient employment and of invariant total output so that the aggregate demand is directly proportional to the quantity of money. To this end, Keynes takes an approach that is totally different from the traditional dichotomy. He does not start off from the quantity of money; he does not assume that the total output and the amount of employment are constant. On the contrary, he starts out on the total output and the amount of employment; he investigates the price level macroscopically, believing that both total output and amount of employment can change and the economy is not always in the state of full employment. Keynes believes that the price level is determined jointly by the prices of various production factors of the marginal cost and the scale of production (amount of employment). However, that is only seen from the angle of supply. One should also consider how changes in the aggregate demand affect the costs and output. So, in his investigation on how to determine the output and the amount of employment, Keynes introduces the money factor that leads to the development of his new theory of money.

Keynes recognizes that the main problem with the traditional quantity theory of money is it did not consider the function of value storage of money, while treating the quantity V in the transaction equation and k in the Cambridge equation as constant and assuming sufficient employment and invariant output so that the price level becomes directly proportional to the quantity of money. In fact, changes in the quantity of money do not directly affect the level of prices. Instead, they only indirectly act on the price level through interactions of such factors as costs, returns on production factors, production scales, interest rates, etc. Keynes believes that the general price level is determined by the wage level and the amount of employment; an increase in the quantity of money could only influence prices through increased wage spending; however, if there is unemployment, increases in the quantity of money would produce no obvious effect on the price level, because the amount of employment is also correspondingly increased. Additionally, Keynes believes that the effective demand does not change with the quantity of money in the same proportion. The quantity of money affects the interest rate, while interest rate in turn influences the speculator demand for money. When the demand for money increases, it also implies that more money is kept on hands, causing the circulation speed of money to decrease so that the storage of money and investment are negatively affected and the realization of sufficient employment is hindered. In order to reduce the liquidity preference, one has to analyze the determinant factors of the demand for money. To this end, Keynes believes that the main factors that influence the demand for money are income and interest rate. So, in order to reduce how much money the public maintains in hands, which is to reduce the demand for money, to promote the circulation of money, and to stimulate consumption and investment, one has to investigate the problem of how to adjust income and interest rate. That is the central problem Keynes is interested in. After analyzing the effect of interest rate, Keynes concludes that although reducing interest rate could increase the speculative demand for money causing reduction in storage and investment, it could also stimulate investment leading to increased income so that the demand for money is increased for the purposes of transactions and prevention for the unexpected. In short, the consequence of reducing interest rate is an increased demand for money while the aggregate demand decreases, because the public holds more cash on hands for the purpose of storing value instead of using the cash in purchases. Hence, it can be seen that reducing interest rate is a macroscopic policy that has the effect of controlling inflation.

In his book "The General Theory of Employment, Interest Rate, and Money", Keynes discards the convention of the classical school of thought that the circulation speed of money is invariant. Instead, he proposes a demand for money theory where interest rate is the key. At that time, because he was on the faculty of Cambridge University, Keynes naturally followed the research method established by his Cambridge predecessors. He further considered the three uses of wealth: the use of consumption, the use of investment, and the use for the prevention of unexpected. Consequently, he developed three motives for people to demand for money: the motive for transactions, the motive for caution, and the motive for investment. Then he more delicately analyzed the factors that affect individuals' decision making than his predecessors. Keynes named his demand for money theory as the theory of liquidity preference.

In terms of the motives of the demand for money, under different circumstances people would have different demands for money due to their varied considerations. All such individual demands constitute the public demand for money. When people demand for money, it is exactly because people need to retain their wealth in the form of money. The greater the proportion of wealth is retained in the form of money, the smaller proportion of the wealth is retained in other forms, such as securities, actual assets, etc. It is because at any chosen moment of time, the total amount of wealth of the society is fixed. When wealth is retained in the form of securities, it brings forward relatively higher returns; when wealth is retained in the form of actual, physical assets, it can directly bring forward the use value for people; when wealth is retained in the form of money, it not only causes losses in interest income but also creates not use value. Then, why do people still like to retain at least a portion of their wealth in the form of money? In other words, why do people still have desire to hold money in hands? According to Keynes, people have three motives for needing money: transactions, measure of caution, and investment.

By the motive of transactions, it means that people need money for the transactions of their ordinary living. Because receiving incomes and various spending occur at different moments of time, people have to maintain enough money to pay for daily needs. The demand for money that is due to this motive is the transaction demand of money. It is determined by levels of incomes, customs, and the system of transaction. Because customs and the system of transaction are stable within short periods of time, they can be seen as constant. So, Keynes believes that the transaction demand for money is mainly determined by income: The high the income, the greater the demand. In other words, the transaction demand for money is positively proportional to the income. In this regard, he agrees with the classical school of thought.

The motive of caution is also known as prevention motive. It stands for the motive that in order to prepare for unexpected payments people need to maintain an amount of money. The demand for money that is due to the motive of caution is then known as precautionary demand or prevention demand. If the synchronization between income and payments is destroyed by the transaction demand, then uncertainties in future income and payments are felt through prevention demand. For example, in order to deal with accidents, unemployment, illness, and other unexpected events, people have to hold a certain amount of money in hands. This amount of money is a part of the prevention demand. As another example, let us assume that a person has had the desire to purchase a dress of a certain fashion. When he/she accidently passes by a store, he discovers that the dress is on sale with 50% off the regular price. If he has the necessary amount of money for such event to occur, then he can make the purchase immediately. Otherwise, he would miss the great opportunity.

Keynes maintains that the prevention demand for money is mainly determined by the outlook people have on their futures and how they expect to make the future transactions. Even so, at the level of the whole society, these future transactions are mostly proportional to incomes so that the prevention demand for money is positively proportional to income. Because of this understanding of the prevention demand for money, Keynes is able to conduct his analysis beyond the framework of the classical theory. At the same time, Keynes recognizes that money possesses the ability to store wealth. He names the need for people to use money to store wealth as speculation motive. In particular, there is a reason behind why people retain their wealth in the form of money instead of those of securities and physical assets: capture the most profitable opportunities to purchase securities. People like to use the changes in interest rate and the prices of securities to speculate. So, for the speculation motive, people need to maintain cash in hands. That is the so-called speculation demand.

Keynes agrees with Marshall, Pigou, and other Cambridge economists, believing that the amount of money held for the motive of storing wealth is directly proportional to income. However, he places more emphasis on the non-ignorable effect of interest rate on this demand for money. Keynes divides wealth into two classes: money and bonds, where bonds create return, while the idling money does not. The price of bonds changes inversely with the interest rate of the bonds: When the interest rate increases, the price of the bonds decreases. When people speculate the interest rate to make profits, they purchase bonds at lower prices and sell the bonds at higher prices, making the profits from the price differences. When the interest rate is relatively low, the price of bonds will be relatively high. So a relatively large amount of bonds are sold. That is, a relatively large amount of bonds is converted into money. That explains when the speculation demand for money is relatively high. Conversely, when the interest rate is relatively high, people use their money to purchase bonds so that the amount of money in hands decreases. That is, the speculation demand for money reduces. In other words, if people expect the interest rate is going to rise, which is equivalent to the expectation of falling bond prices, it means that the current interest rate is low and the current price is high. So, people would sell bonds and convert them into money, leading to increased speculation demand for money. Conversely, if the interest rate is falling, that is, people expect the bonds price to rise, then it means that the current interest rate is high and the price is low so that people would apply their money to purchase bonds, causing the amount of money in hands to drop. That is, the speculation demand for money decreases. Hence, Keynes concludes that increasing interest rate causes the speculation demand for money to drop; the speculation demand for money is negatively correlated to interest rate.

By collectively considering these three motives for people to hold money, Keynes places these motives in his demand formula for money. In doing so, he also strictly separates the nominal and actual amounts of demand for money. If P stands for the price level, then $1/P$ represents the purchasing power, the value, of money. Let M be the demand for money. Then M stands for the nominal amount of demand for money. For example, if the price of a product is doubled, then the same amount of nominal money can only purchase half of the product. Because holding money in hands is for the purpose of retaining wealth in the form of money, Keynes reasons that to reflect the amount of demand for people to hold money for all the three motives, the actual amount of demand should be applied. That is the amount of money M/P, obtained by removing the effect of the purchasing power of money. It is the amount of demand for money that is most affected by wealth Y (the actual income) and interest rate i. That is, it is function $M/P = f(Y, i)$ of the actual income and interest rate. This function satisfies the following properties: $f(Y, i)$ increases with the actual income Y; at the same time, $f(Y, i)$ is a decreasing function of the interest rate i.

By making use of the demand function $f(Y, i)$ of money, the following formula for the circulation speed of money is immediately obtained:

$$V = \frac{PY}{M} = \frac{Y}{f(Y, i)}$$

That is, the circulation speed V is also affected by the actual income and the interest rate, where when the actual income Y changes, $f(Y, i)$ will be different and the change in V will be determined by the elasticity of the actual demand for money against the actual income.

1   If the actual demand for money is not elastic regarding the actual income, then when the actual income Y increases, the actual demand for money will not increase as much as Y does. So, the amount of increase of money held in hands is lower than that of increase in the nominal income, leaving more money in the circulation instead of being held in people's hands. That implies that an increase in the actual income causes the circulation speed of money to rise.

2   If the actual demand for money is elastic regarding the actual income, then it means that when the actual income Y increases, the actual demand for money will increase in a greater magnitude with the amount of hand-held money increasing faster than the actual income. That makes people maintain more money in their hands, causing the circulation speed of money to decrease.

3   If the actual demand for money were elastic regarding the actual income uniquely, then the amount of money held in people's hands would change with income in the same proportion. That means that an increase in Y would not affect the circulation speed of money, because V is equal to the ratio of the nominal income and the quantity of money. This conclusion in fact cannot be shown mathematically. Let E_{MY} represent the elasticity of the real demand for money in terms of the actual income. Then we have

$$E_{MY} = \frac{Y}{f(Y,i)} \cdot \frac{\partial f(Y,i)}{\partial Y} = \frac{Y}{f(Y,i)} \cdot f_Y' \tag{4.34}$$

By computing the partial derivative of the circulation speed V of money with respect to the actual income Y produces

$$\frac{\partial V}{\partial Y} = \frac{f(Y,i)}{[f(Y,i)]^2}\left(1 - \frac{Y}{f(Y,i)}f_Y'\right) = \frac{f(Y,i)}{[f(Y,i)]^2}(1 - E_{MY}) \tag{4.35}$$

So, when $E_{MY} < 1$ (lack of elasticity), V increases with Y; when $E_{MY} > 1$ (there is elasticity), V decreases with Y; and $E_{MY} = 1$ (the elasticity is determined unique by one factor), Y is not affected by V.

In terms of the effect of interest rate i on the circulation speed V of money, we have that when the interest rate i increases, $f(Y,i)$ becomes smaller; when i increases, the circulation speed increases. That implies that the circulation speed of money is positively correlated to interest rate. On the basis of this conclusion we can explain why the circulation speed of money slows down during a recession: The interest rate varies along with the economic cycle, where when the economy expands, the interest rate goes up, and when the economy declines, the interest rate goes down. Hence, the conclusion we just obtained earlier implies that the circulation speed of money decreases during each time period when the economy declines; and during the time of economic expansion, the circulation speed picks up.

From the analysis above it can be seen that Keynes' demand for money theory is mostly the same as the theory of Irving Fisher. The major difference of these two theories on the demand for money is whether or not the circulation speed of money changes. There are two reasons why Keynes objects treating this speed as constant: One is that the circulation speed of money is positively correlated to drastically fluctuating interest rate, and the other is that the expectation of the public about how interest rate would change also affect the demand for money so that the circulation speed of money is influenced.

4.3.3   Further developments of Keynesian theory

After World War II, a group of economists, as represented by William Baumol and James Tobin, further improved and modified Keynes' demand for money theory. They established a more delicate theory to explain the three motives as Keynes proposed for why people need money, where interest rate is considered as the most important factor affecting the demand for money. From the two aspects of income and interest rate, Baumol and Tobin investigate the transaction need of money, in particular, if the transaction need is produced by a time difference between income and payments. If no such difference exists, that is, as soon as income is received, the money is used for payments, then people would not need to hold any cash. For example, assume that a family's income is Y dollars per period and is used to make n even payments, $n > 1$, each of which is in the amount Y/n of dollars. Then the balance of the income is given by

$$Y - \frac{1}{n}Y, \quad Y - \frac{2}{n}Y, \dots, \quad Y - \frac{n-1}{n}Y \tag{4.36}$$

where $Y - (k/n)Y$ stands for the balance of money in hands at the end of the kth spending period. So, the average balance over the pay period is:

$$A = \frac{1}{n-1}\sum_{i=1}^{n-1}\left(Y - \frac{i}{n}Y\right) = \frac{Y}{2} \tag{4.37}$$

That is, the transaction demand is only one half of the periodic income.

If we now look at two specific cases: $Y = \$2000$/month and $Y = 500$/week, then the reasoning above indicates that for the first case, the family needs to maintain $\$1,000$ each month, while for the latter case, the family only needs to maintain $\$250$ each week. That is, if the income is received weekly in the amount of $\$500$ each, where the monthly income still totals to $\$2,000$, the transaction demand for money is merely $\$250$. That is one fourth of that when the income is received only once a month. Now, assume that the income is received not monthly but annually. Then the transaction demand for money of the public will be expanded 12 folds. In short, under the assumption of constant total income, the transaction demand for money decreases as the number of times the income is received in equal amount. In particular, the transaction demand for money is equal to half of the total income divided by the number of times the income is received.

The analysis above indicates that the money held by the public to meet the daily transaction needs constitutes the transaction demand for money, denoted by M_t. That is a function of the nominal income Y:

$$M_t = \frac{1}{2n}Y = kY \tag{4.38}$$

where n stands for the number of times an equal amount of periodic income is received, and Y/n the amount of income received each time, $k = 1/2n$.

In this transaction demand formula, there involves a quantity n. Is it a constant? If it is not a constant, then what factors could affect it? In order to answer this question, let us look at the scenario from a different angle. Denote $x = Y/n$. Then we have $n = Y/x$. Assume that the public consumes the income Y in n periods, spending x dollars per period evenly and gradually. So, the balance of the income is

$$Y - x, \quad Y - 2x, \dots, \quad Y - (n-1)x \tag{4.39}$$

and the average balance is $A = Y/2 = nx/2$. However, the public does not have to keep all of its income in hands. Instead, it only needs to hold some of the income in cash with the rest entirely invested in short-term bond in order to earn interest. Assume that at the start of period 1 the public holds x dollars and uses the rest $Y - x$ dollars to purchase bond. At the start of period 2, the public sells the bond of the value x dollars to cover the daily needs of this period. When period 3 starts, the public once again sells the bond of the value x dollars to cover the daily needs of period 3. Assume this process continues until period n starts, when the public sells its bond for as much as x dollars cash to cover its daily expenses. That is equivalent to the scenario when the total income Y dollars is dispersed n times with x dollars received each period and is evenly and gradually consumed. According to the analysis above, it follows that on the average the amount of money held by the public is $x/2$ dollars. That is to say, there are always $x/2$ dollars that are idling and are not producing any interest income.

Assume that the interest rate of the bond is i. Then the opportunity cost of these $x/2$ dollars to the public is $xi/2$. Additionally, when some of the bond is converted to C dollars cash, there is a transaction fee. Assume that such a fee is b dollars each time. Then the total cost on the transaction fees is nb. So, the total cost to the public when it converts some of its bond into x dollars each period is

$$C = C(x) = \frac{xi}{2} + nb = \frac{xi}{2} + \frac{bY}{x} \tag{4.40}$$

Of course, the public wants the C-value as small as possible. That is, we need to choose an appropriate x so that the total cost C is minimal. As long as x is determined, the number $n = Y/x$ of periods is then also fixed. Evidently, this is a problem of minimization and can be solved by solving the first order condition. That is, let

$$\frac{dC}{dx} = \frac{i}{2} - \frac{bY}{x^2} = 0 \tag{4.41}$$

from which the optimal exchange amount $X = \sqrt{2bY/i}$ is obtained. The corresponding optimal number of periods is $N = Y/X$. So, the transaction demand M_t for money, that is the optimal amount of money the public holds, is

$$M_t = \frac{Y}{2N} = \frac{X}{2} = \frac{1}{2}\sqrt{\frac{2bY}{i}} = \frac{\sqrt{2b}}{2}\sqrt{\frac{Y}{i}} \tag{4.42}$$

This equation indicates that the transaction demand for money is related to not only the nominal income but also the interest rate. Their relationship is: The transaction demand for money increases with income and decreases with interest rate. Symbolically, we have

$$M_t = M_t(Y, i), \quad \text{satisfying } \frac{\partial M_t}{\partial Y} > 0 \quad \text{and} \quad \frac{\partial M_t}{\partial i} < 0 \tag{4.43}$$

The prevention demand for money is produced out of the uncertainty people feel about the future incomes and payments. In order to deal with the potential rise in payments and delay in incomes due to unexpected events, people need to maintain in hands a portion of their assets in the form of cash. It can be reasonably claimed that this portion of cash as prompted by the prevention motive is to a certain degree positively correlated to people's level of income. When the income is relatively high, the prevention demand would be greater and people would also have the ability to hold more cash. It could also be claimed that the prevention demand for money is negatively correlated to the interest rate of the market. When the interest rate increases, people might reduce how much cash they hold to prevent unexpected events while increasing the amount of bond purchase in order to earn higher interest. So, it could be predicted that the prevention demand for money is also a function of income and interest rate. In the following we will specifically analyze how income and interest rate affect the prevention demand for money.

Assume that due to the cautionary motive the amount of money people hold is M. However, because the future is uncertain, whether or not this amount is enough to handle the unexpected events of the future stands for an uncertain event. Even so, we can assume that people could judge the sufficiency of the amount M according to many known events of the past and the present. That is, people could determine a probability $P(M)$ for the amount M to be insufficient. Of course, the greater the amount M of money is held in hands, the smaller the probability $P(M)$ for M to be insufficient. Now the problem becomes: How much cash M has to be held in hand to be appropriate? To this end, let us analyze the situation from the angle of cost and benefit.

For simplicity, assume that there are only two classes of assets: money and bond. Assume the interest rate of the bond is i and people's income is Y dollars. When people hold in hands M dollars for prevention purposes, they give up the opportunity cost of iM dollars. If the amount of money held for the purpose of prevention is insufficient, then a certain amount of loss will be caused. Assume that the proportion of this loss over the income Y is $q > 0$. Denote $y = qY$. That is, y is the amount of loss caused by the insufficient monetary preparation. So, when the amount of cash held is M, the expected loss due to insufficient monetary preparation is $P(M)y = P(M)qY$ dollars. Therefore, the total expected loss L from holding M dollars is

$$L = L(M) = P(M)y + iM \tag{4.44}$$

Note that $P(M)$ is a decreasing function of M. That is, $P'(M) = \frac{dP}{dM} < 0$. Because $L(M)$ stands for the cost of holding M dollars, it should satisfy the law of increasing marginal cost. That is, it should satisfy

$$L''(M) = \frac{d^2L}{dM^2} = P''(M)y > 0 \tag{4.45}$$

That is, we should have $P''(M) > 0$. Because increasing marginal cost is a general law, we can assume $P''(M) > 0$ holds generally.

After knowing the expected total loss, people naturally want the loss to be as small as possible. So, the optimal amount M of cash held in hand for the prevention motive should satisfy the following first order condition:

$$\frac{dL}{dM} = P'(M)y + i = 0 \tag{4.46}$$

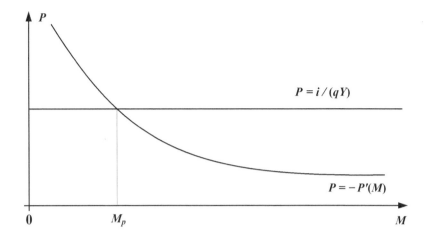

Figure 4.3 The determination of prevention demand for money.

That is,

$$-P'(M) = \frac{i}{Y} = \frac{i}{qY} \tag{4.47}$$

from which the prevention demand for money can be determined. Note that the term on the left hand side is greater than zero. So from $P''(M) > 0$ it follows that $P'(M)$ is an increasing function of M so that $-P'(M)$ is a decreasing function of M. Let the prevention demand M for money, as determined by the previous equation, be M_p, which changes with income Y and interest rate i. When Y increases, M_p become greater; when i goes up, becomes smaller M_p. That is the mechanism that determines the prevention demand, Figure 4.3. So, we have

$$M_p = M_p(Y, i), \quad \text{satisfying} \quad \frac{\partial M_p}{\partial Y} > 0 \quad \text{and} \quad \frac{\partial M_p}{\partial i} < 0 \tag{4.48}$$

Hence, the prevention demand for money satisfies a similar law as the transaction demand. It is because of this reason that people often combine the transaction demand and prevention demand for money into one using the following unified formula:

$$L_T = L_T(Y, i) = M_t(Y, i) + M_p(Y, i) \tag{4.49}$$

where Y is the nominal income, i the interest rate, M_t the transaction demand for money, and M_p the prevention demand for money. And the sum L_T of the transaction demand and the prevention demand is identified as the transaction demand.

Of course, there is also an objective reason for why the transaction demand and prevention demand are combined into one, known as the transaction demand. Along with the rise of income level and constant fluctuation of interest rate, the personal demand that used to be satisfied by the prevention demand can now be met by other ways with little or no cost, such as participation in pension plans, purchase of life insurance, health insurance, property insurance, welfare and security, etc. Therefore, there is no longer much need to hold cash for prevention purposes. In terms of an individual person, because of the introduction of credit cards in the financial market, the amount of cash held for the purpose of prevention has been greatly reduced, because many emergent payments can be dealt with by using credit cards. And because by using credit cards, one can obtain cash, which can, in turn, pay off any kind of expense. Similar to business firms, individual persons also establish their private credit lines to handle emergent payments of large magnitude. A credit line of $1,000 means that this amount of cash can be mobilized according to individual needs. It is as if the person has that amount saved up as his prevention measure in his checking account in his bank. Because of this reason the meaning of the transaction demand for money has been generalized so that it means the demand for money for both the purpose of transactions and the purpose of preventions.

When Keynes analyzes the speculation demand for money, he maintains that when the expected return from bonds is higher than the expected return of money, people would only hold bonds; conversely, when the expected return from bonds is lower than that from money, people would only hold cash; and when the rare circumstance, where the expected returns from bonds and money are equal, occurs, people would hold both bonds and money. However, diversification is a meaningful strategy for selecting assets. So, Tobin develops a model for the speculation demand for money in order to overcome the weakness of Keynes's analysis. Tobin believes that when people decide which asset to hold, they consider not only relative rates of return but also relevant risks of various assets. Tobin assumes that most people are risk averse; they rather hold less risky assets even when their returns are relatively small than much riskier assets even though their returns are much higher. Also, he assumes that there are two classes of assets: money and bonds. The characteristic of money is that the rate of return is fixed, which is assumed to be zero, while the price of bonds fluctuates wildly and sometimes it could even be negative, representing a major risk. So, risk-averse people are willing to use money as their means to store wealth instead of bonds. However, in reality, when people use money to store their wealth, they also wait for opportunities to invest the money to generate speculative profits. So, they purchase bonds, when the price of bonds is low; and they sell bonds when the price is high in order to make profits from the price difference. Now, the problem is: How much money do people need to keep for their speculation purposes? And how can this amount of money be determined?

When money is held in hand, it does not produce interest. So, the return of cash is zero. Assume that the annual rate of return on bonds is i, which represents the return the current bonds holders can definite earn. However, the interest rate of the bond market changes constantly. If R_m stands for the rate of return on bonds, then R_m is uncertain, and is influenced by many random factors. That is, R_m is a random variable. Also, it can be assumed that the expected value of R_m is the annual interest rate i. That is to say, R_m can be written as $R_m = i + g$, where g is an random variable with mathematical expectation zero: $E[g] = 0$. Here, g can be seen as a disturbance variable that affects how the annual interest rate i of the bonds is obtained.

Let α represent the proportion of the total wealth people retain in the form of money, β the proportion of the total wealth people maintain in the form of bonds. Because it is assumed that wealth consists of only two forms: money or bonds, we have $\alpha + \beta = 1$. If we can determine the value of β, then the value $\alpha = 1 - \beta$ follows. That will help to determine how much money people would hold in their hands. So, let us next analyze how to determine β.

Assume that R_β stands for the rate of return from the bonds held that amounts to the proportion β of the total wealth, r_β the expected rate of return of this overall arrangement. That is, $r_\beta = E[R_\beta]$. Let $\sigma_\beta = \sqrt{E[(R_\beta - E[R_\beta])^2]}$ be the standard deviation of R_β. Similarly, let r_m represent the expected interest rate of the bonds; that is, $r_m = E[R_m] = i + E[g] = i$, and $\sigma_m = \sqrt{E[(R_m - i)^2]} = \sqrt{E[g^2]}$ the standard deviation of the interest rate of the bonds. Then we have

$$R_\beta = \beta R_m = \beta i + \beta g \qquad (4.50)$$

$$r_\beta = \beta i + \beta E[g] = \beta i \qquad (4.51)$$

$$\sigma_\beta = \sqrt{E[(R_\beta - r_\beta)^2]} = \beta \sigma_m \qquad (4.52)$$

Therefore, we obtain $\beta = \sigma_\beta / \sigma_m$. Substituting this equation into $r_\beta = \beta i$ produces

$$r_\beta = \frac{i}{\sigma_m} \sigma_\beta \qquad (4.53)$$

This is the relationship between the expected rate of return and the standard deviation of the rate of return when a proportion β of the total wealth is held in the form of bonds. This formula is known as the mean-value and variance constraint of holding bonds.

The standard deviation σ of a random variable expresses the degree the value of the random variable deviates from the mean value. In terms of the random variable of the rate of return, the standard deviation represents the degree of uncertainty of the rate of return, providing an indicator for the magnitude of risk: The greater the standard deviation is, the more uncertain the rate of return becomes and the greater the risk is for actually obtaining the expected rate of return.

Holding cash does not produce interest, while although holding bonds bears return it does involve risk. So, people have to balance between the return and risk. Let r stand for the rate of return and σ the level of risk of obtaining this rate of return. Then the balance between return and risk can be expressed by using a utility function $U(\sigma, r)$ with the following properties: (1) When the risk σ is fixed, the higher the rate r of return, the greater the utility value $U(\sigma, r)$ is; (2) when the rate r of return is fixed, the smaller the risk σ is, the greater value the utility $U(\sigma, r)$ takes. That is, $U'_r > 0$ and $U'_\sigma < 0$.

Let us construct a plane coordinate system $O_{\sigma r}$, where the horizontal axis represents the standard deviation σ and the vertical axis the rate r of return (that is, the mean value). Then, the mean-value and variance constraint for holding bonds is a straight line on this plane with equation given as follows:

$$r = \frac{i}{\sigma_m} \sigma \qquad (4.54)$$

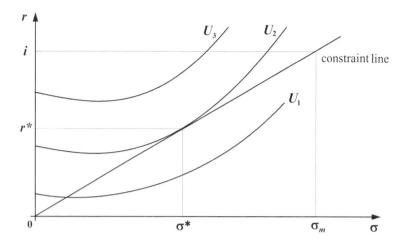

Figure 4.4 Balance between return and risk.

which is referred to as the mean-value and variance line of the bonds held, Figure 4.4. In fact, each point on this line represents a proportional distribution $\beta = \sigma/\sigma_m$ of holding money and bonds. Hence, when selecting how much money and bonds to hold, one would choose such a combination (σ, r) of money and bonds on the mean-value and variance line that makes the utility $U(\sigma, r)$ the maximum. So, the problem of finding the speculation demand for money is transformed into solving the following maximization problem with constraint:

$$
\begin{cases}
\max U(\sigma, r) \\
r = \dfrac{i}{\sigma_m}\sigma
\end{cases}
\tag{4.55}
$$

From the method of Lagrange multipliers, it follows that the solution (σ^*, r^*) to the previous extreme value problem must satisfy the following equation:

$$
\frac{i}{\sigma_m} = -\frac{U'_\sigma}{U'_r}
\tag{4.56}
$$

which is known as the speculation demand for money equation. Its geometric meaning is illustrated in Figure 4.4. The left-hand side of this equation is exactly the slope of the tangent line at point (σ^*, r^*). The tangent point (σ^*, r^*) between the mean-value and variance line and the indifference curve is exactly the point of optimal combination of the mean value and variance. From this point, one can obtain the proportion β^* of bonds within the total wealth. So, the proportion α^* of the speculation demand for money among the total wealth can be obtained as follows:

$$
\beta^* = \frac{\sigma^*}{\sigma_m} \quad \text{and} \quad \alpha^* = 1 - \beta^*
\tag{4.57}
$$

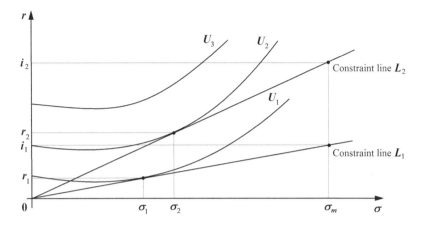

Figure 4.5 How increasing interest rate affects the speculation demand for money.

The β^* value, as determined by the speculation demand for money equation, is a function of the interest rate i and return risk σ_m of the bonds: $\beta^* = \beta^*(i, \sigma_m)$. So, α^* is also a function of i and σ_m:

$$\alpha^* = 1 - \beta^* = 1 - \beta^*(i, \sigma_m) = \alpha^*(i, \sigma_m) \tag{4.58}$$

If we use Y to represent the total income (total wealth), then the speculation demand function M_s for money is given as follows:

$$M_s = M_s(Y, i, \sigma_m) = \alpha^*(i, \sigma_m)Y \tag{4.59}$$

We now consider how changes in income Y, interest rate i, and risk σ_m affect the speculation demand for money. Firstly, let us look at the case when both interest rate and risk stay invariant, while the income Y changes. Evidently, increasing income would lead to increasing speculation demand for money, because in this case the speculation demand is directly proportional to income. Secondly, assume that both income and risk stay constant while interest rate i increases from i_1 to i_2. In this case, the mean-value and variance line moves from L_1 position to L_2 position with steeper slope. So, the tangent point of the mean-value and variance line and the indifference curve moves to the right, as shown in Figure 4.5, from (σ_1, r_1) to (σ_2, r_2) with $\sigma_2 > \sigma_1$. This conclusion implies

$$\beta_2 = \sigma_2/\sigma_m > \sigma_1/\sigma_m = \beta_1 \quad \text{and} \quad \alpha_2 = 1 - \beta_2 < 1 - \beta_1 = \alpha_1$$

that is, the speculation demand for money decreases. Conversely, when the interest rate drops, the speculation demand for money increases. So, the speculation demand for money and interest rate changes in opposite directions (when interest rate increases, people holds more bonds so that less money is held in hands). Thirdly, assume that both income and interest rate stay invariant, while the risk of bonds increases. In this case, the mean-value and variance line moves to positions of smaller slopes. That implies that the tangent point of this line and the indifference curve moves to the lower left direction, causing β^* to decrease and α^* to increase. That makes the speculation demand for money greater. This end indicates that the speculation demand for money is negatively correlated to the risk of bonds (as the risk of bonds increases, people reduce their holding of bonds so that the amount of money on hands increases).

For the general situation, the risk of the bond of a chosen kind is a relatively fixed number. So, σ_m can be seen as a constant and the speculation demand for money becomes a function of only income Y and interest rate i: $M_s = M_s(Y, i)$, satisfying the following properties:

$$\frac{\partial M_s}{\partial Y} > 0 \quad \text{and} \quad \frac{\partial M_s}{\partial i} < 0 \tag{4.60}$$

Let M_d represent the total demand for money, consisting of the transaction demand $M_t(Y, i)$, the prevention demand $M_p(Y, i)$, and the speculation demand $M_s(Y, i)$. Then we have

$$M_d = M_d(Y, i) = M_t(Y, i) + M_p(Y, i) + M_s(Y, i) \tag{4.61}$$

This total demand for money satisfies the following general laws:

$$\frac{\partial M_d}{\partial Y} > 0 \quad \text{and} \quad \frac{\partial M_d}{\partial i} < 0 \tag{4.62}$$

Because the national income Y is budgeted and settled annually, it stays unchanged within short period of time (one year) but varies over long term. So, the total demand for money within a short period of time is only affected by interest rate, while for the long term both national income and interest rate influence the total demand.

4.3.4 Modern quantity theory of money

In 1956, Milton Friedman published an article in which he provided a new explanation for the quantity of money. In this publication, although Friedman often cited results from Irving Fisher's quantity theory of money, his analysis on the demand for money is in fact much closer to the opinions of Cambridge School of thoughts and those of Keynes. Similar to earlier economists, Friedman continued to explore the reason why people hold money. However, what is different from Keynes is that Friedman no longer considers the motives. Instead he broadly believes that factors that affect the demand of other kinds of assets must also influence the demand for money. Hence, Friedman applies the theory of asset demand into the investigation of the demand for money.

Friedman believes that the proportionality k in the theory of cash balances, whose reciprocal $1/k$ is the circulation speed of money, is not a constant; instead it is a variable that is determined by many other economic variables and is a stable function. The demand for money is affected by consumers and producers, where money is wealth in the eyes of consumers, and is capital in the eyes of producers. For consumers, the demand for money is determined by their actual incomes, while incomes are determined by the prices and returns of assets the consumers own. For producers, the demand for money is determined by their returns of production and the cost of holding money. Therefore, at the root level the factors that influence the demand for money are those rates of return or those expected rates of return of those assets that can be used to substitute for money. So, without considering the motives of demand for money, as what Keynes and his followers do, Friedman writes the demand for money function directly as follows:

$$\frac{M_d}{P} = f(Y_p, r_b - r_m, r_e - r_m, \pi_e - r_m) \tag{4.63}$$

where M_d/P stands for the demand for the real balances of money, (which is the quantity of money expressed using physical objects; that is, similar to Keynes, Friedman also admits that people are willing to hold a certain amount of real balances of money), Y_p the permanent income, (which is an indicator Friedman uses to calculate wealth and is the present value of all expected future incomes; in short, it is the average expected value of long term incomes), r_m the expected rate of return of money, r_b the expected rate of return of bonds, r_e the expected rate of return of (common) stocks, and π_e the expected rate of inflation.

Because the demand for an asset is positively correlated to how much wealth people own, and because money is also an asset, the demand for money is also positively correlated to the level of wealth. Friedman uses the concept of permanent income to represent the wealth underneath the demand for money. Its significance is that to a large degree the demand for money does not fluctuate with the economic cycle, because the permanent income varies only slightly within short-period of time. For example, in an economic prosperity, the economic cycle is in the expansionary phase with incomes increasing rapidly. However, some parts of the growth are temporary and transitional, creating short-term changes in the income without major impact on the expectation of long-term income. So, the permanent income does not change much and any change in the permanent income will be smaller than the temporary fluctuations in the income. Similarly, during an economic decline, many parts of the income reduction are temporary and do not affect the average long-term income. So, the permanent income is not affected much. If some, the change in the permanent income is much smaller than the fluctuations in the temporary decreases in the income. So, it follows that permanent income does not fluctuate along with the economic cycle; changes within short periods of time are very minor so that to a large degree the demand for money is not affected by economic cycles. Because the demand for money is directly correlated to the level of wealth, Friedman maintains that the demand for money is directly correlated to permanent income.

Other than using money to store wealth, people can also use other ways to hold wealth. These other forms of wealth Friedman divides into three classes: bonds, (common) stocks, and commodities. The reason for people to hold these three classes of non-monetary assets is their expected rates of return. That is the content of the last three terms in Friedman's demand for money equation. The higher the expected rates of return on these assets, the more wealth people would hold in these assets' forms, and so the less demand for money would be.

The expected rate r_m of return of money appears in all of the last three variables of Friedman's demand for money equation. This rate is affected by two factors:

1   The services banks provide their owners of deposits that are part of the demand for money. These services increase the expected rates of return on the money.
2   The interest income of money balances. The higher the interest is, the higher the expected rate of return for holding money.

The first two variables $r_b - r_m$ and $r_e - r_m$ represent the expected rates of return of bonds and stocks with respect to money. The greater these two values, the smaller the relative expected rate of return of money, and the lower the demand for money will be. The last term $\pi_e - r_m$ stands for the expected rate of return of commodities with respect to money, where the expected rate of return of holding commodities is exactly that obtained when the prices of the commodities go up. So, the expected rate of return of holding commodities is equal to the expected rate π_e of inflation. For example, when the expected rate of inflation is 10%, the prices of commodities are expected to increases 10%. So, by holding commodities, one expected to earn a return of 10%. When $\pi_e - r_m$ increases, the expected rate of return from holding commodities increases with respect to money so that people are more willing to use commodities to store their wealth; consequently the demand for money decreases.

In his investigation, Friedman adopts the same methods as used by Keynes and other economists of Cambridge School. However, he does not closely look at the motives for why people demand for money. On the contrary, Friedman applies the theory of asset demand to illustrate the demand for money, while treating money as an asset. Let us now look at the difference between Friedman's theory and Keynes' theory.

Firstly, Friedman considers that many assets can be used as substitutes for money; so he maintains that in terms of the operation of the whole economy more than one important rate of return exists. In comparison, Keynes classifies all assets, except money, into bonds. He believes that the returns of these other assets often fluctuate synchronously, and because the expected rate of return of bonds is a good indicator of the expected rates of returns of all these other assets, there is no need to list the expected rates of returns of these other assets in the demand for money

Secondly, Friedman treats money and commodities as substitutes of each other. He believes that when people decide how much cash they want to hold, they compare money and commodities. So, the expected rate of return of money is an important factor affecting the demand for money and has to be included in the demand for money function. That is an important difference between Friedman's demand for money function and Keynes' demand for money function. The convention that money and commodities are replaceable of each other indicates that changes in the quantity of money might directly affect the total output.

Thirdly, Friedman believes that interest rate only affects the demand for money slightly, while Keynes maintains that interest rate is an important factor that influences the demand for money. The reason for Friedman to take this point of view is that he believes the expected rate of return of money is not a constant. When the interest rate of the market rises, banks can earn additional profits from loans so that banks will put in more effort to attract deposits.

(1) If there is no interest rate control, then banks would attract deposits by raising interest rate. Because of the intense competition of the banking sector, the expected rate of return of the money held in the form of bank deposits will rise on the back of the increasing rates of return of bonds and loans. The competition in the deposit interest rate among banks will continue until the super profit disappears (the interest rate of deposits rises continuously until it is equal to the rates of loans and bonds). This process of competition diminishes the profit difference between placing deposits and making loans. The ultimate result is that $r_b - r_m$ becomes quite stable, where at the same time when r_b increases, r_m is also rising accordingly.

(2) If there is an interest rate control, then banks cannot pay additional interest on deposits. So, the competition in deposit interest rate will be curbed. However, there will still be a competition among banks in the quantity of services they provide. For example, banks can provide additional services to depositors, including arranging additional tellers, providing automatic payment accounts, installing ATMs in greater territories, etc. Improvements in any of these service areas can help the banks to improve their expected rates of return on the deposits. So, although banks are limited on how much interest they can pay to their depositors, the rising market interest rate still eventually causes the expected rate of return of money to increase majorly, making the difference $r_b - r_m$ relatively stable. In short, no matter whether there is an interest rate control or not, increasing interest rate does not affect $r_b - r_m$ much; so increasing interest rate does not affect the demand for money much.

Friedman also provides a further explanation for why the demand for money is insensitive to interest rate. To him, the reason why the demand for money is insensitive to interest rate is not because the demand for money is insensitive to changes in the monetary opportunity cost of other assets; instead it is because changes in interest rate only bring forward very minor effect on various opportunity costs of the demand for money function. When an increasing interest rate causes the expected rates of return of other assets to increase, it also pushes up the expected rate of return of money accordingly. When these increases cancel each other, the opportunity costs, both $r_e - r_m$ and $\pi_e - r_m$, in the demand for money function stay invariant.

Fourthly, Friedman believes that the demand for money function is of stability. Because there is very little random fluctuation in the demand for money, this demand can be accurately predicted by using the demand for money function. Friedman maintains that in essence permanent income is the main factor that affects the demand for money; so his demand for money function can be roughly written as $M_d/P = f(Y_p)$, where the permanent income Y_p is the average expected value of all long-term incomes. Because many changes in short-term income are temporary, they do not affect the permanent income. So, permanent income fluctuates only slightly within short-term periods and does not change along economic cycles. That is, to a great degree short-term demand for money does not fluctuate with economic cycles and possesses a good amount of stability. However, Keynes does not agree with this convention.

Fifthly, Friedman points out that the circulation speed of money can be predicted accurately. The main basis for him to derive this conclusion consists of two points of view: The demand for money function is quite stable, and the demand for money is insensitive to changes in interest rate. According to Friedman, the demand for money is exactly the cash balance; it is the amount of wealth people hold in the form of money. If Y stands for the total amount of wealth and P the price level, then PY is the nominal total of wealth (the nominal income). So, $M_d/(PY)$ is the proportionality k of the portion of the wealth held in the form of money over the total of wealth and is the reciprocal of the circulation speed V of money. So, we obtain the following formula for the circulation speed of money:

$$V = \frac{1}{k} = \frac{PY}{M_d} = \frac{Y}{M_d/Y} = \frac{Y}{f(Y_p)} \qquad (4.64)$$

Generally, the relationship between Y and Y_p can be predicted. The stability of the demand for money guarantees that the demand for money function $f(Y_p)$ will not experience any obvious displacement. So, the demand for money can be accurately predicted and in turn the circulation speed $V = Y/f(Y_p)$ of money can be accurately determined.

Because of this conclusion about the circulation speed of money, Friedman produces the same result as the traditional quantity theory of money: Money supply is the main factor that determines the nominal income. That is because although the circulation speed of money is not a constant, one can predict the circulation speed for the next time period. So, he can predict the effect of changes in the quantity of money on the total output. That explains how nominal income is still determined by the quantity of money. So, it can be seen that Friedman's demand for money theory is a new version of the traditional quantity theory of money.

We know that Keynes' liquidity preference function (that is Keynes' demand for money function $M_d/P = f(Y, i)$, where the interest rate i is an important independent variable) can explain the pro-cyclical phenomenon of the circulation speed of money as discovered in empirical data. That is, during times of economic prosperity, the circulation speed of money increases, while during times of economic decline, the circulation speed of money decreases. To this end, Keynes's explanation consists of three key components:

1 Interest rate is an important factor that affects the demand for money. When interest rate goes up, the demand for money goes down so that the circulation speed of money increases; conversely, decreasing interest rate causes the demand for money to rise so that the circulation speed of money goes down.
2 Interest rate changes frequently, and fluctuates along with the economic cycle. When the economy is prospering, interest rate goes up; when the economy declines, interest rate goes down.
3 The effect of interest rate on the demand for money and the pro-cyclical change of interest rate explain why the circulation speed $V = Y/f(Y, i)$ of money also changes pro-cyclically.

In comparison, can Friedman's demand for money function explain this phenomenon? To answer this question, the key is to explain how permanent income evolves with the economic cycle, because instead of statistical income what is listed in the demand for money function is the permanent income Y_p. During a period of economic expansion, because most of the growth in income is temporary, the proportion of increase in the permanent income is much smaller than that in income. So, in comparison to the increase in the statistical income Y, the increase in the demand for money, as caused by the growth in the permanent income, is relatively small so that the circulation speed $V = Y/(Y_p)$ of money increases during the economic expansion. Similarly, during a period of economic decline, the magnitude of decrease in the permanent income is much smaller than that in income. So, the amount of decrease in the demand for money, as caused by the decreasing permanent income, is relatively small compared to the decrease in income. Hence, the circulation speed of money decreases. That is to say, from Friedman's demand for money function, one can also explain the pro-cyclical phenomenon of the circulation speed of money.

From the discussions above, it follows that both Friedman's and Keynes' demand for money theory are originated from the classical quantity theory of money. Friedman believes that the modern quantity theory of money is a different presentation of the classical theory, while other economists recognize that it is an in-depth statement of Keynesian demand for money theory. So, the close connection of these three theories can be clearly seen. Most recently, there appears another school of thought, known as the school of rational expectations, as represented by (Robert Lucas, 1972) and (Thomas Sargent, 1971). The main convention of this school consists of the following: Unbiased people can produce rational expectations based on the totality of available information and knowledge. Such rational expectations will make government policies useless. Elastic prices and wages would guarantee equal market demand and supply, where prices would adjust quickly to always make the market in equilibrium state. The theory of rational expectations is developed as a necessity when neither Keynesian nor monetarist could provide an effective strategy to deal with stagflations. Because of its contribution to economics, Lucas won the Nobel economics price in 1995 and Sargent in 2011.

The theory of rational expectations has been crucial in providing explanations for the mechanism underlying financial markets. The earlier expectation theories of the mechanical kind become inappropriate when coming to explaining such speculative markets as stock and bond markets. The theory of rational expectations provides a useful simulation method for the mechanism underlying the determination of interest rate, stock prices, and exchange rates. Such theoretical conventions as that people could rationally expect and that prices are elastic agree with the reality of financial activities. They approximately reflect the phenomenon of speculation widely existing in financial markets.

4.3.5 Empirical tests of money demand

We have seen various theories on the demand for money. They explain differently how money provides various economic functions. Then a natural question is: Which of these theories describes the real world more accurately? This is an important question, and has been the focus of debate on how monetary policies affect the overall economy. To this end, let us consider two basic problems: Is the demand for money sensitive to changes in interest rate? In the long run is the demand for money stable? Investigations of these two problems help to separate the theories of the demand for money and influence the outcome of the inquiry on whether or not the quantity of money is a main determining factor of the total output.

4.3.5.1   Interest rate and demand for money

From the discussions in the previous sections, it follows that if interest rate does not affect the demand for money, then the circulation speed of money could be constant, or at least it could be predicted. So the conclusion of the classical quantity theory that the total output is determined by money supply might be correct. However, if the demand for money is sensitive to changes in interest rate, then the circulation speed of money is not predictable, and the relationship between money supply and total output is no longer clear. And, the more sensitive the demand for money is to interest rate, the more unpredictable the circulation speed of money is, and the less clear the relationship between money supply and total output. When the sensitivity of the demand for money to interest rate reaches the extreme situation, Keynes' liquidity trap will appear. This extreme situation is known as the ultra-sensitivity of the demand for money to interest rate. That is, the elasticity of the demand for money over interest rate is infinite (the interest rate is so low that it cannot go any lower; otherwise the demand for money has to become infinite). In this extreme situation of ultra-sensitivity, the demand for money does not bear any effect on the total output so that monetary policies fail, meaning that monetary policies no longer affect the total output.

Tobin (1956) uses American data to empirically test the relationship between the demand for money and interest rate. He treats the money balance used for transaction purposes as transaction balance (that is the transaction demand), then obverses the average American annual transaction balances and the corresponding average American annual interest rates of commercial papers for the years 1922–1941. What he discovers is that the transaction balances are conversely correlated to interest rates. This discovery indicates that even the money balance used for transaction purposes is very sensitive to changes in interest rate. So, Tobin concludes that the demand for money is sensitive to interest rate.

What Tobin did is one of the earliest empirical studies in this area. After that work, many scholars also conduct empirical tests on the demand for money and provide additional evidence to support what Tobin concludes: The demand for money is sensitive to interest rate.

Following this line of works, economists turn to empirically confirm the validity of Keynes' liquidity traps. That is, they consider whether or not the sensitivity of the demand for money to interest rate is near the ultrasensitive state for monetary policies to lose their effectiveness. As a matter of fact, in his initial publication of his "General Theory" in 1936, Keynes claims that when interest rate is extremely low, there might appear liquidity trap. However, he also confirms that he had never actually seen the appearance of a liquidity trap.

Karl Brunner and Allan Meltzer (1968) employ typical evidence to illustrate why liquidity trap has never appeared before. By observing whether or not the sensitivity of the demand for money to interest rate changes from one time period to another, especially by observing whether or not the sensitivity of the demand for money to interest rate rises during the period of the 1930s when the interest rate is particularly low, they discover that when interest rate goes down, the sensitivity does not exhibit any tendency of increasing. So, they conclude that if there were liquidity traps, then as a result of the existence of such liquidity traps the degree of sensitivity of the demand for money to interest rate should increase for the time period of the 1930s when the interest rate is extremely low. Therefore, the demand for money function, as estimated for this particular time period, cannot be used to accurately predict the demand for money in normal periods of time. However, the reality is not so. The sensitivity of the demand for money to interest rate does not change from one period of time to another; and Brunner and Meltzer discover that the demand for money function, as estimated earlier using the data from the 1930s, can accurately predict the demand for money for the 1950s. This result of the empirical test does not provide any evidence for the existence of liquidity traps during the time period of the Great Depression.

The results obtained by various scholars from their empirical studies are quite consistent: They all support the conclusion that the demand for money is sensitive to changes in interest rate; and almost all the results do not back the existence of liquidity traps, as derived theoretically. That is, we can speech with certainty that the demand for money is affected by interest rate and is sensitive to changes in interest rate; however, there is not any liquidity trap. In other words, the state of ultra-sensitivity of the demand for money to changes in interest rate does not exist.

4.3.5.2 Stability of the demand for money

The stability of the demand for money function is very important in terms of selecting monetary policies. If this function were instable, that is, the demand curve for money could be displaced often, making unpredictable large-scale swings in the demand for money, as claimed by Keynes, then the circulation speed of money would be unpredictable and the amount of money would not be as closely related to total output, as claimed by the modern quantity theory of money so that the central bank would not be using money supply instead of interest rate as its monetary policy. Conversely, if the demand for money were indeed stable, as claimed by the modern quantity theory of money, then the central bank could choose money supply to macroscopically control the total output.

The empirical study on the demand for money by Brunner and Meltzer in fact has touched on the stability problem of the demand for money function. Their empirical analysis indicates that the so-called liquidity trap does not exist; the demand for money function estimated using the data from one time period can be employed to accurately predict the demand for money of the next period. In particular, they apply the data of the 1930s to estimate the demand for money function and employ this function to successfully predict American demand for money for the time period after World War II. This empirical result indicates that the demand for money function is stable within long period of time. Additionally, the fact that the sensitivity of the demand for money to interest rate is relatively stable also implies the stability of the demand for money function.

In the early 1970s, economists employed the narrow money M_1 as the definition of money supply. By using the seasonal American data since after World War II as evidence, the relevant empirical studies on the demand for money totally support the conclusion of stability of the demand for money function. Therefore, the M_1 demand for money function becomes the convention for economists to use. And until 1973, the stability of this function had been accepted as a proven fact.

However, starting in 1974, the conventional M_1 demand for money function seriously overestimates the demand for money for the United States. This phenomenon of instability in the demand for money function is known as the missing money. The appearance of the missing money phenomenon makes it clear that the demand for money function can no longer be used as an effective tool to measure how well monetary policies affect the overall economy. That encourages economists to reconsider the problem of how money affects the economy. Additionally, active works are conducted to resolve the problem of missing money and to establish a stable demand for money function. Out of this massive effort, there appeared two development directions in the research of the stability problem of the demand for money function.

The first direction is to locate the reasons that cause the demand for money function instable. The focus is to see whether an incorrect definition of money could lead to the instability. After 1974, American inflation, high nominal interest rate, and the advancement of computer technology quickly change the payment system and the techniques of cash management. At the same time, the speed of financial innovation quickens with many brand new financial tools introduced. The rapid development of financial innovations makes people doubt about the appropriateness of the traditional definition of money within the new economic environment. People start to identify the financial tools that have been mistakenly excluded out of the traditional definition of money, hoping that by doing so the missing money could be found. One good example in this area of research is the overnight repurchase agreements. They represent riskless overnight loans using treasury bills as collaterals. Those companies that have opened current accounts in commercial banks often use overnight repurchase agreements to make overnight loan against their large sums of deposit balances. These loans are very analogous to money. Economists discover that if these overnight repurchase agreements are included in the money supply, the overestimate of the money supply out of the demand for money function can be greatly reduced so that the demand for money function will be stabilized. However, many studies using more recent data cast doubts on whether or not this method can truly acquire the stability of the demand for money function.

The secondly research direction is to find a new variable for the demand for money function that can make the function stable. Studies along this direction indicate that if the average dividend/price ratio of common stocks (average dividend divided by average price) can be used as a computational value of interest rate, then a stable demand for money function can be obtained. Additionally, some researchers find that if the term structure of all interest rates is used as a new variable of the demand for money function, a stable demand for money function can also be produced. This method of introducing a new variable to stabilize the demand for money function also attracts its share of criticism. For example, these newly designed variables did not accurately reflect the opportunity costs of holding money so that there is not sufficient theoretical reason for them to enter the demand for money function. Later works also point to the inquiry about whether or not this kind of alteration to the demand for money function could truly bring forward stable demand for money function for the future.

4.3.5.3 Circulation speed of money

When the demand for money function overestimated the money demand for the mid period and later period of the 1970s, it also underestimated the circulation speed of money $V = PY/M$, where the actual speed V of increasing in was faster than what was estimated. After entering the 1980s, such weakness of the traditional demand for money function becomes more obvious. In fact, starting in 1982, the circulation speed of M_1 dropped significantly, which was missed by the traditional demand for money function. Although scholars have tried to come up with the reason for why the circulation speed of M_1 dropped, no success has been achieved.

At the same time, what is found is that during the 1980s the circulation speed of M_2 is much more stable than that of M_1. This relatively more stability indicates that employing the demand for money function of M_2 is more advantageous than using that of M_1. And researchers of the Federal Reserve discover that the circulation speed of M_2 is very stable and this speed is closely correlated to the opportunity costs of M_2. However, after entering the 1990s, the circulation speed of M_2 experienced a dramatic drop. Some scholars believe the traditional demand for money function could not explain why this happened. That is, the stability problem of the demand for money is still unsettled.

In short, the empirical studies on the demand for money indicate that the rapid development of financial innovations could be the main cause behind the instability of the demand for money function. However, as of present day, scholars are still searching for evidence to support this claim. Also, as of today, scholars are still looking for truly stable, satisfactory demand for money function.

The near-term instability of the demand for money function challenges the community of economists with the question of whether or not the past empirical studies on the demand for money are sufficient. The demand for money function has been a tool of guidance used by policy makers. However, its instability makes people doubt about its validity. Whether or not the demand for money function is stable is very important in terms of implementation of monetary policies and specifically how these policies are implemented. Because in recent years the demand for money function has become very instable, the prediction of the circulation speed of money has turned into a difficulty. So, rigorously adjusting money supply might no longer be an effective monetary policy method to control the economic total output.

4.4   A FEW FINAL COMMENTS

Based on what is learned in Chapter 2: The Yoyo Model and Its Justification, it seems to suggest that the historically available theories on the supply and demand of money are not completely, to say the least. It is because only with the economic forces of supply and demand interact with each other, other economic events take place. However, the currently available theories have mostly investigated the supply and demand of money separately without systematically looking at the interactions of these two mutually influencing aspects of the money theory.

Another point of note is that the community of economists has been mostly after the idea of establishing a stable demand function of money in order to achieve a centralized control of the economy through the central bank. To this end, there are at least two possible aspects one could look at. One is that in the long run, such desirable stable demand might not exist; the other is that interactions between different economies should have certain bearing on the stability of a particular economy in general and the stability of money demand in particular. The former aspect is intuitively supported by the pattern changes as observed in the dishpan experiment, for details, see Section 3.7: Innovation: Livelihood of National Financial Infrastructure, while the latter aspect is evidenced by economic conditions experienced by regions that have been greatly affected by tourists or by foreign investments.

Chapter 5

Interest: A factor influencing monetary supply and demand

In the previous chapter, interest rate was discussed as a factor that influences the demand for money and the supply of money. That in fact means that interest rate is treated as an exogenous variable. In this chapter, we treat interest rate as a various indogenous to the economic system. By employing interactions of the demand and supply of money, we investigate how interest rate is determined and why interest rate fluctuates, while looking at the problem of how interest rate and total output are determined simultaneously. We will also expand our discussion to why the interest rates of different bonds are different. At the end, we look at how the marketplace reacts respectively to safe and risky investment opportunities.

This chapter is organized as follows. In Section 5.1, we look at loanable funds. In Section 5.2, we study the theory of liquidity preference. In Section 5.3, the relationship between interest rate and total production output is considered. In Section 5.4, the risk structure of interest rates is analyzed, in Section 5.5 the term structure of interest rates is investigated. Different of the existing literature, in Section 5.6, we look at the market behaviors and response to different-scale investment opportunities. This chapter ends with a few concluding words in Section 5.7.

5.1 LOANABLE FUNDS

We have learned that the interest rate and price of bonds are correlated inversely. Assume that we know the mechanism under the determination of bond price and why the price fluctuates, then we would be able to find how the interest rate of bonds is derived and how it would change. Speaking generally, the bond market is a competitive market, where the bond price is determined by the demand and supply. So, let us start by analyzing the demand-supply relationship of bonds in order to find out how interest rate is determined.

First, we assume that other than the interest rate and price of bonds, all other variables stay constant. So, for the interest free bond that matures in one year, if its price P is lower than its face value by discount, then the price P and interest rate i are related as follows:

$$P = \frac{V}{1+i} \quad \text{or} \quad i = \frac{V-P}{P} \tag{5.1}$$

As for the interest-bearing bond that matures in n years, which pays a fixed annual interest C each year, and repays the face value V upon the maturity, the price P and interest rate i are related as follows:

$$P = \sum_{k=1}^{n} \frac{C}{(1+i)^k} + \frac{V}{(1+i)^n} \tag{5.2}$$

In short, the general relationship between the price P and the interest rate i of a bond can be written in the following form:

$$\begin{cases} P = P(i) \\ P'(i) = \dfrac{dP}{di} < 0 \\ P''(i) = \dfrac{d^2P}{di^2} > 0 \\ P(0) = P_0 \\ P(\infty) = \lim_{i \to \infty} P(i) = 0 \end{cases} \tag{5.3}$$

where P_0 stands for the maximum price of the bond, which is the discounted value of the bond when the interest rate is zero. For a bond of any chosen kind, this relationship $P = P(i)$ between the price P and interest rate i is known. So, as long as the relationship between the demand and supply of bonds and the bonds' price is given, the relationship between the demand and supply and the interest rate will be known.

The bond demand curve reflects the relationship between the amount of demand for the bond and the price of the bond. For each possible price level P, the maximum amount $B_d = B_d(P)$ of the bond the public is able to and willing to purchase is the amount of demand for the bond at price P. This functional relationship $B_d = B_d(P)$ is known as the price demand function of the bond, or known briefly as the demand function of the bond. The graph of this function is the price demand curve of the bond, or known briefly as the demand curve. From the general law on the demand for commodity, it follows that demand evolves inversely with the price. Therefore, the demand curve of bond trends lower to the right hand side as shown in Figure 5.1.

By substituting the relation $P = P(i)$ between the price and interest rate of the bond into the demand function $B_d = B_d(P)$ of the bond, we obtain the relationship between the demand for the bond and the interest rate of the bond: $B_d = B_d(P(i)) = \hat{B}_d(i)$. This function is referred to as the interest rate demand function of the bond. Its graph is known as the interest rate demand curve, Figure 5.1. Because the demand B_d for the bond varies inversely with the price P of the bond, and the price P and interest rate i are also inversely correlated, it follows that B_d and i evolve in the same direction.

In order to see how the demand for bond varies in the same direction as the interest rate of bond, let us consider the situation where the public holds their bonds until maturity after they purchase the bonds. For such bonds, which are held until maturity, the expected yield is the yield of the bonds when held to maturity, which is the interest rate of the bonds. So, as a class of assets, the higher the expected yield of the bonds is, the greater the demand. Hence, the higher the bonds' interest rate is, the greater the demand for the bonds will be.

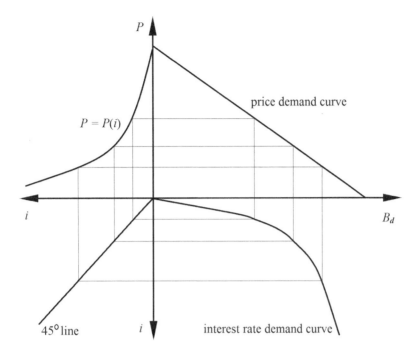

P

$P = P(i)$

price demand curve

i

B_d

45° line

i

interest rate demand curve

Figure 5.1 Bond demand curve.

The supply curve of bonds reflects the relationship between the supply of bonds and the price of the bonds. For each price level P, the maximum amount $B_s = B_s(P)$ of bonds the public is willing and is able to purchase at this price represents the supply for the bonds at price P. This functional relationship $B_s = B_s(P)$ is referred to as the price supply function of bonds, or briefly the supply function of bonds. The graph of this function is known as the price supply curve of bonds, or briefly the supply curve. According to the general law of supply and demand of commodities, the supply curve of bonds is a curve that slants upward to the right hand side, Figure 5.2. That is, the higher the bonds' price is, the greater the supply of the bonds. It also means that the lower the interest rate of the bonds (the higher the bonds' price) is, the greater the supply of the bonds. So, the supply of bonds is inversely correlated to the interest rate of the bonds. In order to specifically describe this inversely varying relationship, we substitute the relationship $P = P(i)$ between the price and the interest rate of the bonds into the supply function $B_s = B_s(P)$, leading to the function $B_s = B_s(P(i)) = \hat{B}_s(i)$, known as the interest rate supply function of the bonds. Its graph is known as the interest rate supply curve, Figure 5.2.

For a fixed price, if the amount of commodities the market participants are willing to purchase is equal to that of the supply of the commodities, the market reaches a state of equilibrium. The particularly fixed price in this case is known as an equilibrium price; and the corresponding market supply is known as an equilibrium supply. To this rule the bond market is no exception. When the demand for bonds is equal to the supply

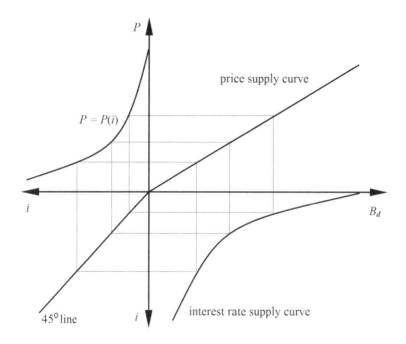

Figure 5.2 Bond supply curve.

of bonds, that is when $B_d = B_s$, the bond market realizes an equilibrium. The bonds' price that makes the bond supply equal to the demand is known as an equilibrium price of the bonds, or the clearing price of the market. The interest rate that corresponds to an equilibrium price is referred to as an equilibrium price of the bonds. With all these terms introduced, the equilibrium model of the bond market can be written in the following two forms:

$$
\begin{cases}
B_d = B_d(P) \\
B_s = B_s(P) \quad \text{(Model of price equilibrium)} \\
B_d = B_s
\end{cases}
\tag{5.4}
$$

$$
\begin{cases}
B_d = \hat{B}_d(P) \\
B_s = \hat{B}_s(P) \quad \text{(Model of interest rate equilibrium)} \\
B_d = B_s
\end{cases}
\tag{5.5}
$$

That is to say, the intersection of the price demand curve and price supply curve determines the equilibrium amount and the equilibrium price of bonds. And the intersection of the interest rate demand curve and the interest rate supply curve provides the equilibrium amount and the equilibrium interest rate of bonds, Figure 5.3.

The equilibrium price and the equilibrium interest rate are truly the market price and market interest rate of bonds. That is because under the current price P, if the demand for bonds is greater than the supply, there will be an excess demand in the

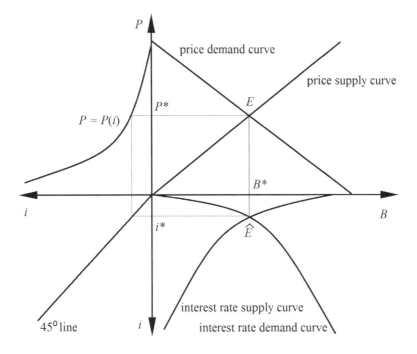

P

price demand curve

price supply curve

$P*$ E

$P = P(i)$

$B*$

i

B

$i*$ $\hat{E}$

interest rate supply curve

$45°$ line i interest rate demand curve

Figure 5.3 Equilibrium price and equilibrium interest rate.

market place so that the amount of bonds people are willing to purchase is greater than the amount of bonds people are willing to sell. That difference causes no trade to be reached at the prevailing price. That is, the excess demand will push the price toward a higher level. Conversely, if the supply of bonds is greater the demand, then there is an excess supply where the amount of bonds buyers are willing to purchase is lower than the amount of bonds sellers are willing to sell. So, at the prevailing price no trade can be entertained; and the excess supply will bring down the price. Only when the supply is equal to the demand, trades can be facilitated. Hence, the equilibrium price is the ultimate price of market transaction, the market price. Similarly, the equilibrium interest rate represents the ultimate rate of the bond market transactions, the true market interest rate.

As for the supply and demand of bonds, there is still another explanation. The company that provides bonds actually obtains its loans from the purchasers of the bonds. So, "supply of bonds" is identical to "demand for loans". Similarly, purchasers of bonds in fact give out loans to the companies that issue bonds so that "demand for bonds" is identical to "supply of loans". If the amount of bonds is renamed as the amount of loanable funds, then the interest rate demand curve of bonds becomes the supply curve of loanable funds, and the interest rate supply curve of bonds becomes the demand curve of loanable funds. The intersection of the supply curve and demand curve of loanable funds determines the equilibrium amount and equilibrium interest rate of the loanable funds, Figure 5.4. Because of this reason, this section's discussion

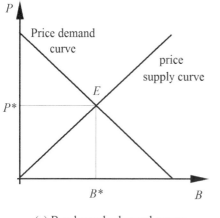
(a) Bond supply-demand curves

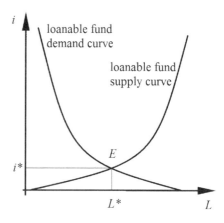
(b) loanable fund demand-supply curves

Figure 5.4 Theory of loanable funds: Determination of equilibrium interest rate.

on how the price and interest rate of bonds are determined is known as the theory of loanable funds.

The theory that the bond supply determines the bonds' interest rate can be employed to analyze the cause underneath the changes of the market interest rate. To this end, let us separate changes in the supply-demand relationship into two scenarios: changes in the amount of supply and demand, and changes in the supply and demand. The so-called change in the amount of demand (respectively, the supply) means the amount of demand (resp., supply) changes along the demand curve (resp., supply curve) with the price. The so-called change in demand (resp., supply) stands for the demand curve (resp., supply curve) experiences displacement, causing the amount of demand (resp., supply) to change at every price level. If the bond market is not in equilibrium, both the amount of supply and the amount of demand of bonds will change and adjust until equilibrium state is reached when the market interest rate and price are determined. So, changes in the amounts of supply and demand along the supply and demand curves are the prerequisite to materializing equilibrium as well as the process of reaching equilibrium without affecting the equilibrium amounts, equilibrium price, and equilibrium interest rate. That is to say, changes in the amounts of supply and demand are not the cause behind the changes in the market interest rate.

On the contrary, changes in supply and demand are different. No matter whether it is the demand curve or the supply curve that experiences displacement, the equilibrium of the bond market will change from one state to another. That is, the intersection of the supply and demand curves moves from one location to another, which evidently causes changes in the equilibrium amount, equilibrium price, and equilibrium interest rate of bonds. Hence, changes in the supply and demand of bonds are the true reason why the market interest rate of bonds varies. In the following, we analyze the factors that affect the positions of the supply and demand curves of bonds. That is, we look at the problem of why the supply and demand of bonds change.

There are four main factors that determine the demand for bonds: wealth held by the public, expected yield of bonds compared to other substitute assets, relative risk scale of bonds in comparison with other substitute assets, and the liquidity of bonds relative to those of substitute assets. When these four factors are fixed, the price of bonds is determined by the demand of the bonds. So, these four factors are the main causes underneath the changes in the demand for bonds and the displacement of the demand curve.

The law of how the demand for bonds changes with wealth is that when the amount of wealth the public holds increases (respectively, decreases), the demand for bonds of the public also increases (resp., decreases) accordingly. That explains that when wealth increases, the demand curve of bonds moves to the right; when wealth decreases, the demand curve moves to the left. Speaking generally, during periods of economic prosperity, the amount of wealth held by the public increases so that the demand for bonds also goes up, while during periods of economic decline, the amount of wealth decreases so that the demand for bonds also accordingly drops. In terms of expected yield,

1   The expected interest rate of bonds affects the expected yield so that it influences the demand for bonds. In fact, for the discount bonds that mature in one year, the expected interest rate from holding the bonds for entire year is equal to the interest rate of the bonds. So, the expected yield of such bonds does not affect the position of the demand curve of bonds. However, for bonds that mature beyond one year period of time, the situation is completely different. The expected yield of these bonds although held for one year often deviates from the interest rate. So, the expected yield affects the demand for bonds greatly. If people expect the interest rate for the next year will go up, that is, people expect the price of bonds will go down for the next year, then they will experience losses if they purchase the bonds this year and sell them out after holding the bonds for one year. So, for long-term bonds, the expected increasing interest rate will cause the demand curve of bonds to move to the left and the demand for bonds to drop. Conversely, if the interest rate is expected to drop, then the price of bonds is expected to go up so that by holding these bonds for one year is expected to earn a positive return. That causes the demand for bonds to increase and the demand curve of bonds to move to the right.
2   Expected inflation also affects the expected yield of bonds so that it influences the demand for bonds. If the inflation is expected to go up, it means that the prices of the substitute assets for bonds is expected to increase so that the expected yield for holding these substitutes is rising. However, as soon as bonds are held, the nominal rate of return of the bonds at maturity will not change so that when inflation is expected to rise, the expected yield from holding bonds in comparison to holding other decreases. So, the demand for bonds goes down, causing the demand curve to move to the left. Hence, when inflation is expected to rise (respectively, drop), the demand curve of bonds will move to the left (resp., the right), and the demand decreases (resp., increases).

As for risk, if the risk from holding bonds increases, then people would not like to hold additional bonds while wanting to reduce the amount of bonds held. So,

Table 5.1   Directional change in the demand for bonds.

Factor	Wealth	Expected interest rate	Expected inflation	Risk	Liquidity
Direction of change in factor	↑	↑	↑	↑	↑
Direction of change in demand	↑	↓	↓	↓	↑

Table 5.2   Direction of change in the supply of bonds.

Influencing factor	Expected return on investment	Expected inflation	Fiscal deficit
Direction of change in factor	↑	↑	↑
Direction of in supply	↑	↑	↑

increasing risk causes the demand curve of bonds to move to the left, and the demand for bonds to drop. In terms of liquidity, the more easily a kind of bond can be converted to cash, the more willingly people like to hold this kind of bond. That is because people prefer the liquidity of assets. That explains that the more liquid a bond is, the more demand for the bond will be. So, increasing the liquidity of bonds can cause the demand curve of bonds to move to the right and the demand to increase. The previous analysis on the factors that affect the location of the demand curve of bonds can be summarized in Table 5.1.

The main factors that cause the supply curve of bonds to move include: the expected profitability of each investment opportunity, expected inflation, and government activities. Specially, the more profitable opportunities a business firm expects, the more willing the firm is to take on loans and increase its amount of debts in order to finance these investments. So, the higher a firm expects its investment returns are, the more willing the firm is to issue additional bonds, causing the supply of bonds to increase and the supply curve to move to the right. Especially during periods of economic prosperity, because the expected profitability of investments is very high, companies tend to issue large amounts of bonds to financie their investments so that the supply of bonds increases tremendously. Conversely, during periods of economic decline, the number of expected profitable opportunities decreases, causing the supply of bonds to decrease and the supply curve to move to the left. If the inflation is expected to rise, the actual cost of taking out loans drops so that companies are more willing to borrow funds. That causes the supply of bonds to increase and the supply curve to move to the right. Government activities also bear great influence on the supply of bonds. When a deficit appears in the fiscal budget, in order to cover the deficit, the government will issue bonds. That surely increases the supply of bond market, making the supply curve of bonds to move to the right. Additionally, when the government implements the monetary policy and the central bank trades securities in the open market, the supply of bonds is also affected, making the supply curve move.

The results of the previous analysis on the factors that affect the location of the supply curve of bonds can be summarized in Table 5.2.

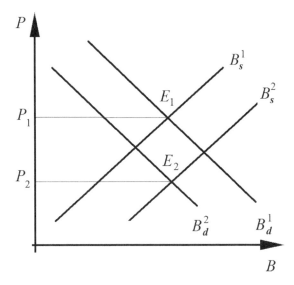

Figure 5.5 Reaction to expected inflation.

Now, let us analyze changes in equilibrium interest rate by using the knowledge of how the position of the supply curve of bonds changes.

1   Fisher Effect

We have seen that inflation bears effect on both the demand and supply of bonds. Now, let us consider how inflation affects the market interest rate. If the inflation is expected to rise, then the demand curve of bonds moves to the left and the supply curve of bonds moves to the right, Figure 5.5. The demand curve moves from position B_d^1 leftward to position B_d^2, while the supply curve moves from position B_s^1 rightward to position B_s^2. The equilibrium point moves from E_1 to E_2, and the bond price drops from P_1 to a relatively lower level at P_2. So, the interest rate of the bonds rises from the original i_1 to the current i_2. That is, when the inflation is expected to rise, the market interest rate goes up. This phenomenon is known as Fisher effect.

Also, we have seen that during periods of economic prosperity, the economy grows rapidly so that the public demand for bonds also increases accordingly, making the demand curve of bonds to move rightward. At the same time, the economic prosperity companies have more profitable investment opportunities. So, companies are more willing to borrow funds and to issue bonds in order to financial their investments. That makes the supply of bonds increase and the supply curve to move to the right. The consequence for both the demand curve and the supply curve to move to the right is that the equilibrium quantity of bonds increase, Figure 5.6. However, whether the equilibrium interest rate and equilibrium price go up or down depends on which one of rightward moves of the demand curve and the supply curve has greater magnitude. If the supply curve moves to the right more than the rightward move of the demand curve,

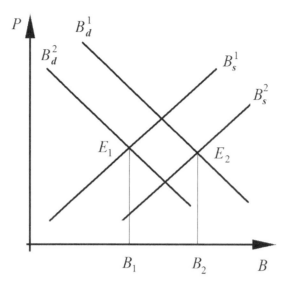

Figure 5.6 Reaction to economic cyclical expansions.

then the equilibrium price falls and the equilibrium interest rate rises. Conversely, if the demand curves moves to the right more than the rightward move of the supply curve, then the equilibrium price goes up and the equilibrium interest rate goes down. If both of these curves move to the right parallelly the same distance, then the equilibrium price and equilibrium interest rate stay invariant.

2   Transaction effect

During periods of economic expansion (periods of either recovery or prosperity), the wealth of the society increase and the level of income rises. That causes the demand for bonds to increase and the demand curve of bonds to move rightward. When seen from the angle of supply, it follows that during periods of economic expansion, various profitable opportunities appear increasingly and the room of profit expands constantly so that companies are optimistic about investment prospects. Consequently, the return of investment is expected to rise. That motivates companies to make more investments, which in turn requires additional financing, causing the supply of bonds to increase and the supply curve of bonds to move to the right. The result for the demand curve and the supply curve of bonds to move right is that the equilibrium quantity of bonds (that is the amount of bonds traded) rises. That is the transaction effect of economic expansion, Figure 5.6. However, whether or not the market interest rate is also rising during periods of economic expansion can not be seen merely from the analysis of the bond market. It is really determined by which of the rightward moves of the demand curve and the supply curve is greater. For the same reason, it follows that during periods of economic contraction (periods of either economic decline or depression), the equilibrium quantity falls. So, bond trading activities (that is the equilibrium quantity of bonds) fluctuates along with the economic cycle.

5.2 LIQUIDITY PREFERENCE

Based on the supply and demand relationship of bonds, the theory of loanable funds analyzes the mechanism that determines the interest rate. That is one theoretical framework for the investigation of how interest rate is determined. Keynes proposes and establishes another theoretical framework, which analyzes the mechanism of how interest rate is determined through the use of the supply and demand relationship of money. This theoretical model is known as the theory of liquidity preference. Although these two frameworks of analysis look different, the analysis of the money market using the liquidity preference is closely related to the theory of loanable funds of the bond market. To a certain degree, these two theories provide the same conclusion regarding how to determine the equilibrium interest rate. Even so, these theories have their different strengths and weaknesses. For example, when analyzing how changes in expected inflation affect the interest rate, the theory of loanable funds is more convenient and readily to apply. When analyzing how changes in income, price level, and supply of money affect the interest rate, the theory of liquidity preferences provides a more straightforward method of analysis.

Keynes divides the assets people use to store wealth into two classes: money and bonds. With this assumption, Keynes believes that the total amount of wealth in an economy is equal to the sum of total quantity of bonds and total amount of money, which is the sum of the bond supply B_s and the money supply M_s. Because the amount of assets people purchase is limited by the total amount of wealth they own, the sum of the amount B_d of bonds and the amount M_d of money people are willing to hold must be the same as the total of wealth. That is, B_d represents the demand for bonds, and M_d the demand for money. Therefore, the following equation holds true:

$$B_s + M_s = B_d + M_d \qquad (5.6)$$

By placing there terms of bonds on the left hand side and by placing the terms of money on the right hand side, the previous equation becomes

$$B_s - B_d = M_d - M_s \qquad (5.7)$$

where $B_s - B_d$ stands for the excess supply of bonds, which people do not hold, and $M_d - M_s$ represents the excess demand of money, which are the quantity of additional money people are willing to have in hands. Equ. (5.7) implies that in the economy the excess supply of bonds is equal to the excess quantity of money people are willing to hold.

It follows immediately from equ. (5.7) that if the money market is in the state of equilibrium, that is, the demand M_d for money is equal to the supply of money, then the excess supply of bonds is zero, that is, the supply of bonds is equal to the demand for bonds and the bonds market is in the state of equilibrium. Similarly, if the bond market is in the state of equilibrium, then the money market is also in the state of equilibrium. Therefore, determining the interest rate by letting the supply and demand of money be equal is no different from the determination of interest rate through forcing the supply and demand of bonds to be the same. So, in this sense, analyzing the theory of liquidity preference of the money market is equivalent to analyzing the

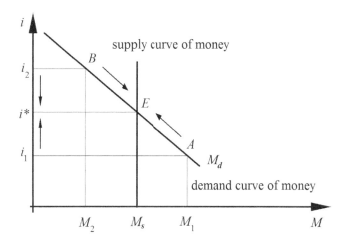

Figure 5.7 Interest rate determined by demand-supply of money.

theory of lonanable funds of the bond market. However, in practice, these two theories are quite different. Evidently, Kaynes categories assets into two classes, while ignoring the impact of the expected returns of such properties as automobiles, houses, etc., on the interest rate. So, the theory on how interest rate is determined using liquidity preference is not complete.

From Keynes' supply and demand theory of money, as discussed earlier, it follows that the demand for money is a function of income Y and interest rate i, and varies in the same direction as income and the opposite direction as interest rate. Now, if the income is assumed to be fixed, then the demand for money becomes a function of interest rate only: $M_d = M_d(i)$, where income only determines the position of the demand curve of money. From the inverse directional relationship between the demand for money and interest rate, the demand curve of money is slanted downward to the right, Figure 5.7. Keynes also assumes that the money supply of the economy is directly within the control of the central bank. Because the income is fixed, the total money supply becomes a variable that does not change with interest rate. Assume that under the control of the central bank the quantity of money supply is M_s, then the supply of money is a constant function $M_s(i) = M_s$ so that the money supply curve is a vertical line perpendicular to the horizontal axis, Figure 5.7.

If for the current interest rate there is an excess demand for money, that is, the demand is greater than the supply, the money supply cannot keep up with the pace of demand, then the market force would push up the interest rate, making the demand for money to drop (the price of bonds drops so that people would increase their holding of bonds, making money less desirable). As reflected in Figure 5.7, point A moves closer to point E, and interest rate i_1 increases. Conversely, if under the current interest rate there is an excess supply of money, that is, the supply of money is more than the demand, then the market force would drag down the interest rate making the demand for money increase (the bond price to go up so that people would sell off their bonds making the demand for money increase). As shown in Figure 5.7, point B moves closer to point E and interest rate i_2 drops.

Such variation in the amount of money held by the public stops when point E in Figure 5.7 is reached, where the demand for money is equal to the supply, so that changes in interest rate also ceases to exist. Only at this moment, market transaction is completed (because before reaching this moment, interest rate still fluctuates constantly so that both sides of the transaction are still negotiating on the interest rate). So, it can be seen that the situation when the demand for money is equal to the money supply represents a relatively static state of equilibrium. It is such a state that ultimately determines the market interest rate. This state is known as the equilibrium of the money market, the corresponding interest rate the equilibrium rate. Hence, the solution i^* of the equation $M_d(i) = M_s$ is the market interest rate determined through money supply, the equilibrium interest rate.

In Keynes' theory of money supply and demand, there are two factors that cause the position of the demand curve to move: income level and price level. No matter which one of these factors goes up, the demand curve of money will move rightward. Keynes believes that during periods of economic expansion, income increases and wealth increases so that people like to have more money to store their values. On the other hand, with the economic expansion, the increasing income motivates people to complete additional transactions. Consequently, people are willing to hold more money. What can be concluded from this observation is that along with the increase of income, the demand for money at each interest rate level increases so that the demand curve of money moves to the right. Keynes also believes that what people care about is the actual quantity of money they hold, which is the amount that can be used to purchase goods and labor. When the price level goes up, the purchasing power of money goes down. The same amount (the nominal amount) of money can purchase less goods and labor after the price level increases. So, increasing price level causes the actual amount of money people hold to decrease. In order to recover the actual amount of money held to the level before the price level increases, people would hold more money. That implies that increasing price level causes the demand for money to rise and the demand curve of money to move to the right.

Figure 5.8 shows that when the demand curve of money moves rightward, the equilibrium interest rate increases. If the rightward movement of the demand curve of money is caused by increasing income, the consequent increase in the equilibrium interest rate is known as the income effect. Similarly, if the rightward movement of the demand curve of money is caused by increasing price level, the consequently increase in the equilibrium interest rate is known as the price effect.

In the earlier chapters, we have learned that the money supply is entirely within the control of the central bank and various factors affect the money supply. In particular, such factors as the amount of wealth in the economy, rate of statutory reserve, the operation of the central bank in the open market (buy and sell of bonds), discount rate (the interest rate of discount loans of the central bank), rate of income tax, bank panics, illegal activities, etc., all cause changes in the money supply. In short, these factors can all cause the position of the demand curve of money to change.

Figure 5.9 shows how the equilibrium interest rate changes when the money supply increases. It can be seen that when the money supply increases, the equilibrium interest rate decreases. This conclusion can also be drawn by using the theory of loanable funds. In fact, the main method for the central bank to increase the money supply is to purchase bonds in the open market. That causes the demand for bonds to increase

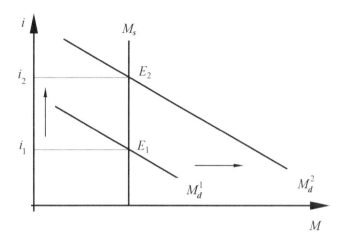

Figure 5.8 When the demand for money increases, the equilibrium interest rate rises.

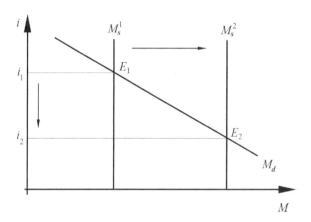

Figure 5.9 When the supply of money increases, the equilibrium interest rate drops.

(or the public's demand for bonds decreases), the demand curve of bonds to move to the right (or the supply curve of bonds to move to the left), the equilibrium price of bonds to increase, and the equilibrium interest rate to decrease. So, when the central bank increases the money supply through purchasing bonds, no matter whether it is from the theory of liquidity preference or from the theory of loanable funds, it can be concluded that the equilibrium interest rate goes down. This conclusion is of the following significant policy implication: Government policy makers often increase the money supply to release the pressure of inflation based on this result. Milton Friedman names this result that increasing money supply causes the market interest rate to drop as the liquidity effect.

Friedman (1969) points out that the liquidity effect appears only when the money supply increases while all other conditions are fixed. He emphasizes that in reality

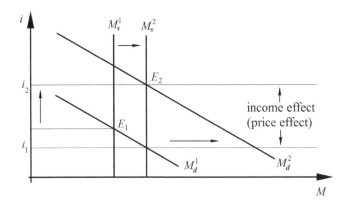

Figure 5.10 The income effect (price effect) of increasing money supply.

increasing money supply has to cause other conditions to vary and brings forward other effects to the economy; if these other effects are great, it is very possible that the market interest rate can go up. So, even though theoretically speaking the liquidity effect is correct, it is problematic when seen practically. In reality, when the money supply is increasing, it is very possible that the market interest rate also goes up. Hence, the policy of pressuring down the interest rate by increasing the money supply might not be effective at all.

Next, let us consider the other effects that lift the equilibrium interest rate when the money supply is increased.

1   The income effect of increasing money supply: Increasing money supply has to bring forward expansionary effect on the economic production, making the levels of both income and wealth increase. When income and wealth grow, the demand curve of money moves rightward, that is, the demand for money rises so that the corresponding equilibrium interest rate also increases. After the money supply is increased, the rising interest rate, as caused by the increasing income, is known as the income effect of the increasing money supply, Figure 5.10. What needs to be noted is that when the supply curve of money moves rightward, the equilibrium interest rate drops; when the demand curve of money moves rightward, the equilibrium interest rate rises. Now, increasing money supply causes the supply curve of money to move rightward. If the income effect of the increasing money supply is greater than the liquidity effect, then the equilibrium interest rate has to go up.

2   The price effect of increasing money supply: Increasing money supply also lefts up the price level, making the demand curve of money to move rightward and the demand for money increases. So, the equilibrium interest rate also correspondingly rises. After the money supply is increased, the increase in the interest rate, as caused by the increasing price level, is known as the price effect of the increasing money supply, its graphical representation is similar to that in Figure 5.10. Similarly, the increasing money supply causes a simultaneous rightward movement of the

supply curve of money. If the price effect of the increasing supply is greater than the liquidity effect, then the equilibrium interest rate has also to rise.

3   Fisher effect of increasing money supply: When money supply is increased, it can make people expect the future price level to increase so that the expected inflation increases. From the Fisher effect of the theory of loanable funds, it follows that when the expected inflation increases, the equilibrium interest rate of bonds goes up. After the money supply is increased, such increasing interest rate, as caused by the increasing expected inflation, is known as Fisher effect of increasing money supply, or the expected inflation effect of increasing money supply. When the money supply is increased, if the liquidity effect is smaller than Fisher effect, then the equilibrium interest rate has also to rise.

Speaking roughly, Fisher effect and the price effect of increasing money supply do not seem to be any different, because inflation is about increasing price level. However, in reality, there is a subtle difference between these two kinds of effects. The time moments for the occurrence and existence of Fisher effects and price effects are different. Assume that a one-time increase is given to the money supply. The consequence is that the price level of the following time period is expected to increase, making the expected inflation increase within the current period. That in turn causes the equilibrium interest rate of the current period to rise. However, because the current prices are not up yet, there does not exist any price effect. In the following period when the price level is increased, the price effect appears. If the price level is of the tendency of increasing continuously, then the expectation of increasing inflation continues to exist so that there is also Fisher effect. If the prices have reached their maximum level and will no longer go any higher, then the expectation of any inflation no longer exists so that Fisher effect disappears, although the price effect, consequent to the rising prices as caused by the increasing money supply, still presents. So, it can be seen that Fisher effects (expected inflation effect) reflect an effect on the current equilibrium interest rate of the future price increases, while price effects represent the effect on the current equilibrium interest rate by the current price increases. These two effects are different.

From the discussions above, it follows that increasing money supply creates four kinds of effects on the interest rate: liquidity effect, income effect, price effect, and Fisher effect. The liquidity effect indicates that increasing money supply causes the equilibrium interest rate to drop. The other three kinds of effects imply that increasing money supply can cause interest rate to rise. So, the direction of liquidity effect is opposite of those of the other three effects. If an effect that brings down the interest rate is expressed by using negative numbers, while an effect that lifts up the interest rate by using positive numbers, then the liquidity effect is negative, while the income effect, price effect, and Fisher effect are all positive. The sum of these four kinds of effects as caused by increasing money supply on interest rate is referred to as the total effect of the increasing money supply.

The total effect could be negative, and could also be positive, the particular situation of which is determined by whether or not the liquidity effect is greater than the sum of all other kinds of effects. That implies that in the operation of monetary policies the method of adjusting the market interest rate by controlling the money supply might not achieve the expected outcome. The impact of increasing money supply on the interest rate is a complicated process, leading to uncertain results that are difficult to predict.

5.3   INTEREST RATE AND TOTAL OUTPUT

Economists and government agencies often need to make predictions for the gross domestic production (GDP) and interest rate, where the IS-LM (Investment-Saving/Liquidity preference-Money supply) model represents one of the widely employed economic prediction models (Hicks, 1937). Under the condition that the price level stays constant, this model explains how to determine the interest rate and the total output (that is the total income) of the economy. Evidently, this theory carries the theories on how interest rate is determined, as discussed in the previous two sections, a few steps forward: it allows both interest rate and total output to vary, while the previous two sections assumed other than interest rate all other factors are fixed. This model is important not only in the area of explaining how interest rate is determined and how economic predictions can be made, but also in helping with our understanding of how economic policies affect economic activities. In this section, we look at this model from the angle of how to determine the interest rate.

Keynes is particularly interested in changes of the total output. He attempts to explain why major economic disasters occur, and to address the problem of increasing employment by using economic policies. Keynes believes that the aggregate demand for the economic output consists of four payments: the payment C for consumption, the payment I for investment, the payment G for government expenditure, and the payment $X - M$ for the net export, where X stands for the total export and M the total import. If Y^d stands for the aggregate demand and Y the total output, then we have

$$Y^d = C + I + G + (X - M) \tag{5.8}$$

Based on his supply and demand analysis, Keynes points out that when the total economic output is equal to the aggregate demand, that is, $Y = Y^d$, the economy reaches its equilibrium, where the producers could sell off all of their products so that the producers will not change their production. After analyzing the factors of various levels that affect the aggregate demand, Keynes is able to explain why the total output is maintained at a certain level, how insufficient demand could cause insufficient employment, and other problems. In his investigations, Keynes assumes that the price level is fixed constant at a certain level. So, when he talks about changes in nominal quantities, he means changes in the actual quantities.

The cost C of consumption stands for the aggregate demand of the economy for goods and labor; it consists of two parts: spontaneous consumption C_0 and derived consumption C_1, where spontaneous consumption C_0 is a various unrelated to the total output Y, while derived consumption C_1 changes directly proportional to the disposable income Y_D: $C_1 = cY_D$. The disposable income Y_D means the balance of the total income (that is the total output) Y after taking out the income taxes, $Y_D = Y - T$. The symbol c stands for the marginal propensity to consume, which is defined to be the amount of increase in consumption when the disposable income grows by a unit. Tax

T is also a function of the total output Y. If t stands for the average rate of tax, then we have $T = T(Y) = tY$. From all these notations in place, it follows that the consumer spending is a function of the total output, known as the consumption function, as follows:

$$C = C_0 + cY_D = C_0 + c(Y - T) = C_0 + c(1 - t)Y \tag{5.9}$$

The term "investment" as used in investment spending stands for planned investments to be made by industrial and commercial firms on new, physical capitals and new houses. It belongs to investment in physical assets instead of those investments in securities. Planned investment also consists of two parts: spontaneous investment I_0 and derived investment I_1, where the spontaneous investment I_0 is a quantity that is not affected by the total income (but it is generally influenced by the interest rate), while the derived investment I_1 is always affected by the total income. When the interest rate i is fixed, the derived investment is directly proportional to the income: $I_1 = vY$, where v stands for the marginal propensity of investment, which is defined to be the amount of increase in the planned investment when the income grows by one unit. That is, when the interest rate i is fixed, this functional relationship between the investment spending I and the total output Y is referred to as investment function: $I = I_0 + vY$. It is worth noticing that investment is negatively correlated to interest rate: The higher the interest rate (of bonds or loans) is, the smaller the planned investment for industrial and commercial firms.

The term "government" in government spending means the aggregate demand of various levels of the government for goods and labor, including military expenditure, expenses for the construction of public facilities, expenditure on education, wages of government workers, expenditure of government operations, spending on relieving efforts, etc. Here, government spending is treated as an exogenous variable so that it does not change with the total output Y.

The total export stands for the aggregate foreign demand for domestic goods and labor. The total import represents the aggregate domestic demand for foreign goods and labor. So, the net export stands for the net foreign demand for domestic goods and labor. Here, we also assume that the situation of international trades is fixed so that the net export is a pre-determined quantity NX.

After explaining all the terms, the aggregate demand can be written as follows:

$$Y^d = C + I + G + NX = (C_0 + I_0 + G + NX) + [c(1 - t) + v]Y \tag{5.10}$$

When the aggregate demand is equal to the total output, the economy reaches its equilibrium. So, the total output can be determined as follows:

$$Y = C + I + G + NX = (C_0 + I_0) + [c(1 - t) + v]Y + (G + NX) \tag{5.11}$$

By solving this equation, the total economic output is obtained as follows:

$$Y = Y^* = \frac{C_0 + I_0 + G + NX}{1 - (1 - t)c - v} \tag{5.12}$$

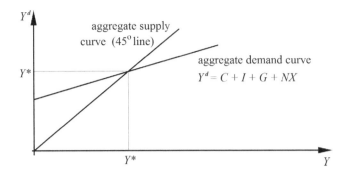

Figure 5.11 The total output when interest rate is fixed.

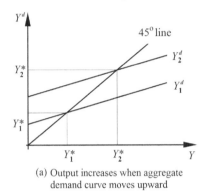

(a) Output increases when aggregate
demand curve moves upward

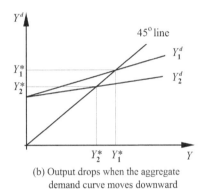

(b) Output drops when the aggregate
demand curve moves downward

Figure 5.12 Changes in the total output.

Figure 5.11 shows the mechanism of how the total output is determined: The inter-section of the aggregate demand curve and the 45° line gives the aggregate demand and the total output when the economy is in its equilibrium state. The 45° line can be understood as the aggregate supply curve of the economy. That is, it is the aggregate supply, the total output, and the total income.

Equ. (5.12) implies that changes in spontaneous consumption, or spontaneous investment, or government spending, or net export help to push the equilibrium total output increase. The reason is that changes in these variables always move the aggregate demand curve to move upward and cause the aggregate demand to increase so that the equilibrium point moves correspondingly, making the total output grow, Figure 5.12(a). However, changes in the tax rate bear different effect on the total output. When the tax rate t is brought upward, the slope of the aggregate demand curve becomes smaller, and the position of the curve moves downward, making the equilibrium point slide along the 45° line so that the total output declines. So, the total output is negatively correlated to tax collection, Figure 5.12(b).

From equ. (5.11), we can solve for the equilibrium total output Y^* for fixed interest rate i. So, we obtain a function $Y = Y^* = Y(i)$ that reflects the relationship between the total output Y and interest rate i. When the total output and interest rate satisfy this relationship, it means that the aggregate demand of the output is equal to the aggregate supply, and that the product market is in equilibrium. Similarly, the equilibrium interest rate, as produced in the previous section, is obtained when the total output Y is unknown. It stands for a relationship between the market interest rate i and the total output Y: $i = i(Y)$. When the interest rate i and the total output Y satisfy this relationship, it means that the money supply is equal to the demand for money, and that the money market is in equilibrium. So, it can be seen that only when both the product market and the money market are in equilibrium, the level of total output and the market interest rate can be truly determined. That is, the interest rate and the total output are determined simultaneously. In the following, we will look into more detailed discussions.

In Keynes' analysis, the fundamental way for interest rate to affect the total output is through its impacts on investment and net export. By effect of interest rate on investment, it means that interest rate influences the spending on the planned investment, which stands for investment in physical assets. Assume that the funds needed for companies to make investment are from taking out loans. Evidently, if the expected return on the investment is higher than the interest cost on the loans, then the companies will make the investment. However, when interest rate is very high, very few investment opportunities would provide returns that are higher than the cost of loans. In this case, the planned investment will be very low. So, by lowering the interest rate, industrial and commercial firms would be more willing to make investments so that planned investment spending would increase. Even if the funds used for investment are excess capital of the companies instead of from loans, the planned investment of the companies is still affected by interest rate. If the interest rate of bonds is high, then the opportunity cost for companies to apply their own funds to the investment of physical assets will be very high. In this case, companies are more willing to purchase bonds from the bond market instead of investing in physical assets. So, the planned investment decreases. Conversely, if the interest rate of bonds is relatively low, then the opportunity cost for companies to invest is also relatively low. Their investment returns from physical assets might be higher than the interest rate of bonds, causing the planned investment to go up. In short, interest rate bears important effect on planned investment, where the latter is negatively correlated to the former: $I = I(i)$ and $I'(i) = dI/di < 0$. That is, when the interest rate rises (respectively, falls), the planned investment decreases (resp., increases).

Assume that the price level is fixed. When the domestic interest rate rises, the deposit of domestic currency is more attractive than that of foreign currency, the domestic currency deposit is more valuable than foreign currency deposit, causing the exchange rate to go up. Increasing interest rate leads to rising value of the modestic currency, which in turn causes the exchange rate to go up. A consequence of such chain reactions is that the prices of domestic products go up in comparison with the foreign products so that export falls, while import rise, leading to reduced net export. So, it can be seen that the net export is negatively correlated to interest rate: $NX = NX(i)$ and $NX'(i) = dNX/di < 0$.

After learning the effect of interest rate on investment and net export, we see that increasing interest rate causes declines in planned investment and net export so that the aggregate demand $Y^d = C + I + G + NX$ falls. Its consequence is the equilibrium level of the total output decreases. Figure 5.13 shows that when the interest rate rises

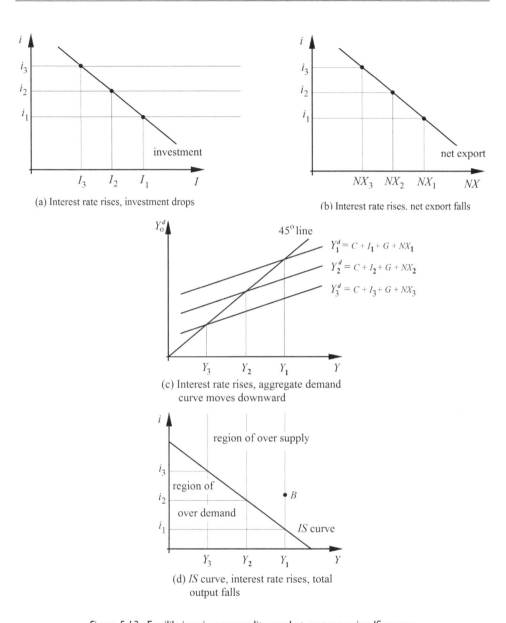

Figure 5.13 Equilibrium in commodity market, as seen using IS curves.

from i_1 to i_2 and then to i_3, how the equilibrium level of the total output changes: The planned investment spending drops from I_1 to I_2, and then to I_3, Figure 5.13(a); the net export falls from quantity NX_1 to NX_2, and then to NX_3, Figure 5.13(b); the aggregate demand curve moves from position Y_1^d down to Y_2^d, and then further down to Y_3^d, Figure 5.13(c); and lastly, the total output slides from Y_1 to Y_2, and then further down to Y_3. The equilibrium total output varies with interest rate. The trajectory of the consequent point (X, Y) of the interest rate and the total output is known as the IS curve, Figure 5.13(d).

The IS curve is an equilibrium curve of the product market. It reflects the law of how the equilibrium of the product market varies with interest rate, and determines the relationship between the equilibrium total output Y and the interest rate i: $Y = Y(i)$ and $Y'(i) = dY/di < 0$. That is, the equilibrium total output is negatively correlated to interest rate and satisfies the following:

1   When the point, consisting of the interest rate and the total output of the economy, falls on the IS curve, the product market is in equilibrium;
2   When the point, consisting of the interest rate and the total output of the economy, falls below the IS curve, Figure 5.13(d), it means that the aggregate demand corresponding to this interest rate and the level of total output is greater than the total output so that there is excess demand. In this case, in order for the product market to be in equilibrium, the level of total output has to be lifted up to reach the IS curve.
3   When the point, consisting of the interest rate and the total output of the economy, falls above the IS curve, such as the location of point B in Figure 5.13(d), it means that the aggregate demand, corresponding to the interest rate and the level of total output, is smaller than the total output so that there is excess supply. In this case, in order for the product market to be in equilibrium, the level of total output has to be lowered to fall on top of the IS curve.

Just as how the IS curve is derived from the equilibrium condition for the product market, from the equilibrium condition of the product market, we can also derive the LM curve. Specifically, when the level of income Y is given, staring off from Keynes' theory of liquidity preference and the equilibrium condition of the money market, the equilibrium interest rate that corresponds to this level Y of income $i = i(Y)$ can be determined.

The theory of liquidity preference claims that the actual demand for money of the public is determined by the income Y and interest rate i, and this demand is correlated to the income positively and to the interest rate negatively. Keynes' demand for money function is $M_d/P = f(Y, i)$, for details see previous Chapter 4. Because the total income is equal to the total output, the demand for money is positively correlated to the total output. For any fixed level of total output Y, let $M_d(Y)$ stand for the demand for money function at this level of total output. Evidently, this demand function only depends on the interest rate i, and is slanted to the lower right hand side.

Figure 5.14 shows the process of how the LM curve is determined from the equilibrium condition of the money market: When the level of total output rises from Y_1 to Y_2, and then to Y_3, then demand for money curve moves up from position M_1^d to position M_2^d, and then to position M_3^d. Correspondingly, the equilibrium interest rate increases from i_1 to i_2, and then to i_3, Figure 5.14(a). Therefore, the trajectory of the moving point (i, Y), consisting of the equilibrium interest rate and the total output, is the LM curve, Figure 5.14(b). Its functional form can be written as $i = i(Y)$.

The LM curve reflects the equilibrium of the money market, determines the relationship between the equilibrium interest rate and the total output. When the point, consisting of the interest rate and the total output, falls on the LM curve, the money

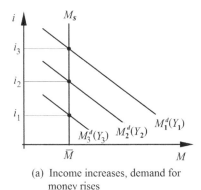

(a) Income increases, demand for
money rises

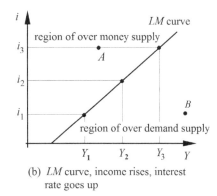

(b) *LM* curve, income rises, interest
rate goes up

Figure 5.14 Equilibrium in money market, as seen using *LM* curves.

market is in equilibrium. When the point, consisting of the interest rate and the total output, falls in the region on the left to the *LM* curve, such as point *A* in Figure 5.14(b), the interest rate is higher than the equilibrium rate at that level of total output, indicating that there is an excess money supply.

When the point, consisting of the interest rate and the total output, falls in the region on the right to the *LM* curve, such as point *B* in Figure 5.14(b), the interest rate is lower than the equilibrium rate at that level of total output, indicating that there is an excess demand for money.

From how the *LM* curve is determined, it follows that when the total output increases, the demand curve of money moves upward, causing the interest rate to rise accordingly. So, the *LM* curve is slanted upward to the right hand side. That is, the equilibrium interest rate is positively correlated to the output: $i = i(Y)$ and $i'(Y) = di/dY > 0$.

By placing the *IS* curve and the *LM* curve, as constructed above, in one graph, Figure 5.15, we obtain the *IS-LM* model that determines simultaneously the interest rate and the total output, where the point consisting of the interest rate and the total output should fall on the *LS* curve as well the *LM* curve. So, the intersection of the *IS* curve and the *LM* curve gives the equilibrium interest rate i^* and the equilibrium total output Y^*, where the money market and the product market are in equilibrium at the same time. When the economic state is away from the intersection point, either the money market or the product market, at least one of these markets, is not in equilibrium. In this case, adjustment will appear so that the economy will move closer to the intersection point. Only at point *E* (Figure 5.15) both the product market and the money market are in equilibrium, where the interest rate is i^* and the total output is Y^*. If the economy falls at any other point, then the market force will act to push the economy toward the equilibrium point *E*.

Although the *IS-LM* model indicates that the economy moves toward a certain level of total output Y^*, there is no reason to believe that at this level of total output sufficient employment would be achieved. If the unemployment rate is too high, then

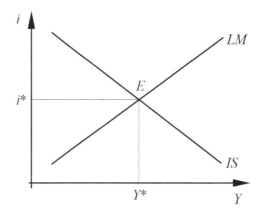

Figure 5.15 Simultaneous determination of interest rate and total output.

the government might try to reduce the unemployment by increasing the total output. The *LS-LM* model well implies that by applying monetary and fiscal policies the government could materially increase the total output in order to achieve the goal of reducing unemployment.

5.4   RISK STRUCTURE OF INTEREST RATES

What has been considered in the previous three sections is on how the interest rate is determined. In fact, there are many kinds of bonds, each of which carries a different interest rate. Analyzing the relationship between various interest rates, and understanding why the interest rates of different bonds are different will provide us a holistic comprehension about interest rates. That is the task of this and the following section. Doing so can also help business firms, banks, insurance companies, and individual investors decide on which bonds to sell and which bonds to buy. This section looks at why even bonds of the same terms of maturity still can bear different interest rates. The relationship between the different interest rates of bonds of the same terms of maturity is known as the risk structure of interest rates.

For bonds of the same maturity terms, variations in their interest rates appear to have the following important characteristics: For any given period of time, different bonds bear different interest rates and the differences between the interest rates vary with time. Then what causes this phenomena to appear? Based on their studies, economists discover that these three factors as risk, liquidity, and rules of income tax play the decisive role of formulating the risk structure of interest rates.

The bond issuer could default, meaning that he cannot afford to pay either the interest or the face value at maturity. That is an inherent risk of bonds. It affects the interest rate of bonds. Companies that suffer from major losses might have to delay the payments of their bonds' interest. The risk of these companies' bonds is very high. Conversely, governments can always pay off their debts either by increasing the tax revenue or by using the method of printing paper money. So, government bonds

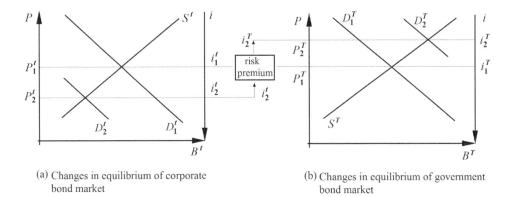

(a) Changes in equilibrium of corporate bond market

(b) Changes in equilibrium of government bond market

Figure 5.16 Effect of default risk of corporate bonds on interest rate.

generally do not have any risk of default. This class of bonds is known as bonds of default risk free. The difference between the interest rates of bonds with and without default risk is known as the risk premium, which is the additional interest to be earned on certain risky bonds. By applying the supply and demand analysis of bonds, it can be concluded that those bonds with default risk generally have positive risk premium, and why the riskier bonds are, the greater the risk premiums are.

In order to observe the impact of default risk on the interest rate of bonds, let us first look at the supply and demand graphs of bonds either without or with default risk. Figure 5.16(a) depicts the supply and demand curves of a company's long-term bond (that possesses default risk), while Figure 5.16(b) shows the supply and demand curves of a treasury bond (that does not have any default risk). Assume that at the start these two classes of bonds do not have any default risk (speaking generally, at the initial insurance and listing bonds do not have any default risk, because companies with low credit ratings are not allowed to be listed and to issue new bonds). That is, at the start, these two classes of bonds have the same attributes (the same risk and the same term), the same initial equilibrium price, and the same initial equilibrium interest rate. So, the risk premium of the company bond is zero. Let i_1^c and P_1^c be the initial equilibrium interest rate and the equilibrium price of the company bond, and i_1^T and P_1^T the initial equilibrium interest rate and the equilibrium price of the government bond. Then we have $i_1^c = i_1^T$ and $P_1^c = P_2^T$ so that the risk premium is zero. That is, $i_1^c - i_1^T = 0$.

Assume that in its business operations the company later on suffers from a heavy loss so that the chance for the company to default is increased. That is, the default risk of the company bond goes up and the expected yield drops, making the yield of the company bond uncertain. So, comparing to the Treasury bond that has no default risk, the expected yield of the company bond is relatively low and the risk is relative high so that fewer investors would be willing to purchase the bond, leading to decreasing demand. Hence, the increasing default risk of the company bond makes the demand curve for the company bond move to the left. Because all other conditions stay the same, the position of the supply curve is fastened so that the equilibrium price of

Table 5.3 Credit ratings of bonds by *Moody* and *Standard and Poor*.

Moody's rating	S & P's rating	Meaning of rating	Classification
Aaa	AAA	Highest quality (minimum default risk)	Investment grade
Aa	AA	High quality (low default risk)	Investment grade
A	A	Above average quality	Investment grade
Baa	Bbb	Average quality	Investment grade
Ba	Bb	Below average quality	Junk bond
B	B	Speculative quality	Junk bond
Caa	Ccc, Cc	Possibly default	Junk bond
Ca	C	Highly speculative	Junk bond
C	D	Worst quality (maximum default risk)	Junk bond

the company bond drops, the equilibrium interest rate rises. At the same time when the equilibrium interest rate of the company bond changes, the expected yield of the default risk free Treasury bond rises relatively, the relative risk becomes smaller so that more investors are willing to purchase this bond, leading to increasing demand and the demand curve moving to the right. The consequence is that the equilibrium price of the Treasury bond goes up, and the equilibrium interest rate goes down.

Figure 5.16 shows the changes in the equilibrium prices and equilibrium interest rates of the company bond and the Treasury bond. After the company suffers from heavy losses, the demand curve of the company's bond moves from the initial position D_1^t leftward to the position D_2^t, the equilibrium price drops from the initial P_1^t to P_2^t so that the equilibrium interest rate rises from the initial i_1^t to i_2^t. At the same time, the demand curve of the Treasury bond is displaced rightward from the initial position D_1^T to position D_2^T, the equilibrium price rises from the initial P_1^T to P_2^T so that the corresponding equilibrium interest rate drops from the initial i_1^T to i_2^T. So, the risk premium of the company's bond rises from the initial zero $(i_1^t - i_1^T = 0)$ to the positive value $i_2^t - i_2^T$ (because $i_2^t > i_1^t$ and $i_2^T < i_1^T$). It can be concluded that all bonds with default risk always carry positive risk premium, and the premium rises with the increase in the default risk.

Because of such important impact of the default risk on the risk premium of default risk bonds, bond purchasers want to know before they buy the company bond indeed how much is the chance for the company to default. *Moody* and *Standard and Poor* are two main investment consulting firms that provide credit rankings of different companies' bonds according to their chances of defaulting on their promises. These consulting firms offer the relevant information on default risks to bond purchasers. Table 5.3 describes the credit ratings these two firms provide to the investing public, where bonds with ratings at or above Baa (Bbb) are considered bonds that have relatively low default risk, known as investment grade bonds; and bonds with credit ratings below Baa (Bbb) are considered as having relatively large default risk, known as junk bonds.

Case study: October 19, 1987, known as the "black Monday", bears great impact on the bond market. It lets many investors start to wonder whether or not the companies that issue junk bonds with low credit ratings are financially sound. The default risk of junk bonds has increased so that no matter what interest rates these companies promise to pay, their bonds are no longer as welcomed as before. Therefore, the demand for junk bonds decreases, and the demand curve moves leftward. That implies that the price of junk bonds falls, and the interest rates rise. What actually happened is indeed like this. On that particular Monday, although the interest rate of American junk bond market rapidly rose 1 percentage point, the default-risk free Treasury bonds of American government became more attractive to the investors, making their demand curve move to the right. This phenomenon is referred by some analysts to as rushing to the high-quality bonds, which helped to lower the interest rate of the government bonds by one percentage point. That is, the difference between the interest rates of junk bonds and the government Treasury bonds was increased to two percentage points, making the original 4% before the black Monday rapidly increased to 6% after that Monday. This real-life scenario shows how default risk bears important effect on risk premium and in turn on interest rate.

From the previous theoretical analysis and the case study we can obtain the following important fact: Default risk is one of the important factors that cause variations in interest rates; the risk premium placed on the bonds with default risk represents the compensation for taking the default risk. The higher the default risk is, the greater the risk premium, and so the greater the difference between the interest rates of bonds.

Another important factor that affects the interest rates of bonds is the bonds' liquidity. The more liquid an asset is, the more the asset is welcomed by the public. In the United States, Treasury bonds are most widely traded. They can be most easily purchased or sold with relatively low costs. In comparison when in emergency, company bonds will be difficult to sell. That explains why the cost of selling company bonds is higher. That is, Treasury bonds have the most liquidity, the amount of transactions of any company bond is smaller than that of Treasury bonds and the liquidity of the former is smaller than that of the latter.

The supply and demand analysis of bonds can be used to explain how decreasing liquidity could affect the interest rate of the bond. To this end, let us provide a different set of explanation for Figure 5.16. Assume that Figure 5.16(a) describes the market situation of a company bond, while Figure 5.16(b) depicts the market supply situation of one particular Treasury bond. Assume that at the start the equilibrium prices of the company bond and the Treasury bond are the same and the equilibrium interest rates are also the same with all other attributes identical: They have the same term, the same liquidity, and the same risk. As the market tradings continue, the trading scope of the Treasury bond incessantly expands and goes way beyond that of the company bond. So, the liquidity of the Treasury bond is much higher than that of the company bond. That is equivalent to saying that the liquidity of the company bond is lowered comparatively. When the company bond's liquidity lowers comparatively, it causes the demand for the company bond to drop, the demand curve to move leftward from position D_1^t to D_2^t; however, the conditions with respect to the aspect of supply do not change. So, the equilibrium price of the company bond falls from the initially relatively higher level P_1^t to the relatively lower level P_2^t, and the corresponding interest rate rises from the initially lower level i_1^t to the relatively higher level i_2^t.

At the same time when the equilibrium of company's bond market changes, the equilibrium of the Treasury bond market also evolves. Because the liquidity of the Treasury bond is relatively high, the public's demand for this bond increases, causing the demand curve to move rightward from the initial position D_1^T to position D_2^T. So, the equilibrium price of the Treasury bond rises from the initially lower level P_1^T

to the relatively higher level P_2^T; and correspondingly, the interest rate drops from the initially relatively higher level i_1^T to the relatively lower level i_2^T.

At the initial time moment, we have $P_1^t = P_1^T$ and $i_1^t = i_1^T$. That is, there is no difference between the company bond and the Treasury bond. However, for now, because of the changes in their liquidities, the interest rate of the company bond goes up, while the interest rate of the Treasury bond goes down, leading to a difference $i_2^t - i_2^T$ in the interest rates, known as the liquidity premium. It can be concluded that the greater liquidity the Treasury bond has (comparatively, the smaller liquidity the company bond experiences), the greater the liquidity premium is, and the greater the interest rate difference is. So, the liquidity premium is a compensation for the smaller liquidity of the company bond.

The previous analysis indicates that the difference in interest rates of different bonds of the same term reflects only the default risks of the bonds, but also the difference in the liquidities of the bonds. So, such difference in interest rates can be decomposed into two parts: the risk premium and the liquidity premium. That is, a decrease in the demand of one bond causes that bond's interest rate to rise and the demand for the other bond to increase and the other bond's interest rate to fall, leading to the appearance of differences in interest rates of the bonds. Essentially, this difference in interest rates is caused by an increase in one bond's default risk and a decrease in the bond's liquidity.

From the analysis on the default risk and liquidity, we can see that in terms of two bonds, if the default risk of one bond is greater than that of the other bond, or the liquidity of the former bond is lower than that of the latter, then the interest rate of the former bond is higher than that of the latter bond. When this theory is employed to analyze municipal bonds and Treasury bonds, there has been an unsettled problem: the default risk of municipal bonds is higher than that of Treasury bonds, and the liquidity of municipal bonds is lower than that of Treasury bonds; however, for a long time it has been the case that the interest rate of American municipal bonds is lower than that of Treasury bonds. This fact disagrees with the previous theory. Next, let us focus on this puzzling problem.

Municipal bonds are not default risk free bonds. In the past some American states and local governments had defaulted on the bonds they issued, which happened during periods of time of economic crises. In the 1983 investigation of cases involving the public energy supply system in Washington State, some default behaviors were uncovered. Additionally, the trading range of municipal bonds is evidently not as wide as that of national debts with much smaller liquidity. If we only considered the situation from the angles of risk and liquidity, the interest rate of municipal bonds would be higher than that of national debts. However, the historical facts of at least 40 years point to the opposite: the interest rate of municipal bonds has always been lower than that of national debts. That forces us to consider the reasons behind the difference in interest rates of different bonds from other angles.

Note that there is an important distinction between municipal bonds and national debts. Up to this point we have not considered its impact on the interest rate. That is, the interest income of municipal bonds is waived from paying national and local taxes, while the interest income of national debts still has to pay tax. To this end, let us employ our imagination: Assume that you have entered a high-income class and for each dollar of additional income you need to pay 40% of income tax. If you hold a treasury bill of face value $1,000 with sales price $1,000 and interest $100. Then after the 40% income tax, your net income from the Treasury bill is merely $60. That is, the rate of after-tax return is 6%. Now, assume that instead of the national debt you hold a municipal bond of the same face value $1,000. Although the interest might only be $80, lower than that of the Treasury bill, your actual return is $80 (because the municipal bond is tax free), which is higher than the net income from the Treasury bill. It is quite clear that although the default risk of the municipal bond is higher and its liquidity is lower, you will still be willing to hold the municipal bond.

When analyzing the situation by using the supply and demand relationship, we can see that the income tax makes the expected yield of the national debts fall, and the expected yield of tax-free bonds rise. When expected yield of bonds rises, the demand for the bonds goes up, making the demand curve move to the right. Conversely, if the expected yield of one bond falls in comparison with another bond, then the demand for the former bond will drop, which makes the demand curve of this bond move to the left. Now, because of the tax-free treatment, the expected yield of municipal bonds goes up and the expected yield of the national debts goes down, the demand curve of municipal bonds moves to the right and the demand curve of the national debts moves to the left. That causes the price of municipal bonds to go up and interest rate to go down, while the price of national debts to go down and interest rate to go up. The resultant consequence is that the interest rate of municipal bonds is lower than that of national debts. In Figure 5.17, at the beginning, the equilibrium interest rate of a municipal bond is equal to that of a Treasury bill. However, the income tax imposed on the income of the Treasury bill causes the demand curve of the municipal bond to move from D_1^t rightward to D_2^t, the price to rise from P_1^t to P_2^t, and the corresponding interest rate to fall from i_1^t to i_2^t, while the demand curve of the Treasury bill moves from position D_1^T leftward to D_2^T, the equilibrium price falls from P_1^T to P_2^T, and the corresponding interest rate rises from i_1^T to i_2^T. So, there appears an opposite different in interest rates: $i_2^T - i_2^t$. More generally, we can say that imposing tax or increasing the tax rate lowers the interest rate of tax-free bonds, while increasing the interest rate of taxable bonds, causing the appearance of an opposite difference in the interest rates of tax-free and taxable bonds.

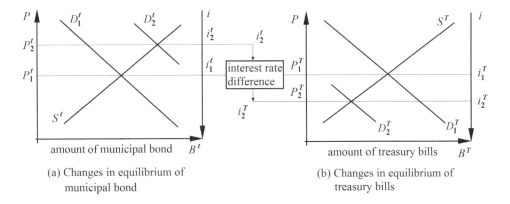

(a) Changes in equilibrium of
    municipal bond

(b) Changes in equilibrium of
    treasury bills

Figure 5.17 Effect of income tax on equilibrium interest rate of bond market.

Let us now summarize what is studied in this section. Such concepts as default risk, liquidity, and imposing the income tax on the earnings of bonds or increasing the existing tax rate explain the relationship of the interest rates of various bonds of the same terms. That is, they provide an explanation for the risk structure of bonds. The risk premium of a bond increases with the rise of the bond's default risk; the liquidity premium of a bond rises with the fall of the liquidity of the bond; and the interest rate of a bond that enjoys the tax-preferred treatment is lowered than that of a bond that is not given any tax preference.

5.5 TERM STRUCTURE OF INTEREST RATES

The term of a bond also affects its interest rate. Bonds with the same risk, liquidity, and tax treatment can also bear different interest rates due to the differences in their terms. The relationship of the interest rates of bonds of different terms is referred to as the term structure of interest rates. This relationship can be often described by using yield curves. A so-called yield curve means such a curve that connects the maturity interest rates of bonds of different terms assuming all other factors, such as risk, liquidity, and tax treatment are all the same. For example, the terms of the treasuries include 3 months, 6 months, 1 year, 2 years, 3 years, 5 years, 7 years, 10 years, and 30 years. These treasuries of different terms can be seen to have the same risk, the same liquidity, and tax treatment. Then, the interest rates of these treasuries of different terms can be connected into a curve, as shown in Figure 5.18, which is known as a yield curve of the treasuries. This curve describes the term structure of the treasuries. Figure 5.18 explains that at different time moment, the yield curve takes a different shape. The yield curve could be slant upward to the right, could also be slant downward to the right, and could indeed be a horizontal line. When the curve is slanted upward to the right, it means that the long-term interest rates are higher than the short-term rates; when the curve is slanted downward to the right, it represents the fact that the long-term rates are lower than the short-term rates; and when the curve is horizontal, it shows that the long-term rates are the same as the short-term rates. In reality, we often see yield curves that are slant upward to the right. But occasionally, we can also see yield curves that are slant downward to the right. In terms of yield curves, there are three

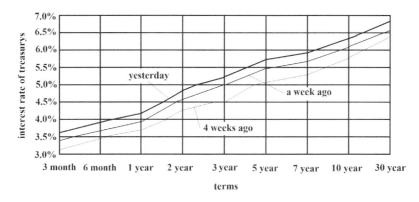

Figure 5.18 Yield curve of treasury bills.

important empirical facts: The interest rates of bonds of different terms fluctuate with time together; (2) if the short-term interest rates are lower, then the yield curve is slanted upward to the right; conversely, if the short-term interest rates are higher, then the yield curve is slanted downward to the right; and (3) yield curves are almost always slant upward to the right.

The term structure theory of interest rates explains not only why yield curves take different shapes at different times, but also the three empirical facts mentioned above. Because of this reason, there have been three kinds of theories or three different explanations for the relationship of interest rates of bonds with different terms: (1) the expectations hypothesis, (2) the market segmentation theory, and (3) the term choice theory. The expectations hypothesis can well explain the first two empirical facts but is incapable of illustrate the third. The market segmentation theory can well explain the third empirical fact, but is incapable of illustrating the first two. So, these two theories complement each other in terms of explaining the empirical facts the other one is not able to explain, while none of them could explain what the other theory could explain. The term choice theory combines both the expectation hypothesis and the market segmentation theory, and aims at provide a better explanation for the term structure of interest rates. It collects the characteristics of the previous two theories and offers a very good explanation for the third empirical fact. That makes the term choice theory well accepted.

In particular, the expectations theory is developed on the so-called proposition of expectations hypothesis: The interest rate of a long-term bond is equal to the average of all the expected interest rates of short-term bonds before the maturity of the long-term bond. For example, assume that people expect the average of short-term interest rates in the next five years is 12%. Then the expectations theory believes that the interest rate of the 5-year long-term bond is also 12%. If people expect the short-term interest rates would rise after 5 years so that the average of the short-term interest rates is 15% in the next 20 years, then the interest rate of the 20-year bond is also 15%. So, because for different time periods, such as future 5 years, 10 years, 20 years, or 30 years, people have different expected short-term interest rates, the interest rates of different terms, such as 5 years, 10 years, 20 years, or 30 years, will be different.

The proposition of expectations hypothesis is a key assumption. It assumes that the bond purchaser does not have any preference over any specific term and only decides on whether or not to purchase a bond based on its expected yield.

When the expected yield of a bond is lower than that of another bond, the public will no longer hold the bond with the lower expected yield. That situation can be seen as that the bond with higher expected yield replaces the other bond with lower expected yield. That is, bonds of different terms can be used to completely replace each other. The complete replacement causes the bond of higher expected yield to be used to replace the bond of lower expected yield. Because the public does not hold any bond of low expected yield, the expected yield of various bonds the public holds must be ultimately identical. Now, let us look at a real-life example by considering the following two bond investment strategies using $R.

1 Purchase one year bond. After the maturity, use the entire funds, consisting of principal and interest, to purchase more one year bond;
2 Purchase two year bond until its maturity.

Let i_t stand for today's interest rate of the one year bond, that is, the interest rate of the tth year, i^e_{t+1} the expected annual rate of following year, that is, the $(t+1)$th year, i_{2t} represent for today's interest rate of the two year bond, that is, the interest rate of the tth year. If the first strategy is employed, that is, purchase the one year bond for \$$R$. Then at the end of two years, the net yield of the investment is $R(1+i_t)(1+i^e_{t+1}) - R$. So, the net yield R_1 of applying the first strategy is

$$R_1 = R(1+i_t)(1+i^e_{t+1}) - R = (i_t + i^e_{t+1} + i_t i^e_{t+1})R$$

Because $i_t i^e_{t+1}$ is very small, it can be ignored so that we have $R_1 = (i_t + i^e_{t+1})R$. That is, the expected yield r_1 of purchasing one year bond two times in two years is $r_1 = R_1/R = i_t + i^e_{t+1}$.

If we apply the second strategy, that is, purchase two-year bond for \$$R$ and hold it until maturity. Then the net yield R_2 of the investment is

$$R_2 = R(1+i_{2t})^2 - R = (2i_{2t} + i^2_{2t})R$$

Because $(i_{2t})^2$ is very small, it can be ignored. So, $R_2 = 2i_{2t}R$. That is, the expected yield r_2 from purchasing the two-year bond is $r_2 = R_2/R = 2i_{2t}$.

If $r_1 < r_2$, then people would not apply the first strategy to invest and would not hold the one-year bond. Conversely, if $r_1 > r_2$, then people would not purchase the two-year bonds. So, as long as these bonds bear different expected yields, then the bond with higher yield will completely replace the other bond with lower yield, forcing the bond of the lower yield to retire from the market entirely. However, in reality these two bonds exist simultaneously in the market place; people hold not only the two-year bond, but also the one-year bond. That implies that the expected yields of these two bonds must be equal or close to each other: $r_1 = r_2$. So, it can be concluded that the interest of the two-year bond is equal to or very close to the average of the expected interest of the one-year bond: $i_{2t} = (i_t + i^e_{t+1})/2$. Similarly, the relationship between the interest rates of an n-year bond and one-year bond can be established as follows:

$$i_{nt} = (i_t + i^e_{t+1} + i^e_{t+2} + \cdots + i^e_{t+n-1})/n$$

where i_t stands for the current year's interest rate (the tth year), i^e_{t+k} the expected interest rate of the $(t+k)$ year, for $k = 1, 2, \ldots, n-1$, and i_{nt} the yield of the nth year at maturity, that is, the interest rate of holding the bond for n years until its maturity. For the expectations theory, this formula is accurate. It means that the interest rate of the one-year bond is equal to the mean of the expected interest rate at the mid-year of the one-year bond.

The proposition of expectations hypothesis well explains the reason behind the difference between the interest rates of bonds of different terms: People have varied expectations for the short-term interest rates of different future terms so that with the changes of the terms of bonds, the interest rates of bonds, which are equal to the mean of expected short-term interest rates, change. Additionally, the yield curve is slanted upward to the right means that the future short-term interest rate is expected to go up; when the yield curve is slanted downward to the right, it implies that the future short-term interest rate is expected to go down; when the yield curve is a horizontal line, it means that the future short-term interest rate is expected to stay constant. That very well explains why at different times the yield curve might take different shapes: The current expectation of the future short-term interest rate is different of that of yesterday, causing the shape of the yield curve to change.

Why are the interest rates of bonds of different terms change simultaneously with time? Based on is proposition, the expectations theory explains that when seen from the history, the short-term interest rate has such a characteristic: If today's short-term interest rate rises, then the future short-term interest rate has the tendency to go even higher. It is because of this characteristic that today's rising short-term interest rate will cause the position of the yield curve to move. That implies that the interest rates of bonds of different terms go up simultaneously. So, the expectations hypothesis well explains the first empirical fact.

For the second empirical fact, the explanation of the expectations hypothesis is as follows: If the current short-term interest rate is low, then people would generally expect the future short-term interest rate would go up so that the interest rate of long-term bonds, as the average of the expected short-term interest rates, would be higher than the current short-term interest rate, causing the yield curve to be slant upward to the right. Conversely, if the current short-term interest rate is high, then people would generally expect the future short-term interest rate to fall so that the interest rate of long-term bond, as the average of the expected short-term interest rates, would be lower than the current short-term interest rate, making the yield curve to be slant downward to the right.

So, it can be seen that the expectations hypothesis provides an elementary yet convincing explanation for the term structure of interest rates, attracting the interest of many economists. However, this theory could not plain the afore-mentioned third empirical fact. That is indeed unfortunate. In fact, according to the expectations theory, the future short-term interest rate could either rise or fall so that the average of the expected short-term interest rates should be fixed at a certain level. That implies that the yield curve should take the shape of a horizontal line. However, in reality what is often seen is that the yield curve is slanted upward to the right. So, the expectations theory could not explain the third empirical fact successfully.

Now, let us look at the market segmentation theory of the term structure of interest rates. This theory maintains that the bond's interest rate of each term is determined by the supply and demand of the bond; the market of bonds of each fixed term should be treated as completely independent from the rest of the bond market. That is to say, bonds of different terms are mutually independent from each other and cannot replace each other. Evidently, this convention is totally opposite of the expectations hypothesis, where bonds of different terms can completely substitute for each other. Hence, the expectations theory and the market segmentation theory represent two extremes: The former believes the total replacement of bonds of different terms, while the latter maintains that bonds of different terms cannot replace each other at all. The reason why this theory proposes this convention of non-replacement is that in the mind of the investor he has a definite time period to hold his bond; that makes him have a strong preference over a certain specific term. That is, the investor prefers the bonds of this particular term over bonds of any other term. If the term of a bond coincides with the preferred term of the investor, then the investor would have a preference toward this term and the expected yield of the bonds of this particular term would be equal to the interest rate (the maturity rate) of the bonds. So, the yield is definite without any risk.

Because the investor is particularly interested in a certain term, he only cares about the expected yield of the bonds of his preference. For example, investors who hold bonds for very short-terms are willing to purchase short term bonds so that they only care about the interest rates of short-term bonds. Conversely, if a person accumulates funds for his young child to go to college after ten years, then in his mind he holds 10 years as his term and he will be willing to purchase 10 year bonds so that he will be more interested in the yield of the 10 year bonds.

The market segmentation theory provides the following explanation for why yield curves take different shapes during different time periods: The interest rates of bonds of various terms are determined by the supply and demand of the bond market; because the supply and demand situation of each bond is different in different time period, that is why the corresponding interest rates are different. That explains why the yield curves in different time periods take on different shapes. That is, the reason behind the different shapes of yield curves is the varied situation of supply and demand.

The explanation of the market segmentation theory for the afore-mentioned third empirical fact (the typical yield curve is slanted upward to the right) is the following: In the ordinary circumstances, people prefer bonds of relatively short terms, and relatively small risk. So, the demand for short bonds is relatively big, while the demand for long-term bonds is relatively small. According to the theory of how the supply and demand determine the interest rate of bonds, it follows that the relatively big demand causes the bond prices to go up and the interest rate to go down, while the relatively small demand causes the prices of bonds to decline and interest rate to rise. So, the interest rates of short-term bonds are relatively low, and the interest rates of long-term bonds are relatively high. That of course makes the yield curve slant upward to the right. Hence, this explanation out of the market segmentation theory seems to be very reasonable. However, the market segmentation theory cannot illustrate the first and the second empirical facts. Because this theory separates the markets of bonds of different terms completely from each other, changes in the interest rate of the bond of one term would not affect the interest rate of the bond of another term. So, it cannot explain why the interest rates of all bonds of different terms fluctuate together (the first empirical fact). Additionally, the market segmentation theory does not provide any description for the relationship between the supply and demand of long-term bonds and those of short-term bonds. It does not explain how changes in the supply and demand of long-term bonds affect the changes in the interest rates of short-term bonds. So, it cannot illustrate the second empirical fact: When the short-term interest rates are low, the yield curve is slanted upward to the right; when the short-term interest rates are high, the yield curve is slant downward to the right.

From the discussions above, it can be seen that both the expectations theory and the market segmentation theory provide explanations for one empirical fact that the other theory cannot explain, and each of these theories cannot elucidate the empirical fact the other theory can enlighten. So, these two theories have their individual strengths and weaknesses. They represent two extremes. A seemingly logical method is to combine the characteristics of these two theories together. That is exactly the idea that is behind the theory of term choice and the closely relevant theory of liquidity premium.

By relating the characteristics of the expectations theory and the market segmentation theory, the term choice theory corrects the practice of going to extremes of these theories. It contemplates that although bonds of different terms cannot be complete substitutes of each other, as what the expectations theory claims, they do possess a certain degree of mutual replacability; and at the same time, bond investors do have certain kinds of preferences over bonds' terms. So, from the conventions of the term choice theory, it follows that although the bond market of each individual term is almost an independent market, it cannot really be an operationally independent market from all bond tradings of other terms. The bond markets of different terms are closely intertwined in which there are replacability and term preferences. That is to say, the expected yield of one bond can affect the expected yields of other bonds of various terms. What investors are more interested in is the yield situations of the bonds of their preferred terms. The simultaneous existence of both replacability and term preference means that investors are also interested in knowing the yields of the bonds that they do not prefer. By summarizing the various points, we have that investors do not allow the expected yield of one bond to be too far off from that of another bond. However, because they do prefer some bonds over others, for those bonds beyond their preferences, only when the expected yields of these bonds are much better, they would potentially consider purchasing these not preferred bonds.

For example, assume that an investor prefers short-term bonds over long-term bonds. Then, even though the expected yields of short-term bonds are slightly lower than those of long-term bonds, the investor will be more willing to hold the short-term bonds instead of long-term bonds. Only when the expected yields of the long-term bonds are much higher than those of short-term bonds, he could potentially consider in invest in the long-term bonds. It means that in order to attract investors to purchase long-term bonds, on the basis of the interest rates of long-term bonds, as implied by the expectations theory, term premiums have to be used. That is, investors have to be compensated for holding bonds they do not prefer to have. So, the formula of bond's interest rate, as proposed in the expectations theory, should be rewritten as follows:

$$i_{nt} = \frac{i_t + i_{t+1}^e + i_{t+2}^e + \cdots + i_{t+n-1}^e}{n} + k_{nt} \tag{5.13}$$

where k_{nt} stands for the term premium in the tth year for the bond of n-year term, that is, in the tth year (the current year), in order for investors to purchase the bond of n-year term, additional interest rate has to be paid to the investors. That is a positive value; it represents the compensation paid to the investors to purchase bonds they do not prefer; and the longer the term, the more compensation is necessary. So, with the bond's term n the value of k_{nt} increases.

The theory of liquidity premium also proposes a similar formula. In a certain sense, the modification to the proposition of expectations hypothesis made by the theory of liquidity premium is more straightforward that that made by term choice theory. The theory of liquidity premium maintains that the liquidity of long-term bonds is obviously lower than that of short-term bonds. In order to encourage an investor to purchase long-term bonds, he is compensated with a positive liquidity premium for his holding of bonds of relatively low liquidity. According to this reasoning, the theory of liquidity premium also derives a formula for determining the interest rate that is similar to that in equ. (5.13). In particular, the theory of liquidity premium assigns the symbol k_{nt} in equ. (5.13) the meaning of liquidity premium, believing that the longer the term of a bond is, the lower the liquidity of the bond so that the greater the liquidity premium (compensation) should be. That is, the value of k_{nt} increases as the term n of the bond becomes longer.

In order to help with the understanding of the term choice theory and the liquidity premium theory, as expressed by equ. (5.13), let us look at an example. Assume that in the coming 5 years, the expected interest rates for a one-year bond are respectively 5%, 6%, 7%, 8%, and 9%, and an investor prefers short-term bonds. That implies that the term premiums for bonds of 1–5 year terms should respectively be 0.00%, 0.25%, 0.50%, 0.75%, and 1.00%. So, from equ. (5.13), it follows that the interest rate of the 2-year bond is

$$(5\% + 6\%)/2 + 0.25\% = 5.75\%$$

The interest rate of the 3-year bond is

$$(5\% + 6\% + 7\%)/3 + 0.5\% = 6.5\%$$

The interest rate of the 4-year bond is

$$(5\% + 6\% + 7\% + 8\%)/4 + 0.75\% = 7.25\%$$

And the interest rate of the 5-year bond is

$$(5\% + 6\% + 7\% + 8\% + 9\%)/5 + 1\% = 8\%$$

By comparing these results with those obtained from using the expectations theory, we discover that because the investor prefers short-term bonds, the yield curve is steeper and is slant upward more obviously than the yield curve of the expectations theory.

Now, let us look at how the term choice theory and the liquidity premium theory explain the term structure of interest rates and the afore-mentioned three empirical facts.

Both the term choice theory and the liquidity premium theory combine the strengths of the expectations theory and the market segmentation theory. They provide a much better explanation for why yield curves take different shapes at different times. On the one hand, the expectation of the future short-term interest rate changes with time. Today's expectation is very likely different from that of yesterday so that the average of expected short-term interest rates changes with time. On the other hand, the interest rate of each individual bond is related to the relationship between the market supply and demand of the particular bond, while the supply situation of the bond is different from one time moment to another so that the interest rate is also different from one time to another. By combining these two aspects of reasons, we see that the yield curve takes different shapes at different times.

As for the first empirical fact, which is the problem of why the interest rates of bonds of different terms changes with time simultaneously, the term choice theory and the liquidity premium theory provide a moch plausible explanation. First of all, today's rise in short-term interest rate can cause future short-term interest rate to go even higher. Because of this reason, today's rise in short-term interest rate makes the future average short-term interest rate of each time period to increase so that the position of the yield curve moves upward. That implies that the interest rates of different terms rise simultaneously. At the same time, because of the shortness of their terms of short-term bonds and the strong liquidity enjoyed by short-term bonds, people always like to hold these short-term bonds. That motivates capital demanders to raise funds through short-term bonds, causing the supply of short-term bonds to increase, the supply curve to move rightward, prices to drop, and interest rates to go higher. That is the main reason for why today's interest rate rises. So, the equilibrium quantity of short-term bonds increases in the bond market. That is, the rising short-term interest rate of the current day brings forward an increasing supply of short-term bonds. And because there is a certain degree of replacability between short-term bonds and long-term debts, when the amount of purchase increases in the short-term bond market, the demand for long-term bonds will fall, the demand curve will move to the left so that the market price of long-term bonds will fall and market interest rate will go up. In short, rising short-term interest rate today causes the expected future short-term interest rate to go up, that in turn causes the demand for long-term bonds to drop, investerest rate to rise. When these two effects are combined, the interest rates of long-term bonds become higher than before. So, the interest rates of bonds of different terms change simultaneously, going up and down together.

For the second empirical fact, the term choice theory explains as follows. If currently the short-term interest rate is low, then people would generally expect the interest rate of the future short-term bonds to rise so that the average expected short-term interest rate of each individual time period would be higher than the current short-term interest rate. Together with the positive term premium, the yield curve is slant upward to the right. Conversely, if the current short-term interest rate is too high, it lets people foresee large magnitude drops in future short-term interest rates. So, the average expected short-term interest rate for a long time will be greatly lower than the current short-term interest rate. That makes that even if there is a positive term premium, the yield curve will be slant downward to the right.

For the third empirical fact (the typical yield curve is slant upward to the right), the explanation given by the term choice theory is that each investor prefers short-term bond so that as the term goes longer, the term premium (the liquidity premium) will increase. That makes the interest rate of long-term bond become greater with the increase of the term. So, under the ordinary circumstance, the yield curve is slant upward to the right, because under the ordinary circumstance, according to the explanation of the expectations hypothesis, the average expected short-term interest rate should be fixed at a constant level.

As for the accidental appearance of yield curves that are slant downward to the right, how does the term choice theory provide an explanation? To this end, both the term choice theory and the liquidity premium theory believe that at some time periods, the expected future short-term interest rate could fall drastically. That makes the average expected short-term interest rate much smaller than the current short-term interest rate. So, even with the positive term premium (the liquidity premium) added, the long-term interest rate would still be lower than the current short-term interest rate. So, the reason for yield curves to occasionally be slant downward to the right is a drastical fall in the expected future short-term interest rate.

The one feature of both the term choice theory and the liquidity premium theory that attract people's attention is that by simply observing the slope of the yield curve,

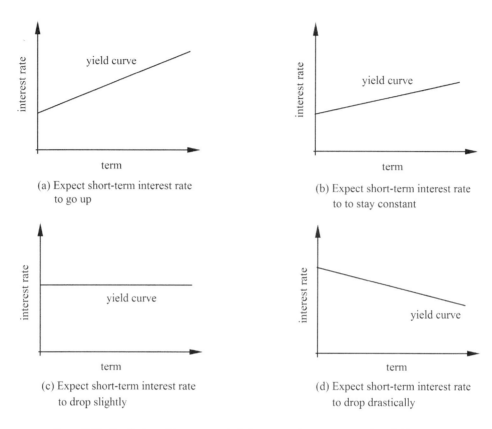

Figure 5.19 Prediction of future trend of short-term interest rate using yield curves.

one can predict the tendency of development of the future market interest rate of short-term bonds, Figure 5.19.

1   If the yield curve is slant obviously upward to the right, it means that the future short-term interest rate is expected to go higher;
2   If the yield curve is slant upward to the right only slightly, it means that the future short-term interest rate is expected to stay constant;
3   If the yield curve is a horizontal line, it means that the future short-term interest rate is expected to drop slightly;
4   If the yield curve is slant slightly downward to the right, it means that the future short-term interest rate is expected to fall majorly.

After checking with the term structure of interest rates of the 1980s in America, some scholars start to doubt the validity of using the slope of yield curve to predict the future tendency of short-term interest rate. They discover that the difference between long-term interest rate and short-term interest rate is not always helpful in the prediction of future short-term interest rate. However, since 1984, with more deliberate and better designed measurements studies find that the term structure of interest rates, that is, the yield curves, indeed contains information about the future development tendency of interest rates (Fama, 1984; Fama and Bliss, 1987; Campbell and Shiller, 1987; 1991).

5.6 MARKET BEHAVIORS AND RESPONSE TO DIFFERENT-SCALE INVESTMENT OPPORTUNITIES

In the previous two sections, we learned that interest rates have both risk structures and term structures. However, according to the analysis of the systemic yoyo model, such structures of interest rates have to be consequences of interactions of different economic forces. Based on this understanding, in this section, we look at how the financial market reacts to these structures. To this end, let us look at two different categories of investment opportunities: riskier opportunities and safer opportunities. Throughout this section, we assume that there is no discount of the future over time. Assume that in time period 1, a commercial company commits I_s of investment to a safer investment opportunity; and in time period 2, the fundamental value of the investment opportunity is v_s for the company when the investment opportunity is fully completed and the benefit of the completed investment is materialized in the marketplace. So, the expected profit from embarking on this safer opportunity is given as follows:

$$(v_s - I_s R_s)P_s, \tag{5.14}$$

where $R_s > 1$ stands for the gross interest spent on the total investment amount I_s, and $P_s \approx 1$ the probability for the safer investment opportunity to be successful.

Similarly, assume that in period 1, the company invests as much as $I_L >> I_s$ for a riskier investment opportunity. In time period 2, the investment is completed and its benefit is fully materialized so that the fundamental value the investment creates for the company is v_L. So, in this period, the expected profit from engaging in this investment opportunity is

$$(v_L - I_L R_L)P_L, \tag{5.15}$$

where $R_L > R_s > 1$ is the gross interest spent on the total investment I_L and $P_L \approx 0$ the probability for the riskier investment opportunity to be successful for the reason that riskier opportunities tend to involve many unexpected factors that may turn out to be detrimental to the success of the investment opportunities. The reason why $R_L > R_s$ is that the latter kind of investment opportunity tends to be much riskier than the former kind of opportunity so that the lender of funds charges a much higher rate of interest.

Now, the second kind investment opportunity is lots riskier with much greater return if successful than the first kind investment opportunity. So, if the company decides on investing in the riskier opportunity, the return has to be proportionally much higher than that of the first kind opportunity. So, we have the following equation:

$$k(v_s - I_s R_s)P_s = (v_L - I_L R_L)P_L. \tag{5.16}$$

where k is the constant of proportionality. By rewriting this equation, we obtain:

$$kv_sP_s = v_LP_L + kI_sR_sP_s - I_LR_LP_L. \tag{5.17}$$

If $P_s = \ell P_L$, for some large real number $\ell > 1$, equ. (5.17) can be rewritten as follows:

$$k\ell v_s = v_L + k\ell I_sR_s - I_LR_L \quad \text{or} \quad k\ell(v_s - I_sR_s) = v_L - I_LR_L. \tag{5.18}$$

Equ. (5.18) provides an analytic explanation for why some decision-makers like to take on riskier investment opportunities, because the success of just one such investment opportunity would produce as much financial profit as $k\ell$ many safer investment opportunity completed one by one. Considering the meanings of the constants k and ℓ, the product $k\ell$ can potentially be a really large number. In particular, it means that if one is successful with riskier investment opportunity just once, twice, or a few times, he would strike it huge in the business world. On the other hand, if one is risk averse and takes on safer investment opportunities only, it may very well take him forever to build his business to a certain respectable magnitude.

The definition of safer and riskier investment opportunities is asset-dependent. That is, when companies have different market capitalizations, their definitions of safer and riskier investment opportunities change from one company to another. A specific investment opportunity could be considered safe for one company and immensely risky for another company. So, equ. (5.18) provides an explanation for why most new start-ups fail because of a lack of funds, since due to their limited financial resources, almost all investment opportunities new start-ups take on would be considered riskier. In this case, if they do not make it with just one opportunity, the companies would be over permanently.

Now, assume that a company has a choice of either taking on relatively safe investment opportunities or one relatively riskier investment opportunity in time period 1. If the company chooses the option of safe opportunities, then its stock is traded at p_s ($<, =$, or $> v_s$) in time period 1, where v_s is the fundamental share-value of the company. In this time period, an investor buys $n(p_s)$ shares of the stock at the market value p_s a share. His total cost of investment is $I_s = n(p_s)p_s$. Assume that in time period 2, the trading price equals the fundamental value v_s. So, the profit of this investor is given as follows:

$$n(p_s)v_s - I_sR_s = \frac{I_sv_s}{p_s} - I_sR_s = I_s\left(\frac{v_s}{p_s} - R_s\right), \tag{5.19}$$

where as before $R_s > 1$ is the gross interest spent on the total investment I_s.

On the other hand, if the company takes on the choice of one risky investment opportunity, the same investor buys in period 1 $n(p_L)$ shares of the company stock at p_L a share, where $p_L <, =$, or $> v_L$ with v_L being the fundamental share-value of the company stock. So, the total cost of investment to the investor is $I_L = n(p_L)p_L$. In time period 2, assume that the investment opportunity is completed with its consequence known so that the fundamental share-value v_L is materialized. So, the total profit for the investor is given as follows:

$$n(p_L)v_L - I_L R_L = \frac{I_L v_L}{p_L} - I_L R_L = I_L\left(\frac{v_L}{p_L} - R_L\right), \tag{5.20}$$

where $R_L > R_s > 1$ is the gross interest spent on the total investment I_L.

If the investor wants to produce more return on his investment of the riskier investment-opportunity option, then from equs. (5.19) and (5.20), we have

$$I_s\left(\frac{v_s}{p_s} - R_s\right) < I_L\left(\frac{v_L}{p_L} - R_L\right). \tag{5.21}$$

From the assumption that $I_s = I_L$, meaning that the investor allocated the same amount of funds to each option of investment, we have

$$\frac{v_s}{p_s} < \frac{v_L}{p_L} - (R_L - R_s) < \frac{v_L}{p_L}. \tag{5.22}$$

From the assumption that the safer choice of investment opportunities contains a huge number of relatively safe investment opportunities completed one by one so that the totality of the these investment opportunities does not become a riskier investment, we can take $\frac{v_s}{p_s} \approx 1$. That is, the market read on the potential values of the relatively safe investment opportunities in time period 1 is very close to that of the fundamental value materialized in period 2. So, we have from equ. (5.22) that

$$\frac{v_L}{p_L} > 1. \tag{5.23}$$

This end implies that for the option of taking on one risky investment opportunity, the marketplace tends to underestimate the benefit the potential success the riskier investment opportunity would create for the company. That is, investors value a large number of relatively safe projects completed one by one more than one single riskier project.

The modeling approach, as presented in this section, has been employed in different ways in (Lin, 2008; Lin and Forrest, 2011).

5.7 SOME CONCLUDING WORDS

What is presented in Section 5.6 suggests that interest rates are indeed a product produced as a consequence of the interaction of different economic forces. These interactions not only determine the interest rates, but also provide explanations for why interest rates fluctuate over time and influence the total domestic output.

Chapter 6

Monetary policy: Another factor influencing monetary supply and demand

In the development of the market economy, to avoid certain undesirable economic difficulties national macro regulation has become gradually visible and has been playing an ever increasingly important role, where the focus of macro regulation is about financial regulation. The so-called financial regulation means that through introducing and implementing monetary policies by the central bank, the nation materializes its adjustment and control over the entire spectrum of socio-economic activities. The effect and influence of financial regulations are not simply about some aspects of the socio-economic activities. Instead, they permeate each and every corner of the national economy. By making use of the particular position occupied by the central bank and the tools of monetary policies available to the central bank, the nation regulates such relevant variables as the scale of credit, the quantity of money in circulation, etc. By varying the money supply, the government effectively adjusts the overall level of economic activities so that the needs of national management and necessary intervention of social and economic activities are met. There are multiple standards for evaluating the goals and effects of financial regulation, including such comprehensive criteria as price stability, economic growth, sufficient employment, and balance in international payments. In order to materialize these goals, monetary policies often need to be employed jointly with fiscal policies. Because monetary policies are the core contents of financial regulation, this chapter will focus on monetary policies while discussing other relevant topics.

The chapter is organized as follows. In Section 6.1, we will look at the tasks, contents, and goals of monetary policies. In Section, we focus on the medium-term goals of monetary policies. Then the available tools necessary for carrying out the established monetary policies are studied in Section 6.3. The particular mechanism of how monetary policies affect the economy is discussed in Section 6.4. Relevant empirical evidence on the importance of money is provided in Section 6.5. In Section 6.6, we will look at how the *IS-LM* model can be employed to the understanding of money and fiscal policies, their effects on the total output and interest rates, and the crowding-out effect. Section 6.7 considers how inflation is caused, how inflation is related to the prices of goods, and how inflation can be dealt with. At the end, Section 6.8 concludes this chapter by exploring how systemic structure actually underlies monetary policies.

6.1 GOALS OF MONETARY POLICY

Monetary policies are introduced for the purpose of realizing certain goals of economic development. They are the general term for a series of policies, guidelines and control measures regarding financial activities. Financial regulation is materialized through monetary policies, which are introduced and implemented by the central bank. As the sole currency issuer, the central bank controls the supply of money and maintains the circulation stability of money through employing a series of effective measures so that a financial environment that is conducive for economic development can be created.

6.1.1   Tasks and contents of monetary policy

The basic task of monetary policies is to correctly regulate the money supply and to stabilize the money circulation. Only when the money supply agrees with the objective demand for money of the economic development, the available resources can be sufficiently utilized, the circulation of money can be stabilized, and the economy can be developed coordinately. Conversely, if the money supply is not adequate, it will lead to chaos in the circulation of money, hindering the concerned development of the economy. In particular, a monetary policy, as an important national macro-economic policy, needs to accomplish the following three tasks:

A.   Maintain appropriate money supply so that the normal economic activities will not be adversely affected by inadequate money supply;
B.   Create a conducive and stable monetary, financial environment for the healthy development of national economy; and
C.   Play the role of eliminating the interfering effects of other economic factors. That is, when the economy expands, the task of monetary policy is to reduce the money supply in order to tame the excessive demand for money so that the economy can evolve smoothly. When the economy contracts, the task of monetary policy is to increase the money supply in order to stimulate the aggregate demand so that the economy can be brought out of the "trough".

In general, monetary policies address problems in three different directions: the goals of the policies, the tools of the policies, and the implementation effects of the policies. Because from the time a monetary policy is introduced, to the time when it is practically implemented, and to the time when the desired goals are materialized, there involves a whole series of processes, monetary policies also have to address intermediate objectives and policy transmission mechanisms.

The content of monetary policies specifically includes

1   Credit policy: It consists of a series of guidelines and measures the central bank adopts in order to manage credit. Through implementing credit policies, the central bank, on the one hand, regulates the total amount of credit available to the society in order to make it appropriate to the economic development, and, on the other hand, controls the structure and direction of the available credit so that the use effect of capital can be optimized.
2   Interest rate policy: It is a series of guidelines and principles the central bank adopts in order for the bank to control and regulate the market interest rate. Specifically, the role of an interest rate policy is reflected in two aspects: the regulation on the level of interest rate, and the regulation on the structure (both the risk structure and the term structure) of interest rate.

3   Exchange rate policy: The central bank regulates the foreign exchange rate market through exchange rate policies in order to control the flow of international capital and to maintain adequate foreign exchange reserves so that a stable exchange rate can be preserved and that the normal economic development can be protected from excessive fluctuations in exchange rate.
4   Other policies: Other than introducing credit policies, interest rate policies, and exchange rate policies, the central bank also introduces a whole series of other policies relevant to financial activities in order to maintain the order and stability of the financial markets.

6.1.2   Goals of monetary policy

By the goal of monetary policies, it means the ultimate goal to be achieved by the central bank through its introduction and implementation of monetary policies and its financial regulation over the national economy. The establishment of such a goal is closely related to the status of the socio-economic development of the specific time period; it needs to be adaptive to the macro-economic target of the nation. The goal of monetary policies has experienced a long evolutionary process from the initial single goal before the 1930s of stabilizing the value of currency to the current comprehensive goal of stabilizing prices, maintaining sufficient employment, stimulating economic growth, and sustaining balance of international payments.

Since the time when central banks initially appeared until the time before the 1930s, the Western countries had commonly employed the gold standard system, where the currency possessed its ability to automatically adjust the economy. So, stabilizing the value of currency naturally became the goal of monetary policies of the central banks. Under the circumstances of the time, each of these nations highly praised the markets for their ability to automatically adjust the economy, while the governments still could not apply any macro regulation on the economies. That explains why there was not any other goal for monetary policies and why each of these nations had chosen stabilizing the value of their currencies as the sole goal of their monetary policies.

The major economic crisis that broke out in the 1930s throughout the capitalistic world caused drastically increasing unemployment, which severely affected the economic development and social stability. So, from the year 1930 to the mid-1940s, how to resolve the problem of unemployment became the overriding central task of the time. Along with the rise and acceptance of the Keynesian economics of state interventions, the major capitalistic nations decided one by one in the form of law that maintaining sufficient employment is a main goal of monetary policies of the central bank. So, the original single goal of monetary policies evolved into the dual objectives of stabilizing the value of currency and maintaining sufficient employment. After the 1950s, the economic development of different nations became very uneven. In order to preserve their individual national economic strengths and political status in the international arena, many nations considered economic development and stimulating economic growth as the emphasis of their national monetary policies.

After entering the 1970s, because of the long period of implementation of Keynesian policy of macro regulations, different nations had experienced varying degrees of inflation, which severely affected these nations' international payments. Especially, since the 1970s, the United States constantly experienced international trade deficit. So, the U.S. dollar based international monetary system started to rattle, leading to two U.S. dollar crises one after another. That caused some nations to treat the balance in international payments as an important issue, and to list it as an objective of monetary policies. As of this point in time, the goal of monetary policies has evolved into such a comprehensive system of objectives that includes the value stability of currency, sufficient employment, economic growth, and balance in international payments. Later on, the international monetary system further evolved so that gold lost its functionality as the value carrier of currencies. Consequently, many nations changed the goal of stabilizing their currency values to that of stabilizing prices.

Summarizing the development up to the present day, the goal of monetary policies includes the maintenance of price stability, the preservation of sufficient employment, the stimulation of economic growth, and the upkeep of the balance of international payments. Other than all these objectives, the central bank, as the highest level financial management institution of the nation, also has objectives in the areas of the national financial system and the relevant management. In particular, the central bank is charged with the obligation of supervising and guiding banks and other financial institutions to pursue business activities along healthy and sound tracks, and that of maintaining order and stability of the financial markets.

After experiencing the economic throes of inflation, economists as well as government officials have recognized the socio-economic price of inflation so that they have been more inclined to include the maintenance of price stability as one objective of monetary policies. For example, after Germany suffered through the hyperinflation of the time period from 1921 to 1923, German central bank has bared without any hesitation the strongest sense of obligation to maintain price stability. Consequently, policy makers of other European nations also took maintaining price stability as a basic objective of their central banks. The Maastricht treaty, as signed by the European Union in December 1991, clearly lifted the importance of maintaining price stability as an objective of monetary policies. The treaty suggests establishing a European Central Bank System with functions similar to those of the Federal Reserves of the United States. The constitution of the European Central Bank System treats maintaining price stability as its basic objective and declares that only when the overall economic policies of the European Union are not in conflict with the objective of maintaining price stability, the European Central Bank System will support the economic policies.

The reason why people like to maintain price stability is because continuously increasing prices cause economic uncertainties. For example, if the price level rises constantly, then the information contained in commercial goods and labors will become more difficult to comprehend, making it more difficult for consumers, business entities, and government agencies to make relevant decisions. The event related to extreme instability of prices is hyperinflation. For example, during the latter two years of the German hyperinflation from 1921 to 1923, the continued price increase severely shocked German economy, causing the gross domestic production to fall drastically, while the hyperinflation made it difficult for people to plan for the future. In particular, in an inflationary environment determining how much savings would be needed for children's future college education becomes an impossible task. Additionally, hyperinflations could create tensions for a nation, causing its citizens to lose confidence in their currency. At extremes, the national currency could become such a currency that is so "hot" that people do not want to keep it for any length of time. When that happens, social and economic turmoil would ensue in that particular nation.

Now, a natural question is: At which level should the central bank maintain the price in order to be considered stable? The answer is dependent on the specific circumstances of the nation and the capability of the people involved. Even so, it is advantageous to the development of national economy to retain price increases within the minimum possible magnitude.

Both the Employment Act of 1946 and the Full Employment and Balanced Growth Act of 1978 committed the U.S. government to the dual task of improving sufficient employment and maintaining price stability. There are two reasons for why the problem of sufficient employment is important. Firstly, high unemployment causes a lot of sufferings to the people: household hardships, loss of individual self-esteems, and increase in the number of social crimes. Secondly, high unemployment not only creates a large number of idle workers, not also results in a large number of idle economic resources, such as closed factories and unused equipment, so that the total economic output drops greatly. Hence, people aspire to sufficient employment.

However, what is considered as sufficient employment? At the first sight, it seems that sufficient employment would mean that no one is unemployed. However, such explanation ignores one fact: Some unemployment belongs to frictional unemployment, which is beneficial to the economy. For instance, people often voluntarily leave their current works to pursue other activities. When these people decide to re-enter the labor market and look for jobs, it generally takes them for a while to find an appropriate jobs. During this period of time, these workers are considered unemployed. However, such unemployment is temporary and belongs to frictional unemployment. The existence of some unemployment to a certain degree is beneficial to the economy. It is similar to the situation where the non-zero idle rate of guest rooms in hotels is advantageous to the tourist industry. So, the concept of sufficient employment should be understood as follows: Each willing worker can find an appropriate job within a short period of time. That is, sufficient employment is not equivalent to having unemployment rate equal to zero. It should allow for the existence of a non-zero natural unemployment rate. At the level of this natural unemployment rate, the supply of labor (that is willing and able) is equal to the demand for labors so that the existence of this natural unemployment rate is beneficial to the economy.

Although such understanding of the concept of sufficient employment sounds simple and trustworthy, the reality is so straightforward. It leaves open the following question: What unemployment rate should be considered the natural rate of unemployment? During the Great Depression, the unemployment rate in the United States was over 20%. That surely was not the natural rate of unemployment, because it is obviously too high. During the early part of the 1960s, American economists commonly considered 4% unemployment rate to be reasonable, which with 20-20 hindsight is seen as too low. It is because such a low unemployment rate caused the acceleration in the increasing inflation of the time. Currently, the estimate for the natural unemployment rate is around 6%. However, this estimate is not certain either. In short, the specific level of the natural unemployment rate should be estimated based on specific circumstances involved and for the purpose of stimulating economic development.

By economic growth, it means the increase in the aggregate national total of produced goods and labor within a certain time period; or it means the increase of the per capita GDP of the nation. Economic growth is the objective of all economic policies so that it is also an objective of monetary policies. It reflects the level of a nation's economic development, and is closely related to the objective of sufficient employment. When the unemployment rate is very low, business firms are more willing to invest in capital equipment in order to increase productivity, leading to economic growth. On the other hand, when unemployment rate is relatively high, factories would be idling, companies would not be willing to throw money at adding new work spaces and capital equipment. Consequently, economic growth is greatly hindered. Currently, to measure the indicators of economic growth, eliminating the effect of inflationary factors is often considered.

By international payment, it means the totality of all accounts payables and receivables as caused by political, economic, and cultural exchanges between one nation and the rest of the world within a certain period of time, which ordinarily means one year. It stands for an important component of the national economy of the nation. It reflects the international range of the nation's economic activities, scale, and characteristics as well as the nation's status and role in the global economy. By balance of international payments, it means that the nation's international income is roughly equal to its international payment.

If in a certain year an imbalance in international payment appears, it is not necessarily a bad thing. For example, if a trade deficit appears in one year, it could mean that during that year more foreign machines, equipment, advanced technology, and other goods are purchased to promote the domestic economic growth. Of course, people prefer to see the appearance of surplus within a year. However, if either deficit or surplus appears year after year, it definitely means there is a problem. Consecutive yearly deficit could exhaust the national capability to offset the deficit; and consecutive yearly surplus could lead to continued increase in the nation's foreign exchange reserve, leading to waste of the idling foreign currency assets. So, the ideal situation is to maintain a roughly balanced payment in the international trades for the nation without much deficit and surplus. Because of this reason, each nation has treated the balance in international payment as a major objective of monetary policies.

Other than the afore-mentioned four objectives of monetary policies, these policies also have objectives in the areas of financial mechanism and financial management, which is the maintenance of the stability of the financial system and the financial markets. These objectives are very important in terms of establishing an orderly and operational monetary financial environment that can stimulate the economic development. The reason why a nation would establish its central bank and authorizes it to regulate and to manage the national financial system is for the purpose of supervising and guiding the banks and other financial institutions to conduct their business activities along a healthy and sound trajectory, and the purpose of preserving the stability of the financial system and the financial markets.

The method the central bank uses to maintain the stability of the financial system is through playing its role of the ultimate lender to prevent financial panics. To maintain the stability of the financial markets, the central bank tries to preserve the stability of interest rate. Fluctuations in interest rate can cause a great deal of uncertainty for financial institutions. For example, increasing interest rate could lead to large capital losses of long-term bonds and pledged assets. These losses in turn could cause the financial institutions that hold the relevant assets to go under. In recent years, large fluctuations in interest rate have caused severe difficulties for American Savings and Loan Association and Mutual Savings Banks, especially when these organizations experience fiscal hardships at the same time. Therefore, keeping interest rate stable is extremely important for the stability of the financial markets.

These different objectives of monetary policies are related to each other both consistently and contradictorily. When seen from long term, they are consistent with each other and mutually supplementary to each other. However, when seen in the short-term perspective, these objectives are contradictory and conflicting with each other. Very few nations have consistently dealt with these short-term objectives well.

Over the long term, all the objectives of monetary policies are consistent without suffering from any conflict. They are interdependent, mutually promoting, and are not faced with the problem of giving up one objective in order to accomplish the others. Economic growth is the material foundation of other objectives; stability in prices is the prerequisite of economic growth. Sufficient employment could promote economic growth. Balance in international payment is beneficial to the realization of other objectives. And, stability of financial markets provides a good financial environment. In particular, economic growth could help expand the aggregate supply of the society; improvement in employment rate and creation of additional work opportunities strengthen the capability of imports and exports. Specifically, only with continued, stable, and coordinated economic growth, equilibrium between the demand and supply of the product market and the money market can be potentially reached so that price stability, interest rate stability, and the financial markets stability could be warranted. Sustained, stable, and coordinated economic growth could be materialized only under the condition of appropriate economic structures, while to have the appropriate economic structures, there must be appropriate price structures and accurate information guidance. Only with stable price levels, the price structure of the economy could be appropriate and provide the necessary, accurate price information for the economy. Hence, stable prices are the prerequisite of economic growth. Sufficient employment means sufficient usage of recourses, and the willingness for business firms to invest in capital equipment in order to increase productivity, leading to economic growth. Conversely, when massive unemployment occurs, many factories are shut down, while many others are no longer willing to invest capital in factory buildings and equipment. That would surely and severely hinder the economic growth. So, it can be seen that sufficient employment is closely related to economic growth and that economic growth is a prerequisite of sufficient employment. On the other hand, sufficient employment in turn promotes economic growth.

Balance in international payment is beneficial to the growth of the domestic economy and beneficial to the sufficient usage of international resources so that the domestic productivity is enhanced, the domestic economic structure is improved, and the economic growth is promoted. It is also beneficial to the stability of interest rate and exchange rate so that the stability of the financial markets and foreign exchange markets is boosted. When the financial markets and financial system are stable, a good monetary, financial environment is created for domestic production and investment. Within this environment, prices of goods are stable, because the interest rate is stable. So, the prices of other assets and goods are also unwavering. When the sources of financing for business production and investment are stable, stead economic growth is more or less guaranteed, which in turn promotes higher employment. That leads to

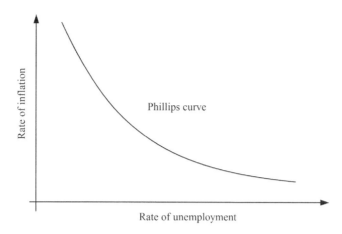

Figure 6.1 The relationship between unemployment and inflation.

a stable increase in the domestic productivity, a rising capability of import and export, and the balance of international payment.

In terms of the short-term perspective, practical implementations of monetary policies often lead to conflicts between the objectives of the policies. The situation of reaching one objective while violating some other objectives appears frequently. It has been shown to be difficult to coordinate all these objectives well. First of all, there is conflict between price stability and sufficient employment. The classical Keynesian economics maintains that inflation is caused by continued increase in demand after the economy has materialized sufficient employment. So, unemployment and inflation cannot exist at the same time. That is, the rate of inflation that accompanies unemployment is zero; the rate of unemployment that accompanies inflation is also zero. This end explains how price stability and sufficient employment conflict with each other. However, this fact was not noticed before the 1950s. By drawing a curve, now known as the famous Phillips curve, that slants downward to the right, (Phillips, 1958) concludes that between the rate of unemployment and the price level is there a relationship where when one variable goes up, the other variable goes down. This work made people realize the first time in history that there is a relationship between unemployment and inflation as shown in Figure 6.1. Both economists and government officials have treated Phillips curve highly because this curve tells people that between unemployment and inflation the society could make a choice and could find the optimal combination of the two. For example, further studies on American Phillips curves reveal that fact that in order to guarantee the unemployment rate in the United States to be not more than 3% in the early 1969s, the rate of inflation has to be between 4%–5%. However, after entering the 1970s, the situation changed, when the Western nations experienced the phenomenon of "stagflation", where high unemployment coexisted with severe inflation. However, in terms of the understanding of stagflation, economists still do not have a commonly accepted theory.

Secondly, although both price stability and economic growth are the core contents of monetary policies, these two objectives often conflict with each other within short terms. For example, when the economy is declining, expansionary monetary policies are introduced and implemented to stimulate demand, to promote economic growth, and to reduce unemployment. However, doing so often causes the quantity of money in circulation greater than the demand for money that is appropriate for the development of the economy, leading to increasing prices. Conversely, when the economy is expanding, in order to subdue the inflation and to maintain price stability, tightening monetary policies are often introduced and implemented in order to reduce the money supply. That has been frequently hindering the economic growth and causing reduced amount of employment opportunities. So, it can be seen that there is conflict between price stability and economic growth. How to determine the optimal combination of these two objectives has been an important, unsettled problem in the selection of monetary policies.

Thirdly, there is conflict between price stability and balance of international payment. Along with the current globalization of the world economy, if one nation maintains its price stability while another nation experiences inflation, then it would cause the exported goods to rise in price, leading to reduced export and increased import of the former nation. This change would create a deficit in the international payment to appear or make the existing deficit to rise. So, it can be seen that maintaining price stability in one nation could not guarantee the nation's balance of international payment. Instead, it might very likely lead to imbalances in international payment.

Fourthly, there is conflict between the stability of prices and that of financial markets. The stability of financial markets is mainly materialized through interest rate stability. And the latter is realized by the central bank through its operations in the open market. For example, when the economy is expanding, to control the inflation and to maintain the price stability, the central bank sells bonds in the open market in order to reduce the supply of money. However, doing so increases the supply of bonds and lowers the price of the bonds. That causes the interest rate to rise, leading to instabilities of the financial markets. Conversely, when the economy is declining, to suppress the prices of goods and to maintain price stability, the central bank introduces and implements expansionary monetary policies, while purchasing bonds in the open market in order to increase the supply of money and to stimulate the domestic demand. However, doing so causes the demand of the bond market to increase, making the price of bond to go up and interest rate to go down, causing instabilities for the financial markets. That is, between price stability and financial markets stability are there inconsistency and conflicts.

Fifthly, there is conflict between sufficient employment and balance of international payment. When the number of employed people increases, the level of income goes up. That leads to rising import of foreign goods and falling export of domestic goods, which enlarges the deficit of international payment. To reduce the deficit, tightening monetary policies are introduced and implemented in order to suppress the domestic demand, which in turn causes the number of employment opportunities to fall and unemployment is rise. So, in the short-term perspective, when employment is sufficient, the international payment is likely not balanced; and when the international payment is balanced, there might be a large number of unemployed workers. So, there are conflicts between the objectives of sufficient employment and balance of international payment.

Sixthly, there is conflict between economic growth and balance in international payment. Economic growth and sufficient employment are two consistent objectives. However, from the discussion above it has been seen that there are conflicts between sufficient employment and balance of international payment. So, it is natural for conflicts to exist between economic growth and balance of international payment. Economic growth inevitably increases the demand for imported goods, including foreign machines and equipment, advanced technology and raw materials, etc. At the same time, expanding export cannot be materialized within short periods of time. So, economic growth generally leads to rising deficit in international payment, causing imbalances in international payment.

The conflicting relationships between the objectives of monetary policies represent one of the most difficult problems facing each and every national government and economist. The monetary policy implemented to realize a specific objective might very well delay the realization of another objective or destroy the desired degree of realization of another objective. So, the task facing financial regulation is to make optimal selections of these both contradictory and consistent objectives so that best possible outcomes could be expected.

6.2   MEDIUM-TERM GOALS OF MONETARY POLICY

Speaking holistically, the essence of monetary policies is about price stability and economic growth, which is such an economic growth that is accompanied with no inflation. That is also the ultimate goal of macro-economic policies. So, in implementing monetary policies, the central bank needs to know the consequent effects of the policies. To this end, quantifiable and operable economic indicators are designed. These indicators are the bridges connecting the monetary policies and the results of their implementations. That is why they are also known as medium-term or intermediate goals or objectives of monetary policies. How to select intermediate objectives represents an important medium step in the implementation of monetary policies. The accuracy of these indicators determines whether or not the ultimate objectives of the policies can be eventually materialized.

When the central bank selects the intermediate objectives for its monetary policies, it uses three criteria: each intermediate objective must be quantitatively computable, it must be controllable by the central bank, and it must possess the capability to predict the ultimate objectives of the monetary policies. Here, by controllability, it means that the central bank can effectively regulate and adjust the intermediate objective variables and can accurately control the changes and tendency of changes of these variables through using various policy instruments. If the central bank cannot control an intermediate objective variable, then even if the central bank knows that the variable is not on the right trajectory, it has no means to adjust the variable so that it will fall back into the needed trajectory. For example, some economists suggest applying the nominal GDP as an intermediate objective. However, because the central bank has very little to do with the nominal GDP, the nominal GDP cannot really provide much help to the central bank in terms of how the central bank should arrange its monetary policy tools. Therefore, the nominal GDP is not appropriate to be used as an intermediate objective variable. Conversely, the central bank can exert effect control over the total quantity of money in circulation and the interest rate from many different angles, these two variables are appropriate to be employed as intermediate objective variables of monetary policies. The so-called predictability stands for the computability of the intermediate objective variables. Specifically, it contains following two aspects of implications. Firstly, the central bank can quickly collect data for the intermediate objective variables; and secondly, the central bank can make accurate measurements, analyses, and predictions on these variables by using the collected data.

It is very necessary for the central bank to be able to make quick and accurate measurements of the intermediate objective variables. Whether or not an intermediate objective variable is appropriate is mainly dependent on if the variable can emit the needed signal quicker than the policy objective when the policy of the variable is off the track. And the prerequisite for such a variable to quickly emit signals is the possibility for the central bank to quickly and accurately take measures for the variable. In other words, the central bank has to be able to quickly collect data on the variable and to make accurate measurements and analyses. So, a necessary condition for a variable to be adopted as an intermediate objective variable is whether or not it can be quickly and accurately measured. For example, assume that the central bank plans to realize a 4% increase in money M_2. However, if the central bank cannot quickly and accurately measure M_2, then what goodness would this plan have? As another example, we know that GDP data are not as accurate as those of the total quantity of money and interest rate. So, putting attention on the total quantity of money or interest rate will provide much clearer signal on the state of monetary policies than placing attention on GDP.

Each intermediate objective has to be closely related to the ultimate objectives of monetary policies. By controlling and adjusting the intermediate objectives, the central bank can promote the realization of the ultimate goals. The close relationship here reflects the influencing power of the intermediate objectives on the ultimate goals. The closer the relationship is, the greater the influencing power and the more effective the central bank is in terms of controlling the ultimate goal variable through adjusting the intermediate objective variables. What is more important is that the influence of intermediate objectives on the ultimate goal must be measurable and predictable so that its strength of influence can be known ahead of time or predicted reliably. For example, the central bank can quickly and accurately calculate the price of tea and is able to control the price. However, such variable is useless, because through adjusting the tea prices the central bank cannot influence the domestic price level of other common goods, the total output, and the situation of employment. However, such variables as the quantity of money supply and interest rate are closely related to price level, total output, and the situation of employment. So, by adjusting these two variables, the central bank will be able to exert its influence on price level, total output, and the situation of employment.

Based on the afore-mentioned three criteria – predictability, controllability, and relevancy, and combined with the accustomed tradition of various central banks, the variables commonly accepted as intermediate objectives of monetary policies by central banks ate of the following three kinds:

1   Interest rate: Market interest rate is closely relevant to economic lives and exerts important effects on the aggregate demand and supply of the society. Selecting

market interest rate as an intermediate objective variable satisfies all the three criteria of intermediate objective variables. Firstly, interest rate possesses a strong measurability. The central bank can obtain the data of market interest rate almost instantaneously. Beyond that, the data of market interest rate are quite accurate with little need for revision. Secondly, interest rate is of strong controllability. Based on the specifics of the national economy, the central bank can affect the market interest rate by altering the demand and supply of the bond market through selling and buying bonds in the open market so that the market interest rate would move along the desired trajectory. Thirdly, interest rate is strongly relevant. From the discussion in the previous chapter on how interest rates are determined, it follows that interest rate directly influences the total economic output, and the market interest rate and the total output are determined simultaneously. Therefore, through adjusting the market interest rate, the central bank materializes the goal of controlling the total output by affecting economic investments and savings.

However, there are also shortcomings with applying interest rate as an intermediate objective variable of monetary policies. Although the central bank can quickly and accurately measure interest rate, the interest rate that can be so measured is the nominal interest rate. It cannot explain the actual cost of borrowing (the actual interest rate), while it is the actual cost of borrowing that can relatively accurately illustrate in what amount the GDP will change. According to the discussion in Chapter 5 on how the actual interest rate is measured, it follows that the actual interest rate (the actual cost of borrowing) is equal to the nominal interest rate minus the expected inflation rate. Unfortunately, there is no direct method to measure the predicted inflation rate. So, it is extremely difficult to compute the actual interest rate. That also makes it difficult for the central bank to control the changes in the actual interest rate. It is exactly because of this reason that when interest rate is applied as an intermediate variable of monetary policies, there are also matters that make it inappropriate.

2   The supply of money: Also, the supply of money satisfies all the three criteria of intermediate objectives of monetary policies. Firstly, the supply of money is of strong predictability. No matter whether it is M_1 money or M_2 money, it is contained in the balance sheets of the central bank and commercial banks. That makes it convenient to collect the statistics of money supply and to analyze the collected data. For example, Federal Reserve of the United States announces the data of money supply once every two weeks. Secondly, the supply of money is of very strong controllability. From the previous discussion on the supply of money, it follows that the central bank indeed can exert powerful influence on the supply of money through such methods as manipulations in the open market, adjusting the rate of statutory reserves, altering the rate of re-discount, etc. Additionally, the rights of issuing currency and injecting new currency into circulation are completely within the control of the central bank. Hence, the central bank can effectively control the amounts of supply of M_1 and M_2 money. Thirdly, the amount of money supply is closely related to economic activities, such as the total output, the level of employment, the level of prices, etc. So, when the amount of money supply is used as an intermediate objective of monetary policies, it possesses a very strong relevancy. For example, in economic prosperity, reducing money supply can suppress inflation and maintain price stability; during a time of economic decline,

increasing money supply can stimulate the aggregate demand and increase employment opportunities so that economic growth (that is, the total output is increased) is promoted. So, a relatively strong relevancy between money supply and the objectives of monetary policies appears. However, there are still controversies regarding the relationship between these two ends. Currently it is still not clear whether it is interest rate that is more closely related to the objectives of monetary policies or it is the money supply that is more closely related to the objectives of monetary policies.

3   The base money: The base money also satisfies all the three criteria of intermediate objectives of monetary policies. Firstly, it is more convenient for the central bank to measure the base money than to collect data for money supply and interest rate. The central bank can directly calculate the total liabilities from its balance sheet so that it knows the amount of base money. That is definitely more convenient than for the central bank to compute the total sum of all demand deposits of commercial banks from around the nation. Secondly, the base money is the money that is directly under the control of the central bank, which holds the right to issue money. Thirdly, any change in the base money affects the changes of the money supply. So, it affects the realization of the objectives of monetary policies. Therefore, between the base money and the objectives of monetary policies is there a relatively strong relevancy.

Selecting the base money as an intermediate objective of monetary policies might be better than selecting interest rate or money supply, because of the following three reasons: (1) The base money is the currency of the central bank and is equal to the total liabilities of the central bank. It is recorded directly on the balance sheet of the central bank and is under the direct control of the central bank. (2) The base money is the foundation for the commercial banking system to create deposit currency. By adjusting the base money, the central bank can change the amount of deposit currency created by commercial banks so that the total money supply of the society is changed. (3) When the base money is selected as an intermediate objective, it has a better operability than selecting interest rate or money supply. The central bank can manipulate the base money in order to achieve its desired outcome by simply analyzing its own balance sheet. Additionally, by changing the base money, the central bank can cause changes to interest rate and money supply. So, if speaking in this sense, the base money is an objective variable that can be more conveniently manipulated. Hence for the central bank, it has stronger operability than interest rate and money supply.

6.3 TOOLS OF MONETARY POLICY

Financial regulation achieves its goals mainly through adjusting such intermediate objectives of monetary policies as the money supply, interest rate, and the base money. As for adjusting the intermediate objectives of monetary policies, it is materialized by making use of certain methods, which are known as the tools of monetary policies. The central bank employs the tools of monetary policies to influence the intermediate objectives of the monetary policies, and materializes the ultimate goal of financial regulation (that is, the goals of the monetary policies) through the changes of the intermediate objective variables. The tools of monetary policies widely employed by the Western nations consist of two major classes. One class contains the general tools that can produce commonly felt impacts on the entire national economy; and the other class includes all other tools, including selective monetary policy tools, direct and indirect credit control tools, and others.

In terms of the general monetary policy tools the central bank uses to control the money supply, they are of three different kinds: open market operations, rate of statutory reserves, and rate of discount.

Now, let us look at the tool of open market operations. This is the most important monetary policy tool. It represents the basic factor that influences the changes of the base money and is the main reason underneath the variance of money supply. When purchases are made on the open market, they expand the base money so that the money supply is increased. When sales are made on the open market, they help to reduce the base money so that the money supply is decreased. Through its operations in the open market, the central bank influences the base money so that it materializes the control of the money supply. The operations in the open market are classified into two groups. One group represents initiative operations in the open market with the purpose of varying the level of reserves and the base money. The other group contains defensive operations in the open market with the purpose of cancelling the effects of the changing base money or other factors. The object of the open market operations is government bonds in general and Treasurys in particular. These bonds have the most liquidity and enjoy the greatest magnitude of trading. And through trading these bonds, the central bank does not cause any damaging fluctuation to the marketplace. In particular, in the United States the objects of open market operations are Treasury bonds and government agency bonds. The decision-making authority is the Open Market Committee; the particular operations are handled by the trading department of New York's Federal Reserve Bank, where the manager of the trading department supervise the bonds' buying and/or selling of his traders. To understand how this operation is carried out, the best way is to walk through the activities of a normal day within the trading department.

The day of the department manager starts with his reading the last-night report on the total amount of statutory reserves of the banking system. This report helps the manager decide with what magnitude of change in the reserves the desired level of money supply can be reached. After completing the report, the manager discusses with several traders of government bonds in order to be updated with the current price trend of the bond market. After the meeting, the manager receives a detailed research report on what short-term factors might influence the base money. This report also estimates how much currency is held by the public. Based on this report and his understanding of the bond market, the manager of the trading department will draft a detailed plan of defensive operations in the open market. For instance, if the report predicts that the amount of cash in transit will be reduced because of the accelerated clearance of checks, then the manager will decide to use defensive operations in the open market to eliminate the predicted drop in the base money as caused by the reduced amount of cash in transit. If the report predicts that the amount of currency held by the public is going to rise, then the manager will decide to purchase bonds in the open market in order to increase the bank's reserves and to prevent a potential drop in money supply. After making the plan of defensive operations in the open market, the department manager telephones the U.S. Treasury Department to get an update on the prediction of that department's deposits project. Then the manager compares this prediction of the Treasury Department with those made by his department workers.

Then this manager of the trading department receives an instruction from the Open Market Committee of Federal Reserve regarding what indicators of the quantity of total money supply need to be materialized. Based on this instruction, the manager plans on how to actualize these indicators through using initiative operations in the open market. At the end, the manager combines the drafted plans of initiative and defensive operations to formulate the action plan of operation in the open market for the day. The afore-mentioned process of decision making generally is completed by 11:15 am when the manager of the trading department holds the daily telephone conference with the members of the Open Market Committee of Federal Reserve to summarize the content of his action plan. As soon as the plan is approved at around 11:30 am, the manager orders his trader to quote the standing selling prices in increasing order from low prices to high prices. Then Federal Reserve System purchases Treasury bonds starting with low standing selling prices and gradually going higher until the planned amount of bonds is successfully purchased. This process of collecting price quotes and making purchases is generally completed by 12:15 pm. After then, although the trading department calms down, the traders still need to continuously monitor the money market and the movement the bank reserves. In rare occasions, the trading department might decide that further trading is necessary.

Compared to other monetary policy tools, operations in the open market have the following advantages. (1) These operations are actively performed by the central bank. So their magnitudes are within the total control of the central bank; (2) these operations are flexible and delicate, and the trading scales can be adequately managed. No matter whether a small vibration to the base money is desired or a major change is wanted, the central bank can achieve it through operations in the open market; (3) these operations in the open market are reversible and can be reversed easily. When errors appear, the central bank can immediately correct them by using this tool in the opposite direction. For example, if the central bank concludes that the excessive growth of money supply is caused by its over purchase in the open market, then the central bank can lower the speed of the excessive growth by immediately selling in the open market; and (4) these operations in the open market are generally speedy without delay. When the central bank decides to modify the quantity of the base money, the bank only needs to place an order to purchase or sell bonds with the floor trader and the trade is carried out and completed immediately.

So, it can be seen that operations in the open market represent an ideal monetary policy tool for the central bank to conduct its daily financial regulation. However, there are some prerequisite conditions for this tool to effectively function. Firstly, the central bank has to have such a financial strength that is powerful enough to interfere and to control the entire financial market. Secondly, the domestic financial market has to be normalized and working properly. Thirdly, there must be other monetary policy tools to work jointly and coordinately. For example, if there were no deposit reserve system in place, it would be difficult for this tool to function right.

Secondly, let us look at the rate of statutory reserves. By adjusting the rate of statutory reserves, the central bank can cause the monetary multiplier to change so that the amount of money supply is affected. Indeed, the rate of statutory reserves is a powerful tool that influences the money supply. Its influence on the money supply is so huge that the central bank generally does not use this tool to regulate the money supply.

The main advantage of applying the rate of statutory reserves is that its influence on all banks is equal while the money supply is greatly affected. However, as a powerful monetary policy tool, it might have more weaknesses than advantages. Its weaknesses are mainly shown in two aspects. Firstly, it is difficult to employ the rate of statutory reserves to make minor adjustment to the money supply, because the proportion of demand deposits is relatively high among all bank deposits. Secondly, banks with very low excessive reserves might immediately experience the problem of liquidity if the rate of statutory reserves is increased. In short, the method of changing the rate of statutory reserves as a monetary policy tool bears more potential problems than expected. That explains why this tool has not been applied very often.

In recent years, the central banks of many nations from around the world have either lowered or eliminated the requirement of statutory reserves. The United States jettisoned the requirement of statutory reserves for time deposits in December 1990, and lowered the rate of statutory reserves for check deposits in April 1992 from 12% to 10%. The steps taken by Canada are even greater. The financial market regulations issued in April 1992 abandoned the requirement of statutory reserves for all deposits with terms of more than two years. The central banks of Switzerland, New Zealand, and Australia have completely done with the requirement of statutory reserves. Considering what has been happening, if the simple deposit multiplier formula were employed without necessary modification, the amount of bank deposits would expand infinitely. However, that cannot be any further from the fact. In the process of deposit creation there are many leakages that are not considered in the development of the formula. So, if these leakages are included, a more delicate formula for the deposit creation will be established, where eliminating the requirement of deposit reserves will not lead to infinite growth in bank deposits. Additionally, because the reserves commercial banks deposited at the central bank do not bear interest, they create a high level of opportunity costs for the commercial banks. That means that comparing to the nonbanking financial institutions that are not required to maintain statutory reserves at the central bank, the competitiveness of commercial banks has suffered a great deal. Due to this reason, lowering or eliminating the requirement of statutory reserves has no doubt fortified the strength the commercial banks in their market competition and maintained the stability of the banking system, while the money supply will not increase infinitely.

Thirdly, let us look at the rate of discount. By altering the rate of discount, the central bank can influence the quantity of discount loans so that it can influence the amount of base money and money supply. The facility through which the central bank issues discount loans to commercial banks is known as the discount window. In the following, let us look at how the discount window works and the roles played by the central bank through the discount window.

In terms of how the discount window works, the central bank generally influences the magnitude of discount loans through one of two methods. One is to alter the discount rate, the interest rate of discount loans; and the other is to apply administrative management to the discount window.

1　Altering the discount rate: The specifics of influence on the magnitude of discount loans by altering the discount rate are that: When the discount rate is lifted higher, the number of discount loans will decrease. When the discount rate is lowered, the number of discount loans will increase. This end has been discussed earlier. So, all the details will be omitted here.

2   Types of discount loans: In order to learn how the central bank influences the number of discount loans through managing the discount window, let us first look at the types of discount loans and how these loans are issued. There are three classes of discount loans that the central bank lends to commercial banks: adjustment loans (also known as primary credit), seasonal loans (also known as seasonal credit), and sustaining loans (also known as secondary credit).

The central bank gives out adjustment loans to assist commercial banks to overcome short-term liquidity need due to temporary outflows of deposits. These are the most common discount loans and can be acquired by one telephone call. However, they need to be repaid quickly. For large banks, these loans need to be repaid by the end of the next business day. Seasonal loans are issued by the central bank to those banks located at resorts or in crop production areas that have seasonal characteristics and experience seasonal needs. Sustaining loans are discount loans the central bank lends to those commercial banks that suffer from severe liquidity problems due to continued outflow of deposits. They do not have to be repaid quickly. To obtain a sustaining loan from the central bank, the needy commercial bank must hand in an application with clear explanation for why such a loan is desirable along with a plan on how the bank is going to regain the necessary liquidity.

3   The management of the discount window: The reason why the central bank manages the discount window and restrains the discount loans is because it wants to prevent commercial banks from abusing their discount loans. In recent years, the discount rate in the United States has often been lower than the market interest rate. That of course creates a great speculation opportunity for commercial banks if they could take out discount loans from the central bank at a lower rate and then loan the money out at higher rates or purchase bonds that pay higher rates of interest. It is this potential scenario that the central bank tries to manage and restrain. In order to prevent such scenario from occurring, the central bank establishes a number of rules, limiting how frequently a bank can obtain discount loans. If a certain commercial bank applies for discount loans too frequently, the central bank can and will refuse to give any loan to this bank a warning to remind the bank that the discount window is a special the central bank provides commercial banks instead of a right commercial banks have. This management rule the central banks uses to manage the discount window is known as a moral advice.

When a commercial bank applies for a discount loan, it will permit the central bank to investigate its credit. If it has applied for loans too frequently, future application might be rejected. After obtaining a discount loan, the commercial banks will have to pay interest to the central bank. All of this constitutes the cost of discount loans and affects the number of discount loans. The higher the cost is, the fewer discount loans there are.

In terms of the lender of the last resort, giving out discount loans by the central bank plays the role of preventing financial panics. So, the central bank can be righteously seen as the lender of the last resort, which used to be seen as the most important role played by the central bank. When a bank experiences difficulties before it loses control of the situation, the central bank could provide reserve capital to the bank system in order to prevent bank panics and financial panics. During the time of a bank crisis, discount loans are a particularly effective way to inject reserve capital into the banking system. Through this method, the reserve capital can be delivered directly and immediately into the hands of those banks that are in urgent need. So, the discount tool can be used to avoid financial panics by the central bank as it plays the role of the lender of the last resort. That represents an extremely important requirement for successfully introducing monetary policies.

As the lender of the last resort, although the central bank plays the important role of preventing financial panics, there still involve cost. Because commercial banks know that when they are in crises, the central bank will lend its help by providing them discount loans, the commercial banks would generally be motivated to take more risky actions and movements in their daily operations, leading to such problems as moral hazards of the deposit insurance. Such moral hazards are more clearly shown with large banks. These banks believe with a degree of certainty that they belong to the class of such banks that are too large to fail without causing major chain reactions to the society. They are confident that as soon as they fall into a difficult situation, the central bank will lend a hand to save them in order to avoid bank panics caused by failures of these large banks. So, when the central bank applies its discount tool to prevent financial panics, it also considers the moral hazard problem it brings with it as the lender of the last resort by balancing the potential moral hazards and the prevention of financial panics and by limiting itself to frequently play the role of the lender of the last resort.

The most important advantage of the discount policy is that the policy makes the central bank be the lender of the last resort and play the role of preventing financial panics from happening. This role has shown its importance in the history. Another advantage is that the discount policy can be utilized as a piece of information that signals the intention of the future monetary policies of the central bank. If the central bank decides to rise the interest rate (reduce the money supply) in order to slow down the speed of economic growth, it can show its policy intention in the form of increasing the discount rate. From such a showing of intention through the discount policy, the public can draw the conclusion that the future monetary policy is less expansionary. So, the signal sent out by lifting the discount rate is helpful for slowing the economic growth.

However, the discount policy also suffers from two major shortcomings. Firstly, the central bank's announcement of adjusting the discount rate could be misunderstood by the public in terms of the intention of the monetary policy of the central bank. That is because the purpose for the central bank to raise the discount rate could also be about limiting the total amount of money used for discount loans instead of intentionally employing a less expansionary monetary policy. But, the public could very likely understand the central bank's announcement of raising discount rate as the national intention of moving toward a tightening monetary policy. With this misunderstanding consequences opposite to what is intended will be resulted. Secondly, when the central bank fastens the discount rate at a certain specific level, the difference $i - i_d$ of the market interest rate and the discount rate will fluctuate along with the market interest rate. These fluctuations will cause large oscillations in the scale of discount loans and in the money supply of non-policy intensions, making it more difficult to control the money supply.

Other than the two afore-mentioned major shortcomings, the number of discount loans is not under the complete control of the central bank through its adjustment of the discount rate, where the behaviors of commercial banks carry quite good amount of weight on the number of discount loans, see previous chapters for more in-depth discussion. Additionally, the discount policy is not as effective as open market operations, either. Therefore, economists have proposed different ideas to modify the discount policy as a monetary policy tool.

(Friedman, 1969) suggests that the central bank should do away with discount loans in order to develop a more effective mechanism of money supply. His reason is that because of the existence of the Federal Deposit Insurance Corporation, it has already eliminated the possibility of bank panics. Hence, there is no longer any need for the central bank to play the role of the lender of the last resort in order to prevent banks panics from happening. So, it is no longer necessary to employ the discount policy. When discount loans are done away with, because the consequent unintended fluctuation of money supply, as caused by the number of discount loans, will disappear, the mechanism of money supply will become more effective. Some other economists suggest bundling market interest rate together with discount rate in order to eliminate the difference between these rates. By doing so the central bank can continue to employ the discount tool and play the role of the lender of the last resort, while eliminating the origin of rate difference, as caused by changes in the number of discount loans. That of course exterminates the chance for commercial banks to make profit from the discount window by using rate differences. Additionally, changes in discount rate no longer deliver any useful signal so that no public misunderstanding would be led to. However, the central bank does not go along with such idea of reform. Firstly, the central bank does not want to lose this authority it has been holding. Secondly, the central bank maintains that when market interest rate changes, holding discount rate constant is beneficial for reducing the fluctuation of the market interest rate. Especially when the market interest rate rises, this method will increase the number of discount loans, causing banks' reserves to go up and deposits to multiple. That to a certain degree can stop the market interest rate from going high. That is, the liquidity effect of increasing money supply will appear. Although the central bank dissents the idea, the economists' suggestion has produced a good deal of impact. The central bank has followed a discount policy that is not too far from the reform economists suggested. The purpose of the new discount policy is to keep the discount rate not too different from the market interest rate so that the number of discount loans would be under control.

All of the three types of general monetary policy tools, as afore-described, materialize the ultimate goals of monetary policies through adjusting the intermediate objective of money supply, which is our focus of discussion. Other than what has been discussed, the central bank can also either directly or indirectly exert its control over the credit of specific areas. So, there have appeared other types of monetary policy tools. In the following, let us look at some of these tools.

1   The control of consumer credit: This means the control the central bank exerts on the financing of various durable consumer goods other than real estate properties. Its main content includes: (i) the central bank regulates the minimum amount

of the first payment when durable goods are purchased using installments; (ii) the central bank stipulates the maximum term of consumer credit loans; (iii) the central bank specifies the particular kinds of consumer goods on which consumer credit loans are allowed; and (iv) for different kinds of durable consumer goods, the central bank identifies the corresponding terms of consumer credit loans.

2   The control of stock market credit: This means the limitation the central bank imposes on various loans that are related to stock trades in order to contain excessive speculation in the stock market. The content includes: (i) the amount of the first payment when securities are purchased using loans; and (ii) adjust the rate of security deposit either higher or lower depending on the circumstances of the financial market.

3   The control over real estate credit: (i) the maximum loan amount; (ii) the longest term of real estate loans, and (iii) the minimum down payment.

4   Preferential interest rates: This means the loans with preferential interest rates the central bank gives to particular industries that nation wants to foster.

5   The import/export security deposits: The central bank requires importers and exporters to put up with their security deposits that are proportional to their total amounts of import and export. The purpose of this security deposit system is to control the speed of growth in import and export.

By direct credit control, it means that the central bank directly limits the credit activities of commercial banks in the form of executive orders or other forms. There are following types of direct credit controls: (i) the maximum interest rate: According to the law, the central bank specifies the interest rate ceiling payable to time deposits and savings deposits attracted by commercial banks; (ii) credit quotas: Based on the need of the financial market and the need of the economy, the central bank rationally allocates and restrains credit scales to commercial banks after it weighs priorities; (iii) Proportions of liquidity: The central bank stipulates for commercial banks the weights of liquid assets among all bank assets in order to limit the credit expansion of commercial banks; and (iv) direct interference: It means that the central bank interferes with the credit business, the range of loans, and other activities of commercial banks.

In terms of indirect credit control, the following are the main methods the central bank uses: (i) moral advice: when necessary, the central bank provides commercial banks and other financial institutions with either oral or written advice in order to influence their numbers and directions of lending so that they coincide with the intention of financial regulation; (ii) window guidance: Based on the market condition, the trend of price change, and the movement of the financial market, the central bank makes provisions for commercial banks and requires action from the commercial banks regarding how their loan amounts need to be seasonally adjusted.

6.4   CONDUCTION MECHANISM OF MONETARY POLICY

The reason why the central bank can adjust the macroeconomic objectives through using monetary policies is because changes in the monetary policy tools can influence the behaviors of economic participants. The so-called conduction mechanism of monetary policies stands for how applications of monetary policies cause changes in economic activities so that the pre-determined goal of the policies is materialized. The conduction of monetary policies is a complex process. Although there are many different understandings regarding this process, one point is basic. That is, the mechanism of interest is the basis for economic participants to react to changes in the tool of monetary policies. That is the basic starting point of all different approaches of analyzing the conduction problem of monetary policies. Also, starting on this starting point, this section will analyze the conduction mechanism of monetary policies.

Regarding the transmission process of monetary policies, Keynesian theory and monetary theory are two rival systems. The former believes that money supply has relatively minor effect on the level of income, and that the mechanism for changing money supply to play its role of adjusting the macro economy is not due to the existence of a stable close relationship between money supply and the level of income. On the other hand, the monetary theory maintains that money supply is an extremely important factor that affects the level of income, and that there is a stable close relationship between money supply and level of income. One major contribution of the monetary theory is its explanation of the transmission process of money supply of the 1950s and 1960s by using the systems theory of adjusting the amount of priced securities. In particular, when money supply is altered, through a complicated relationship of substitution between financial assets and physical assets, changes in the national level of income are gradually seen. Both of these theories describe the adjustment process of the amount of priced securities similarly. However, they differ on the substitution problems between money and other financial assets and between money and physical assets.

The rate of return on assets is the prerequisite for explaining transmission processes. Speaking generally, each kind of asset produces return; assets offer returns to their owners according to their rates of return. This opinion can be supported readily by various cases. For example, the rates of return of factory buildings and durable production equipment can be easily calculated; the rates of return of financial assets, such as securities, can be obtained from daily papers of finance or television programs; the rates of return of consumer goods, such as daily necessities, household appliances, etc., and money, such as cash, securities, etc., can be easily felt from the services they provide to their owners, although these rates cannot be clearly written as those of capital assets.

Other than increasing the print of currency, another commonly employed method for the central bank to increase the money supply is to purchase national debts through its operations in the open market. When the central bank purchases Treasury bills, it pushes the prices of the Treasuries higher so that the rates of return go down. If the prices of other securities stay constant temporarily, then there will appear an arbitrage process where the owners of wealth adjust their proportions of different securities: they sell the bullish bonds and purchase those whose prices stay unchanged. That process makes the overall return of financial assets go down. And with all other factors constant, the initial purchase of the central bank increases the reserves of commercial banks so that the commercial banks will acquire additional securities and provide more loans. These security purchases by commercial banks make the return of these assets under greater pressure; the increased amount of capital available for lending in turn pushes the service burden of the borrowers lower. When people recognize that such trades of converting money into financial assets are no longer profitable, the initial imbalance caused by the increasing quantity of money in securities will be corrected, and the arbitrage process is completed.

The afore-described process ignores one fact: At the beginning, the expected rates of return of various physical assets do not change because of the increasing money supply and the overall fall in the returns of financial assets. Because of the existence of this fact, in the afore-described process, although at the start the wealth owners want to use other financial assets to substitute for his money, sooner or later they will apply physical assets to replace financial assets. That will cause the demand for physical assets to increase. The rising demand in turn leads to the prices of the physical assets to go higher, while the rising prices stimulate the production of these assets to surge. There is still another aspect to this whole picture. The adjustment process of the amount of securities also directs the wealth owners to expand their demand for the capital goods of consumer type. The increasing money supply induces a whole series of substitution activities among securities. As soon as these substitution activities are completed, it means that the total amount of assets of the durable consumer type, other kinds of physical assets, and financial assets is greatly increased.

Speaking differently, adjustments to the amount of securities represent the consequence of relative changes in the returns of different assets. Throughout this process of adjustment the demand for physical assets rises, or in other words, appear substitution activities that convert financial assets into physical assets. And converting into physical assets is a consequence of the lowering expected return from financial assets when compared to physical assets. This conclusion stands for a basic proposition of economics. It touches on the key difference between Keynesian and monetary theories. Keynesians believe that through changes in interest rate or in rate of return the money supply affects the demand for physical assets so that changes in money supply indirectly affect the national level of income. Although monetarists, as represented by Friedman, also established an adjustment process of the amount of securities, similar to what is discussed above, and discussed the effects of changes in money supply, they do not believe that changes in interest rate or in the rate of return are the prerequisite for changes in the demand for goods and labor. When the money supply is increased, there might appear an adjustment in the amount of securities, which includes movements from money directly into goods. (Friedman and Meiselman, 1963, pp. 221) points out that the ultimate results do not have to come from changes in interest rate; they can also be from changes in price level or income level. It is similar to the situation that an increasing amount of water can flow through a lake without causing the water level of the lake to be any higher than the ordinary.

So, it can be seen that the monetarism in fact believes that changes in money supply will directly cause the national level of income to change; and there is a stable connection between the two variables. According to the description of Keynesians, decreasing rate of return of government bonds will cause the return of other securities to go down due to arbitrage processes; on a relatively low level of securities returns, the substitutions needed for the amount of securities to reach its equilibrium involve not only a relatively good amount of money and a relatively small amount of securities, but also a relatively good amount of goods and a relatively small amount of securities. When there is no profitable change, there will be no reason for increasing demand for goods. Hence, Keynesians believe that there is not any direct stable connection between money supply and income level as claimed by monetarists.

Whether or not there is a direct and stable connection between money supply and income level is the main distinction between monetarism and Keynesian. Even for those who strongly oppose monetarism, they might not deny the possibility that such a connection could truthfully exist. However, they might think that the connection could involve more than simply money supply. Instead it could be that changes in money supply cause changes in the total wealth of the public; without changes appearing to the total wealth, a change in money supply would not lead to any direct effect on the demand for goods. Therefore, whether or not changes in money supply could lead to changes in wealth becomes an important problem. Its answer is expected to be that certain changes in money supply would lead to the described consequences, while other changes in money supply would not.

Increasing money supply causes interest rate to go down, which in turn lifts the present value of income flows of capital. The value of an asset is determined by the capitalization of the income flows the asset generates. When the capitalization is done with a low interest rate, the value of the asset is higher. And because the market values of desired assets, as seen by the public, rise through capital gains, it can be expected that the demand for goods and labor will also go up. One form of connection, which is the so-called wealth effect arising from interest, is to establish a certain connection between money supply and demand without relying on the relative rate of interest.

Changes in money supply inevitably cause changes in income level. With regard to this fact do both Keynesian and monetarism not have any disagreement. Their difference is only about how changes in money supply affect the income level and whether or not there is a close, stable connection between money supply and income level. Keynesians believe that such a relation is loosely defined and experiences large variations from time to time. On the other hand, monetarism believes that this relation is tightly defined with only minor modifications needed once in a while. This end poses a natural question to us: What is the origin of the difference between Keynesian and monetarism? In the following, let us discuss this question.

Although both Keynesians and monetarists describe the adjustment process of the amount of securities in a similar fashion, they hold their individual positions on the substitution problem between money and physical assets. Keynesians believe that money and financial assets are close substitutes of each other. For two goods that are close substitutes of each other, when the price of one good rises while the price of other stays the same, then the demand for the other good will increase. According to this principle, if the prices of other financial assets rise, as caused by an increase in the money supply, then the demand for money will also go up. When other financial assets are relatively higher priced while carry relatively low returns, it means that the public would prefer to hold more money and less other financial assets. Because money plays the role of generally equivalent objects, it means that the public would like to own more physical assets. When the demand for physical assets rises, it stimulates the production of goods to increase so that the national level of income goes up. However, this relationship between changes of money supply and variations in income level is loosely determined, where the goods and physical things the public uses to make substitutions represent only a part of the change in money supply, while this part can vary frequently. From the relevant elasticity (the elasticity of money demand with respect to interest rate), an increase in money supply could lift the income level significantly.

Monetarists, at least Friedman, believe that close substitutions for money are physical assets instead of financial assets. An increasing money supply promotes an adjustment of the amount of priced securities. However, among all the substitutions that take place during the adjustment, there is no monetary substitute that can be used to replace financial assets. Although the rate of return of other financial assets decreases, the public does not decide to hold more money and less other financial assets. The demand for money does not have the elasticity of interest rate. So, what the public decides to do is to own more physical assets (or goods) and less money. In other words, physical assets will be used to substitute for money. The rising demand for physical assets stimulates increasing production of the physical assets, causing the national level of income to go higher. This process of substituting goods for money continues until the productivity of the goods and the national level of income reach such a height that the actual amount of money held by the public is equal to a fixed proportion of the income. At this time, the total quantity of money the public holds is equal to the money demand for the media needed to handle business transactions at the higher level of income. So, the relationship between money supply and income level is direct and stable.

Based on the discussion above, it can be seen that the origin of difference between Keynesian and monetarism is on the substitution problem of assets. This problem is essentially significant, from which there appears the monetary quantity theory of money and the opposing attitude of Keynesians. If we agree with that physical assets are closer substitutes for money than other financial assets, then we will walk toward the quantity theory of money. If we do not think so, then we will be walking toward Keynesian or to the opposite side of the quantity theory of money.

The traditional viewpoint of Keynesian conduction mechanism of monetary policies can be summarized as follows. When the money supply M increases, interest rate i drops, investment expense I goes up so that the total expenses Y rise. That is, $M\uparrow \to i\downarrow \to I\uparrow \to Y\uparrow$. The empirical tests of the monetary school on the importance of money cause people to explore the conduction mechanism of monetary policies freely and generally. Once these different conduction mechanisms are combined into the structural model of Keynesian school, the resultant models will be able to more accurately reveal the impacts of money on economic activities. The consequence will be a cognitive convergence of Keynesians and monetarists on the importance of money. As of this writing, Keynesian school has accepted the monetary opinion on the importance of money. But, it does not support the monetary position that only money is the most important. The structural model of Keynesian school is still employed as sound tool for reasoning and powerful evidence for introducing fiscal policies.

The exploration on the conduction mechanism of monetary policies can be classified into three categories: Monetary policies conduct and play their roles (i) through investment spending, (ii) through consumer spending, and (iii) through international trades.

1   Investment spending

Keynesian school emphasizes on the effect of investment on the fluctuations of economic cycles. That is why the early investigations on the conduction mechanism of monetary policies first focused on the investment spending.

Economists discover that monetary policies affect investment spending through influencing stock prices. Here, there are two points that need to be illustrated. Firstly, how do monetary policies affect stock prices? According to Keynesian school, increasing money supply could lower the interest rate of bonds so that stocks are more attractive to investors than bonds. Along with the increasing demand of stocks, the price of stocks goes up. Another straightforward explanation without Keynesian color is that when the money supply increases, the public discovers that they hold more money than they need so that they would spend the extra money somehow. One of the places the public could spend the money is the stock market. That is, the public would increase their demand for stocks so that the price of stocks goes up. In short, no matter which explanation is applied, it tells who monetary policies affect the stock prices, where increasing money supply raise stock prices.

Secondly, how do stock prices affect investment spending? To address this problem, (Tobin, 1969) develops a theory on the connection between stock prices and investment spending. This theory is known as Tobin's q theory, where Tobin defines q as the ratio of the company's market capitalization value divided by the replacement cost of capital. When q is very high ($q > 1$), the company's market capitalization value is higher than the replacement cost of capital so that the capital of new factory buildings and new equipment would be lower than the company's market value. Under this circumstance, the company could issue stocks and receive from the stocks a somewhat higher price than that it is paying for facilities and equipment. Because the company could issue relatively fewer shares of its stocks for its purchase of new investment goods, its investment spending would increase. Conversely, when q is very low ($q < 1$), because the company's market value is lower than the cost of its capital, the company would not purchase new investment goods so that its investment spending (the purchase of new investment goods) would be very small.

Tobin's q theory provides a very good explanation for the phenomenon of minimum rate of investment spending during the Great Depression in the United States. The higher (lower) the stock prices are, the greater (smaller) Tobin's q is. During the Great Depression, the stock prices slumped. In 1933, the stocks were worth only about one tenth of the value of late 1929. Tobin's q reached unprecedented lower levels so that the rate of investment spending also fell to a very low extreme.

Now, if the previous two explanations are combined together, the following conduction mechanism of monetary policies is naturally obtained: When the money supply M is raised, the stock price P_s goes up, Tobin's q increases, the investment spending grows, and the total spending Y rises. That is, we have the following:

$$M\uparrow \rightarrow P_s\uparrow \rightarrow q\uparrow \rightarrow I\uparrow \rightarrow Y\uparrow \tag{6.1}$$

When the central bank purchases bonds in the open market, it increases the supply of bank loans. Because banks are the most important source of external financing for business firms, bank loans play a special role in economic activities. So, the increasing bank loans will cause a rise in investment spending, leading to increased total spending. That gives us the following conduction mechanism of monetary policies from the point of view of credits:

$$\text{Purchase in open market} \rightarrow M\uparrow \rightarrow \text{bank loans}\uparrow \rightarrow I\uparrow \rightarrow Y\uparrow \tag{6.2}$$

One important meaning of the credit point of view is that the effect of monetary policies on small firms is greater than that on large firms. That is because small firms depend more on bank loans, while large firms could raise investment capital through the stock and bond markets directly.

From how asymmetric information could affect the financial market, it follows that the higher a firm's net value is, the smaller the adverse selection and moral hazards are. Relatively high net value means that the borrower in fact has more guarantees for his loans and the chance for the borrower to take on risky activities is quite small. Increasing net value also reduces the problem of adverse selection so that it encourages financing loans for investment spending. On the other hand, the relatively high net value also means that the shareholders of the company have invested more capital in the company that in turn reduces the problem of moral hazards. The more capital investment there is, the less willing the shareholders are to take risky activities and the less possible the shareholders would invest capital in projects that are beneficial to personal gains than to increasing the company's profits. When the borrower pursues less risky activities and spends less on personal gains, that makes it more likely to have the loans repaid. So, an increasing net value of a firm will lead to rising amount of loans and rising amount of investment spending. Furthermore, increasing stock price raises the net value of the company. And because of a reduction of the adverse selection and moral hazard problems, relatively higher level of investment spending is resulted. Because monetary policies affect stock price P_s, the concept of asymmetric information provides us another conduction mechanism of monetary policies:

$$M\uparrow \rightarrow P_s\uparrow \rightarrow \text{firm's net value}\uparrow \rightarrow \text{adverse selection/moral hazard}\downarrow$$
$$\rightarrow \text{loan}\uparrow \rightarrow I\uparrow \rightarrow Y\uparrow \tag{6.3}$$

2   Consumer spending

The simplified analysis of the empirical tests of the monetary school also believes that there might be a direct connection between monetary policy and consumer spending, where consumer spending consists of two parts. One part is the spending on durable consumer goods; and the other part is the spending on non-durable consumer goods. Durable consumer goods include automobiles, refrigerators, televisions, air conditioners, furniture, etc. The consumer spending on these goods are mostly from borrowed funds, while the spending on non-durable goods is often supported by increased wealth. The early works on the relationship between monetary policy and consumer spending focus on the possible effect of interest rate on the spending of durable consumer goods. Then researchers consider the decision-making problem of how the balance sheet of assets affects the person's consumer spending. These works develop the concept of wealth effect of monetary policies, and consider the relationship between monetary policy and spending on non-durable consumer goods. It is discovered that because the liquidity of durable consumer goods is very poor while the liquidity of stocks is relatively strong, the stock market also affects how a person decide to spend his money. Consequently, the concept of liquidity effect of monetary policies is introduced.

Because the spending on durable consumer goods is often supported by borrowed funds, the developers of the early structural model of Keynesian school look for effects of interest rate on the spending of durable consumer goods. When interest rate is lowered, the cost of borrowing is reduced so that consumers would borrow more to purchase durable consumer goods, leading to increased spending on these goods. The way how money supply affects the total spending can be described as follows:

$$M\uparrow \rightarrow i\downarrow \rightarrow \text{spending on durable consumer goods}\uparrow \rightarrow Y\uparrow \qquad (6.4)$$

This fashion of influence through durable consumer goods is also applicable to spending on houses, because the spending on houses, similar to that on durable consumer goods, is also often supported by borrowed funds. So, when interest rate goes down, the cost of borrowing is reduced, which arouses people's desire to purchase new houses. Hence, we obtain another explanation of how money affects the total spending:

$$M\uparrow \rightarrow i\downarrow \rightarrow \text{spending on new houses}\uparrow \rightarrow Y\uparrow \qquad (6.5)$$

However, the reality shows that such influence is very minor. For example, in recent years, the interest rate in the United States has been kept at nearly zero. However, the impact of the low interest rate on spending on durable consumer goods and new houses has not been clear. In order to explain why monetary policies can influence consumer spending, we must also consider other ways of impact.

(Franco Modigliani and Brumberg, 1954) consider the decision-making problem of how a person's balance sheet of assets affects his consumer spending by making use of their life cycle hypothesis of consumption. The basic premise of Modigliani's theory is that consumers arrange their consumptions evenly according to time. Based on this assumption, the consumer spending as Modigliani talks about stands for the spending on non-durable consumer goods, where the factor that determines the spending is the lifelong riches of the consumer instead of the current income. An important component of the consumer's lifelong riches is financial wealth, while a major portion of the financial wealth is common stocks. When stock prices go up, the value of financial wealth increases so that the consumer's lifelong riches also increase consequently. That leads to an increased spending on non-durable consumer goods. Because we have seen that increasing money supply causes stock prices to go up, we now obtain yet another conduction mechanism of monetary policies:

$$M\uparrow \rightarrow P_s\uparrow \rightarrow \text{wealth}\uparrow \rightarrow \text{lifelong riches}\uparrow$$
$$\rightarrow \text{spending on non-durable consumer goods}\uparrow \rightarrow Y\uparrow \qquad (6.6)$$

From his investigations, Modigliani discovers that this mechanism is very powerful for transmission through the money market; it makes the effect of monetary policies greatly magnified.

Stock market affects the consumer spending of not only non-durable consumer goods but also durable consumer goods. Now, let us see how stock market affects the consumer spending of durable consumer goods.

Durable consumer goods do not have liquidity. Let us imagine that you have to sell some of your durable goods, such as your car, because you suddenly need a certain amount of cash. Then you can expect that you will suffer from great losses. In such an urgent situation where a sale has to be done, the value of a durable consumer good in general cannot be fully materialized. Conversely, if what you hold is a piece of financial asset, such as a bank deposit, some stocks, bonds, etc., then you can quite readily convert the asset into cash according to the market value of the asset. The high level of liquidity of financial assets and the illiquidity of durable consumer goods force consumers to consider such a question: If expecting a forthcoming financial difficulty, should then the consumer hold more financial assets or more durable consumer goods? Evidently, the consumer would be willing to hold those financial assets that have good liquidity. So, we can say that if the chance for a financial difficulty to occur increases, then consumers would cut down their spending on durable consumer goods; if the possibility for a financial difficulty to appear decreases, then the spending on durable consumer goods will increase. That is the liquidity effect.

Next, we like to ask: What factors could cause the possibility for a consumer to have financial difficulties to change? Evidently, the answer to this question has something to do with the balance sheet of the consumer's assets and debts. In particular, the greater the difference between the consumer's financial assets and his debts is, the lower possibility for the consumer to experience financial difficulties, so the more he will spend on durable consumer goods. However, the value of financial assets is related to the stock market. When stock prices go up, the value of financial assets also goes up. Correspondingly, the value of the consumer's financial assets in comparison with his debts also increases so that the chance for him to experience financial difficulties will go down, leading to increased spending on durable consumer goods. With this discussion, we derive the following conduction mechanism of monetary policies:

$$M\uparrow \rightarrow P_s\uparrow \rightarrow \text{value of financial assets}\uparrow$$
$$\rightarrow \text{chance of having financial difficulties}\downarrow$$
$$\rightarrow \text{spending on durable consumer goods}\uparrow \rightarrow Y\uparrow \tag{6.7}$$

The previous discussion is also appropriate for the analysis of spending on housing, because houses are also durable consumer goods, and belong to the class of real estate properties without much liquidity. In particular, when stock prices go up, it improves the balance sheet of the consumer so that the chance for the consumer to suffer from financial difficulties is lowered. So, his desire to purchase a new home is heightened. By writing out this analysis, we have the following path through which money affects spending:

$$M\uparrow \rightarrow P_s\uparrow \rightarrow \text{value of financial assets}\uparrow$$
$$\rightarrow \text{chance of having financial difficulties}\downarrow$$
$$\rightarrow \text{spending on new homes}\uparrow \rightarrow Y\uparrow \tag{6.8}$$

3   International trades

Along with the economic globalization and the implementation of floating exchange rate system, the impact of exchange rate on net export has become an important conduction mechanism of monetary policies. When the domestic interest rate falls, assuming that the rate of inflation stays constant, the attraction of domestic deposits when compared to foreign deposits decreases; and the amount of foreign currency deposit one unit of the domestic currency deposit can exchange for decreases. That means that the exchange rate e falls, and the value of the domestic currency goes down so that domestic goods becomes cheaper compared to foreign goods. That causes the net export NX to increase and the total output Y to go up. Therefore, the following conduction mechanism of monetary policies is derived through international trades:

$$M\uparrow \rightarrow i\downarrow \rightarrow e\downarrow \rightarrow NX\uparrow \rightarrow Y\uparrow \tag{6.9}$$

6.5   EMPIRICAL ANALYSIS ON THE IMPORTANCE OF MONEY

The relationship between money and economic activities is the key for us to investigate the conduction problem of monetary policies. The reason why Keynesian school and the monetary school hold different stances on the conduction problem of monetary policies is mainly because they hold their respectively different understandings on the relationship between money and economic activities (total output, total income). That is, they have different views on the importance of money. In this section, we will discuss the relationship between money and economic activities from the angle of empirical analysis. This discussion not only helps us to understand the importance of money on economic activities, but also provides us with a new way to evaluate some of the seemingly unresolvable debates in economics.

In economics, there are two different types of empirical analyses: Empirical analyses using structural models and empirical analyses using simplified forms. In an empirical analysis using structural model, the available data are used to construct a model, which is then used to explain how one variable affects another variable and whether or not one variable affects another variable at all. In an empirical analysis using simplified forms, the relationship between two variables is directly observed so that the fact of whether or not one variable affects another variable is detected. The difference between these two methods of empirical analyses can be illustrated vividly by using the problem of whether or not drinking coffee can cause heart disease. When the approach of structural model is employed to analyze this problem, an empirical model first needs to be constructed. Then this model is used to analyze the data on how coffee causes metabolic changes in the body, how coffee affects the operation of the heart, how the effect of coffee on the heart could lead to a heart disease, and other relevant processes. On the other hand, if the approach of empirical analysis using simplified forms is employed, direct observations are made to compare whether more coffee drinkers or more non-coffee drinkers suffer from heart problems.

From different types of model analyses, different conclusions can be derived. It is exactly because Keynesians and monetarists employed different approaches of empirical analyses that they arrive at different conclusions and formulate different

Money supply $M \longrightarrow$ Interest rate $i \longrightarrow$ Investment payment $I \longrightarrow \longrightarrow$ Total payment Y

Figure 6.2 Description of Keynesian structural model.

opinions. Keynesians focus on empirical analyses using structural models, which lead to the opinion that money is not important. Because monetarists employ the empirical analyses of simplified forms, they discover that money is extremely important in economic activities so that they pose a major challenge to Keynesian opinion. In order to comprehend the difference in the opinions on the importance of monetary policies, there is a need for us to look at the properties, advantages and weaknesses of these two approaches of empirical analyses.

Keynesian school focuses on how money supply affects economic activities by establishing a structural model in its investigation of the effect of money on economic activities. In this model, to describe economic operations is a series of equations introduced for the behaviors of business firms and consumers. With these equations is then the problem of how monetary and fiscal policies affect the economic total output and total spending explained. A structural model of Keynesian school can contain several equations that describe the working of monetary policies. A systemic depicture of such a Keynesian model is shown in Figure 6.2.

The conduction mechanism as described by this model is that the money supply M affects the interest rate i, while interest rate i affects the investment spending I, whereas the investment spending I produces effects on the total output and the total spending Y. That is, by considering the path of monetary effects, Keynesian school investigates the relationship between money supply M and total output (total spending) Y.

By employing empirical analyses of structural models, we can understand how the economy operates if the models are correct. That is to say, if the employed model contained all the conduction mechanisms of monetary and fiscal policies that affect economic activities, then this empirical analysis would have the following three advantages:

First, more evidences can be obtained on whether or not monetary policies carry important effects on economic activities. For example, if we discover money has important effects on economic activities, then we can speak with certainty that changes in money supply will cause changes in economic activities and that there is a causal relationship between money supply and total output.

Second, more accurate predictions can be made on how money supply affects the total output. For example, when the interest rate is relatively low, we might discover that the effect of expanding money supply is quite minor; when the interest rate is relatively high, we would be able to predict that expanding money supply have quite major effect of the total output (when compared to the time when interest rate is low).

Third, from understanding how economy operates, it might be possible to predict how a change in the economic system would affect the relationship between money supply and the total output. For example, if the control on the interest rate payable to savings deposits is removed, then when the interest rate is raised higher, consumers

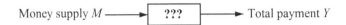

Figure 6.3 The relationship between currency and economic activities.

will be able to acquire more benefits from their savings so that the amount of savings will increase. Additionally, because of the rapid development of financial innovations in recent years, it becomes more important to be able to predict how changes in systems could affect the relationship between money supply and total spending (total output).

The previous advantages of the structural model approach indicate that the quantity of structural model analysis depends on the quantity of the structural model employed. Only under the condition that all conduction mechanisms are sufficiently comprehended, employing structural analysis is the best choice. However, if the structural model employed misses one or two relevant conduction mechanisms of monetary policies, then structural model analysis could greatly underestimate the effect of money supply on total spending (total output). That is exactly what worries the school of monetarists: Many of Keynesian structural models might have missed some important conduction mechanisms of monetary policies. For example, if a conduction mechanism of monetary policies has everything to do with consumer spending while it has nothing to do with investment spending, then the Keynesian structural model that emphasizes on investment spending $(M \rightarrow i \rightarrow I \rightarrow Y)$ might underestimate the importance of money on economic activities. So, the monetary school believes that Keynesian school specifies too particularly how money functions so that the importance of monetary policies cannot be sufficiently comprehended. That is indeed the shortcoming of the structural model analysis.

As for the monetary school, it does not have any concrete way to describe how money supply affects the total output. Instead, it observes whether or not changes in total output are highly correlated to changes in money supply and draws the conclusion on the effect of money on economic activities. That is the empirical analysis using simplified forms of the monetary school. In other words, when analyzing the effects of money supply on the total output, the economy is seen as a black box whose inner function is not visible to the man. A systemic representation of the empirical analysis on the relationship between money and economic activities using simplified form is shown in Figure 6.3.

The advantage of the analysis of simplified forms over that of structural models is that the former does not impose any condition on the form of how monetary policies affect economy. So, even though we do not know all the conduction mechanisms of monetary policies, it is still possible for us to identify all the effects of monetary policies on the total output through observing whether or not changes in total output are highly correlated to changes in money supply.

The reason why the monetary school encourages the use of empirical analysis of simplified forms is because the members of the school believe that the specific ways of how money supply affects the total output can take many different forms and change from one circumstance to another constantly; and in fact it is difficult for people to know all conduction mechanisms of monetary policies. Hence, employing structural models to make analysis can easily underestimate the importance of money. So, to analyze the relationship between money and economic activities, the approach of analysis using simplified forms is more advantageous than that of structural models.

However, the analysis using simplified forms also suffers from shortcomings. In particular, when changes in the total output are not caused by money supply, the analysis using simplified forms could lead to misunderstanding. If the relationship between two variables is not causal, no matter how closely they are related, none of the variables could be treated as the cause, while the other as the consequence. For example, when the number of criminal activities is on the rise, there will be more police officers patrolling the streets. Even though the number of criminal acts and the number of police officers on the streets are correlated, we cannot conclude that because of the patrolling of police officers criminal activities are led to. So, one weakness of the analysis using simplified forms is that it suffers from the problem of causal relationship, where we really do not know which variable is the cause and which variable is the consequence.

The other side of the causality problem is that there might be another unknown factor that is behind the variables that change in concert. For example, coffee drinking might be related to having heart disease. However, this relationship cannot be explained as coffee drinking causes the heart disease. Instead, it is because many of the coffee drinkers are under heavy pressures in life. It is these heavy pressures that cause the heart problems. So, reducing coffee drinking cannot really decrease the onset of heart problems. Similarly, if there is a unknown factor that causes both money supply and the total output change coordinately, then simply controlling the changes in money supply will not help to improve the control on the total output.

The theoretical framework Keynes develops in 1936 for analyzing the overall economic activities only gains acceptance in the 1950s and early 1960s by the majority of economists. The research of Keynesian school on the importance of money on economic activities during this period of time shows clearly the following tendency: changes in money have nothing to do with changes in total output. As for the present day, the recognition of Keynesian school on the importance of money is different of that of the earlier time. They accept the criticism of the monetary school and recognize that money plays a very important role on economic activities.

The earlier Keynesian school believes the ineffectiveness of monetary policies. They construct three structural models to empirically support this claim. Firstly, the earlier Keynesian school believes that monetary policies only affect nominal interest rate, while the nominal interest rate influences investment spending, which moves the total output. During the time of Great Depression, the interest rate of American Treasurys fell to the lowest level in the history. For example, the interest rate of three-month Treasury fell below 1%. Such low interest rate during the Great Depression indicates that the monetary policies were loose, because they encouraged investment. So, there was no tightening effect during this period of time. So, the monetary policies cannot be used to explain why the worst economic contraction ever happened. Based on this phenomenon, Keynesian school obtains the conclusion that: Changes in money supply do not affect the total output. In other words, money is not relevant. Secondly, the early empirical research also discovers that there is no correlation between the nominal interest rate and investment spending. Because the early Keynesian school treats the relationship between nominal interest rate and investment spending as a path through which money supply affects the aggregate demand, when the relationship between the nominal interest rate and investment spending is very weak, changes in money supply will produce almost no effect on the total output. Thirdly, the early Keynesian school surveys the investment of industrial and commercial firms on physical assets. That study indicates that the decision of the firms on the amount of investments on new physical assets is not affected by the market interest rate. This evidence further supports that the relationship between interest rate and investment spending is quite weak, which further backs the conclusion that money is irrelevant. Because of the previous empirical explanation on the importance of money of the early Keynesian school, most economists did not give enough attention to monetary policies until the 1960s.

Just when Keynesian economics was at its heyday, Friedman and A. J. Schwartz in 1963, for more details see (Friedman and Schwartz, 1971), did an empirical investigation on the monetary policies of the United States during the Great Depression. They discovered that during that period of time, a large number of American banks went bankrupt, causing the money supply to decrease and an economic contraction of an unprecedented magnitude in history. In other words, monetary policies can be used to explain the reason behind the economic contraction in the United States during the Great Depression. So, the monetary school consequently developed its opinion completely opposite of that of the early Keynesian school: The monetary policies during the Great Depression were not loose; the Great Depression cannot be singled out to show the ineffectiveness of monetary policies during this period of time; so, it cannot be excluded that during this period of depression the monetary policies had played an important role. And the fact was indeed so: no monetary policies were ever tighter than those employed during the Great Depression. Because of the criticism of the monetary school, many economists had to reconsider their positions regarding whether or not money is important.

The monetary school also disagrees with the claim of the early Keynesian school that the relationship between nominal interest rate and investment spending is not strong so that investment spending is not affected by monetary policies. Instead, the monetary school believes that although the relationship between nominal interest rate and investment spending is relatively weak, that does not exclude the possibility that there is a relatively strong correlation between the actual interest rate and investment spending. In fact, when nominal interest rate is employed to illustrate the dynamics of the actual interest rate, many misunderstanding can be induced. Such misunderstandings exist not only for the time period of the Great Depression, but also other time periods of normal economic development. In fact, nominal interest rate cannot truthfully reflect the real costs of borrowing, and low nominal interest rate not necessarily means low borrowing cost. For example, if the price level drops 10%, even if the nominal interest rate is zero, the actual cost of borrowing is still as high as 10%. So, only the actual interest rate can truthfully reflect the real cost of borrowing. Even so, the relationship between the real interest rate and investment decision is stronger than that between nominal interest rate and investment decision. So, although the empirical analysis of early Keynesian school indicates that nominal interest rate has very minor effect on investment spending, it cannot exclude that changes in money supply have very strong influence on investment spending and in turn on the aggregate demand.

At the same time, the monetary school also believes that the effect of interest rate on investment spending can only be one of the many ways through which monetary policies affect the aggregate demand. Even as what the early Keynesian school claims as that interest rate has very minor effect on investment spending, monetary policies can still produce large impacts on the aggregate demand through other paths. So, money is important on economic activities instead of irrelevant.

In the early 1960s, on the basis of empirical analysis using simplified forms, Friedman and his colleagues conducted a large amount of research on the importance of money and developed a whole set of theory and evidence on why money is important on economic activities. The empirical analysis using simplified forms can be roughly divided into three types: timing confirmation, statistical test, and historical verification. A timing confirmation observes whether or not a variable often changes ahead of another variable. Consequently a causal relationship between the variables is derived. A statistical test uses the normal statistical tests to verify the statistical relationship between two variables. A correlation relationship between two variables is resulted. A historical verification observes specific historical periods in order to check whether or not changes in one variable can cause changes in another variable. In the following, staring with these three types of empirical analyses using simplified forms we will see how the monetary school analyzes the importance of money.

Firstly, the timing confirmation of the monetary school mainly considers how the rate of increase in money supply changes with respect to the economic cycle. (Friedman and Schwartz, 1963) discover that within each of the economic cycle of the past nearly one hundred years, the rate of increase in money supply has always dropped before the total output falls. When using average, it is found that the peak rate of increase in money supply appears about 16 months ahead of the peak of the total output. Such lead time is variable, ranging from several months to over two years. Based on this fact, Friedman and Schwartz conclude that changes in money supply cause the cyclic movement of the economy, while the effect of the former on the latter has a long-term variable time delay.

The basis of timing confirmation is the following philosophical principle: If one event appears after another event, then the second event is caused by the first event. In fact, this principle only holds true under the following condition: The first event is an exogenous event. That is, this event is the result of some independent action, where the independent action can be caused neither by any event that appears after the action nor by any exogenous factor that might affect the action and other events that appear after the action. According to this principle, if the first event is exogenous while the second event occurs after the first event, then the first event is the cause, and the second event is the result. For example, a chemist mixes two chemicals together in his lab experiment. Unexpectedly, a sudden explosion occurs and the chemist is killed. From this we can firmly say that the cause of death of this chemist is that he mixed the particular chemicals together. In scientific experiments, this philosophical principle is very useful.

However, the study of economics is not as exact as that of hard sciences. In general, it is not possible to determine whether or not an economic event, such as the falling of the growth rate of money, is an exogenous event. Changes in the growth rate of money can be caused by exogenous factors; but they can also be induced by the events expected to happen if the growth rate of money varies. So, it is generally difficult or impossible to determine whether or not an economic event is the cause of another economic event. Hence, timing confirmation clearly possesses the characteristics of analysis using simplified forms. That is, it directly investigates the relationship between two variables. To this end a natural question is: there might be a reversed causal relationship. For example, if the growth of money supply occurs before the grow in the total output, however, the growth in money is caused by the growth in the total output, then this is exactly the opposite of the causal relationship as described in the previous philosophical principle. So, it follows that timing confirmation can very easily lead to misunderstandings. Furthermore, in order to discover the desired event, people might very likely focus their attention on certain variable so that some untruthful relationship is obtained. For definite causal relationships, timing confirmation might be a dangerous method of research.

Secondly, the statistical tests the monetary school uses employ the normal statistical methods to uncover the relationship between money and economic activities. For instance, (Friedman and Meiselman, 1963) tested monetary models and Keynesian models. According to Keynesian theories, they treated investment spending and government spending as the reason for the fluctuation seen in the total demand. So they introduced a Keynesian autonomous expenditure variable A, which is equal to the sum of investment spending and government spending. After that, they characterized the Keynesian model as follows: A should be highly correlated to the total spending Y, while the money supply M would not be highly correlated to the total output Y. At the same time, they also established a monetary model, where money supply M is seen as the cause for the total output Y to fluctuate so that M and Y should be highly correlated, while A is not highly correlated to Y. Now, which of these two models should be correct? One logical method to answer this question is to compare M and A and see which one of them should be more highly correlated to Y. To this end, Friedman and Meiselman used the available American data collected for different time periods. Their work discovered that the correlation between M and Y is higher than that between A and Y. So, they concluded that in terms of studying how to determine the total spending, the monetary analysis method is better than the Keynesian analysis method.

However, regarding empirical analysis of Friedman and Meiselman there are also some criticisms. Summarizing what is known in the literature, these criticisms can be grouped into the following three areas. Firstly, the employed empirical tests using simplified forms might suffer from reverse causal relationships and the possibility that there might be such an exogenous factor that is behind both variables of concern. Secondly, such tests are not fair, because the description of the Keynesian model is too much simplified. Generally speaking, a Keynesian structural model contains over one hundred equations, while the Keynesian model Friedman and Meiselman established has only one equation. So, this simplified model is not enough to reflect various effects of autonomous spending. Additionally, Keynesian models generally also contain effects of other variables. As soon as the effects of these additional variables are ignored, the effect of monetary policies could be overestimated, while the effect of autonomous spending is underestimated. Thirdly, the construction of the statistical index A, as introduced by Friedman and Meiselman for autonomous spending, might not be appropriate. That hinders the Keynesian model to play its roles effectively. For instance, (Ando and Modigliani, 1965) deliberately constructed the variable of autonomous spending and made careful measurements. Their conclusion was the exact opposition: Keynesian models are advantageous than monetary models. For the question of which method of defining autonomous spending is more appropriate, (Poole and Kornblith, 1973) did a post-event experiment. Their work did not provide clearly indications on whether Keynesian models are better than monetary models or monetary models are better than Keynesian models.

Thirdly, (Friedman and Schwartz, 1971) did an empirical investigation on the American monetary policies employed during the Great Depression. They discovered that the Americans employed monetary policies during the Great Depression that were not loose. Instead, the bank panics of the time caused a drastic fall in the money supply. This book also explained by using detailed historical documents that the monetary growth appears ahead of economic cycle, and the rate of monetary growth falls each time before the economic depression sets in. Among all the historical time periods considered in this book, almost all changes in money supply were exogenous events. These time periods behaved almost exactly like controlled experiments. So, in the empirical tests of these time periods, the philosophical principle that "the first is the cause and the second is the result" may very well be tenable: If during these historical periods as soon as the rate of change in money supply falls, the level of the total output follows the suite, then the conclusion that monetary growth is the acting force behind economic cycles is backed with a strong empirical test. One very good example of such historical time period is that during 1936 to 1937 the increasing rate of statutory deposit reserves caused the growth rate of money supply to fall drastically. At the time, for the purpose of improving its control over monetary policies, the Federal Reserve System of the United States raised the rate of the statutory deposit reserves (Note that this decision was not made to deal with a particular economic situation). So, from the existence of a reverse causal relationship between the total output and money supply, it follows that the decline in the money supply of this historical period can be treated as an exogenous event of the characteristic of controllable experiments. Soon after the occurrence of this exogenous event, the severe economic depression appeared during 1937 and 1938. So, it can be concluded with certainty that the decline in the money supply, as caused by the increasing rate of the statutory deposit reserves, was indeed the reason for the economic depression to follow.

(Friedman and Schwartz, 1971) also did a historical survey of the bank panics that occurred in 1907 and other times. They discovered that the relevant falls in money supply were also exogenous events. So, the frequent appearance of economic declines following falls in money supply provides strong evidence that changes in money supply indeed bear quite important effect of the total output.

The precious discussions on the empirical tests of money tell us that because there might be reverse causal relationships and the effects of exogenous factors, people do not feel certain about the conclusions produced out of timing confirmation and statistical tests. However, some historical verification can indeed provide powerful supports for opinions of the monetary school. So, if the three approaches of empirical studies – timing confirmation, statistical tests, and historical verification – are combined together and each produces the same conclusions, then such results could be seen as having sound bases. The empirical tests of the monetary school on the importance of money materialized the unification of timing confirmation, statistical tests, and historical verification. So, it is sound to believe that money is very important to economic activities. This conclusion of the empirical tests of the monetary school shocked the community of economists of the time, because at that time most economists followed the opinions of the early Keynesian school, believing that money is irrelevant to economic activities. From then on, the research of economics has evolved in two directions. One is checking the importance of money on economic activities by using more complicated models of simplified forms of the monetary school; and the other continues the method of structural models to explore other ways of how money affects the total spending without limiting on the study the effect of interest rate on investment.

6.6   MONEY AND FISCAL POLICIES IN THE IS-LM MODEL

Monetary and fiscal policies represent the two major policies through which the government regulates the economy under the condition of modern market economy. If the decision makers of the government decide to increase either the money supply or the government spending, then the IS-LM model will be helpful to them to analyze the effects of these policy tools on the total output and interest rate. That is, the IS-LM model can help us understand the effects and effectiveness of monetary and fiscal policies on economic activities.

Which of these two classes of policies is better? Which one is more effective? The answers to these questions are relative and conditional. In practical applications, monetary policies and fiscal policies can substitute for each other and complement each other. When they are combined together appropriately, they can greatly help to materialize the desired economic goals. This section will address these questions by applying the method of structural model analysis on the basis of the IS-LM model. We will include various kinds of conduction mechanisms of monetary policies into the structural model in order to reveal how money affects the total output in different fashions.

6.6.1   Movement of IS curve

For each chosen interest rate IS curve depicts the equilibrium point of the commodity market, which is the various combination of the total output and interest rate under the equilibrium condition that the total output is equal to the aggregate demand. When an autonomous factor that has nothing to do with the interest rate changes independently of the total output, the position of the IS curve moves. And the displacement of the IS curve causes the changes in the equilibrium output. Now, we analyze the five autonomous reasons, which are those factors that have nothing to do with interest rates, that can make the position of the IS curve move: change in consumer spending, change in investment spending, change in government spending, change in tax collection, and change in net export.

Firstly, let us look at changes in autonomous consumer spending. Speaking generally, the reason why autonomous consumer spending increases is because the consumer expects a good economic future. For example, the discovery of a new oil field, agricultural harvest year after year, the prosperity of the market place, etc., are all signs of expecting a good economic future. In such circumstances, the autonomous consumer spending generally increases. When the autonomous consumer spending goes

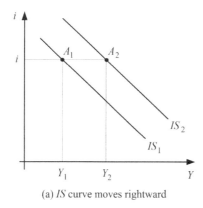

(a) *IS* curve moves rightward

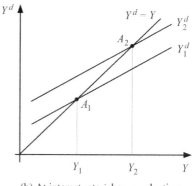

(b) At interest rate *i*, how production
output market is affected

Figure 6.4 Effect of changes of autonomous factor on the location of the IS curve.

up, it causes the aggregate demand curve to move upward so that the IS curve moves
rightward. In order to explain how such movement occurs, let us assume that the
initial position of the IS curve is located at the IS$_1$ position in Figure 6.4(a). Now let
us look at the situation of movement of the equilibrium total output at each level of
interest rate. Assume that *i* is an arbitrarily chosen level of interest rate and that at
this level of interest rate, the initial position of the total demand curve is located at
the position of Y_1^d as shown in Figure 6.4(b), where the equilibrium total output is
at the level Y_1. Assume that the initial consumer spending is C_1, and that the increase
in the autonomous consumer spending causes the consumer spending goes to C_2. So,
the total demand changes from $Y_1^d = C_1 + I + G + NX$ to $Y_2^d = C_2 + I + G + NX$, the
aggregate demand curve moves from the position of Y_1^d to that of Y_2^d so that the equilibrium total output increases from the level Y_1 to the level Y_2. That means that when
the autonomous consumer spending increases, it makes the equilibrium total output
at each fixed level of interest rate rise so that the IS curve moves rightward.

In the discussion of the conduction mechanisms of monetary policies in the previous section, it is pointed out that consumer spending *C* is also related to interest rate *i*.
That is to say, consumer spending consists of two parts: the autonomous spending that
is related to interest rate and the other autonomous part that has nothing to do with
interest rate. When interest rate rises, on the one hand, due to the increasing opportunity cost of the current consumer spending, the current consumption goes down.
On the other hand, the interest rate of deposits also rises. So, each one dollar current spending means more spending the consumer can enjoy in the future, which also
causes the current spending to fall. With this consideration in place, the consumption
function under the fixed interest rate can be completely written as follows:

$$C = C_0 + c(Y - T) - \alpha i \tag{6.10}$$

where *T* stands for the total tax, *Y* the total income, $C_0 - \alpha i$ can be explained as
autonomous spending, and C_0 the autonomous, spontaneous spending.

Secondly, let us look at changes in autonomous investment spending. Changes in interest rate can affect the spending of planned investments so that the equilibrium total output is also affected. However, such changes represent those along the same IS curve without affecting the location of the IS curve. In comparison, changes in autonomous investment spending are different. These changes are not affected by interest rate and belong to changes of exogenous factors. So, these change can affect the location of the IS curve. In other words, along with changes in autonomous investment spending, the equilibrium total output at each chosen level of interest rate also changes. Because investment spending consists of two parts – the autonomous investment spending and the investment spending that is related to interest rate, the autonomous investment spending is exogenous, it does not change with interest rate, while the investment spending that is related to interest rate is endogenous. The earlier discussion on the determination of the total output with fixed interest rate in fact holds true only when the autonomous investment spending is predetermined. Now, because changes occur to this pre-determined condition, the IS curves both before and after the change are different. Speaking strictly, the investment function, considering the autonomous investment spending should be written as

$$I = I_0 + vY - \beta i \tag{6.11}$$

That is, I changes in the same direction of the total income Y and in the opposite direction of the interest rate i. In this equation, $I_0 - \beta i$ can be seen as the spontaneous investment as mentioned earlier. Then I_0 is the autonomous, spontaneous investment; it has nothing to do with either the income or the interest rate. It is a completely exogenous factor. The quantity $I_0 + vY$ can be seen as the autonomous investment, that is, the investment when the interest rate (of loans) is zero. It has nothing to do with the interest rate. If the economic conditions are expected to be good, then the autonomous investment spending will increase. And an increasing autonomous investment spending will cause the aggregate demand curve $Y^d = C + I + G + NX$ to move upward as shown in Figure 6.4(b), so that the IS curve moves rightward, Figure 6.4(a).

Thirdly, let us look at changes in government spending. When the government spending increases, it will make the aggregate demand curve at any given level of interest rate move upward, Figure 6.4(b), so that the IS curve moves to the right, Figure 6.4(a). Conversely, when the government spending falls, it will make the IS curve move leftward.

Fourthly, let us look at changes in the amount of tax collection. When the government increases its tax collection, it makes the disposable income of the public decrease so that the consumer spending falls ($C = C_0 + c(Y - T) - \alpha i$), where C decreases when T increases. That makes the aggregate demand curve at any level of interest rate moves downward and the equilibrium total output fall. Consequently, the IS curve moves leftward. Conversely, reducing tax collection can make the aggregate demand curve at any given level of interest rate move upward, as shown in Figure 6.4(b), the equilibrium total output rise, and the IS curve move rightward, as shown in Figure 6.4(a).

Tax is the main source of income for the government. That is, the operation of the government is mainly covered by tax income. Speaking generally, tax income changes simultaneously with the government spending. When tax income increases, the government spending goes up; when tax income decreases, the government spending goes down. According to the previous analysis, increasing tax income makes the IS curve move leftward. So increasing government spending makes the IS curve move rightward.

Next, let us address the problem of whether changes in tax collection or changes in government spending bear more effect. To this end, assume that at the same time when the government increases its tax collection, it also increases its spending in the same amount. Let the increase in tax income T is ΔT. So, the government spending G also increases in the amount of $\Delta G = \Delta T$. In this case, at any given level of interest rate i and given level of total output Y, the following describes the situation of change in the aggregate demand $Y^d = C_0 + c(Y - T) - \alpha i + I + G + NX$:

$$\Delta Y^d = \Delta G - c\Delta T = (1 - c)\Delta T \tag{6.12}$$

Because c is the marginal propensity to consume, satisfying $0 < c < 1$, we have $\Delta Y^d > 0$. That is to say that although the increasing tax income makes the aggregate demand curve to move downward as much as $c\Delta T$, the equivalent amount of increase in government spending makes the aggregate demand curve move upward as much as ΔG, where $\Delta G > c\Delta T$. So, the net result is that the aggregate demand curve moves upward in the amount of $(1 - c)\Delta T$. That causes the IS curve to move rightward. That is, the equilibrium total output at any given level of interest rate has been increased.

In the following let us compute how much the IS curve has moved rightward. To this end, let i be the given level of interest rate. In this case, the equilibrium equation of the commodity market is

$$Y = C_0 + c(Y - T) - \alpha i + I_0 + vY - \beta i + G + NX \tag{6.13}$$

Hence, the equilibrium total output is:

$$Y = \frac{C_0 - \alpha i + I_0 - \beta i + G + NX}{1 - c - v} + \frac{G - cT}{1 - c - v} \tag{6.14}$$

Now, because only G and T are changed (they are increased by the same amount) while all other variables stay constant, the change in the equilibrium total output is given as follows:

$$\Delta Y = \frac{\Delta G - c\Delta T}{1 - c - v} = \frac{1 - c}{1 - c - v}\Delta T \tag{6.15}$$

That is the magnitude of how much the IS curve moves rightward. Notice that $(1 - c)\Delta T$ stands for the magnitude of the upward movement of the aggregate demand curve. Since $0 < 1 - c - v < 1$, it follows that the magnitude of the rightward movement of the IS curve is greater than the magnitude of the upward movement of the aggregate demand curve.

What is analyzed above is of important policy significance: If the government increases its tax collection and at the same time increases its spending in the same amount in such areas as maintaining the normal economic operation, safe-keeping national security, constructing public projects, relieving poverty, etc., such a fiscal policy of "from the people and giving back to the people" can expand the domestic demand of the nation and stimulate the improvement of the national total output.

Fifthly, let us look at changes in autonomous net export. As what has been concluded earlier, interest rate affects the net export. However, the change in the net export as caused by changes in interest rate can only make the equilibrium total output move along the IS curve without causing the IS curve to relocate. Within the change in net export, there is such a part that is caused by external factors. For example, if a particular product is welcomed by foreign buyers, then the increased production of that particular product can help increase the net export of the nation. That part of net export increase has nothing to do with interest rate. It belongs to the so-called autonomous increase. The consequence of such autonomous increase in the net export is that the aggregate demand curve at each given level of interest rate moves upward and the equilibrium total output rises, making the IS curve move rightward. After we have considered this exogenous factor behind the autonomous change in net export, the net export function can be written as follows: $NX = NX_0 - \gamma i$, where NX_0 stands for the autonomous net export, which is equal to the net export when the interest rate is zero. Now, the autonomous factor (an exogenous factor) that affects the location of the IS curve can be inserted into the equation that determines the equilibrium total output of the commodity market:

$$Y = (C_0 + cY - cT - \alpha i) + (I_0 + vY - \beta i) + G + (NX_0 - \gamma i) \tag{6.16}$$

Therefore, the equation of the IS curve is given as follows:

$$Y = \frac{C_0 + I_0 + NX_0 + G - cT}{1 - c - v} - \frac{\alpha + \beta + \gamma}{1 - c - v} i \tag{6.17}$$

where C_0, I_0, NX_0, G, and T are exogenous factors, c the marginal propensity to consume, v the marginal propensity to invest, satisfying that $c > 0$, $v > 0$, and $0 < c + v < 1$. When the exogenous factors change, the magnitude of displacement of the IS curve, which is the amount of change in the equilibrium total output at any given level of interest rate, is given as follows:

$$\Delta Y = \frac{\Delta C_0 + \Delta I_0 + \Delta NX_0 + \Delta G - c\Delta T}{1 - c - v} \tag{6.18}$$

6.6.2   Movement of LM curve

The LM curve describes how the equilibrium interest rate of the money market changes with the total income, which is the various combinations of total income and interest rate that make the money supply equal the demand for money. There are only two factors that can cause the LM curve to move. One is the autonomous changes in the demand for money, and the other changes in money supply.

By autonomous demand for money, it means the part of the demand for money that has nothing to do with the price level, the level of total output (total income), and the interest rate. The reason why people have autonomous demand for money is because people recognize the fluctuation and risk associated with the securities market so that

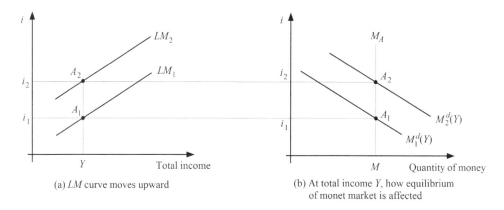

(a) *LM* curve moves upward

(b) At total income *Y*, how equilibrium
of monet market is affected

Figure 6.5 The upward movement of the LM curve when the autonomous demand for money
increases.

holding certain amounts of cash in hands can always make people feel relaxed. When
the autonomous demand for money increases, it causes the demand-for-money curve
at every total income level Y to move rightward as shown in Figure 6.5(b), from the
original location $M_1^d(Y)$ rightward to the new location $M_2^d(Y)$ so that the equilibrium
interest rate at each total income level rises. That is, the LM curve moves upward, as
shown in Figure 6.5(a), from the original location LM_1 to the new location LM_2.

After considering the autonomous demand for money, the demand-for-money
function now can be simply written as follows:

$$M^d = M^d(Y, i) = M_A + aY - bi \tag{6.19}$$

where M_A stands for the autonomous demand for money. When this demand increases
an increment ΔM_A, the amount of rightward movement of the demand-for-money
curve is $\Delta M^d = \Delta M_A$, and the amount of upward movement of the LM curve is
$\Delta M_A/b$, which is the amount of rise in the equilibrium interest rate. It is because
according to the equilibrium equation of the money market

$$M_A + aY - bi = M \tag{6.20}$$

it follows that for any given level of total income Y, the equilibrium interest rate i is
given by $i = (aY + M_A - M)/b$. So, when M_A experiences a change in the amount of
ΔM_A while all other factors stay constant, the amount of change in the equilibrium
interest rate is $\Delta i = (\Delta M_A)/b$, which is exactly the amount of upward movement of
the LM curve.

When money supply is increased, it directly causes the vertical demand-for-money
curve to move rightward, as shown in Figure 6.6(b), from the original location M_1 to
the new location M_2 so that the equilibrium interest rate at every level of income falls.
That is, the LM curve moves downward, as shown in Figure 6.6(a), from the original
location LM_1 downward to the new location LM_2. If M stands for the amount of
money supply and ΔM for the amount of increase in the supply, then the amount of

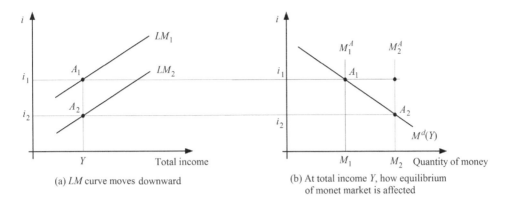

(a) *LM* curve moves downward

(b) At total income *Y*, how equilibrium
of monet market is affected

Figure 6.6 The downward movement of the LM curve when the money supply increases.

change in the equilibrium interest rate $i = (aY + M_A - M)/b$ at every level of the total income is $\Delta i = -(\Delta M)/b$. So, when the money supply increases, the LM curve moves downward in the amount of $(\Delta M)/b$.

6.6.3 Movement of total output and interest rate

Now, let us analyze the reactions of the equilibrium total output and the equilibrium interest rate to monetary and fiscal policies by using the knowledge on what causes the IS and LM curves to move. For this purpose, let us first combine the equilibria of the commodity market and the money market together to formulate the necessary equation in order to determine the equilibrium interest rate and the equilibrium total output. That is, we like to first develop the equation needed to determine the interaction point of the IS and LM curves.

From solving the system of equ. (6.17) of the IS curve and equ. (6.20) of the LM curve, the equilibrium total output Y and the equilibrium interest rate i can be obtained

$$Y = \frac{b(C_0 + I_0 + NX_0 + G - cT) + (\alpha + \beta + \gamma)(M - M_A)}{a(\alpha + \beta + \gamma) + b(1 - c - v)} \tag{6.21}$$

$$i = \frac{a(C_0 + I_0 + NX_0 + G - cT) - (1 - c - v)(M - M_A)}{a(\alpha + \beta + \gamma) + b(1 - c - v)} \tag{6.22}$$

The effect of monetary and fiscal policies on the total output and interest rate can be analyzed by using these two equations.

Assume that initially both the commodity market and the money market are at equilibrium where the economy is situated at the equilibrium state E_1 with the level of total output at Y_1 and the level of the market interest rate at i_1, as shown in Figure 6.7(a). However, at this moment there appears unemployment in the economy so that the current level of the total output Y_1 becomes too lower so that the monetary author-ity decides to reduce the unemployment through increasing the money supply and the output. The increased money supply makes the LM curve moves downward from the

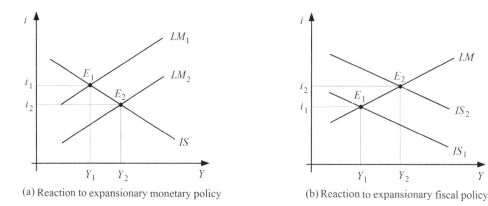

(a) Reaction to expansionary monetary policy   (b) Reaction to expansionary fiscal policy

Figure 6.7   Reactions of total output and interest rate to monetary and fiscal policies.

original location LM_1 to LM_2, the point that makes both the commodity market and the money market simultaneously reach equilibrium moves from the original location E_1 to the new location E_2. That causes the interest rate to fall to i_2, and the total output increases to the new level Y_2, materializing the desired effect of the monetary authority by applying the monetary policy. Here, the specific monetary policy has played the role of stimulating the economic growth and maintained sufficient employment.

In order to fully understand why the total output rose and why the interest rate fell, let us carefully consider that happened during the process when the economy changed from E_1 to E_2. When the economy was still situated at the original state E_1, the monetary authority injected new money into the economy so that the money supply was increased. That broke the initial state of equilibrium and caused the market interest rate to fall due to the additional money supply. The falling interest rate in turn stimulated the consumer spending, investment spending, and net export, making the spending in these areas to rise. That conversely helped to lift the aggregate demand so that the total output was increased. As long as the excessive money supply in the money market does not disappear, the interest rate will continue to fall, consumer spending, investment spending and net export will continue to rise, making the aggregate demand to increase continuously so that the total output will continuously to go up. When the economy reached the point E_2, the excessive money supply disappeared, at the same time there was not excessive demand for money. So, changes in consumer spending, investment spending, and net export were also stopped, making the change in the aggregate demand also cease. So, the increase in the total output was stopped. At this point in time, the growth in the total output reached its minimum magnitude. That is the conduction path of the monetary policy.

The formulas in equs. (6.21) and (6.22) can be used to make quantitative analysis of the implementation effects of monetary policies. Assume that the increment in money supply is ΔM and that all other exogenous factors stay invariant. So, from equ. (6.21) it follows that the increment in the total output is

$$\Delta Y = \frac{\alpha + \beta + \gamma}{a(\alpha + \beta + \gamma) + b(1 - c - v)} \Delta M \tag{6.23}$$

From equ. (6.22) it follows that the amount of fall of the interest rate is:

$$\Delta i = \frac{1 - c - v}{a(\alpha + \beta + \gamma) + b(1 - c - v)} \Delta M \tag{6.24}$$

The roles played by the consumer spending, investment spending, and net export in the conduction of the monetary policy are reflected in these two quantitative formulas through the coefficients α, β, and γ. What is more important is that these formulas indicate that the total output is positively correlated to money supply, and that the market interest rate is negatively correlated to money supply.

Next, let us consider the functions of fiscal policies. Assume that when the economy is situated at the equilibrium state E_1, the employment is not sufficient and the government does not want to reduce the unemployment by stimulating the economic growth through increasing money supply. Then can the government increase the total output and reduce the unemployment through adjusting the government spending and increasing the tax revenue? The IS-LM model indicates that this is possible. Figure 6.7(b) describes how the total output and interest rate react to expansionary fiscal policies. When the government adopts an expansionary fiscal policy, it is about either increasing the government spending or reducing the tax revenue. The discussion on what factors cause the IS curve to move its location has already indicated that both increasing government spending and reducing tax revenue can make the IS curve move rightward so that the total output goes up and the unemployment goes down. In Figure 6.7(b), the IS curve moves from its original location IS_1 to the new location IS_2, the simultaneous equilibrium point of the commodity market and the money market moves from the original location E_1 to the new location E_2. Correspondingly, the total output increases from the originally lower level Y_1 to the new higher level Y_2, and the interest rate also rises from i_1 to i_2. It can be seen that in terms of the objective of stimulating economic growth (that is, lifting the level of total output), the effect of expansionary fiscal policies is the same as that of expansionary monetary policies. However, these two kinds of policies bring forward with different effects to the market interest rate, where expansionary monetary policies cause the interest rate to go down, while expansionary fiscal policies make the interest rate to go up. This different is worth our attention.

In order to fully understand why increasing government spending or reducing tax collection can cause the total output and interest rate to go up, let us analyze the conduction of fiscal policies. When the government increases its spending, it directly increases the aggregate demand. When the tax collection is reduced, it increases the disposable income of the public so that consumer spending is increased. That also causes the aggregate demand to increase. The caused increase in the aggregate demand (the aggregate demand curve moves upward) in turn makes the total output to increase (that is, the level of the equilibrium total output is lifted upward). Now, the relatively high level of the total output in turn causes the demand for money to rise (that is, the demand-for-money curve moves rightward), creating an excessive demand for money in the money market. That stimulates the interest rate to go higher. As long as the excessive demand for money exists, the interest rate will be adjusted continuously higher. Only when this process of adjustment pushes the economy to reach its new state of equilibrium, the process of adjustment will cease to exist. Under the circumstances of the new state of equilibrium, both the total output and the interest rate are at higher levels than before.

The effect of contracting fiscal policies is completely the opposite of that as described above. Decreasing government spending or increasing tax collection will reduce the aggregate demand so that the IS curve will move leftward. That will cause both the total output and interest rate to fall to lower levels. In short, both the total output and the interest rate are positively correlated to fiscal policies. This kind of positive correlation relationship can also be analyzed quantitatively by using equs. (6.21) and (6.22). When the government spending increases an amount ΔG while all other factors stay invariant, then the increment in the total output ΔY and the increment in interest rate Δi can be respectively written as follows:

$$\Delta Y = \frac{b}{a(\alpha + \beta + \gamma) + b(1 - c - v)}\Delta G \tag{6.25}$$

$$\Delta i = \frac{a}{a(\alpha + \beta + \gamma) + b(1 - c - v)}\Delta G \tag{6.26}$$

When the tax collection is reduced by the amount ΔT while all other factors are kept constant, then the increment ΔY in the total output and the increment Δi in the interest rate are respectively given by

$$\Delta Y = \frac{bc}{a(\alpha + \beta + \gamma) + b(1 - c - v)}\Delta T \tag{6.27}$$

$$\Delta i = \frac{ac}{a(\alpha + \beta + \gamma) + b(1 - c - v)}\Delta T \tag{6.28}$$

When the government spending and tax collection are increased simultaneously with the same amount ΔB, then if all other factors stay constant, then the increment ΔY in the total output and the increment Δi in the interest rate are respectively given as follows:

$$\Delta Y = \frac{b(1 - c)}{a(\alpha + \beta + \gamma) + b(1 - c - v)}\Delta B \tag{6.29}$$

$$\Delta i = \frac{a(1 - c)}{a(\alpha + \beta + \gamma) + b(1 - c - v)}\Delta B \tag{6.30}$$

So, the overall effect is till that both the level of the total output and the level of interest rate are lifted upward. That means that the expansionary effect of the fiscal policy of increasing the government spending is greater than the contraction effect of the policy of increasing tax collection. When the effects of these policies are equal, the overall consequence of fiscal policies is still expansionary with somewhat less effect efficiency.

When a budget deficit appears, the government generally increases its tax collection to offset the short fall. However, the fiscal policy of increasing tax collection is contractionary, which will suppress the economic growth. The government does not wish to see such outcome. So, it adopts at the same time expansionary monetary policy to stimulate the economy in order to maintain the necessary economic growth. So, it can be seen that it is very important to employ both fiscal and monetary policies jointly. Combinations of these two kinds of policies that are appropriate to the need can exert their influence on the magnitude of the deficit. In the following we will look at the problem of how to combine these two kinds of policies together from three angles: substitution, complementation, and consistency.

Firstly, the ultimate goals of fiscal and monetary policies are the same, serving the macro economy, promoting sufficient employment, stimulating the economic growth, and maintaining the economic stability. Because of the uniformity in their goals, these policies possess a degree of mutual substitutability. For example, using either fiscal policies or monetary policies can reach the goal of reducing demand. So these policies are substitutable of each other. When a fiscal contractionary policy, either increasing the tax collection or reducing the spending, is applied by itself, the goal of reducing demand can be materialized with short delay and good effect. When a contractionary monetary policy is employed alone, the goals of lifting interest rate, reducing investment, and suppressing demand can be accomplished with a long delay and slowly showing effect. If these policies are utilized jointly, the same goals can be materialized with increased strength and more satisfactory effect.

Secondly, although the same results can be accomplished by applying either fiscal policies or monetary policies, these policies have their individual strengths and weaknesses and can be used to complement each other. If these two kinds of policies are employed jointly, better effects can be expected because the strengths and weaknesses of the policies are complementing each other. For example, if the government spending is increased from the fiscal aspect, then it will stimulate investment and increase the national income. However, if at the same time a monetary policy is also considered to appropriately raise the interest rate, then both investment and income will be suppressed. When both of these policies are combined, one is loosening and the other tightening; by balancing and gradually adjusting the trade-offs the optimal effect can be eventually accomplished.

Thirdly, the substitutability and complementability of fiscal and monetary policies are the foundation that underlies the possibility of jointly applying these policies together. However, between the two kinds of policies is there a degree of inconsistency. Because of the existence of the inconsistency, it requires more attention to be paid to how these policies should be coordinated when used jointly in order to avoid conflicts and to overcome the unavoidable conflicts. Speaking generally, the main objective of monetary policies is to stable the financial market, while the main goal of fiscal policies is to develop service to the public. Because of the difference in their objectives, a series of inconsistencies or conflicts often appear. For example, to satisfy the public needs, the government incurs a large financial deficit, which needs the central bank to help resolve. Through its operations in the open market the central bank purchases a large amount of government bonds in order to provide the government with the needed cash. That in turn causes inflation. So, the central bank has to implement tightening monetary policies to raise interest rate in order to suppress inflation. However, the increasing interest rate will escalate the financial burden of the government, causing its financial deficit to rise. Such vicious cycle can easily strain the relationship between the treasury department and the central bank. In practical applications, in order to avoid and to overcome the potential conflicts arising from jointly employing fiscal and monetary policies, the central bank often yields to the political pressure and does not independently implement monetary policies.

Fourthly, through using equs. (6.21) and (6.22) the effect of jointly applying fiscal and monetary policies can be quantitatively analyzed. In particular, when the government spending increases an amount ΔG, the tax collection grows ΔT, and the money supply surges ΔM, the increments in the total output ΔY and the interest rate Δi are respectively given as follows:

$$\Delta Y = \frac{b\Delta G - bc\Delta T + (\alpha + \beta + \gamma)\Delta M}{a(\alpha + \beta + \gamma) + b(1 - c - v)} \tag{6.31}$$

$$\Delta i = \frac{a\Delta G - ac\Delta T - (1 - c - v)\Delta M}{a(\alpha + \beta + \gamma) + b(1 - c - v)} \tag{6.32}$$

The effect of both loosening policies: When both loosening policies are adopted, we have $\Delta G > 0$ (or $\Delta T < 0$) and $\Delta M > 0$. So, the amount of increase in the total output Y is very large. However, at the same time when an expansionary fiscal policy causes the interest rate to go up, the expansionary monetary policy has the effect to bring the interest rate down. So, the resultant increase in the interest rate will not be large and the stability of the interest rate can be maintained. This effect of both loosening policies is exactly the basis for Keynesians to advocate the employment of expansionary fiscal and monetary policies. In the 1960s, the Western nations adopted both loosening policies based on the suggestion of Keynesians. However, after entering the 1970s, these nations experienced the situation of stagflation. That is, both stagnated production and inflation co-existed at the same time. That declared the bankruptcy of Keynesian theory and forced people to reconsider the problem of inflation from new directions. That also got rid of the laws that inflation is a price economic growth can accept and that inflation can stimulate sustained economic growth.

The effect of one loosening policy and one tightening policy: When an expansionary fiscal policy and a tightening monetary policy are implemented, we have $\Delta G > 0$ (or $\Delta T < 0$) and $\Delta M < 0$. So, the magnitude of increase in the total output Y is not large. That can help to stabilize the total output. However, in such circumstances, the interest rate will climb hugely higher, causing instabilities in the financial market and sliding in the stock market. In the 1980s, the Reagan administration of the United States replaced the earlier either both loosening or both tightening policies by a combination of one loosening and one tighten policies. That experiment indeed brought forward with a visible effect, and helped American economy talk out of the stagflation and enjoyed six years of prosperity. However for these results paid were big prices: huge financial deficit and slow economic growth.

6.6.4 The crowding-out effect

Our discussion up to this point indicates that both monetary policies and fiscal policies can affect the level of total output. This conclusion holds true entirely on the basis that the money supply is sensitive to changes in interest rate. From the discussion on the demand of money in Chapter 4, it follows that Friedman believes that the demand for money is not sensitive to changes in interest rate and in fact he believes that the

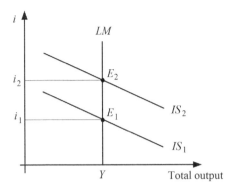

(a) Reaction of total output and interest rate
to expansionary fiscal policy

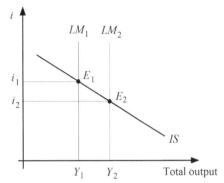

(b) Reaction of total output and interest rate
to expansionary monetary policy

Figure 6.8 Effects of monetary and fiscal policies when the demand for money is not affected by interest rate.

demand for money is irrelevant to changes in interest rate. If the situation is indeed so, can either monetary or fiscal policies affect the level of total output? Next, let us address this problem.

Assume that the demand for money has nothing to do with interest rate. Then the coefficient b in the demand-for-money function $M^d = M^d(Y, i) = M_A + aY - bi$ should be zero. So we have $M^d = M_A + aY$ and the equilibrium equation of the money market $M_A + aY = M$. From this end, we obtain the equation $Y = (M - M_A)/a$ for the LM curve. It means that the LM curve is a vertical line, as shown in Figure 6.8(a). Because $b = 0$, it follows that the total output and interest rate when the commodity market and the money market are both at equilibrium are respectively

$$Y = \frac{(\alpha + \beta + \gamma)(M - M_A)}{a(\alpha + \beta + \gamma)} = \frac{M - M_A}{a} \tag{6.33}$$

$$i = \frac{a(C_0 + I_0 + NX_0 + G - cT) - (1 - c - v)(M - M_A)}{a(\alpha + \beta + \gamma)} \tag{6.34}$$

Equ. (6.33) tells us that at this particular moment in time, implementing fiscal policies will not affect the total output so that the fiscal policies become invalid in terms of stimulating economic growth and increasing employment, Figure 6.8(a), and that expansionary fiscal policies can only cause the interest rate to go up. However, equ. (6.34) indicates that at this moment of time, monetary policies are effective, as shown in 6.8(b).

The, what is causing this phenomenon to appear? The reason is that when expansionary fiscal policies are implemented, the IS curve moves rightward, causing the interest rate to rise. However, because the LM curve is vertical and no monetary policy is adopted, no displacement occurs to the LM curve so that the rightward movement of the IS curve can only cause the interest rate to go up without making the total output to rise. In other words, the rightward movement of the IS curve causes the interest rate to rise, the consumer spending, investment spending, and the net export to drop so that the total spending actually is lowered. The amount of decrease is exactly equal to the increased spending of the expansionary fiscal policy. This effect of expansionary fiscal policies is known as the crowding-out effect. That is, the increased government spending from implementing expansionary fiscal policies cancels the totality of consumer spending, investment spending, and net export. This phenomenon of implementing fiscal policies without producing any effect on the total output is commonly known as complete crowding out. Under the general circumstance, the LM curve is slant to the upper right direction instead of being perfectly vertical. So, expansionary fiscal policies always suffer from some degree of crowding-out effect. However, it is not a complete crowding out. This phenomenon is known as partial crowding out.

Additionally, equs. (6.21) and (6.33) also tell us that the more insensitive to changes in interest rate the demand for money is (that is, the smaller b is, the greater the crowding-out effect, the less effective fiscal policies are, and the greater monetary policies are so that the more effective monetary policies are in comparison to fiscal policies.

The analysis above indicates that the degree of sensitivity of the demand for money to changes in interest rate is very important to decision-makers when it comes to a time to determine whether it is appropriate to adopt monetary policies or to adopt fiscal policies to interfere with the economic activities. It is exactly because of this importance that economists have done a great deal of research on the problem of how the demand for money is related to interest rate, leading to many debatable conclusions.

6.7 INFLATION

Inflation represents one of the major problems studied in modern economics. From the angle of money, this section discusses how monetary policies could cause inflation by using the analysis of aggregate supply and demand, and then considers the monetary and fiscal policy measures that could be applied to prevent inflation.

6.7.1 The phenomenon

When the quantity of money in circulation is greater than what the economy actually needs, then the money devalues, causing the general price level to go up comprehensively and sustainably. That is referred to as inflation. The percentage change in the price level from one time period to another is known as the rate of inflation. Therefore, the effective interest rate = the nominal interest rate – the rate of inflation. Generally, inflation is an economic phenomenon caused by money. When it is seen for the short term or statically, inflation stands for the amount at which the money supply is greater than the necessary quantity of money determined by the supply of actual goods. When it is seen from the long term or dynamically, inflation represents the quantity of money by which the actual money supply is more than the objective demand for money as determined by the rate of natural economic growth. Inflation can be shown either as continuous, persistent price increase, or as tenacious strengthening of non-price signals.

Based on their strength (the speed of increase), inflation is classified as mild (when the annual price level rises within 10%), severe (10%–100% annually), and hyper (more than 100%). Based on their manifestations, inflation is classified as open (as reflected in the statistics of prices), concealed (not reflected in the statistics of prices), and suppressed (as reflected through various indirect ways under the condition of price control, such as the lines formed in store fronts, searches, ranting, etc.). Based on the degree of expectations, inflation is classified as unexpected (prices gone up more than expected), and expected (prices change along a pattern). Based on their effect of prices, inflation is classified as balanced (the prices of different goods go up with the same proportion) and imbalanced (the prices of goods go up in individual proportions). Based on their causes, inflation is classified as demand pull, supply side, and structural, where the demand-pull inflation is caused by the swelling demand for money triggered by excessive money supply, the supply-side inflation consists of two kinds: cost push, which is caused by the rising prices of the upstream products (such as the petroleum) and wage rates, and profit push, which is caused by high profits of monopoly that actually exists due to insufficient mechanism of competition, and the structural inflation is caused by imbalanced or partially imbalanced supply and demand due to changes in the production structure. The conduction mechanism is the price rigidity and price comparison.

6.7.2   Price levels of goods

As of this location in the book, the discussion on how the total output is determined is done on the premise that the level of prices is fixed. That is, we have not considered the effect of price level on the total output. Evidently, such a theory cannot be complete. Without further developing that theory, it will be difficult for us to employ that theory in the explanation of the phenomenon of inflation. The analysis of aggregate demand and supply in this subsection will be done by using the IS-LM model while considering the determination of the total output along with the price variable.

The IS-LM model, as established in the previous section, did not consider the effect of changing prices on the economy. That is, it was assumed that the price level is constant. Speaking strictly, the equilibrium total output produced out of this IS-LM model with fixed price level of course is a function of the price level. When the price changes from the original level to a new level, it means that a variable, which is exogenous to the IS-LM model, has been changed. So the level of total output obtained out of this model will of course change accordingly. Therefore, it can be seen that the level of equilibrium total output that makes both the commodity market and money market be at equilibrium is really a function of the price level. Since at the sense of equilibrium the total output is the same as the aggregate demand, we will refer to the functional relationship between the total output (the level of aggregate demand) that makes both the commodity market and money market reach their equilibria at the same time and the price level as the aggregate demand function. The trajectory that consists of the corresponding price levels and total outputs is referred to as the aggregate demand curve. This curve is slant to the lower right hand side, as shown in Figure 6.9(b).

A natural question at this junction is: Why is this aggregate demand curve always slanting to the lower right hand side? To this end, let us use the IS-LM model to address this question. When the price rises from the original level P_1 to the new level P_2, because

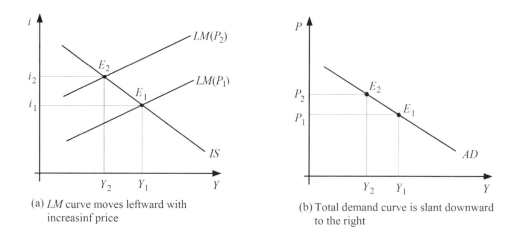

(a) *LM* curve moves leftward with
increasinf price

(b) Total demand curve is slant downward
to the right

Figure 6.9 The derivation of the aggregate demand curve.

the money supply we consider is the real money supply that has deducted the effect of
price changes, which is equal to the nominal money supply divided by the price level,
when the prices increases, it causes the actual amount of money supply to drop and the
(vertical) money supply curve to more leftward. That naturally causes the equilibrium
interest rate of the money market at every level of the total income to rise so that the
position of the LM curve moves upward as the price level increases. The consequence is
that the level of the total output that makes the commodity market and money market
reach their equilibria simultaneously falls as the price level increases, Figure 6.9(a).
That explains why the aggregate demand curve that reflects the relationship between
the total output and price level is always slant downward to the right.

Evidently, other than changes in the price level, which can only cause the total output to move along the aggregate demand curve, any other factor that can make either
the LM curve or the IS curve move can also make the aggregate demand curve travel.
Firstly, if a certain factor causes the LM curve move rightward, then the interaction of
the LM curve and the IS curve at any price level will travel along the IS curve upward
to the right. So, the level of total output at any price level is lifted upward. That implies
that the aggregate curve has also moved rightward. Similarly, any factor that causes
the IS curve to move leftward will also cause the aggregate demand curve to move
leftward. So, the IS curve and the aggregate demand curve travel in the same direction.
Secondly, if a certain non-price factor causes the LM curve moves rightward, then the
LM curve at any price level will move to the right so that its intersection with the IS
curve will accordingly move downward to the right. That illustrates that at any price
level the total output is lifted so that the aggregate demand curve travels rightward.
Similarly, any factor that causes the LM curve moves leftward also causes the aggregate demand curve travels leftward. So, both the LM curve and the aggregate demand
curve also move in the same direction.

When talking about aggregate supply while assuming the price level stays constant, we can only say that whatever the output is the supply will be. However, it
is impossible for the price level to stay invariant. With variant price the aggregate

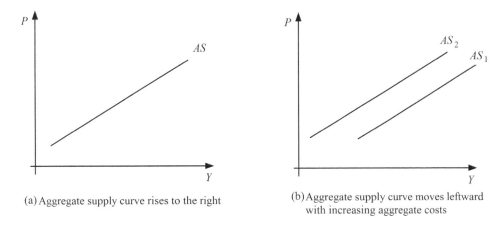

(a) Aggregate supply curve rises to the right

(b) Aggregate supply curve moves leftward with increasing aggregate costs

Figure 6.10 The aggregate supply curve.

supply of output products will be affected by the price level. These two variables enjoy a special relationship: The aggregate supply and the price level change in the same direction. This relationship between the aggregate supply and price level is known as aggregate supply function. When various price levels and corresponding aggregate supplies are plotted, the resultant curve is known as the aggregate supply curve. This curve is slanted upward to the right, as shown in Figure 6.10(a).

Then, why is the aggregate supply curve slant upward to the right? In order to understand the answer to this question, let us analyze the factors that cause the quantity of output supply to change. We know that business firms aim at maximizing their profits. So, the quantity of output supply is determined by the magnitude of profit acquired from each unit product. If the profit of each unit product increases, then the quantity of output supply also rises. Conversely, if the profit of each unit product decreases, then the quantity of output product also goes down.

The profit of each unit product is equal to the selling price minus the production cost. Within a short period of time, many factors of production are relatively fixed. For example, the wage of labor is invariant within a period of time according to the work contract; the purchasing price of raw materials stays constant within short periods of time based on the long-term agreement. Because these production costs do not change within short periods of time, when the price level rises, the price of the products will increase in comparison to the production cost. So, the profit of each unit product goes up, causing the company to increase its production so that the aggregate supply of the output is increased. So, the aggregate supply curve moves upward to the left.

In the previous explanation, the phrase of "short term" is used. That implies that the aggregate supply curve might change with time. Indeed, although within a short period of time the cost of producing a unit product does not change much. However, the situation is different for long periods of time. As time goes forward, the cost of each unit product can change drastically. When the cost goes up, it makes the profit per unit product go down at each price level. So, the output supply at each price level decreases accordingly. That means that when the production cost increases, the

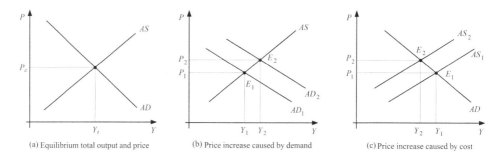

Figure 6.11 The equilibrium of the total supply and total demand.

aggregate supply curve moves leftward and that when the production cost decreases, the aggregate supply curve moves rightward, as shown in Figure 6.10(b).

At the point where the demand for the total output products is equal to the supply, there appear the equilibrium total output and equilibrium price. As shown in Figure 6.11(a), the aggregate demand curve AD intersects the aggregate supply curve AS at point E, where the economy reaches its equilibrium state. At the time moment, the level of the total output Y_t is known as the equilibrium total output. The corresponding level of price Y_ε is known as the equilibrium price. In the analysis of aggregate demand and aggregate supply, having located the equilibrium point of the economy does not mean the end of the research. That is because such equilibrium state is temporary, many factors of the economy can destroy the ready reached equilibrium so that the economy will have to readjust toward a new equilibrium. And only when the economy indeed evolves toward equilibrium, the concept of equilibrium becomes meaningful.

As what has been mentioned above, any factor that causes the IS curve or the LM curve to move can also cause the aggregate demand curve to move in the same direction. As shown in Figure 6.11(b), when the aggregate demand curve moves to the right (that is, when the aggregate demand increases), both the equilibrium total output and equilibrium price level increase. Such price increase as caused by the right-ward movement of the aggregate demand curve is known as price increase pulled by demand. The previous analysis on the reason why the aggregate supply curve could move indicates that any increase in the production cost will cause the aggregate supply curve to move leftward. As shown in 6.11(c), when the aggregate supply curve moves to the left, the level of the equilibrium total output will fall and the equilibrium price level will rise. Such price increase as caused by the leftward movement of the aggregate supply curve (due to an increasing cost in production) is known as price increase pushed by cost.

6.7.3 Causes of inflation

(Friedman, 1969) pointed out that: No matter when and where inflation is a monetary phenomenon. This proposition has been firmly supported by empirical evidence from nations that have suffered from lasting high levels of inflation. That is, in a nation that experiences a lasting high level of inflation, the rate of monetary increase is also high.

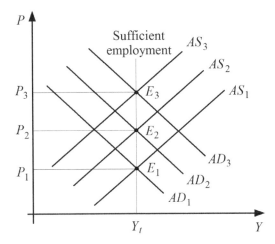

Figure 6.12 The consequence of constantly increasing money supply.

So, the following conclusion can be seen: high levels of monetary increase cause high degrees of inflation. In the following, we will first employ the analysis on the aggregate demand and aggregate supply to illustrate why significant and sustained rise in price level can only occur at the time when the money supply continues to go up, and then consider the government policies that commonly cause inflation.

Assume that the economy is initially situated at the equilibrium point E_1 as shown in Figure 6.12, where the aggregate supply is equal to the aggregate demand and the employment is sufficient. Now, the monetary authority decides to implement an expansionary monetary policy by increasing the money supply. That, as we have discussed earlier, will cause the money supply curve to move rightward so that the interest rate at any given level of the total output will fall, leading to a rightward (downward) movement of the LM curve. From the previous discussion on the movement of the aggregate demand curve, it follows that the increasing money supply also causes the aggregate demand curve to move rightward. A consequence of that movement is that the levels of the total output and prices are lifted at the same time. The rise in the total output strengthens the demand for labor. However, the economy has been since the start at the state of sufficient employment. So, even though the current rate of unemployment is lower than the natural rate, there is still a shortage of labor. So wages have to rise.

Along with the rising wages, the per-unit production cost increases, causing the profit to slide. So, business firms decide to reduce the level of production and the supply of goods. That in turn causes the aggregate supply curve to move leftward again and again. Because under the condition of sufficient employment companies cannot receive additional labors to increase their production, the just lifted total output has to fall back to the original level. However, the rightward movement of the aggregate demand causes the total output to rise. So, the consequent leftward movement of the aggregate supply curve once again makes the level of the total output fall back to the original level. The result of such moves – increase and increase again in the price level – is maintained. Hence, when the economy reaches sufficient employment, increasing money supply cannot make the total output to increase. Instead it will cause the price level to go up majorly. As shown in Figure 6.12, when the money supply is increased, the aggregate demand curve moves from the original location AD_1 rightward to AD_2, and the economy evolves from the state E_1 to the state E_1'. Following that, the aggregate demand curve moves from the original location AS_1 leftward to the location AS_2, and the economy evolves from E_1' to E_2, and the total output recovers to its original level. However, the price level has been lifted upward from the original level P_1 to a higher level P_2.

If the government continued its implementation of an expansionary monetary policy, then under the condition of sufficient employment the price level will keep on rising while the actual level of total output will stay invariant: the aggregate demand curve keeps on moving rightward from AD_1 to AD_2, then to AD_3, ...; the aggregate supply curve keeps on moving leftward from AS_1 to AS_2, then to AS_3, ...; and the price level continues its upward trend from P_1 to P_2, then to P_3, ... This consequence of implementing expansionary monetary policies is inflation. To this end, both Keynesians and monetarists have the same conclusion, and agree with what Friedman said: no matter when and where inflation is a monetary phenomenon, believing that the root cause of high inflation is in the high rate of increase in the money supply.

Many people believe that inflation bears adverse effect on the economy. If it is so, why then does inflation still exist? And why does the government still adopt inflationary monetary policies? Addressing these problems will help us understand why there is inflation. In the following, let us consider those government policies that commonly cause inflation.

Most national governments pursue after high levels of employment as their number one objective. That often causes inflation. In order to promote high levels of employment, the adopted economic policies can lead to two kinds of inflation: the demand-pull inflation and cost push inflation. The former is caused the government's adoption of such economic policies that cause the aggregate demand curve to move rightward, while the latter is originated from the opposite shock of the supply, that is, workers demand for higher wages. In other words, it is caused by the aggregate demand curve moves to the left. Now, let us employ the analysis of the aggregate supply and aggregate demand to see how high levels of employment can lead to these two kinds of inflation.

Sufficient employment is not equivalent to zero unemployment. The rate of unemployment at the time of sufficient employment is known as the national rate of unemployment. It is a number greater than zero. If the government officers that make decisions on the rate of unemployment aim at achieving a rate of unemployment lower than the natural rate, then the goal of a high rate of employment will create a condition for high increase rate of money supply so that inflation will follow naturally. Figure 6.13 illustrates how all of this happens, where the level Y_t of output stands for that with sufficient employment. That is, it is the objective level of the total output at the natural rate of unemployment.

After having made the decision of reaching a high rate of employment that is lower than the natural rate of unemployment, the decision makers will then try to make the total output to reach an objective Y_T that is higher than the level Y_t at the natural rate. In order to materialize this objective, they increase the money supply – a policy that stimulates the aggregate demand. As a consequence, the aggregate demand

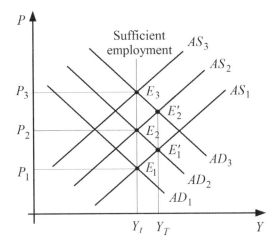

Figure 6.13 Inflation caused by demand: unemployment is too low.

curve moves rightward from the initial location AD_1 to AD_2 so that their desired objective of higher rate of employment is achieved. However, during this process, the price level is also floated upward. Remember that at the output level Y_T, the rate of unemployment is too low. That causes wages to rise and production cost to increase, leading to a leftward movement of the aggregate supply curve. Initially the decision makers thought as soon as they materialized the objective level Y_T of output they could stop without expecting that the aggregate supply could fall short. When the aggregate supply curve moves leftward from AS_1 to AS_2, it makes the level of total output fall back to the original level Y_t and the unemployment recover to the initial state. That is, the state of high rate of employment did not get maintained. Even so, the price level has been raised to a higher level, reaching P_2.

Because of the recently achieved high rate of employment was not maintained, the decision makers once again increase the money supply in order to stimulate the aggregate demand and to reach the level Y_T of the total output in their materialization of high employment. However, the result of this action, just like what has happened before, first makes the economy reach the level Y_T of total output while realizing the desired high employment. But, that achievement is temporary. Very soon everything returns to their original state: the high level Y_T of total output falls back to the original state Y_t of the sufficient employment, while the price level has been brought up further from P_2 to P_3, where at first the aggregate demand curve moves rightward from AD_2 to AD_3, and then the aggregate supply curve moves leftward from AS_2 to AS_3. The consequence is that both the total output and the unemployment rate recover to their original levels, while the prices have experienced two rounds of large rises.

If the decision makers continue to pursue their objective of nigh employment by implementing expansionary monetary policies without recognizing that they could not maintain their desired high employment that is lower than the natural unemployment rate, then the result has to be that they see continuously climbing price levels and

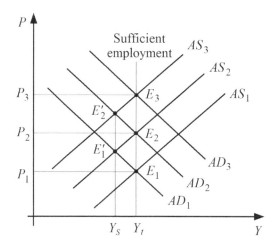

Figure 6.14 Inflation caused by cost: the rate of wage rises.

the invariant levels of total output and employment. That explains how demand pull inflation is created. Its root cause is the continued increase in money supply.

As for the analysis of cost push inflation, we still assume that the economy is initially situated a state of sufficient employment, where the aggregate supply is equal to the aggregate demand. If this economic state is maintained for long without raising the workers' wages, that then will make the workers unhappy and will demand for better benefits and higher salaries. If business firms do not meet the demand, the labor unions will coordinate strikes. Under such unions' pressure and demand for higher wages and better benefits, companies will have to accept the requirements. However, increasing wages will cause the production cost to rise and the total output to fall. That is, the aggregate supply curve will move leftward from the initial location AS_1 to AS_2, as shown in Figure 6.14. The consequence of this change is that the level of total output falls to Y_s and more unemployment is caused so that the unemployment rate is higher than the natural rate, while the price level has been lifted upward.

The increasing unemployment catches the attention of the government. In order to maintain the objective of sufficient employment, the government adopts the policy of stimulating the aggregate demand in order to promote production – that is, increasing the money supply. This move of the government indeed is effective. It makes the aggregate demand curve moves rightward from AD_1 to AD_2, the economy materialize its new equilibrium, and the total output and employment recover to their initial states of sufficient employment. However, such policy causes the price level to move further upward to P_2.

People's desires are endless. Now, with milk to drink and bread to eat, people demand for more: They want to own a house, a car, etc. The result is that the workers once again demand for higher wages. Under pressure, companies meet the demand. Then the outcome is the same as discussed before: The aggregate supply curve first moves leftward from AS_2 to AS_3; then in order to maintain sufficient employment, the government adopts expansionary monetary policies, causing the aggregate demand curve to move rightward from AD$_2$ to AD$_3$. That makes the economy reach its new equilibrium at E$_3$. Now, the prices climb from the already high level P$_2$ to another higher level P$_3$.

No matter how high wages are, people's desires will never be satisfied. So, such demand for higher wages will not stop, and companies will continue to make compromises, causing the production cost continuously to go higher. Pushing by such increasing production cost, the price level will be raised step by step so that cost push inflation appears. From this analysis, it follows that the root reason behind the cost push inflation is also the constantly increasing money supply.

As for the demand pull inflation and the cost push inflation, the following points need to be noted:

1 The process for demand pull inflation to appear is: The AD curve moves rightward → the AS curve moves leftward → the AD curve moves rightward → the AS curve moves leftward → ... The process for cost push inflation to appear is: The AS curve moves leftward → the AD curve moves rightward → the AS curve moves leftward → the AD curve moves rightward → ...

2 The root reason for these two kinds of inflation to occur is the constant increase in money supply.

3 The demand pull inflation generally appears at the time when the unemployment is lower than the natural rate, while the cost push inflation generally appears at the time when the unemployment is higher than the natural rate.

4 In real life, these two kinds of inflation are difficult to separate. It is because firstly it is difficult to tell correctly whether it is the aggregate demand curve that moves rightward first or the aggregate supply curve that moves leftward first. Secondly, in order to tell which kind the inflation that has ready appeared is, one must know for the current moment whether the unemployment is lower than the natural rate or high. However, it is difficult to calculate the natural rate of unemployment. To this end, no good method has been established. As for using 6% as the natural unemployment rate, people have had many doubts. As for indeed what is the natural unemployment rate? It is still an unsettled problem of economics.

In terms of budget deficit and inflation, empirical tests and analyses indicate that another possible reason for the government to adopt an inflationary monetary policy is budget deficit. When there appears a budget deficit, the government can employ one of two methods to borrow money to offset the deficit: sell public debts and create money (that is the so-called printing money). When public debts are sold to the public, it does not affect the base money, because the money supply is not directly affected. However, printing money is different. It makes the bas money to increase so that the money supply is raised. That leads to the aggregate demand curve to move rightward, which could cause inflation.

What needs to be explained is that the key for budget deficit to cause inflation is a persistent existence of the deficit. If the deficit is temporary, then one time increase in the money supply will not cause the price level to rise quickly and continuously so that no inflation will follow. However, if the deficit exists persistently, where the printing of money offsets the deficit in one period and the deficit still exists in the following period, then if the government continue to print money to offset the deficit, then inflation will appear unavoidly.

From combining what has been discussed above, it follows that for the following two situations budget deficit might become a root reason for inflation to appear: (i) the budget deficit exists persistently instead of temporarily; (ii) the government offsets the deficit by using the method of creating additional money instead of selling public debts.

Because it is known that dealing with deficit by using the method of creating additional money can lead to inflation, then why does the government still adopt such method? The reason is that in those nations where the money market and capital market are not well developed, it is difficult to sell the government bonds or national debts. For these nations, the only method to deal with huge budget deficit is to print money. A second natural question at this junction is: Will the method of dealing with budget deficit by selling public debts not cause inflation to appear? To this end, we would say that this method does implicitly suffer from such a possibility. When the government raises funds to offset its budget deficit through selling national debts to the public, it might create a force to lift up the interest rate. It is because the risk of the public debts is low and if the interest rate is comparable, the public will rush to buy the public debts. That will cause the demand for company bonds to fall so that the prices of these bonds will rise and interest rate will increase. In order to prevent the market interest rate from rising too high, the central bank will come to rescue by purchasing some amount of bonds in order to release the pressure for the interest rate to go higher. This movement of the central bank causes the money supply to income so that the price level will grow higher. If the government's budget deficit exists persistently and continuously sells public debts, then the central bank will have to make purchases constantly in the open market. That of course will cause the price level to continue its way upward, causing inflation to appear. So, no matter which method is employed, as long as the deficit exists continuously, inflation will possibly appear. The only difference is that if the method of printing money is used to offset the deficit, then inflation will be created directly; and if the method of selling public debts is applied, it might indirectly cause inflation through creating a hidden danger of inflation.

6.7.4 Game theory explanation of stagflation phenomena

In the 1950s, the US government adopted Keynesian fiscal and monetary policies, leading to two financial crises in 1951–1954 and 1957–1958 with greatly slowed economic growth. Facing this situation, Keynesian school of thought suggested that expansionary monetary and fiscal policies should be employed not only during the periods of decline and economic recessions but also during the periods of recovery in order to stimulate growth and sufficient employment. This suggestion was adopted by the Kennedy and Johnson administrations and led to noticeable successes in the first half of the 1960s. However, the good time did not last. At the end of the 1960s and the start of the 1970s, American economy experienced a situation of stagflation of both production stagnation and inflation. Facing the stagflation, economists could not come up with any effective prescription until neoclassicists investigated the behaviors of monetary policy decision makers using game theory. They provided not only explanations for the stagflation but also treatment to the problem. For our purpose here, let us use a classical example to illustrate the particularly useful thinking of game theory in the explanation of stagflations.

Neoclassicists noticed that in the past economic studies the government was not treated as an independent agent. However, like an agent, the government also has its goals, and its ability to make decisions is limited by certain constraints. Generally speaking, the goal of the government is to optimize the societal welfare, while individual agents pursue after the optimization of personal interests. When the government is seen as a rational agent, it can easily get involved in conflicts with the public. So the behaviors of the government and the public influence each other. To describe and investigate such mutual influences that are accompanied with conflicting decision-making behaviors and interest, game theory is an appropriate theory.

To this end, the monetary policy making can be seen as a game played by the government and the labor union. To achieve the goal of pacifying inflation and providing sufficient employment, the government needs to apply influence over the agreement on wages. However, the final outcome in turn is determined by how the union predicts and reacts to the government's decision making. The union has two choices: either request an increase in wages or not, while the government can decide on whether to increase the supply of money or not. In this game, the payoff table of the government is given in Table 6.1, while the payoff table of the union in Table 6.2.

The payoffs in these tables can be understood as follows: At the best possible scenario where the union does not request any increase in wages and the government does not plan to increase the money supply, which represents the state of economy with low inflation and low unemployment, the payoffs of both sides are respectively 3. If the union did not propose to raise wages, while the government increased the money supply, then the actual wages would have declined, then the payoff of the union drops from 3 to 0. At the same time, because of the increased money supply, the government helped to improve the employment situation so that the payoff of the government jumped one unit and became 4. On the other hand, if the union requested

Table 6.1 The payoffs of the government.

Union	Government	
	Increase in money supply	No increase in money supply
Increase in wages	1 (high inflation, high unemployment)	0 (low inflation, high unemployment)
No increase in wages	4 (high inflation, low unemployment)	3 (low inflation, low unemployment)

Table 6.2 The payoffs of the union.

Union	Government	
	Increase in money supply	No increase in money supply
Increase in wages	1 (high inflation, high unemployment)	4 (low inflation, high unemployment)
No increase in wages	0 (high inflation, low unemployment)	3 (low inflation, low unemployment)

to raise wages while the government did not increase the money supply, then the wages have actually improved so that the payoff of the union jumped one unit and became 4. At this same time, because of the increasing unemployment rate, the payoff of the government dropped to 0. If the union requested to raise wages and, to satisfy such demand, the government increased the money supply, then the rates of inflation and unemployment are consequently elevated higher while the actual wages did not materially increase much. Therefore, both the union and the government experience losses so that their payoffs drop 2 units and become 1.

Intuitively, the best option is that the union does not request any increase in wages and the government does not increase the supply of money. In this case, both sides earn 3 units of payoffs, while the economy experiences low inflation and low unemployment. However, by using a simple analysis, we can see that this option is instable. In particular, if the union does not propose to increase the wages while the government provides additional money supply, then because of the decline in the actual incomes the payoffs of the union becomes zero. If the union does request an increase in wages, then no matter whether the government increases the money supply or not, the union's payoff increases by one unit. Similarly, if the government does not increase the money supply while the union requests an increase in wages, then the government's payoff will drop to 0. If the government increases the money supply, then no matter whether the union requests an increase in wages or not, the government will gain one unit in its payoff. So, it can be seen readily that driven by maximizing the payoff, the optimal option for the union is to request wage increase; and the best option for the government is to increase the money supply. Therefore, both the union and the government can only obtain the payoff of one unit instead of the most possible 3.

In short, the key of this game between the government and the union is that each side is influenced by additional gains in payoffs so that neither side cares about what choice the other party takes. The optimal solution of this game is (request wage increase, increase money supply), which is different of the ideal solution (not request wage increase, not increase the money supply). That is to say, the outcome of (increase wages, increase money supply), which implies stagflation, is the Nash equilibrium of the game. So, the phenomenon of stagflation is a consequence of the game between the government and the union, reflecting their pursuit after their respective short-term interest. Because both the government and the union only go after their short-term interests, stagflation becomes a definite result.

When the game described above can be repeated infinitely many times, the situation becomes different, where the outcome (no wage increase, no money supply increase) of the game becomes a 'feasible conclusion". It represents the long-term interest of the government and the union so that both parties would pursue after such a conclusion. In particular, when the afore-described game is played repeatedly without end, in order to arrive at the feasible conclusion of (no wage increase, no money supply increase), each player would signal its 'goodwill' to the opponent. That is, the government would signal its desire of not attempting to increase the money supply, while the union the desire of not requesting wage increase. If the union single-mindedly pursues after wage increase without reacting to the government's "goodwill", then the government would lose its patience and would forever implement the strategy of increasing money supply. Similarly, if the government adopts the strategy of increasing money supply without answering the union's "goodwill", then the union would from then on forever lose its confidence in the government and implement the strategy of requesting wage increases. As a consequence of not answering the signal of the other player, each side would lose its long-term benefit for the gain of short-term payoffs. One result that is led to based on such a reasoning is that the current move of a player will be answered sooner or later by its opponent; in other words, future moves of the opponent might be dependent on what move the player takes currently.

For the sake of convenience of our communication, let us donate the union's move of not requesting wage increase after receiving the signal of "goodwill" from the government as the strategy of "collaboration", and the government's move of not increasing money supply after learning the signal of "goodwill" from the union as the strategy of "collaboration". Let us now look at whether or not the situation of (collaboration, collaboration) can appear as equilibrium of the game played repeatedly between the government and the union. To this end, we consider two possibilities: the game is played finitely many times, and the game played infinitely many times.

First we assume both the government and the union know the game will be played a predetermined number of times, say 5 times. Before the last round of the play, both players know that is the last match so that the standard reasoning of equilibrium would be applied, leading to the outcome of each player choosing 'not to collaborate''. Next, let us look at the match right before the last round. Here, each player seems to value collaboration so that such collaborative spirit could be continued into the next round. However, because in the last round both players would not collaborate with each other, it implies that in this second to the last round, collaboration does not seem to have any advantage. Therefore, either player would choose not to collaborate. By applying such a reasoning continuously, the backward induction shows that the strategy of collaboration in no round of the game play before the last round would produce long-term benefit, while believing that the opponent in the last round would simply take the strategy of not collaboration. So, applying current "goodwill" to influence the outcome of the next round game play would be a waste of time. Therefore, in the game that is repeated finitely many times, the outcome of each round is the identical (no collaboration, no collaboration). That is, the appearance of stagflation is the guaranteed outcome of finitely repeated game plays.

Secondly, we look at what happens when the game is repeated infinitely many times. Because at each round of the play both players know the game will be repeated more than once, the strategy of collaboration has a chance and arriving at the long-term benefit is still possible. So, the strategies of the government and the union are sequences of functions, where for each player its decision of whether to collaborate or to betray at every round is dependent on the game history prior to this current round. Each player's payoff of the game is the discounted sum of all rounds payoffs. In particular, assume that the payoff of a player at the t-th round is u_t, $t = 0, 1, 2, 3, \ldots$ Then, the discounted sum of the player's payoffs is $\sum_{t=0}^{\infty} u_t / (1 + t)^t$, where r stands for the discount rate. As long as the discount rate is not very high, the equilibrium outcome of each round play is (collaboration, collaboration), where each player would see the long-term benefit of being collaborative. In order to illustrate this conclusion, let us use the data given In Tables 6.1 and 6.2.

Assume that both players, the government and the union, had collaborated in every round of the play until the T-th round when one player decided not to collaborate. Then the other play's payoff for this round would become 0 so that this player would forever in the following rounds take the strategy of no collaboration in order to punish the opponent. As a consequence, the player, who was first to not collaborate, from then on had to apply the strategy of no collaboration, because its payoff of collaboration would be 0. Therefore, although for the T-th round play, this player gained 4 unit of payoff, it would only gain 1 unit of payoffs in each and every following rounds of play. This player's discounted payoff is

$$R_T = \sum_{t=0}^{T-1} \frac{3}{(1+r)^t} + \frac{4}{(1+r)^T} + \sum_{t=1}^{\infty} \frac{1}{(1+r)^{T+t}} = \frac{3(1+r)}{r} + \frac{r-2}{r(1+r)^T} \qquad (6.35)$$

On the other hand, if neither player betrayed the opponent, then each player would have gain the following discounted payoff

$$R = \sum_{t=0}^{\infty} \frac{3}{(1+r)^t} = \frac{3(1+r)}{r} \qquad (6.36)$$

By comparing equs. (6.35) and (6.36), it follows that as long as the discount rate $r < 2$, one has $R_T < R$. This end implies that as long as the discount rate is not very high, by taking the strategy of forever not to collaborate, the player who was first to apply this strategy will suffer from gaining greatly reduced payoff. Therefore, both players would be better off if they collaborate forever. In other words, when the discount rate is not very high, the equilibrium in the infinitely-many-times repeated game play is that both the government and the union collaborate at round: the government does not increase the money supply and the union does not request raise in wages.

6.7.5 Dealing with inflation

After having comprehended the reason why inflation appears, we will be able to resolve the problem of inflation starting from its root level and to suppress the appearance of inflation. To this end, a natural question arises: Why do we want to suppress inflation? The Keynesians once suggested using mild inflation to stimulate demand so that the problem of employment is revolved and economic development is enhanced. And after World War II, the Western countries commonly adopted this Keynesian suggestion. However, during the late 1960s and early 1970s, these countries suffered from the problem of stagflation, where the economic growth stagnated and inflation became severe day by day. Especially, the increasing severity of inflation lowered people's real income, causing uneven distribution of income, distorted price signals, and worsening pattern of resource distribution. Eventually the economy was forced into the situation of recession. The community of economists began to reconsider about Keynesian beliefs, the governments of these Western nations changed their logic of thinking. They shifted the main objective of macroeconomic regulation from revolving employment problem to preventing and curing inflation, where inflation was seen as the number 1 major problem of the government. In terms of how to handle the inflation, different national governments adopted their different methods. To sum up, all the methods employed can be seen as jointly applying monetary and fiscal policies with differences in how specifically they were combined to deal with particular circumstances. In the following, let us use the United States as an example to illustrate the effect of jointly applying monetary and fiscal policies in the area of preventing and curing inflation.

During the late 1960s and early 1970s, in order to control the inflation, the government of the United States alternatively employed both loosening and both tightening monetary and fiscal policies. In other words, during the time period when inflation was severe, both loosening policies were utilized to tighten the fiscal budget and to tighten the money supply (through tightening credit and increasing interest rate). During the time period of economic decline, loosening fiscal and monetary policies were employed, where credit was loose, fiscal spending was high, and interest rate was lowered so the deficit and expansionary policies were applied to stimulate the economy. The effect of the double tightening policies was obvious. The inflation was under control quickly. However, what followed was a fast decline of the economy. Similarly, the effect of the double loosening policies was noticeable. It caused the inflation once again to rise rapidly. The consequence of implementing the double tightening and loosening policies was the vicious cycle of alternating economic decline and high inflation, making the economy experience non-stopping large magnitude swings back and forth. For the entire 1970s, the interest rate in the United States floated up and down while rising gradually and definitely. That caused the inflation spirally higher and economic growth slowed significantly.

In order to escape from the stagflation, the Federal Revers of the United States announced that it would use money supply instead of interest rate as its objective to deal with inflation. That indicated that the American government had adopted the opinion of monetarists and recognized that no matter when and where inflation is forever a phenomenon of money, and the only method of handling inflation is to slow down the increasing rate of money supply. Since 1980, the American government gradually lowered the magnitude of increase of money. For the years of 1982, 1983, and 1984, the rate of inflation in the United States was respectively 6.2%, 3.2%, and 4.3%. That means that by using the method of lowering the increasing rate of money supply, the United States had achieved its initial success in terms of handling the inflation.

Since the start of the 1980s, the Reagan administration adopted an economic policy that combined the opinions of the monetary school and the supply school by employing a loosening and a tightening policy, where the fiscal side was loose and the monetary side was tight, or vice versa. Such collocation of monetary and fiscal policies with one loose and the other tight provided the American government with four methods to choose in its fight against inflation: (i) loose monetary and loose fiscal policies; (ii) tight monetary and tight fiscal policies; (ii) loose monetary and tight fiscal policies; and (iv) tight monetary and loose fiscal policies. Based on the particular circumstance and specific characteristics of the inflation and economic development, selecting the appropriate combination to apply monetary and fiscal policies had achieved magnificent effects. That helped the American economy walk out of the difficult stagflation and realize a period of six years of economic prosperity. During this time period, the rate of inflation dropped consecutively year after year until reaching the acceptable level.

Since the mid 1990's, the main objective of macro regulation for the United States is to maintain economic growth at an appropriate speed, to control the inflation index, and to balance the two at the same time by applying monetary and fiscal policies. When the economy is too cold, appropriately easing up the money to heat up the economy so that the speed of economic development would pick up. When the economy is too hot and inflation seems to rise, tightening the money causes the economy to cool and inflation to be tamed. For example, at the start of 1994, the economy of the United States was beginning to heat up and inflation rebounding. In order to suppress the swelling inflation, the Federal Reserve raised the rate of federal funds, which is the interest rate of overnight lending between commercial banks, seven times from February 4, 1994, to February 1, 1995, from 3% to 6%. This movement effectively suppressed the increasing prices and made the actual rate of economic growth stabilize at around 2.5%. And in June 1999 the Federal Reserve took note that inflation in the United States was about to rise. So, it raised the prime rate slightly in order to have the then not yet swelled inflation under control. Because of the preemptive action, the Federal Reserve avoided the future need of major economic surgery to have the situation under control by utilizing a quite minor remedy.

It is more difficult to tame a hyperinflation, which generally requires a major economic surgery beyond merely applying monetary and fiscal policies. It is because for such a circumstance money has almost entirely lost its functionalities as the media of trades, standard of value, and storage of wealth and the monetary system can no longer operate normally. The government has to implement a monetary reform and drastically reduce the money supply along with other political and economic methods. For more severe situations, the government cannot help but abolish the old currency and issue new currency, because the old currency has lost its functionality of money completely.

6.8   SYSTEMIC STRUCTURE OF MONETARY POLICY

By combining what has been discussed in this chapter with the systemic yoyo model of economies, each of which could be seen as a field of spinning fluids, one can see the following systemic structure of monetary policy: Within the national boundary, the economic "fluid", consisting of people, business entities, economic activities, information, various supply and demand forces, etc., spins around a core, consisting of various profit opportunities, where the term "profit" can be in the traditional terms of money or other relevant significance, such as good wills, welfare for certain sectors of the society, etc. In order to maintain the desired order of the society, a central bank is organized and artificially placed within the core of all business activities.

In order to achieve certain goals of economic development, monetary policies are introduced to coordinate, to regulate, and to control financial processes and activities through the functions of the central bank. The ultimate control over the national finance is materialized through implementing monetary policies. The relevant monetary policies are made and carried out by the central bank. As the only national organization that can print and issue money, the central bank controls the supply of money through a series of effective procedures in order to maintain a stable circulation of currency and to create and sustain an ideal financial environment for economic development.

If we theoretically imagine a national economy as a pool of spinning fluids, as in the case of Hide's dishpan experiment (Hide, 1953), for more details, see Chapter 2, then what happens here is that the inner cylinder of the rotational dish spins at an adjustable speed in order to exert influence on how the fluid between the inner and

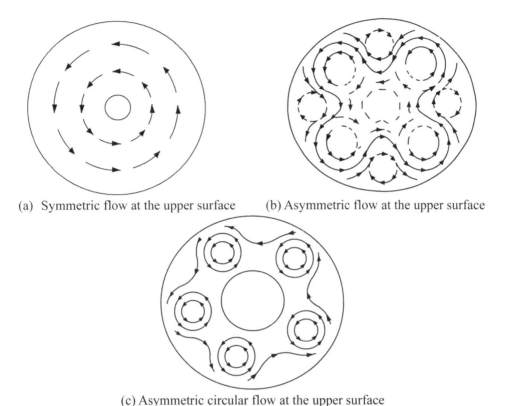

(a) Symmetric flow at the upper surface   (b) Asymmetric flow at the upper surface

(c) Asymmetric circular flow at the upper surface

Figure 6.15 Patterns observed in Hide's dishpan experiment.

outer cylinders flow and consequently what patterns of movement the fluid creates. For the sake of convenience of our communication, let us reproduce Figure 6.15 here.

More specifically, when Hide's dishpan is employed to simulate a national economy, the originally solid outer and inner borders of the fluid should have become blurred. That is, they are realistically there in nature; however, when one wants to quantitatively pinpoint where they are located, the person will have trouble to do so.

Without the central bank or any central body to coordinate the development of the fluid (economic activities), the alterations of the flow patterns, as described in Figure 6.15, will be chaotic due to the fact that not all geographic regions provide the same policies toward economic development. In terms of the language of the dishpan experiment, it means that the cylinders are not perfectly circular, the bottom of the dish is not smooth, and the density of the fluid located in between the cylinders is not even and varies from one location to another. Therefore, the periodic alteration of flow patterns as depicted in Figure 6.15 will be not as predictable as in the dishpan experiment, where a perfect symmetry is a guarantee. In fact, when such a perfect symmetry does not exist, the alteration of the expected flow patterns becomes unpredictable as documented in the studies of economic cycles.

On top of the afore-mentioned, seemingly chaotic alterations of flow patterns, a central body, such a central bank, is introduced for the purpose of maintaining the movement of the economic fluid in a desirable pattern. Here, different forms of governments prefer different patterns of economic flows. For instance, for a nation with its powerful central government, such as ex-Soviet Union, the desired pattern of the economic flow would be something close to that in Figure 6.15(a), where the powerful central government wants all corners of the national economy stay in pace with what the head of the government dictates. On the other hand, for a democratic nation, such as the United States of America, the government prefers a pattern of economic flows that resembles closely to that in Figure 6.15(b). It is because American people prefer to have a certain degree of orderly autonomy. To this end, what is depicted in Figure 6.15(c), when combined with non-existing symmetry, stands for economic independence for certain geographic areas. In general, no politically solvent nation likes to maintain such a pattern of economic flow.

Now, in Hide's dishpan experiment let us identify the smaller, inner cylinder as the core, the totality of all potential profit opportunities and the central bank of a specific nation, and the larger, outer cylinder as the boundary of the nation's economy. Then the observed liquid patterns in Figure 6.15 would be caused by uneven distribution of economic forces acting on the liquid particles (economic regions) located in different distances from the core. Here, differences in reinforcing the monetary policies from one geographic region to another and differences in the intensities of economic activities across the entire nation strengthens the uneven distribution of the interacting forces acting on the economic fluid. The observed regular period of transition of flow patterns in the Hide's dishpan experiment is determined by the perfect symmetry of the liquid (no impurities) and the cylinders about the axis of rotation. Now, because the earthly conditions, distributions of natural resources, and the varied effects of the core in different directions are not perfectly fair and even, the orientation of the circular eddy motions in Figure 6.15(b) and (c) would not be perfectly in line with the development of the national economy. This fact explains why economic centers travel from one location to another. Additionally, this fact provides us a means to study the effectiveness of the core by carefully analyzing the paths along which economic centers have travelled.

Because, as analyzed above, the economic core tends to "spin" at an adjustable speed, that is the core at one time moment might be 'spinning' faster and at other times slower than the outer boundary of the national economy, it explains why monetary policies could have respectively positive and negative effects. Here, by spinning faster it means that certain monetary policies are introduced to prevent certain predicted undesirable events from happening, while by spinning slower it represents those scenarios where monetary policies are introduced to alter certain existing development trends.

By combining the fact that the core spins at different speed from the rest of the pool of economic activities with the Hide's dishpan model of national economies, it can be seen that policy mistakes could be more easily made during the time when the economic liquid in between the inner and outer cores flows relatively symmetrically and uniformly with respect to the center of rotation (Figure 6.15(a)). At this time moment, the meridian field that holds the systemic yoyo structure of the national economy together is the weakest, since no sub-eddy, those regional spinning pools in Figure 6.15(c), exists to create additional meridian forces. In this case, if the inner core rotates slightly faster or slower than the rest of the economic pool, the uneven distribution of spinning forces acting on the liquid particles located at different distances from the inner core would possibly reverse from that as shown in Figure 6.15(b) and (c). So, consequently, the chain of identical eddy motions (in Figure 6.15(b) and (c)) formed in the area between the inner core and outer boundary would be spinning in the opposite direction. Such direction change in the sub-eddies will reverse the direction of the overall economic development (the direction of the meridian field). Because of the non-homogeneous impacts of the inner core, the outer boundary, and the distribution of natural resources, and the influences from other economic bodies, the reversals of economic development have seemed to be random and difficult to predict. This yoyo model also explains why economic downturns most likely to occur when the strength of the meridian field of the national economy is the lowest, because that is the time when the economic fluid in the region between the inner core and the outer boundary flows relatively evenly and symmetrically with respect to the center of rotation (Figure 6.15(a)) without any existing eddy leaf as shown in Figure 6.15(b) and (c). This is a systems-scientific explanation for the appearance of seemingly unpredictable economic cycles widely observed in market economies.

The mechanism beneath the flow pattern changes in the dishpan experiments is the uneven distribution of the forces acting on the fluid particles. As observed in the laboratory, two factors affect this distribution of acting forces: The rotation speed of the overall dish and the difference in temperature between the core and the outer periphery. When the national economy is simulated as a spinning field, the speed of rotation of the field can be modeled by the level of economic activities, while the temperature difference by that between the orderliness of the central bank and high-level policy makers in their thinking and planning for achieving the desired state of economic development and actual pursuit of various activities after profit making, where the former stands for order while the latter randomness and chaos.

When systemic yoyo model is introduced to model an economy, as what is done above in this section, we can organically thread all the topics of investigation learned in this chapter into a whole and see to where more attention should be directed.

Chapter 7

Portfolio of assets

As the last chapter of this part on the domestic financial system, seen as a closed system, let us look at the studies on the portfolio of assets. In other words, in this chapter we will learn as an individual, such as a person, a fund manager, an investment firm, how he could select financial assets to invest his available money in order to maximize his return.

The theory of asset demand is the theoretical foundation of all financial investments. It focuses on how a person, a fund manager, or an investment firm could increase his return in order to maximize his utility (satisfaction) by selecting appropriate financial assets from a huge pool of various available choices.

This chapter is organized as follows. Section 7.1 addresses the problem of why investment opportunities always exist from the angle of the systemic yoyo model. Section 7.2 presents the concepts of interest rates and yields. Different factors that affect the investment portfolio returns are listed in Section 7.3. In terms of balancing the risk and yields, Section 7.4 investigates the need of diversity of investment, the concept of mean variance utility, and matters related to the determination of the portfolio proportion. Then this chapter is concluded by Section 7.5, where we study the expected yields of portfolios, demands of assets, and the advantages of diversity.

7.1 CONSTANT EXISTENCE OF INVESTMENT OPPORTUNITIES

In this section, we look at in terms of the systemic yoyo model why there are different opportunities of investment by looking at how companies operate differently when their financial recourses are either limited or unlimited. We first intuitively model each commercial firm as an economic yoyo. Then, on the basis of the evolution of flows of such spinning structures, we can see how these economic yoyos interact with each other through combinations and breaking-ups. This end shows why there exist companies of different sizes in each economic sector or industry. Similar analysis shows why at any chosen moment of time in history, there are investment opportunities of various scales. With such a dynamic systemic analysis in place, we establish a simple profit-maximization model to study large and small firms and their difference in areas of production, determination of product selling prices, and the cost basis of their products. Among what we find is that when a firm has limited resources, its business has a glass ceiling for its potential maximum level of profits.

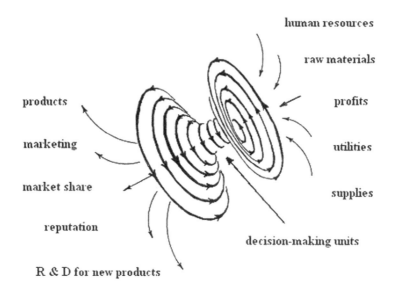

Figure 7.1 The yoyo model for a viable commercial company.

In comparison, it is found that financially resourceful companies have a large array of advantages over those companies that are limited by their resources. One example of such advantage is that the former companies bring in handsome profits in two-dimensions – the dimension of their products and that of their personnel, while the later companies could only produce their profits from the single dimension – the products – with an invisible glass ceiling. This end provides the long-sought-after explanation for the stable investment opportunities that exist over time and across national borders.

7.1.1 Economic yoyo flows

Let us model a commercial company as a spinning yoyo as shown in Figure 7.1, where the black-hole side sucks in all the basic supplies to sustain the vitality of the company, such as all needed raw materials, utilities, human resources, various services, and most importantly, the profit. The big-bang side emits the company's products. Through marketing effort, the company establishes its reputation for quality and occupies a certain percentage of the consumers' market. With the help of its R & D efforts, the company continually eyes on expansions in new areas of business. Various internal decision-making units, including the board of directors, the CEO and other executives, etc, connect the two sides of the spinning yoyo (Gillan, 2006).

The force necessary for such a yoyo to exist is the continued inflow of profits at and above the minimum sustainable level. Without the profits or the ability to generate the adequate profits, the company will fail and its economic yoyo will disappear. When the opportunity for earning adequate profits exist in the marketplace, many entrepreneurs will enter the same sector of the economic actions to fight for their shares of the potential pool of profits. Therefore, an industry or an economic sector starts to emerge. If we imagine all the possible spinning yoyos, each of which stands for a company, of an industry, floating against each other, we then can see the fact that some yoyos will have the tendency to combine into greater yoyos and some other yoyos may break into smaller yoyos or simply disappear, depending on the yoyos' spinning directions, speeds, and the angles of tilt. For details on how spinning yoyos could interact with each other, please consult Chapter 2 or (Lin, 2007). As long as the flow of demands is unstable, the economic yoyos will continue to fight for their own

1   Existence,
2   Survival, and then
3   Possible expansion.

When the picture of demands begin to crystallize, that is, when the acting and reacting forces start to stabilize, the intensive struggles for market shares between the economic yoyos will decline to a minimum and the entire economic sector begins to stabilize. This description is based on studies of fluid motions (OuYang, 1994) without using any of the terminology.

Now, if each economic yoyo, representing an individual company, in the previous imagined picture is replaced by such a bigger economic yoyo that represents an economic sector or industry, then a similar evolution of struggles for market share or the available resources and profit opportunities between industries or economic sectors would exist. This evolution of struggle would be dynamic if we place it in the flow of time. In particular, some yoyos, representing economic sectors or industries, will have the tendency to combine into greater yoyos and others may break apart into smaller yoyos or simply disappear. This evolution process is similar to that of a current rushing down a high land. If the boundary conditions are complicated, the state of the current flow pattern will be difficult to model using the traditional mathematics. It is because the overall state of motion is moving downward mixed with local variations, such as jet streams, whirlpools of different sizes, strengths and directions, etc. So, if we take a snap shot for the evolution of economic power struggle between the economic yoyos of various industries, in the still momentary picture, we can expect to see spinning yoyos of different sizes, meaning some industries (yoyos) suck in more profits and human resources than others. It is reasonable to expect that industries, whose yoyo structures suck in more profits and other available resources, behave differently than other industries.

The qualitative analysis of economic yoyos as above can continue along a line parallel to the evolution process of a fluid motion and a corresponding mathematical analysis can be introduced to support such a qualitative analysis. In the following, we will study a simple analytic model based on the premise of the yoyo model to see the analysis above actually works out mathematically.

7.1.2 A simple model for perfect capital markets

Assume that a retailer sells a specific product for $\$p_s$ each unit. The total cost for the entire process of acquiring the product, shipping and handling, insurance, storage, salesmen's salary, etc., is $\$p_p$ each unit. Let $n = n(p_s)$ be the total number of units sold at the price $\$p_s$ per unit. Then, for this particular retailer, his profit from this single line of product is given by

$$P = \text{profit} = n(p_s)(p_s - p_p). \tag{7.1}$$

If the capital markets are perfect, meaning that the retailer can borrow as much funds to purchase or produce as many units of the product as he needs to at any desirable time, then he would determine such a selling price $\$p_s$ so that his profit P in equ. (7.1) will be maximized. The first order condition for this maximization problem is given as follows:

$$\frac{\partial P}{\partial p_s} = n'(p_s)(p_s - p_p) + n(p_s) = 0. \tag{7.2}$$

For each fixed p_s- and p_p-values, we have

$$n'(p_s) = -\frac{n(p_s)}{p_s - p_p}.$$

This is a separable differential equation, its solution is

$$n(p_s) = \frac{C}{p_s - p_p}, \tag{7.3}$$

where C is the integration constant. If at the price level p_{s0} the initial market demand for the product is $n(p_{s0})$ units, then equ. (7.3) implies that $C = n(p_{s0})(p_{s0} - p_p)$. Substituting this C-value and equ. (7.3) into equ. (7.1) leads to

$$P = \text{profit} = n(p_{s0})(p_{s0} - p_p). \tag{7.4}$$

Equ. (7.3) implies that if the retailer plans to make a profit on each unit of his product, meaning $p_s > p_p$, then the constant C will be positive and $n(p_s)$ increases indefinitely as the selling price $p_s \to (p_p)^+$. If for reasons like competition or clearing out the specific product line, the retailer could choose to get rid of the product by selling it at a price p_s below the cost p_p. In this case, the constant C will be less than 0. If the retailer is involved in a competition to occupy a greater market share for his product, then equ. (7.3) also implies that he has to keep the unit price p_s as close to the cost basis p_p as much as possible in order to maximize the marker demand $n(p_s)$.

To maximize his profit P, equ. (7.4) indicates that if the difference $(p_{s0} - p_s)$ stays constant, the lower the p_{s0}-value the greater the demand $n(p_{s0})$ and the greater the total profit P. So, to generate as much profit as possible, the retailer has to keep his per-unit cost p_p as low as possible. Since the costs like shipping and handling, insurance, storage, salesmen's salaries, etc., are exogenous to the retailer and are quite robust, the retailer's efficient strategy to drastically reduce the cost basis p_p is to locate a manufacturer who can massively produce the needed product at a price below all other competitors. There are many ways to achieve this goal. For example, the manufacturer can hire a labor force at a below-the-market price, purchase his raw materials at extremely low prices, or introduce a revolutionary technology to drastically increase the labor's productivity.

Since raw materials and energies are international commodities mostly traded on global markets and due to the advances of information technology, all competing manufacturers can quite easily catch up with technological innovations, to purchase products at below-the-market prices, the retailer would naturally go to places where the labor quality is adequate and the labor costs are as low as possible. When many retailers look for such ideal labor markets, manufacturing businesses naturally start to relocate in different geographic locations with as ideal labor bases as possible. That is, the sea of economic yoyos starts to flow, caused by the uneven force of competition. This end explains why the current trend of moving manufacturing operations from industrialized nations to the third world countries does not seem reversible in the foreseeable future as long as the international transportation costs stay low and the global economic system stays open and competitive. In fact, this end is simulated well by the fluid patterns found in the dishpan experiment.

Assume in equ. (7.1) that the selling price p_s drops an increment $\Delta p_s > 0$ per unit. If the retailer does not pick up any additional market demand for his product at this reduced price $(p_s - \Delta p_s)$, then he would have a loss in the amount of $n(p_s)\Delta p_s$. So, if at the lower unit price level, the retailer expands the market demand $n(p_s)$ to that of $n(p_s - \Delta p_s)$ and if the profit increment $P(p_s - \Delta p_s) - P(p_s)$ is greater than the theoretical loss $n(p_s)\Delta p_s$, that is,

$$n(p_s - \Delta p_s)(p_s - \Delta p_s - p_p) - n(p_s)(p_s - p_p) > n(p_s)\Delta p_s, \tag{7.5}$$

then the retailer would rather reduce his unit selling price p_s to maximize his profit. Equ. (7.5) is equivalent to the following:

$$\frac{dP(p_s)}{dp_s} = \lim_{\Delta p_s \to 0^+} \frac{P(p_s - \Delta p_s) - P(p_s)}{-\Delta p_s} < -n(p_s). \tag{7.6}$$

Now, if instead of reducing the per-unit selling price p_s by an increment $\Delta p_s > 0$, the unit cost basis p_p is increased by an increment $\Delta p_p > 0$. If the retailer does not pick up any additional demand for his product with his increased cost basis, then he would experience a loss in the amount of $n(p_p)\Delta p_p$. However, if such an increase in his cost basis produces a new market demand $n(p_p + \Delta p_p)$ and the change in profit P satisfies

$$P(p_p + \Delta p_p) - P(p_p) > n(p_p)\Delta p_p, \tag{7.7}$$

then the retailer would rather increase his unit cost basis by as much as necessary in order to maximize his profit. Now, equ. (7.7) is equivalent to the following

$$\frac{dP(p_p)}{dp_p} = \lim_{\Delta p_p \to 0^+} \frac{P(p_p + \Delta p_p) - P(p_p)}{\Delta p_p} > n(p_p). \tag{7.8}$$

Combining equs. (7.6) and (7.8) provides the following conclusions.

Proposition 7.1. *Assume that the capital markets are perfect. If at a unit selling price* $p_s = p_{s0}$ *and at a unit cost basis* $p_p = p_{p0}$, *the total profit* $P(p_s, p_p)$ *from selling the product satisfies*

1. $\left.\dfrac{\partial P}{\partial p_s}\right|_{p_s=p_{s0}} < -n(p_{s0})$, then the retailer can reduce his unit selling price p_{s0} to increase his total profit; or

2. $\left.\dfrac{\partial P}{\partial p_p}\right|_{p_p=p_{p0}} > n(p_{p0})$, then the retailer can increase his unit-cost basis p_{p0} to reap in additional profits; or

3. $\left.\dfrac{\partial P}{\partial p_s}\right|_{p_s=p_{s0}} < -n(p_{s0})$ and $\left.\dfrac{\partial P}{\partial p_p}\right|_{p_p=p_{p0}} > n(p_{p0})$, then the retailer can both reduce his unit-selling price p_{s0} and increase his unit-cost basis p_{p0} to boast his total profit from his specific line of product. $\qquad\square$

What is implied in the previous analysis includes the fact that if by reducing his unit profit $(p_s - p_p)$ the retailer can greatly increase his total profit by gaining additional market demand, he will try all he can to accomplish that goal. More specifically, he might

1 Launch an aggressive commercial campaign where p_s is increased in order to expand the market demand;
2 Acquire larger number of units of the product at a much lower price where p_p is decreased so that p_s can be accordingly lowered to attract more customers;
3 Lower per-unit shipping and handling costs with increased volume of business;
4 Lower per-unit insurance costs with increased volume of business so that the savings can be passed on to the customers to create a healthier market demand; and
5 Increase the salesmen's wages. With such a potential of drastically increasing take-home pays, the salesmen would work harder and smarter so that the market demand is consequently pushed to new extremes.

7.1.3 A simple model when capital markets are imperfect

In this section, we analyze the simple model established in the previous section for the case of imperfect capital markets. That is, the profit P in equ. (7.1) is subject to the following constraint:

$$n(p_s)p_p = I, \tag{7.9}$$

where I is the total available funds for the retailer to invest in his line of product and $n(p_s)$ seen as the size of the inventory. The first order conditions for maximizing the profit P subject to the constraint in equ. (7.9) are given by

$$\frac{\partial P}{\partial p_s} = n'(p_s)(p_s - p_p) + n(p_s) = \lambda n'(p_s)p_p \tag{7.10}$$

and

$$\frac{\partial P}{\partial p_p} = n'(p_s)\frac{dp_s}{dp_p}(p_s - p_p) + n(p_s)\left(\frac{dp_s}{dp_p} - 1\right) = \lambda\left[n'(p_s)\frac{dp_s}{dp_p}p_p + n(p_s)\right], \quad (7.11)$$

where λ is the Lagrange multiplier. Substituting equ. (7.10) into equ. (7.11) leads to $(1 + \lambda)n(p_s) = 0$.

Equ. (7.9) implies that $n(p_s) \neq 0$, which means $\lambda = -1$. So, equ. (7.10) becomes $n'(p_s)p_s = -n(p_s)$.

Solving this equation for $n(p_s)$ gives us

$$n(p_s) = \frac{n(p_{s0})p_{s0}}{p_s}, \quad (7.12)$$

where $n(p_{s0})$ is the initial market demand when the product is sold for $\$p_{s0}$ per unit. So, the retailer's profit P is given by

$$P = \text{profit} = \frac{n(p_{s0})p_{s0}}{p_s}(p_s - p_p) \quad (7.13a)$$

$$= n(p_{s0})p_{s0}\left(1 - \frac{p_p}{p_s}\right). \quad (7.13b)$$

Equ. (7.12) indicates that to expand the market demand, the retailer has to decrease his unit selling price. Since the capital markets are imperfect, it means that the retailer has a limited resource to invest in his product. That is, he cannot afford to compete with a retailer who has unlimited resources. Similar to the situation of a financially powerful retailer, our small retailer can also increase his profit by reducing his selling price p_s, if he can keep the unit profit $(p_s - p_p)$ constant. However, unlike the powerful retailer in Section 12.2, our small retailer has a cap that is $n(p_{s0})p_{s0}$ (equ. (7.13b)) on how much he can expand his potential of total profit. This comparison tells us the following two facts:

1   When financial resources are limited, any venture will have a glass ceiling for its maximum level of profits.
2   Because of their limited resources, small retailers or poorly funded ventures do not have many opportunities to locate extremely low-priced manufacturers. One reason is that they do not have the ability to place large orders and two because they do not have the financial strength to create their own low-priced manufacturing operations to strengthen their ability to compete.

By comparing equs. (7.13a) and (7.13b) to conclusions (1)–(5) in the previous section, we can see the following facts:

1   While financially powerful companies are promoting their product to expand their market share and appearance, companies with limited resources cannot afford to devote much of their scarce resources to do so. One reason is that they do not have much money to allocate for the purpose of promotion. Another reason is that, as equ. (7.13b) indicates, excessive amount of spending will keep their unit selling

price p_s high. To increase their profit potential, companies with limited resources have to control their spending so that their profit can be maximized by lowering their unit-selling price p_s.

2   While financially powerful companies are placing large orders at much reduced wholesale prices, companies with limited resources just cannot take such opportunities. Similarly, other volume related savings are not available to ventures with limited resources.

In the rest of this chapter, we will study how an individual person's investment portfolio can be optimally constructed using various analytically models.

7.2   INTEREST RATES AND YIELDS

The index that can accurately measure the interest rate of bonds is the yield to maturity of the bonds. When economists use the phrase "interest rate", they mean the yield to maturity. The concept of yield is related to the concept of interest rate; the yield earned from bonds can be different from the interest rate earned from bonds. That explains why from the interest rate of a bond one cannot really judge whether or not investing in this bond is not good decision or not.

7.2.1   Present values

In the financial market, credit instruments carry different terms and requirements for repayments, which might even though look similar. To determine which instrument can bring forward with more return to the investor, one has to apply the concept of present value. In particular, the value of one dollar one year from now is different from the value of one dollar today. The former is lower than that of the latter. When the present one dollar is deposited in a bank account, the account balance after one year will be more than one dollar. The concept of present value is such a basic knowledge. In general, the asset invested in a certain project will receive $R total return after n years. If the interest rate is i, then when the future $R is brought back to the value of today, then the present value (PV) of the $R is given by

$$PV = \frac{R}{(1+i)^n} \tag{7.14}$$

which is known as the simple discount formula. This process of calculating the present value of some future amount is known as discounting the future. This computation provides the discount of one future value. In reality, some investment projects provide annual returns each year for n years into the future. Assume that the return of the kth year is R_k, $k = 1, 2, \ldots, n$, and the interest rate is i. then the formula for computing the present value of the returns of this investment is

$$PV = \sum_{k=1}^{n} \frac{R_k}{(1+i)^k} = \frac{R_1}{1+i} + \frac{R_2}{(1+i)^2} + \cdots + \frac{R_n}{(1+i)^n} \tag{7.15}$$

which is known as the discount formula with fixed terms and fixed amount. What this formula means is that if one invests PV dollars now at interest rate i, then after receiving R_1 in the first year, R_2 in the second year, ..., and R_n in the nth year, the entire investment will be returned entirely. As a special case, assume that the interest rate of an investment project is i, and it pays $\$R$ back each year. If in n years the entire investment is recovered, then the present value of this investment project is worth

$$PV = \sum_{k=1}^{n} \frac{R}{(1+i)^k} = \frac{R}{i}\left[1 - \frac{1}{(1+i)^n}\right] \tag{7.16}$$

Another specific scenario is: The interest rate of an investment project is i, each year the project pays $\$R$, and after doing so continuously for n years there are still $\$J$ left. Then the present value of this investment project is worth

$$PV = \sum_{k=1}^{n} \frac{R}{(1+i)^k} + \frac{J}{(1+i)^n} \tag{7.17}$$

The general scenario is that the interest rate of an investment project is i, the project pays back $\$R_1$ for interest and principal for the first year, $\$R_2$ for the second year, ..., and $\$R_n$ for the nth year with $\$J$ remaining in the project. Then the present value of this investment project is worth

$$PV = \sum_{k=1}^{n} \frac{R_k}{(1+i)^k} + \frac{J}{(1+i)^n} \tag{7.18}$$

7.2.2 Yield to maturity

The so-called yield to maturity of a bond stands for the interest rate that makes the present value of the return from investing in the bond equal to the value of the present investment. For example, an investor invests $\$V$ in a bond. After n years, he can receive one amount of $\$R$ as his total return. Then the interest rate i that makes the present value PV of these $\$R$ is equal to the value V of present investment should be determined from the formula $PV = R/(1+i)^n = V$. That is, $i = \sqrt[n]{R/V} - 1$. This interest rate of the bond is the yield to maturity of the bond. If an interest-bearing bond cannot be sold for its face value, then there will be a selling/purchasing price P. So, in the computation of the yield to maturity, the discounted value will have to be the same as the purchase price P, which is the value of the present investment, instead of the face value. In this case, the yield to maturity of the interest-bearing bond is given by the following formula:

$$P = \sum_{k=1}^{n} \frac{C}{(1+i)^k} + \frac{V}{(1+i)^n} \tag{7.19}$$

where the yield to maturity i is unknown, while the face value V of the interest-bearing bond, the annual interest C, the purchase price P, and the term n of the bond are known.

This yield to maturity is the most accurate index for interest rate. As economists talk about interest rate, they mean this yield to maturity. In order to avoid computational complications, there are two methods to approximate interest rate: the current period yield and discounted yield, where the current yield of a bond represents the ratio of the interest of the current period over the current price of the bond. Symbolically, assume that the current price of the bond is P, and the interest for the current period is C. Then the yield i_c of the current period is $i_C = C/P$. As for an interest-bearing bond of face value V and price P, the current period yield C is equal to the (annual) interest of the bond. So, from the formula that determines the yield to maturity i of an interest-bearing bond

$$P = \sum_{k=1}^{n} \frac{C}{(1+i)^k} + \frac{V}{(1+i)^n} = \frac{C}{i}\left(1 - \frac{1}{(1+i)^n}\right) + \frac{V}{(1+i)^n} \tag{7.20}$$

it follows that the relationship between the current period yield i_c and the yield to maturity i is

$$i_C = \frac{C}{P} = \frac{1 - \dfrac{V/P}{(1+i)^n}}{1 - \dfrac{1}{(1+i)^n}} \tag{7.21}$$

Therefore, it can be seen that

1 When the price P of the bond is equal to its face value V, the current period yield i_c will be the same as the yield to maturity i;
2 When $P \to V$, we have $i_c \to i$. That means that the closer the bond price is to its face value, the closer the current period yield to the yield to maturity;
3 When $n \to \infty$, we have $i_c \to i$. That indicates that the longer the remaining term the bond has, the closer the current period yield will be to the yield to maturity; and
4 The current period yield is positively correlated to the yield to maturity.

All these facts tell us that the current period yield is an approximate value of the yield to maturity, and that changes in the current period yield always signal the same directional changes in the yield to maturity. So, that explains why people tend to use the current period yield to replace for the yield to maturity due to the convenience of computation.

The fiscal year used by the Treasury Department of the United States consist of 360 days, while the actual teams of bonds are calculated using 365-day years. So, the interest of one-year bond should also consider the effect of the 5 extra days. That is to say that

Face value of bond – purchase price – bond interest

$$= \text{face value} \times \text{bond's interest rate} \times \frac{365}{360} \tag{7.22}$$

The concept of interest rate as defined in this sense is known as the discount yield, written as i_d. For the bond of one-year term, its discount yield is given by

$$i_d = \frac{\text{face value} - \text{purchase price}}{\text{face value}} \times \frac{360}{365} \tag{7.23}$$

More generally, if V stands for the face value of the bond, P the purchasing price, and T the number of days until maturity, then the discount yield i_d is given as follows:

$$i_d = \frac{V - P}{V} \frac{360}{T} = \frac{\text{face value} - \text{purchase price}}{\text{face value}} \times \frac{360}{\text{number of days until maturity}} \tag{7.24}$$

This method of calculating interest rate is of two clear characteristics: (1) It uses the percentage yield $(V - P)/V$ of the bond's face value instead of the percentage yield $(V - P)/P$ of the bond price; and (2) it computes the yield using 360-day year instead of 365-day year. These two characteristics determines the fact that the discount yield is smaller than the yield to maturity.

Additionally, the longer the term a bond carries, the greater difference between the bond's face value and the bond's price so that the greater the difference between the discount yield and the yield to maturity. In short, the relationship between the discount yield i_d (of one-year bond) and the yield i to maturity can be integrated into the following formula:

$$i_d = \frac{P}{V} \times \frac{360}{\text{number of days until maturity}} \times i \tag{7.25}$$

Because $P < V$, it follows that $i_d < i$, and that the longer the term is, the smaller P/V is, and the greater the difference between i_d and i.

Up to this point in the book, when we have talked about interest rate, we have not mentioned the factor of inflation. So speaking more strictly, the interest rate we have been talking about should be known as the nominal rate. When the expected inflation is subtracted from the nominal rate, what is obtained is the actual interest rate. The reason why we use expected inflation is because what is represented by the nominal interest rate is the future return. Let i stand for the nominal interest rate, and π_e the expected inflation. Then the actual interest rate is given by $i_r = i - \pi_e$. When compared to the nominal interest rate, the actual interest rate can better reflect the dynamics of the lending activities, and can more accurately indicate the tightness of money of the financial market.

7.2.3   Prices, rates of return, and yields of bonds

The interest rate of bond, as computed by using the yield to maturity, in reality reflects the discount rate of all future incomes from the bond, is the rate of return from the bond when it matures. It is only relevant to the initial purchasing price of the bond without anything to do with the price fluctuations of the market during the entire term of the bond. So, the yield to maturity is a rate of return internal to the bond. That is, it is a rate of return determined totally by the bond without being affected by the price factors of the market. It is exactly because of the internality of this interest rate that any increase in the interest rate does not really bring any benefit to the bond investor. In fact, the quantity that can accurately measure how much return the investor gains within a certain period of time from holding bonds or other priced securities is the yield, which is a concept of external rate of return.

The bond holder can sell his bond at the market price before the maturity date of the bond. The time period from the time an investor bought the bond to the time when he sold the bond is known as the holding period of the bond, or simply the holding period. Let P_b stand for the purchase price of the bond, P_s the selling price, and C the interest of the holding period, which is the interest the bond holder receives from the bond. Then the yield r of holding the bond for the holder is given by

$$r = \frac{C + P_s - P_b}{P_b} \tag{7.26}$$

the ratio of the sum of the interest and the price difference of buying and selling over the buying price. This index r is known as the yield of the bond, or more accurately the yield of the holding period.

Speaking generally, an investor purchases bond in one period and sells in another period so that he makes profits from the difference of the market prices. Assume that the investor purchases the bond in period t for the price P_t and sells the bond in period $t + 1$ for the price P_{t+1}. If the interest of the bond for each period is C, such as the annual interest of an interest-bearing bond, then the yield of the bond is

$$r = \frac{C + P_{t+1} - P_t}{P_t} \tag{7.27}$$

the ratio of the sum of the interest and the price difference between period t and period $t + 1$ over the price of the earlier period. If we separate the right hand side into the sum of two terms $r_C = C/P_t$ and $g = (P_{t+1} - P_t)/P_t$, where the first term is the current period yield and the second term the interest rate of the investment, then the yield of the bond is equal to the sum of the current period yield and the interest rate of the investment: $r = r_C + g$.

Let us now look into this equation a little further in details. When the interest rate of the bond changes, it means that the interest rate of the first period is different of that of the second period. When we say that the interest rate rises, it means that the price of the bond in the second period is lower than that of the first period, that is, $P_{t+1} < P_t$. So in this case, the interest rate of the investment is negative. Because r_c is computed by using the price p_t of the first period, the rising interest rate does not affect the current yield r_c in the computation of the yield. That implies that rising interest rate means that the price of the bond goes down so that the yield falls, which might lead to negative yield. That confirms our earlier statement that when the interest rate of a bond rises, it does not necessarily mean that the bond investor actually benefits from it.

The closer the bond is to its date of maturity, the closer the bond price to its face value, the smaller the price fluctuates, and the smaller the corresponding interest rate goes up and down. Conversely, the further the bond is away from its date of maturity, the greater fluctuation the bond price and the corresponding interest rate are. At the same time, the fluctuations in interest rate in turn cause the yield to go up or down accordingly. So, the longer the bond's term is, the more severe the bond price and interest rate fluctuate, and the more austere the yield swings. That explains such a fact that the price and yield of long-term bond are more unstable than those of short-term bond. So, fluctuations of interest rate tremendously increase the risk of long-term bond. Such risk of the return of assets that is caused by fluctuations in interest rate is known as interest rate risk. Controlling interest rate risk represents a great concern of managers of financial institutions.

What needs to be noted is that interest rate risk only exists with long-term bonds that are still far from their maturity. If a bond is already near its date of maturity, then the price of the bond will be close to its face value with very minor fluctuation so that the yield is almost the same as the yield to maturity without much interest rate risk. That illustrates that although long-term bonds experience high interest rate risk, short-term bonds do not have much such risk at all.

Like interest rate, yield can also be divided similarly: the nominal yield and actual yield. The yield without considering the effect of inflation is known as the nominal yield, which is the yield we have been discussing as of this point. When inflation is subtracted from the nominal yield, we obtain the actual yield, which represents the quantitative magnitude of additional goods or labor that can be purchased from the investor's holding of securities.

7.3 FACTORS AFFECTING PORTFOLIO CHOICES

When a person decides whether or not he should purchase and hold a certain financial asset, and whether he should purchase this kind of asset or that kind, he must consider several factors: wealth, expected return, risk, liquidity, and expected utility.

By wealth, it means all resources including all kinds of assets the person owns. When the person's wealth increases, he has more resource to purchase assets so that his demand for assets goes up. Of course, we cannot exclude those assets for which the demand decreases as the amount of wealth increases. Such assets are known as low quality. In the study of economics, it is generally assumed that the demand for assets increases as the amount of wealth goes up. That is to say, wealth is an important factor that affects the demand for assets.

There are many different kinds of assets. Different assets react to changes in the amount of wealth differently. When the level of wealth rises, the amount of demand for certain assets increases less than the magnitude of increase in wealth, while the amount of demand for some other assets increases more than the magnitude of increase in wealth. So, let us divide assets into two classes by using wealth elasticity of asset demand, which is defined by the amount of increase in asset demand (written in percentage form) divided by the amount of increase in wealth (written in percentage form):

$$\text{wealth elasticity of asset demand} = \frac{\text{percentage change in asset demand}}{\text{percentage change in wealth}} \quad (7.28)$$

When the wealth elasticity E_ε of an asset is less than 1, the asset is referred to as lack of wealth elasticity. When $E_\varepsilon = 1$, the asset is referred to as an elasticity unit of wealth. By using the magnitudes of wealth elasticity, we can divide assets into two classes: necessary assets and luxury assets, where an asset is necessary if its wealth elasticity is less than or equal to 1, and an asset if luxury if its wealth elasticity is greater than 1. For example, cash and deposits in checking accounts are necessary assets. In order to handle daily transactions people have to hold a certain amount of cash and maintain a certain amounts of balance in their checking accounts. On the other hand, stocks and bonds belong to luxury assets. Only when people's income reaches a certain level, they will apply the remnant capital to invest in stocks and bonds.

We know that the yield on assets measures the magnitude of profit for people to hold an asset; it is related to the price of the asset, especially the future price. When a person purchases an asset in the current period, if the future price of this asset will be higher than the purchase price of the current period, then holding the asset will produce profit and the higher the future price will be, the higher rate of return. However, if the future price is lower than the purchase price, then purchasing and holding the asset will lead to losses. However, the future price of the asset is influenced by many random (uncertain) factors, whether it is higher or lower than the purchase price is unknown in the current period. To this end, people can only make estimates or form expectations. Because of this reason, the profit situation of purchasing and holding an asset is unknown in the current period and can only be estimated. If people expect the yield of one asset is relatively high, then in the current period, the demand for this asset will increase. So, the expected yield of an asset is another important factor that affects the demand for the asset.

An index that measures the magnitude of expected yield from purchasing and holding an asset is the expected rate of return of the asset. For example, if people estimate that the common stock of a petroleum company will bring 15% of return within one half of the next one year, and 5% within the other half of the year, then the expected rate of return from holding the common stock of this petroleum company is

$$r_e = 0.5 \times 15\% + 0.5 \times 5\% = 10\%$$

where 0.5 stands for the probability for the stock to earn 15%, while the probability to earn 5% is also 05.

More generally, assume that the return of an asset is influenced by many random factors so that people estimate that the rate of return of this asset has m possibilities with the rate of return r_i for the ith possibility, which could happen with probability p_i, ($p_i \geq 0, i = 1, 2, \ldots, m; p_1 + p_2 + \cdots + p_m = 1$), then the expected rate of return r_e of this asset is

$$r_e = p_1 r_1 + p_2 r_2 + \cdots + p_m r_m = \sum_{i=1}^{m} p_i r_i \qquad (7.29)$$

That is, the expected rate of return is the mathematical expectation of all possible rates of return.

If the probability for the rate of return of a certain asset to be a fixed number is 100%, then this asset is known as a safe asset or riskless asset. Generally, government bonds are considered safe assets, because a currently well run government does not fall in the foreseeable future. So, the return from investing in governments' bonds is stable without risk. Its expected return is equal to the bonds' interest rates (the yields to maturity). On the other hand, if the rate of return of an asset is not stable, and cannot reach a fixed number with 100% chance, then this asset is known as risky asset. For example, stocks and bonds of companies are assets without any fixed yield. So, they are risky assets.

In the previous discussions, we have pointed out two important factors that affect the demand for an asset: wealth and expected rate of return. However, these factors do not affect the demand curve of the asset. Studies of microeconomics tell us that each change in the demand for a good includes two meanings: One is a change in the quantity of demand, and the other a change in the location of the demand curve. Next let us look at the factors that affect the location of the demand curve of an asset.

One of such factors is people's attitude to risk. There is a degree of risk involved in financial investment. That is, the rate of return of investment cannot be really certain. For example, for the previously mentioned common stocks of the petroleum company, its rate of return for half of the coming year is 15%, and for the other half of the year is 5%. So, the rate of return is not fixed, meaning that investing in the stock of this company is risky. Now, there is another asset, one-year government bond, which is riskless asset with 10% of rate of return (yield to maturity). This guaranteed rate of return is equal to the expected rate of return of the company stock and is not affected by any random factor. So, when faced with the problem of selecting whether the stock of the petroleum company or the government bond to invest, how will an investor make his decision? That will depend on his attitude toward risk.

Some people believe that although there is a 50% chance for the company stock to earn more than 15%, there is still another 50% chance for the stock to earn less than 5%. If investing in this stock, they might very likely earn 5% instead of 15%. So, investing in this stock may very well produce a loss for the investor. Because the rate of return from the government bond is the same as the expected rate of return from the company stock, and because the rate of return of the government bond is stable without risk, investing in the government bond is more stable and better than that in the company stock. The number of people holding such beliefs is not small. In fact, most people reason and think this way. That is to say, most people do not like to take risk and are known as risk aversers. There are also such people who believe in that although investing in the government bond can guarantee the same rate of return as the expected rate of return of the company stock, investing in the government bond misses out the opportunity of earning as high as 15% of return. So, to chase after the 15% high yield, these people are willing to risk the chance of earning only 5% of return by investing in the company stock. These people are known as risk lovers. They enjoy taking risks.

The previous analysis indicates that when people have different attitudes toward risk, they will make different selections of assets in terms of investments, as shown in how they subjectively evaluate various assets. For risk aversers, when the expected rates of return are the same, the smaller the risk is the better, while for risk lovers, the riskier the better if the expected rates of return are the same. The difference in the attitudes toward risk determines the personal preference of assets so that it in turn defines the location of his demand curve of assets.

Another important factor that determines the demand curve of an asset is the liquidity of the asset. That is the ability to convert the asset into cash. The reason why people are fond of cash instead of any other non-cash forms of assets is because cash is the media of transactions and represents the asset with the highest level of liquidity. So, the more liquid an asset enjoys, the more welcomed it is. That is the liquidity preference, which governs how people evaluate assets and determines the location of the demand curve of an asset.

7.4 BALANCE BETWEEN RISK AND YIELDS

Tobin's analysis on the speculative demand for money, Chapter 4, has indeed improved Keynesian theory. However, it also suffers from weaknesses. For example, it does not clearly point out whether or not the speculative demand for money indeed exists. If the rate of return of bonds is very high, then is it still necessary for people to hold cash in order to speculate or store wealth? The answer to this question is negative, because in this case holding the bonds will be more cost effective than keeping cash.

7.4.1 Diversity of investment

Most investment opportunities involve risky assets, where investors cannot tell how much return they can earn for the coming year. For example, the price of an office building can fall and the price of a particular company stock may rebound. However, the specific situations of fall and rise cannot be known ahead of time, leading to uncertainties in the return of the assets. Additionally, inflation is also uncertain and causes the actual rate of return uncertain. Even so, rational investors make their decisions by using their expected rates of return formed on the basis of the available information and relevant knowledge. After making comparisons between different expectations, these investors select their assets of investment among all the risky assets.

That is to say, the yield of any risky asset is influenced by random factors and is a random variable. If m stands for a risky asset, and R_m the rate of return of this asset, then R_m is affected by many random factors $\omega \in \Omega$, where Ω is a random variable defined on a space of natural states. Assume that the distribution function of R_m is F. Then the expected rate of return on the risky asset m is given by

$$E[R_m] = \int_{-\infty}^{+\infty} x dF(x) = \int_{-\infty}^{+\infty} x f(x) dx \tag{7.30}$$

For a particular year, the actual rate of return of the risky asset could be much higher or much lower than the expected rate of return, However, after a relatively long period of time, the average rate of return will be roughly the same as the expected rate of return.

In fact even a safe asset can be seen as risky, a degenerated risky asset, where the yield is not affected by any random factor and so a constant. If R_f stands for the rate of return of a certain safe asset, then R_f can be seen as a degenerated random variable: For any natural state $\omega \in \Omega$, $R_f(\omega) = R_f = a$ constant. The expected rate of return of this safe asset is equal to its yield: $E[R_f] = R_f$.

Table 7.1 Investment risks and returns in the United States (1926–1991).

	Actual rate of return (%)	Risk (standard deviation) (%)
Common stocks	8.8	21.2
Long-term company bonds	2.4	8.5
Short-term treasuries	0.5	3.4

Different assets have different rates of return. Table 7.1 lists the yields and risks of common stocks, long-term company bonds, and short-term treasuries of the United States during 1926–1991.

From Table 7.1, it follows that the yield of stocks is nearly 9%, while the actual rate of return on short-term treasuries is not even 1%. However, there were still a lot of people who purchased and held the short-term bonds with very low expected rate of return instead of investing in the common stocks which carried a much high expected rate of return. So, a natural question is *why*? The answer is that the demand for assets is not merely dependent on the expected rate of return; it is also dependent on the risk of the asset. To this end, the risk of the common stocks reached as high as 21.2%, while the risk of short-term treasuries was merely 3.4%. What this table reflects is that the higher the expected rate of return of an asset class carries, the riskier the asset class is. Evidently, when making investment decisions, one has to balance the return and risk of the assets of interest.

When a consumer has m risky assets to consider, what proportion of his savings should he use in the investment of risky assets? Next, let us concentrate on the study of this question. To this end, let us assume that there are only two assets: one risky asset m and safe asset f, and the consumer will invest his savings in either the risky asset or the safe asset, or a combination of the two. Let β stand for the proportion of his savings he invests in the risky asset. Then the proportion of his savings he invests in the safe asset f is $1 - \beta$. Now, the question is: How does the consumer decide on this proportion β?

Assume that the rate of return of the risky asset m is R_m, which is a random variable, the expected rate of return of m is r_m, that is, $r_m = E[R_m]$, and the variance σ_m^2 of is R_m. Similarly, we let R_f represent the rate of return of the safe asset f, which is a constant, and r_f the expected rate of return of f. Then $r_f = E[R_f] = R_f$. Only when the expected rate of return r_m of the risky asset m is greater than the rate of return r_f of the safe asset f, there will be some risk aversers to invest in the risky asset. Otherwise, they would invest their entire savings in the safe asset.

Let R_β stand for the rate of return of investing proportion β of his savings in the risky asset m, and proportion $1 - \beta$ of his savings in the safe asset f, r_β the expected rate of return of this asset combination, that is, $r_\beta = E[R_\beta]$, and σ_β^2 the variance. Then we have

$$R_\beta = \beta R_m + (1 - \beta)R_f = R_f + \beta(R_m - R_f) \qquad (7.31)$$

$$r_\beta = \beta r_m + (1 - \beta)r_f = r_f + \beta(r_m - r_f) \qquad (7.32)$$

$$\sigma_\beta = \sqrt{E[(R_\beta - r_\beta)^2]} = \beta\sqrt{E[(R_m - r_m)^2]} = \beta\sigma_m \qquad (7.33)$$

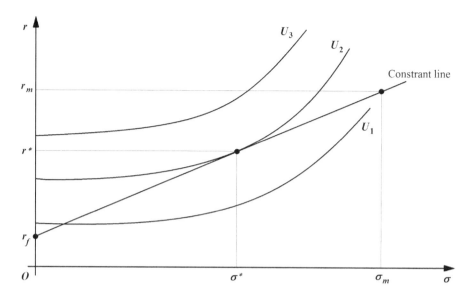

Figure 7.2 Choice between return and risk.

From these equations we obtain

$$\beta = \sigma_\beta/\sigma_m \quad \text{and} \quad r_\beta = r_f + \beta(r_m - r_f) = r_f + [(r_m - r_f)/\sigma_m]\sigma_\beta.$$

the second of which is referred to as the budget line equation. It expresses the trade-off between the risk and return of the assets in the combination, where r_f, r_m, σ_m are known constants, while r_β and σ_β are unknown, and need to be determined by the investor. What the budget line equation tells us is that as the risk (variance) of the combination increases the expected rate of return of the combined assets becomes greater, as shown in Figure 7.1. Because of this reason, the slope $(r_m - r_f)/\sigma_m$ of the budget line equation is referred to as the price of risk. It spells out the degree of risk the investor has to take in order to acquire a higher expected return.

7.4.2 Mean variance utility

The purpose for an investor to select assets for his investments is to bring in additional income in order to gain greater satisfaction. So, the quality of a particular portfolio of assets should be evaluated by using the investor's utility function. Assume that $u(x)$ is such a utility function of the investor, where when the investor acquires x units of income, he obtains $u(x)$ units of utility.

When the investor puts his money with a project of risk, his return will be uncertain, which stands for a random variable. Assume that ξ is a random return variable. It represents an investment activity of risk. Then, the investor's utility $u(\xi)$ is also random.

In this case, to evaluate the quality of the random return variable ξ, one would look at the magnitude of the expected utility $E[u(\xi)]$. The greater the expected utility is, the better the investor would think of his investment activity.

Based on this understanding, we can generalize the return utility function $u(x)$ of the investor to the set X of all random variables as follows: For any $\xi \in X$, define $Eu(\xi)$ as the expected utility of ξ; that is, $Eu(\xi) = E[u(\xi)]$. The reason why this function $Eu(\xi)$ is seen as a generalization of the return utility $u(x)$ is because for any level x of return, it can be treated as a random variable $\xi_x = x$. Therefore, we have $Eu(\xi_x) = E[u(\xi_x)] = u(x)$. Because x can be identified as a random variable ξ in X, will no longer distinguish x and ξ_x so that we have $Eu(x) = u(x)$. This general function $Eu(\xi)$ is known as the expected utility function of the investor.

If we apply this expected utility function to the asset proportion β of the investor's portfolio, then $Eu(\xi)$ could be employed as a criterion about whether or not the proportion of his asset selection is appropriate: The greater the expected utility of the portfolio the better the corresponding portfolio proportion. In the following we will see how particularly such an evaluation can be conducted. For the sake of convenience of communication, the income the investor plans to use in his investment is seen as a unit; and the variable x in the return utility function $u(x)$ stands only for the net return. So, x stands for the yield or the return of the unit investment. Additionally, we can assume $u(0) = 0$.

With these assumptions of simplification, let us once again look at R_β. It stands for the yield of the investment with $1 - \beta$ of the portfolio invested in safe assets and β of the portfolio in risky assets. In the rest of this section, we will use R_β to represent both such a portfolio composition and its yield. Because R_β is a random variable, let us use r_β to denote its expected yield (the mathematical expectation) and σ_β its risk (standard deviation), and $Eu(R_\beta)$ the expected utility. Then, we can rewrite R_β by using σ_β and r_β as follows:

$$R_\beta = r_f + \beta(R_m - r_f) = r_f + \frac{R_m - r_f}{\sigma_m}\sigma_\beta$$

$$= r_f + \frac{r_m - r_f}{\sigma_m}\sigma_\beta + \frac{R_m - r_f}{\sigma_m}\sigma_\beta$$

$$= r_\beta + \frac{R_m - r_f}{\sigma_m}\sigma_\beta \tag{7.34}$$

That implies that for any $\sigma \geq 0$ and $r \geq 0$, we can construct such an investment portfolio $\xi(\sigma, r)$ such that σ is the risk and r the expected yield of the portfolio:

$$\xi(\sigma, r) = r + \frac{R_m - r_f}{\sigma_m}\sigma \tag{7.35}$$

It can be confirmed that the expected yield of this investment portfolio $\xi(\sigma, r)$ is r, that is, $E[\xi(\sigma, r)] = r$, the risk is σ, that is, $\sigma(\xi(\sigma, r)) = \sigma$, and $\xi(\sigma, r)$ stands for the yield of the investment portfolio with safe assets of yield $r_f + r - r_\beta$ and risky assets of yield $R_m + r - r_\beta$ according to the proportion $\beta = \sigma/\sigma_m$:

$$\xi(\sigma, r) = r + (R_m - r_m)\frac{\sigma}{\sigma_m}$$

$$= r + \beta(R_m - r_m)$$

$$= r + \beta(r_f - r_m) + \beta[(R_m + r - r_\beta) - (r_f + r - r_\beta)]$$

$$= r_f + r - [r_f + \beta(r_f - r_m)] + \beta[(R_m + r - r_\beta) - (r_f + r - r_\beta)]$$

$$= (r_f + r - r_\beta) + \beta[(R_m + r - r_\beta) - (r_f + r - r_\beta)] \quad (7.36)$$

It needs to be noted that the investment portfolio as described by $\xi(\sigma, r)$ stands for an ideal case, because there is guarantee that the investor has the choice of having such safe assets of yield $r_f + r - r_\beta$ and risky assets of yield $R_m + r - r_\beta$. Even so, this theoretical analysis provides the convenience for us to investigate the portfolio selection: $R_\beta = \xi(\sigma_\beta, r_\beta)$ is a particular portfolio $\xi(\sigma, r)$. Additionally, for different (σ_1, r_1) and (σ_2, r_2), we have $\xi(\sigma_1, r_1) \neq \xi(\sigma_2, r_2)$. Hence, the first quadrant R_+^2 of the σ-r plane can be bijectively mapped onto the subset $X = \{\xi(\sigma, r) : (\sigma, r) \in R_+^2\}$ of X through $\xi: R_+^2 \to X$. That is, each point (σ, r) of the first quadrant of the σ-r plane can be seen as the investment composition $\xi(\sigma, r)$, where σ stands for the risk of the portfolio, while r the expected yield. However, only the investment portfolios located on the budget line are practically possible.

With the previous understanding of the first quadrant R_+^2 of the σ-r plane, we will call R_+^2 the risk-yield plane. Through the expected utility function $Eu(\xi)$ we can derive the following so-called mean variance utility function U of the risk-yield plane:

$$U(\sigma, r) = Eu(\xi(\sigma, r)), \quad \text{for any } (\sigma, r) \in R_+^2 \quad (7.37)$$

Theorem 7.1. Assume that the return utility function $u(x)$ is twice differentiable. Then the mean variance utility function $U(\sigma, r)$ satisfies:

1   For a risk averser (that is, $u'' < 0$), $U(\sigma, r)$ is a concave function. That is, for any (σ_1, r_1) and $(\sigma_2, r_2) \in R_+^2$ such that $(\sigma_1, r_1) \neq (\sigma_2, r_2)$ and any real number $\alpha \in (0, 1)$, the following holds true:

$$\alpha U(\alpha_1, r_1) + (1 - \alpha)U(\alpha_2, r_2) < U(\alpha(\alpha_1, r_1) + (1 - \alpha)(\alpha_2, r_2)) \quad (7.38)$$

2   For a risk lover (that is, $u'' > 0$), $U(\sigma, r)$ is a convex function. That is, for any (σ_1, r_1) and $(\sigma_2, r_2) \in R_+^2$ such that $(\sigma_1, r_1) \neq (\sigma_2, r_2)$ and any real number $\alpha \in (0, 1)$, the following holds true:

$$\alpha U(\alpha_1, r_1) + (1 - \alpha)U(\alpha_2, r_2) > U(\alpha(\alpha_1, r_1) + (1 - \alpha)(\alpha_2, r_2)) \quad (7.39)$$

3   For a person who is risk neutral (that is, $u'' = 0$), $U(\sigma, r)$ is a linear function. That is, for any (σ_1, r_1) and $(\sigma_2, r_2) \in R_+^2$ such that $(\sigma_1, r_1) \neq (\sigma_2, r_2)$ and any real number $\alpha \in (0, 1)$, the following holds true:

$$\alpha U(\alpha_1, r_1) + (1 - \alpha)U(\alpha_2, r_2) = U(\alpha(\alpha_1, r_1) + (1 - \alpha)(\alpha_2, r_2)) \quad (7.40)$$

Proof. For a risk averser, from $u''(x) < 0$ for any level x of income it follows that $u(x)$ is a concave function of x. That means that for any x_1 and x_2 such that $x_1 \neq x_2$ and any real number α satisfying $0 < \alpha < 1$, the following holds true:

$$\alpha u(x_1) + (1 - \alpha)u(x_2) = u(\alpha x_1 + (1 - \alpha)x_2) \tag{7.41}$$

Therefore, for any (σ_1, r_1) and $(\sigma_2, r_2) \in R_+^2$ such that $(\sigma_1, r_1) \neq (\sigma_2, r_2)$ and any real number $\alpha \in (0, 1)$, we have

$$
\begin{aligned}
\alpha U(\alpha_1, r_1) + (1 - \alpha)U(\alpha_2, r_2) &= \alpha E[u(\xi(\alpha_1, r_1))] + (1 - \alpha)E[u(\xi(\alpha_2, r_2))] \\
&= E[\alpha u(\xi(\alpha_1, r_1)) + (1 - \alpha)u(\xi(\alpha_2, r_2))] \\
&< E[u(\alpha\xi(\alpha_1, r_1) + (1 - \alpha)\xi(\alpha_2, r_2))] \\
&= E[u(\xi(\alpha(\alpha_1, r_1) + (1 - \alpha)(\alpha_2, r_2)))] \\
&= U(\alpha(\alpha_1, r_1) + (1 - \alpha)(\alpha_2, r_2)) \tag{7.42}
\end{aligned}
$$

That is, for a risk averser, the mean variance utility function is concave. Similarly, statements (2) and (3) can be proven. QED.

Theorem 7.2. Assume that the return utility function $u(x)$ is twice differentiable and $u'(x) > 0$ for all x. Then the following hold true:

1 For a risk averser, $\partial U(\sigma, r)/\partial r > 0$ and $\partial U(\sigma, r)/\partial \sigma < 0$ for all $(\sigma, r) \in R_+^2$ so that low-risk and high-yield investment portfolio is better than high-risk and low-yield portfolio.
2 For a risk lover, $\partial U(\sigma, r)/\partial r > 0$ and $\partial U(\sigma, r)/\partial \sigma > 0$ for all $(\sigma, r) \in R_+^2$ so that high-risk and high-yield investment portfolio is better than low-risk and low-yield portfolio.
3 For a person who is risk neutral, $\partial U(\sigma, r)/\partial r > 0$ and $\partial U(\sigma, r)/\partial \sigma = 0$ for all $(\sigma, r) \in R_+^2$ so that high-yield investment portfolio is better than low-yield portfolio regardless what level of risk is involved.

Proof. From the definition it follows that $U(\sigma, r) = E[u(\xi(\sigma, r))] = E\left[u\left(r + \dfrac{R_m - r_m}{\sigma_m}\sigma\right)\right]$. So, we have

$$
\begin{aligned}
\frac{\partial U(\sigma, r)}{\partial r} &= E\left[u'(\xi(\sigma, r))\frac{\partial \xi(\sigma, r)}{\partial r}\right] \\
&= E[u'(\xi(\sigma, r))] > 0 \tag{7.43}
\end{aligned}
$$

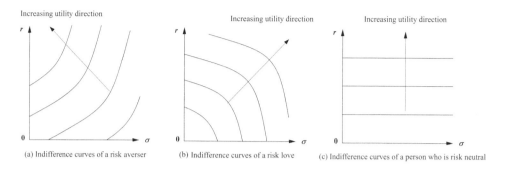

(a) Indifference curves of a risk averser   (b) Indifference curves of a risk love   (c) Indifference curves of a person who is risk neutral

Figure 7.3 The indifference curves of the mean variance utility functions.

and

$$\frac{\partial U(\sigma,r)}{\partial \sigma} = E\left[u'(\xi(\sigma,r))\frac{\partial \xi(\sigma,r)}{\partial \sigma}\right]$$

$$= E\left[u'(\xi(\sigma,r))\frac{R_m - r_m}{\sigma_m}\right]$$

$$= \mathrm{Cov}\left(u'(\xi(\sigma,r)), \frac{R_m - r_m}{\sigma_m}\right) + E[u'(\xi(\sigma,r))]E\left[\frac{R_m - r_m}{\sigma_m}\right]$$

$$= \mathrm{Cov}\left(u'(\xi(\sigma,r)), \frac{R_m - r_m}{\sigma_m}\right) \quad (\text{because } E[(R_m - r_m)/\sigma_m] = 0) \qquad (7.44)$$

where Cov stands for covariance operator $\mathrm{Cov}(\xi,\eta) = E[(\xi - E\xi)(\eta - E\eta)]$.

From $\xi(\sigma,r) = r + [(R_m - r_m)/\sigma_m]\sigma$, it follows that $\xi(\sigma,r)$ is positively correlated to $(R_m - r_m)/\sigma_m$. Therefore, for a risk averser, $u'(\xi(\sigma,r))$ is negatively correlated to $(R_m - r_m)/\sigma_m$ and so $U'_\sigma(\sigma,r) < 0$; for a risk lover, $u'(\xi(\sigma,r))$ is positively correlated to $(R_m - r_m)/\sigma_m$ and hence $U'_\sigma(\sigma,r) > 0$; and for a person who is risk neutral, $u'(\xi(\sigma,r))$ is a constant so that $u'(\xi(\sigma,r))$ is not correlated to $(R_m - r_m)/\sigma_m$ so that $U'_\sigma(\sigma,r) = 0$. QED.

These two theorems provides us with the clear characteristics of the mean variance utility function. Figure 7.3 depicts respectively the shapes of the indifference curves of the mean variance utility functions for risk aversers, risk lovers, and those who are risk neutral.

7.4.3   Determination of the portfolio proportion

Based on the characteristics of the mean variance utility and $r_m > r_f$, it follows that under the constraint of the investor's budge, the optimal choice for risk lovers and those who are risk neutral must be investing all of their available funds in risky assets without considering safe assets. That is, the optimal portfolio proportion for these people is $\beta = 1$. In the following let us look at the strategy of risk aversers.

For a risk averser, his objective is to select an appropriate investment portfolio under the budget constraint, that is selecting an appropriate point (σ^*, r^*), so that his mean variance utility reaches its maximum. This maximization problem with constraints can be written as follows:

$$\max U(\sigma, r)$$

$$\text{s.t. } \sigma > 0, r > 0$$

$$r = r_f + \frac{r_m - r_f}{\sigma_m} \sigma \tag{7.45}$$

To solve this problem, assume that (σ^*, r^*) is the solution of this maximization problem. Then there is a Lagrange multiplier λ such that the following system of equations holds true:

$$\begin{cases} U_r'(\sigma^*, r^*) = \lambda \\ U_\sigma'(\sigma^*, r^*) = \dfrac{-\lambda(r_m - r_f)}{\sigma_m} \\ r^* = r_f + \dfrac{\sigma^*(r_m - r_f)}{\sigma_m} \end{cases} \tag{7.46}$$

Because $U_r'(\sigma, r) > 0$, the Lagrange multiplier also satisfies $\lambda > 0$. In fact, this system of equations is also a sufficient condition for (σ^*, r^*) to become the maximum point of $U(\sigma, r)$ under the investment budget constraints. That is, we have the following result:

Theorem 7.3. Assume that an investor is risk averse, that is, $u'' < 0$, $\sigma^* > 0$ and $r^* > 0$. Then a sufficient and necessary condition for (σ^*, r^*) to be the maximum point of the mean variance utility function $U(\sigma, r)$ under the constraint $r = r_f + \sigma(r_m - r_f)/\sigma_m$ is $(r_m - r_f)/\sigma_m = -U_\sigma'(\sigma^*, r^*)/U_r'(\sigma^*, r^*)$ and $r^* = r_f + \sigma^*(r_m - r_f)/\sigma_m$.

Proof. According to the discussion above, it follows that the necessity of this theorem is guaranteed by the method of Lagrange multiplier. So, it suffices to show the sufficiency only. To this end, assume that $(r_m - r_f)/\sigma_m = -U_\sigma'(\sigma^*, r^*)/U_r'(\sigma^*, r^*)$ and $r^* = r_f + \sigma^*(r_m - r_f)/\sigma_m$.

Now, the maximization problem of $U(\sigma, r)$ under the constraint $r = r_f + \sigma(r_m - r_f)/\sigma_m$ is equivalent to the maximization problem of $g(\sigma) = U(\sigma, r_f + \sigma(r_m - r_f)/\sigma_m)$ without constraint. From Theorem 7.1, it follows that $U(\sigma, r)$ is a concave function so that $g(\sigma)$ is also a concave function. So, $g''(\sigma) < 0$ holds true for all $\sigma > 0$. By computing the first order derivative of $g(\sigma)$, we have

$$g'(\sigma) = U_\sigma'\left(\sigma, r_f + \frac{\sigma(r_m - r_f)}{\sigma_m}\right) + U_r'\left(\sigma, r_f + \frac{\sigma(r_m - r_f)}{\sigma_m}\right) \frac{\partial(\sigma(r_m - r_f)/\sigma_m)}{\partial \sigma}$$

$$= U_\sigma'\left(\sigma, r_f + \frac{\sigma(r_m - r_f)}{\sigma_m}\right) + U_r'\left(\sigma, r_f + \frac{\sigma(r_m - r_f)}{\sigma_m}\right) \frac{(r_m - r_f)}{\sigma_m} \tag{7.47}$$

Because $r^* = r_f + \sigma^*(r_m - r_f)/\sigma_m$, $\quad g'(\sigma^*) = U_\sigma'(\sigma, r^*) + U_r'(\sigma, r^*)(r_m - r_f)/\sigma_m = 0$. Combined with the fact that $g(\sigma)$ is a concave function, we have $g''(\sigma^*) < 0$. From

$g'(\sigma^*) = 0$ and $g''(\sigma^*) < 0$ it follows that σ^* is a maximum point of $g(\sigma)$. Notice that for any concave function, if it has a maximum point, then that point exists uniquely. Therefore, the maximum value of the concave function must also be the absolute maximum. So, σ^* is the absolute maximum point of $g(\sigma)$. That implies that (σ^*, r^*) is the absolute maximum point of the mean variance utility function $U(,)$ under the constraint $r = r_f + \sigma(r_m - r_f)/\sigma_m$. QED.

This theorem tells us that the optimal selection point (σ^*, r^*), which stands for the optimal investment proportion, of a risk averser is determined by the following system of equations:

$$
\begin{cases}
\dfrac{r_m - r_f}{\sigma_m} = -\dfrac{U'_\sigma(\sigma^*, r^*)}{U'_r(\sigma^*, r^*)} \\[4mm]
r^* = r_f + \dfrac{\sigma^*(r_m - r_f)}{\sigma_m}
\end{cases}
\tag{7.48}
$$

Let us now look at the significance of this optimal solution and how to judge whether or not an investment portfolio is optimized. We have seen that the slope $(r_m - r_f)/\sigma_m$ of the budget line stands for the price of risk. That is the amount of increase in the expected yield when the risk is increased by one unit in the investment portfolio. So, the price of risk $(r_m - r_f)/\sigma_m$ represents the return of risk, which is an objectively existing return and cannot be artificially altered or ignored. Next, let us look at the meaning of $-U'_\sigma(\sigma^*, r^*)/U'_r(\sigma^*, r^*)$.

Because the investor analyzed is risk averse, when the risk increases, if the expected yield does not rise accordingly, then the level of his utility has to drop. Only when the expected yield increases above such a level that is equivalent to the original level, the investor would achieve a psychological balance so that he would possibly accept the increased risk. Such a rate of increase in the expected yield that accompanies the increasing risk and the stabilizing utility can also be seen as the investor's demand on the return of increasing risk. Let us use the phrase "rate of desired return" to express this phenomenon.

At an arbitrarily chosen point (σ, r), the rate of desired return of risk, denoted $DR(\sigma, r)$, stands for the amount of increase in the expected yield that is necessary to maintain the stability in the mean variance utility when risk is increased by one unit. Assume that at point (σ, r) the risk increases a small increment $d\sigma$ and the expected yield increases in the amount of dr. In order to keep the mean variance utility constant, we must have

$$
dU = U'_\sigma(r - r)d\sigma + U'_r(r - r)dr = 0
\tag{7.49}
$$

So, we have $dr/d\sigma = -U'_\sigma(\sigma^*, r^*)/U'_r(\sigma^*, r^*)$. Based on the definition of the rate of desired return, $DR(\sigma, r) = dr/d\sigma$ should follow. This end illustrates that $-U'_\sigma(\sigma^*, r^*)/U'_r(\sigma^*, r^*)$ is exactly the rate of desired return $DR(\sigma, r)$ of the investment portfolio at the point (σ, r):

$$
DR(\sigma, r) = -\frac{U'_\sigma(\sigma^*, r^*)}{U'_r(\sigma^*, r^*)}
\tag{7.50}
$$

Besides, $-U'_\sigma(\sigma^*, r^*)/U'_r(\sigma^*, r^*)$ is also the slope of the tangent line at the point (σ, r) of the indifference curve. Therefore, the geometric meaning of the rate $DR(\sigma, r)$ of desired return of risk is the slope of the indifference curve at the point (σ, r).

It should be pointed out that the rate $DR(\sigma, r)$ of desired return is really dependent on the subjective judgment of the investor. Although it also measures the return of risk, it is a subjective rate that represents the investor's demand on the return in front of increasing risk. That is exactly the difference between the price of risk and desired return: The former stands for an objectively existing return, while the latter a subjective demand of return for taking additional risk. Because of this difference, let us simply refer the price of risk $(r_m - r_f)/\sigma_m$ as the rate of objective return (of risk), while the rate of desired return $DR(\sigma, r)$ as the rate of subjective return (of risk). In the following, let us consider the points (σ, r) on the budget line of the investment portfolio: $r = r_f + \sigma(r_m - r_f)/\sigma_m$.

When $(r_m - r_f)/\sigma_m > DR(\sigma, r)$, the slope of the budget line is greater than the slope of the indifference curve at the point (σ, r). So, the rate of objective return of risk is higher than that of subjective return. That is, when risk increases, the realistic yield of the investor's portfolio is higher than his subjective demand of increase. That implies that the increasing risk makes the investor more satisfied than he desires. Hence only increasing risk can help to increase the utility. So, it can be seen that when the price of risk is higher than the desired return, the investor should allow higher level of risk in his portfolio. That is, adjust the investment portfolio along the budget line in the direction of increasing risk, see Figure 7.4 for more details.

When $(r_m - r_f)/\sigma_m < DR(\sigma, r)$, the slope of the budget line is smaller than the slope of the indifference curve at the point (σ, r). So, the objective return of risk is lower than that of the subjective return. That is, when the risk increases, the actual increase in the investor's yield is lower than the demanded subjective return. It implies that the increasing risk makes the investor's utility decrease. On the other hand, decreasing risk lifts the investor's utility to a higher level. Hence, when the price of risk is lower than the desired return of the risk, the investor should make the risk of his portfolio

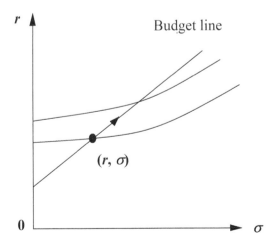

Figure 7.4 Increasing risk can help raise utility.

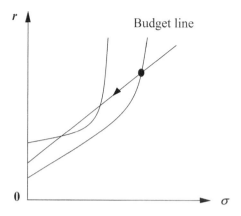

Figure 7.5 Decreasing risk can help raise utility.

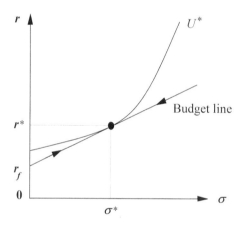

Figure 7.6 Optimal solution: maximum utility.

as small as possible. That is, adjust his portfolio along the budget line in the direction of smaller risk. See Figure 7.5 for details.

In short, as long as $(r_m - r_f)/\sigma_m \neq DR(\sigma, r)$, one can adjust his investment portfolio to raise his utility accordingly based on the particular comparison of his price of risk and desired return of risk. When the adjustment reaches the moment when $(r_m - r_f)/\sigma_m = DR(\sigma, r)$, the utility function has reached its maximum. That is when on the budget line (σ, r) moves to the maximum point (σ^*, r^*), which is the tangent point of the portfolio budget line and the indifference curve, Figure 7.6. This is the significance of the optimal solution (σ^*, r^*): Balance the risk and yield so that the utility is maximized.

From the relationship $\sigma_\beta = \beta \sigma_m$ between σ_β and σ_m of the investment portfolio, it follows that when the maximum utility point (σ^*, r^*) on the investment budget line is determined, the ratio $\beta^* = \sigma^*/\sigma_m$ should be the optimal portfolio proportion of

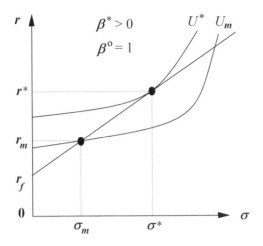

Figure 7.7 Impossible optimal portfolio composition.

the investment. That stands for the optimal proportion of risky assets in the overall portfolio. Therefore, let us refer this ratio as the theoretical optimal ratio. However, β^* could be greater than 1, because when we solve for the optimal solution (σ^*, r^*), we did not require $\sigma^* \leq \sigma_m$ other than imposing the condition that the point (σ^*, r^*) is on the budget line $r = r_f + \sigma(r_m - r_f)/\sigma_m$. When $\beta^* > 1$, it means that the investor needs to not only spend all his funds on risky assets but also convert some of his holdings of safe assets f into risky assets. In other words, the cash needed for the over-investment $\beta^* - 1$ in risky assets is supplied by selling the safe assets already contained in the portfolio. If the portfolio no longer contains any safe asset to be converted to cash and the investor cannot borrow funds without paying interest, then $\beta^* > 1$ means that the investor can not materialize his ideal, optimal asset selection, even though he invests all his available funds in risky assets m. That is to say, when $\beta^* > 1$, the optimal portfolio proportion under the allowable circumstances is $\beta^o = 1$, Figure 7.7. Hence, when the investor locates his optimal portfolio point (σ^*, r^*), his truly operable optimal portfolio proportion within the particular circumstance is determined by $\beta^o = \min\{1, \sigma^*/\sigma_m\}$. This proportion β^o is referred to as operable optimal ratio.

Note that β^o is indeed a kind of optimal portfolio proportion. In fact, the portfolio proportion β^o corresponds to the maximum value of the utility function $U(\sigma, r)$ on the line segment that connects $(0, r_f)$ and (σ_m, r_m):

$$Eu(R_{\beta^o}) = \max\{Eu(R_\beta) : 0 \leq \beta \leq 1\} \tag{7.51}$$

Under what conditions is the theoretical optimal proportion the same as the operable optimal proportion? Based on what is discussed above on the interpretation of the optimal solution (σ^*, r^*), to answer this question, we only need to check whether or not the price of risk is lower than the desired return, as shown in Figure 7.8:

$$(\beta^* = \beta^o) \Leftrightarrow (\beta^* \leq 0) \Leftrightarrow \left(DR(\sigma_m, r_m) \geq \frac{r_m - r_f}{\sigma_m}\right) \tag{7.52}$$

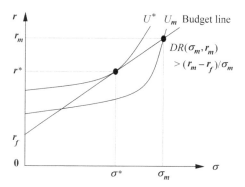

Figure 7.8 Possible optimal portfolio composition.

So now the natural question is: What does the condition $DR(\sigma_m, r_m) \geq (r_m - r_f)/\sigma_m$ really say? To address this problem, let us apply the computational formulas of the partial derivatives of the utility function $U(\sigma, r)$ as derived in the proof of Theorem 7.2 as follows:

$$U'_r(\sigma, r) = E[u'(\xi(\sigma, r))] > 0 \tag{7.53}$$

$$U'_\sigma(\sigma, r) = E\left[u'(\xi(\sigma, r))\frac{R_m - r_m}{\sigma_m}\right] = \frac{E[u'(\xi(\sigma, r))R_m] - r_m E[u'(\xi(\sigma, r))]}{\sigma_m} \tag{7.54}$$

Therefore, the rate of desired return of risk is

$$DR(\sigma, r) = -\frac{U'_\sigma(\sigma, r)}{U'_r(\sigma, r)} = \frac{r_m - \dfrac{E[u'(\xi(\sigma, r))R_m]}{E[u'(\xi(\sigma, r))]}}{U'_r(\sigma, r)} \tag{7.55}$$

By applying this computational formula on (σ_m, r_m), while noticing that $\xi((\sigma_m, r_m) = R_m$, we have

$$DR(\sigma_m, r_m) = \frac{r_m - \dfrac{E[u'(R_m)R_m]}{E[u'(R_m)]}}{\sigma_m} \tag{7.56}$$

So, the condition that $DR(\sigma_m, r_m) \geq (r_m - r_f)/\sigma_m$ is equivalent to that of $E[u'(R_m)R_m] \leq E[u'(R_m)]r_f$.

From probability theory, it follows that

$$E[u'(R_m)R_m] = \text{Cov}[u'(R_m), R_m] + E[u'(R_m)]r_m. \tag{7.57}$$

Hence, the condition that $E[u'(R_m)R_m] \leq E[u'(R_m)]r_f$ is equivalent to that of

$$-\text{Cov}[u'(R_m), R_m] \geq E[u'(R_m)](r_m - r_f). \tag{7.58}$$

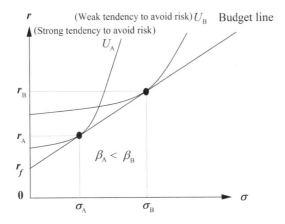

Figure 7.9 Portfolio proportion and tendency of risk avoidance.

Because $u'(R_m)$ and R_m are negatively correlated, the condition (7.58) requires a relatively high degree of negative correlation between $u'(R_m)$ and R_m: the absolute value of the covariance is greater than $E[u'(R_m)](r_m - r_f)$.

It can be seen that the more concave the utility function $u(x)$ is, that is, the greater $-u''(x)$ is, the faster $u'(x)$ decrease and the higher the negative correlation between the marginal utility $u'(R_m)$ and the income R_m. So, it means that the more concave the utility function $u(x)$ is, the more risk averse the investor is. Therefore, equ. (7.58) in fact requires the investor to be quite sensitive and averse to risk. With this understanding in place, the meaning of $DR(\sigma_m, r_m) \geq (r_m - r_f)/\sigma_m$ becomes clear: In order for the theoretical optimal proportion β^* not to be greater than 1, the investor's risk averseness cannot be too weak and needs to be so strong that equ. (7.58) is satisfied.

What can be concluded from this result is that the more risk averse the investor is, the smaller his portfolio proportion β^* is, while the less risk averse an investor is the greater the proportion β^*. That is, the optimal portfolio proportion β^* reflects the level of how risk averse the investor is. Figure 7.9 shows the portfolio selections of two investors with very different levels of risk averseness, where investor A is more risk averse than investor B so that investor A's optimal portfolio composition point (σ_A, r_A) is located on the left hand side of B's optimal portfolio composition point (σ_B, r_B) satisfying that $\sigma_A < \sigma_B$ and $r_A < r_B$.

7.4.4   Estimate of the β coefficient

The optimal portfolio proportion, either β^* or β^o, has been known as the β coefficient of the portfolio. In this subsection, let us see how to understand this β coefficient from the practical point of view.

First of all, the relationship $R_\beta = R_f + \beta(R_m - R_f)$ between R_β and R_m tells us that the β coefficient can be seen as the regression coefficient between the random variable R_β of the yield of the portfolio and the random variable R_m of the yield of risky assets.

As a matter of fact, we can compute the covariance $\mathrm{Cov}(R_\beta, R_m)$ of R_β and R_m based on the formula $R_\beta = R_f + \beta(R_m - R_f)$ and obtain

$$
\begin{aligned}
\mathrm{Cov}(R_\beta, R_m) &= E[(R_\beta - r_\beta)(R_m - r_m)] \\
&= \beta E[(R_m - r_m)^2] = \beta \sigma_m^2
\end{aligned}
\tag{7.59}
$$

That is, $\beta = \mathrm{Cov}(R_\beta, R_m)/\sigma_m^2 = \mathrm{Cov}(R_\beta, R_m)/\mathrm{Var}(R_m)$. This implies that β is the linear regression coefficient of R_β with respect to R_m; it measures the effect of the changes in the yield of the particular risky assets m on the yield of the investment portfolio.

With this understanding in place, we can practically estimate the β coefficient through using the available financial investment data. The particular procedure is described as follows. Because the investor is rational (he maximizes his utility), we can believe that each actual financial investment is done after the investor balances the risk involved and the expected yield (so that the β coefficient is implicitly considered). After having selected a typical risky asset m, such as a stock index, let us assume that the investor's financial behavior is about how to combine this risky asset and a kind of safe asset f, such as a short-term government treasury bill. Let us use R_p to stand for the yield of the selected investment portfolio (that is, R_p also represents the optimal assets ratio of the investor), and R_m the yield on the risky asset m. Surely enough, both R_p and R_m can be naturally seen as random variables. Collect and organize the historical data relevant to R_p and R_m, and then estimate the linear regression coefficient by using the model $R_p = \alpha + \beta R_m$. The resultant β-value stands for an estimate of the optimal portfolio proportion.

If the collected historical data of R_p and R_m are respectively $R_p^1, R_p^2, \ldots, R_p^n$ and $R_m^1, R_m^2, \ldots, R_m^n$ and their corresponding mean values $\bar{R}_p = \sum_{i=1}^{n} R_p^i/n$ and $\bar{R}_m = \sum_{i=1}^{n} R_m^i/n$, then the α-value and β-value, as estimated using the linear regression model, are given by

$$
\begin{cases}
\alpha = \bar{R}_p - \beta \bar{R}_m \\
\beta = \sum_{i=1}^{n} (R_p^i - \bar{R}_p)(R_m^i - \bar{R}_m) \Big/ \sum_{i=1}^{n} (R_m^i - \bar{R}_m)^2
\end{cases}
\tag{7.60}
$$

7.4.5 Portfolio proportions

Now, let us consider the situation of multiple risky assets. Assume that an investor prepares $\$v$ to invest in the asset market; his utility function of monetary income is u. The market has one safe asset and n risky assets, where the safe asset is numbered by 0, and the risky assets numbered $1, 2, \ldots, n$, respectively. The rate of return of the safe asset, such as deposit in a bank, is a constant. Because it is not affected by any random factor, we will call it interest rate, denoted by θ. The rates of return of the risky assets are random variables that are influenced by random factors. Let R_j stand for the rate of return of asset j, and r_j the expected rate of return of asset j, that is, $r_j = E[R_j](j = 0, 1, \ldots, n)$. Then $r_0 = R_0 = \theta$. Next we analyze how the investor selects among these $n + 1$ assets for his investment portfolio.

Let a_j represent the amount of money the investor invests in asset j, $j = 0, 1, 2, \ldots, n$. Then $\sum_{j=0}^{n} a_j = w$, his amount of money invested in the safe asset is $a_0 = w - \sum_{j=1}^{n} a_j$, the total amount of money invested in risky assets is $\sum_{j=1}^{n} a_j$, and the total income from the investment portfolio is

$$W = \sum_{j=0}^{n} a_j(1 + R_j) = \left(w - \sum_{j=1}^{n} a_j \right)(1 + \theta) + \sum_{j=1}^{n} a_j(1 + R_j) \tag{7.61}$$

The objective of the investor is to select an appropriate risk combination $(a_1, a_2, \ldots, a_n)$ of investment so that the utility function $E[u(W)]$ is maximized.

Assume that $a^* = (a_1^*, a_2^*, \ldots, a_n^*) >> 0$ is the risk combination of investment that makes $E[u(W)]$ reach its maximum value. From the first order condition of maximum values, it follows that

$$\frac{\partial}{\partial a_j} E[u(W^*)] = E[u'(W^*)(R_j - \theta)] = 0, \quad j = 1, 2, \ldots, n \tag{7.62}$$

where $W^* = (w - \sum_{j=1}^{n} a_j^*)(1 + \theta) + \sum_{j=1}^{n} a_j^*(1 + R_j)$. The point $(a_1^*, a_2^*, \ldots, a_n^*)$ that makes $E[u(W)]$ reach its maximum is known as the demand for risky assets under the total amount w of investment and interest rate θ, denoted A or $A(w, \theta)$. That is,

$$A = A(w, \theta) = (a_1^*, a_2^*, \ldots, a_n^*) = (A_1, A_2, \ldots, A_n)$$

Equ. (7.62) is known as the demand equation for risky assets that determines A or the marginal equation of the asset combination A. The function $A_j = A_j(w, \theta)$ is known as the demand function for asset j. Evidently, interest rate plays a decisive role in the determination of the demand for risky assets.

Assume that the investor's utility function u is strictly increasing. That is, the first order derivative of u is positive everywhere. As for an investor who is risk averse or risk neutral, that is $u''(x) << 0$, if he decides to invest in each and every one of the assets, then the actual situation has to be: the rate of return of each risky asset is not less than the interest rate. So, in order "not to put all eggs in one basket", he scatters his funds among all the safe asset and risky assets in order to reduce risk and earn relatively high return. If he places all his funds in the safe asset, the actual situation has to be different from what is described above: none of the risky asset warrants an expected rate of return greater than the interest rate. So, only when at least one of the risky assets carries an expected rate of return greater than the interest rate, he will consider investing in this risky asset. In the following we will prove these conclusions.

Assume that $a^* = A(w, \theta)$ is the investor's demand for risky assets. Then we have

$$\forall x \in (-\infty, +\infty)(u'(x) > 0 \ \& \ u''(x) \le 0)$$

If he invests in each of the available assets, then the following have to be true: $a^* > 0$, $\sum_{i=1}^{n} a^* < w$ and $E[u'(W^*)(R_j - \theta)] = 0, j = 1, 2, \ldots, n$. Because $u'(W^*) > 0$, the probability $P(R_j - \theta \ge 0) > 0, j = 1, 2, \ldots, n$, that is, the rate of return from each of the risky asset is not lower than the interest rate θ. That proves the first part of the previous conclusions.

In order to prove the second part of the conclusion, assume that all available funds are invested on the safe asset. That is, $a_j^* = 0, j = 1, 2, \ldots, n$. Notice that u is a concave function, that u is strictly increasing, and that a^* is the point where $E[u(x)]$ reaches it maximum value. So, we obtain

$$E[u'(W^*)(R_j - \theta)] \leq 0, \quad j = 1, 2, \ldots, n \tag{7.63}$$

In this case we have

$$\begin{aligned} E[u'(W^*)(R_j - \theta)] &= E[u'(w(1 + \theta))(R_j - \theta)] \\ &= u'(w(1 + \theta))E[R_j - \theta] \\ &= u'(w(1 + \theta))(r_j - \theta), \quad j = 1, 2, \ldots, n \end{aligned} \tag{7.64}$$

Therefore, $u'(w(1 + \theta))(r_j - \theta) \leq 0$, $j = 1, 2, \ldots, n$. Because $u'(w(1 + \theta)) > 0$, it follows that $r_j = E[R_j] \leq \theta$, $j = 1, 2, \ldots, n$. That is, none of the risky asset carries an expected rate of return greater than the interest rate. That ends the proof of the second part of the conclusion.

The previous conclusion of scattering investment selections indicates that when the interest rate θ is so high that it is possible to be not lower than the expected rate of return of each of the risky assets, people will not take the risk of pursuing after higher returns by investing in risky assets; the safe and secure method is to deposit their money in bank accounts in order to earn stable interest income so that the demand for risky assets is zero. When the interest rate is relatively high but there is such a chance for it to be less than the expected rate of return of at least one of the risky assets, then people might invest in some of the risky assets so that the demand for such risky asset is created. If the interest rate is relatively low so that there is such a chance that the expected rate of return of each of risky assets is not lower than the interest rate, then people will invest in various risky assets so that the amount of money invested in risky assets is greatly increased. That is the law of how the (investment) demand for risky assets changes with interest rate: When interest rate increases, the demand for risky assets decreases; when interest rate goes up, the demand for risky assets goes down. Symbolically, that means $\partial A / \partial \theta < 0$.

7.4.6 Asset pricing

Let us make some changes to the demand equation for risky assets in equ. (7.62). The following is equivalent to the equation $E[u'(W)(R_j - \theta)] = 0$

$$E[u'(W)R_j] = \theta E[u'(W)], \quad j = 1, 2, \ldots, n \tag{7.65}$$

As for the random variables X and Y, because $\mathrm{Cov}(X, Y) = E[XY] - E[X]E[Y]$, that is, $E[XY] = E[X]E[Y] + \mathrm{Cov}(X, Y)$, we have

$$E[u'(W)R_j] = E[u'(W)]r_j + \mathrm{Cov}(u'(W), R_j) \tag{7.66}$$

So, equ. (7.62) becomes

$$(r_j - \theta)E[u'(W)] = -\text{Cov}(u'(W), R_j) \tag{7.67}$$

That is,

$$r_j = \theta + \frac{\text{cov}(u'(W), R_j)}{E[u'(W)]}, \quad j = 1, 2, \ldots, n \tag{7.68}$$

The second term on the right hand side of this equation is known as the rate of return on risk. It is determined by the covariance of the marginal utility of income and the yield of assets. So, the expected rate of return of any risky asset is equal to the sum of the interest rate of the safe asset and the rate of return on risk.

When the rate of return R_j of an asset is positively correlated to the income of wealth W, for a risk averser, the marginal utility $u'(W)$ decreases with the income of wealth W so that the rate of return R_j of this asset j has to be negatively correlated to the marginal utility $u'(W)$ and that there must be one expected rate of return r_j that is greater than the interest rate θ that offsets the risk taking. Conversely, when the rate of return R_j of an asset j is negatively correlated to the income of wealth, for a risk averser, the rate of return R_j of this asset j must be positively correlated to the marginal utility $u'(W)$ so that the expected rate of return r_j of this asset has to be lower than the interest rate of the safe asset. Intuitively speaking, an asset that is negatively correlated to the income of wealth is such an asset that it is especially valuable in terms of reducing risk so that people are willing to hold this asset by giving up the expected return.

7.5   PORTFOLIO SELECTION

For the sake of convenience of our communication, let us assume that in the financial market there are $n + 1$ assets: named asset 0, asset 1, asset 2, ..., asset n, and that the rates of return of these assets take many different possible values, and each of the rates is affected by two mutually independent random events (that is, these two events cannot occur at the same time). Assume that the probability for the jth event to occur is $p_j, j = 1, 2, \ldots, m$, satisfying $\sum_{j=1}^{m} p_j = 1$; and the rate of return of asset i when the jth event occurs is $r_{ij}, i = 1, 2, \ldots, n, j = 1, 2, \ldots, m$. Assume that the rate of return of asset i assumes its value in the set $\{r_{i1}, r_{i2}, \ldots, r_{in}\}, i = 1, 2, \ldots, n$. We also assume that among these $n + 1$ assets there is only one riskless asset. The reason we make such assumption is because the rates of return of all riskless assets are the same. If there were two riskless assets of different rates of return in the market place, then who will be willing to purchase the asset of lower rate of return because the rates are fixed? So, the riskless asset with higher rate of return expels the asset of lower rate of return out of the market place so that all riskless assets in the market place pay the same rate of return. Because of this reason, for our purpose here, we can assume there is only one riskless asset in the market place. For convenience, let us assume the riskless asset is number 0. That is asset 0 is the safe asset, and all the other n assets are risky. The rate of return r_0 of asset 0 is a constant that is not affected by any random factor. It represents the yield to maturity (interest rate). So, we have $r_{01} = r_{02} = \cdots = r_{0m} = r_0$.

7.5.1 Expected yields of portfolios

By asset portfolio selection, it means that the investor selects among various kinds of choices a quantitative combination of assets. If an asset is not among his choice, then in the portfolio the quantity of that asset is zero. Now, among all the $n+1$ assets if the investor decides to purchase x_i units of asset i, $i = 1, 2, \ldots, n$, then the $n+1$ dimensional vector $x = (x_0, x_1, \ldots, x_n)$ represents the asset portfolio of the investor, known as the portfolio vector or simply portfolio. If the investor does not purchase asset i, then $x_i = 0$. So, the zero vector represents that the investor does not purchase any asset.

The expected rate of return r_{ie} of asset i is defined to be mathematical expectation of the rates of return of asset i:

$$r_{ie} = \sum_{j=1}^{m} p_j r_{ij}, \quad i = 1, 2, \ldots, n \tag{7.69}$$

Because asset 0 is riskless, its rate of return is a fixed number r_0 so that $r_{0e} = \sum_{j=1}^{m} p_j r_{0j} = r_0 \sum_{j=1}^{m} p_j = r_0$.

The expected yield R_e from selecting the portfolio $x = (x_0, x_1, \ldots, x_n)$ is defined to be the expected value of all possible yields under this portfolio selection:

$$R_e = R_e(x) = R_e(x_0, x_1, \ldots, x_n) = \sum_{i=0}^{n} x_i(1 + r_{ie}) \tag{7.70}$$

The purpose of investment is to acquire more return of wealth, while the degree of satisfaction obtained from pursuing after wealth is dependent on his subjective evaluation. This evaluation can be expressed by using the utility function $u(w)$ on the return of wealth, where $u(w)$ stands for the utility the investor obtains when he receives w units of return of wealth. Here, the magnitude of the return of wealth is measured by using the standard value as offered by money. Generally, people love wealth. When that is reflected by using the utility function $u(w)$, it means that $u(w)$ is an increasing function of the wealth w. That is, the first order derivative of u is positive everywhere.

However, in terms of financial investments, because the return of an investment is uncertain, it causes the utility an investor could obtain to be also uncertain. So, to evaluate the quality of an asset portfolio one has to make use of the concept of expected utilities. The so-called expected utility $EU(x)$ of an asset portfolio $x = (x_0, x_1, \ldots, x_n)$ is defined as the mathematical expectation of all the possible utilities the portfolio x could bring to the investor:

$$EU(x) = \sum_{j=1}^{m} p_j \cdot u \left(\sum_{i=0}^{n} x_i(1 + r_{ij}) \right) \tag{7.71}$$

where $\sum_{i=0}^{n} x_i(1+r_{ij})$ stands for the yield of the portfolio x when the jth event occurs, denoted by $W_j(x)$. That is we have

$$W_j(x) = \sum_{i=0}^{n} x_i(1+r_{ij}), \quad j = 1, 2, \ldots, m \tag{7.72}$$

So, $u(W_j(x))$ is the quantity of utility the investor obtains from the portfolio x when the jth event occurs. By summing the products of the utilities attainable by x under various possible circumstance with their corresponding probabilities, we obtain the expected utility of the portfolio x as follows:

$$EU(x) = \sum_{j=1}^{m} p_j u(W_j(x)) \tag{7.73}$$

For two different portfolios x and y, when $EU(x) > EU(y)$, the investor will think that x is better than y so that he is will to select x instead of y. When $EU(x) = EU(y)$, the investor will think that the portfolio x and portfolio y are identical without any difference. That is the criterion investors use to evaluate asset portfolios when selecting financial investments.

Each asset portfolio x has its expected yield $R_e(x)$, from which we can compute its quantitative utility $u(R_e(x))$. This utility is known as the utility of the expected yield of the asset portfolio x. However, this utility does not have the significance of expected utility. What it shows is the amount of utility the investor obtains when he acquires $R_e(x)$ units of wealth. So it is a utility in the sense of determinacy. As for a risk averser, he would realize that the expected utility of an asset portfolio x is smaller than the expected return utility of x, because although the expected return of x is equal to $R_e(x)$, there is risk for him to acquire such a return of x with average value $R_e(x)$. Since there is no risk involved when he obtains $R_e(x)$ directly from the safe asset, he would then feel that $EU(x) < u(R_e(x))$. When such attitude of the risk averser regarding risks is reflected in the utility function u, this function $u(w)$ will be a concave function of the return of wealth. That is, for any $0 < p < 1$ and any two returns of wealth w_1 and w_2, the following holds true:

$$pu(w_1) + (1-p)u(w_2) < u(pw_1 + (1-p)w_2) \tag{7.74}$$

This equation can be interpreted as follows: The investor obtains wealth w_1 with probability p, and wealth w_2 with probability $1 - p$; so even though the expected return of this investment is $pw_1 + (1-p)w_2$, it is not as good as the riskless investment that is safe, stable and produces the same return as $pw_1 + (1-p)w_2$.

Conversely, for a risk lover, he would realize that the expected utility of the asset portfolio x is greater than the utility of the safe return $R_e(x)$, because when his investment is done according to x, he would acquire an opportunity to earn a higher return. When this is reflected in the utility function u, it means that the utility function $u(w)$ is a convex function of wealth w. That is, for any $0 < p < 1$ and two returns of wealth w_1 and w_2, we have $pu(w_1) + (1-p)u(w_2) < u(pw_1 + (1-p)w_2)$.

The previous analysis indicates that the investor's attitude toward risk determines the shape of the curve of his utility function. The utility function $u(w)$ of a risk averser is a concave function such that $u''(w) < 0$. That is, the first order derivative function $u'(w)$ is decreasing. Because most people are risk aversers, their marginal utility of wealth is a decreasing function. That is the reason why in financial analysis, it is often assumed that the utility function of the return of wealth is an increasing concave function, which implies that its first order derivative is greater than 0 and its second order derivative function is less than 0.

7.5.2   Demand of assets

The investor compares various investment programs and selects his investment portfolio by using expected utilities in order to maximize the expected utility. However, in his selection, he has to be constrained by how much wealth he owns. That is to say, he cannot make investment beyond the level of his wealth. Assume that the investor plans to use W units of wealth to purchase financial assets. When he selects what to invest among the $n + 1$ assets, the wealth constraint of the asset portfolio $x = (x_0, x_1, \ldots, x_n)$ can be written as follows:

$$x_i \geq 0, \quad i = 0, 1, 2, \ldots, n, \quad \text{and} \quad \sum_{i=0}^{n} x_i = W \tag{7.75}$$

This wealth constraint limits the range X from which the investor can select his asset portfolio:

$$X = \left\{ (x_0, x_1, \ldots, x_n) \in R^{n+1} : \forall 0 \leq i \leq n(x_i \geq 0 \ \& \ \sum_{j=0}^{n} x_j = W) \right\} \tag{7.76}$$

which is referred to as the wealth constraint set. The asset portfolio within X that makes the expected utility of the investor reach the maximum is known as the investor's demand vector of assets, denoted by $\xi = (\xi_0, \xi_1, \ldots, \xi_n)$. Notice that this vector is obtained by maximizing the expected utility from the utility function u with known wealth W and various possible rates of return r_{ij} of all the assets. So, speaking strictly, ξ should be written as follows:

$$\xi = \xi(W, r_0, r_{11}, \ldots, r_{nm}) \tag{7.77}$$

$$\xi_i = \xi_i(W, r_0, r_{11}, \ldots, r_{nm}) \tag{7.78}$$

for $i = 0, 1, 2, \ldots, n$, where ξ_I stands for the investor's demand for asset i, and ξ the investor's demand mapping of assets.

In order to solve for the demand vector of assets, we only need to find the optimal value of the expected utility function $EU(x)$ subject to the wealth constraint. Note that for those asset portfolios x that satisfy the wealth constraint, we have

$$x_0 = W - \sum_{i=1}^{n} x_i = W - (x_1 + x_2 + \cdots + x_n) \tag{7.79}$$

$$W_j(x) = \sum_{i=0}^{n} x_i(1 + r_{ij}) = W(1 + r_0) + \sum_{i=1}^{n} x_i(r_{ij} - r_0), \quad j = 1, 2, \ldots, m \tag{7.80}$$

$$EU(x) = \sum_{j=1}^{m} p_j u(W_j(x)) = \sum_{j=1}^{m} p_j u\left(W(1 + r_0) + \sum_{i=1}^{n} x_i(r_{ij} - r_0)\right) \tag{7.81}$$

$$\frac{\partial W_j(x)}{\partial x_i} = r_{ij} - r_0, \quad i = 0, 1, 2, \ldots, n, \; j = 1, 2, \ldots, m \tag{7.82}$$

$$\frac{\partial EU(x)}{\partial x_i} = \sum_{j=1}^{m} p_j \frac{\partial u(W_j(x))}{\partial x_i} = \sum_{j=1}^{m} p_j u'(W_j(x)) \frac{\partial W_j(x)}{\partial x_i}$$

$$= \sum_{j=1}^{m} p_j u'(W_j(x))(r_{ij} - r_0), \quad i = 0, 1, 2, \ldots, n \tag{7.83}$$

So, the maximization problem of $EU(x)$ subject to the wealth constraint is transformed into the maximization problem of $EU(x)$ without any constraint over the region

$$X_n = \left\{ (x_1, x_2, \ldots, x_n) \in R^n : \forall 1 \leq i \leq n(x_i \geq 0 \; \& \; \sum_{j=1}^{n} x_j \leq W) \right\} \tag{7.84}$$

According to the first order condition of optimizing functions, it follows that if $x = (x_0, x_1, \ldots, x_n) \in X$ is the investor's demand vector of assets, that is, $x = \xi(W, r_0, r_{11}, \ldots, r_{nm})$, then

1 When $(x_1, x_2, \ldots, x_n)$ is in the interior of the region X_n, that is, $x_i > 0$, $i = 1, 2, \ldots, n$, satisfying $\sum_{i=1}^{n} x_i < W$, $\partial EU(x)/\partial x_i = 0$, $i = 1, 2, \ldots, n$, so that the demand for assets $x = \xi$ is determined by the following system of equations:

$$\sum_{j=1}^{m} p_j u'(W_j(x))(r_{ij} - r_0) = 0, \quad i = 1, 2, \ldots, n \tag{7.85}$$

2 When $(x_1, x_2, \ldots, x_n)$ is located on the boundary of X_n while satisfying $\sum_{i=1}^{n} x_i < W$, (that is, each x_i still has space to increase), $\partial EU(x)/\partial x_i \leq 0$, $i = 1, 2, \ldots, n$. That is, we have

$$\sum_{j=1}^{m} p_j u'(W_j(x))(r_{ij} - r_0) \leq 0, \quad i = 1, 2, \ldots, n. \tag{7.86}$$

3   When $x_k > 0$ (that is x_k still has space to decrease), for $1 \le k \le n$, no matter whether $(x_1, x_2, \ldots, x_n)$ is in the interior of X_n or on the boundary of X_n, the following always holds true:

$$\sum_{j=1}^{m} p_j u'(W_j(x))(r_{kj} - r_0) \ge 0 \tag{7.87}$$

What is obtained here tells us the following two facts regarding asset selections:

(i)   In order to have the investor invest in each asset, there must be a chance for the rate of return of each risky asset to be greater than the interest rate of the safe asset. That is because when the investor invests in each of the assets, each component of the demand vector x for assets is greater than zero so that $(x_1, x_2, \ldots, x_n)$ is located within the interior of the region X_n. From item (1) above, it follows that for each asset i, the following has to be true

$$\sum_{j=1}^{m} p_j u'(W_j(x))(r_{ij} - r_0) = 0$$

Because $u'(w) > 0$ and $r_{ij} = r_0$ will not hold true for each $j = 1, 2, \ldots, m$, there must be $j \le m$ such that $r_{ij} > r_0$ (because otherwise the left hand side of the demand equation would be smaller than 0, a contradiction). That implies that for each asset i there is a chance for its rate of return to be greater than the interest rate of the safe asset. Additionally, item (3) above tells us that in order to attract investment to a certain risky asset, there must be a relatively large chance for the rate of return of this risky asset to be greater than the interest rate of the safe asset.

(ii)   In order to have the investor put all his available funds on the safe asset, it must be the case that the expected rate of return of each of the risky asset is not higher than the interest rate of the safe asset. That is because when all the available funds are invested in the safe asset, the demand vector for assets $x = (x_0, x_1, \ldots, x_n) = (W, 0, \ldots, 0)$ satisfies the condition in item (2) above and $W_j(x) = W(1 + r_0)$, $j = 1, 2, \ldots, m$, so that $\sum_{j=1}^{m} p_j u'(W_j(x))(r_{ij} - r_0) = u'(W(1 + r_0))(r_{ie} - r_0) \le 0$. That explains that $r_{ie} \le r_0, i = 1, 2, \ldots, n$, because $u'(w) > 0$. That is, there is not a single risky asset whose expected rate of return is greater than the interest rate of the safe asset.

7.5.3 Advantages of diversity

The afore-mentioned two facts are very important. They indicate that in investment of financial assets, as long as a risky asset has a chance to earn greater rate of return than the interest rate of the safe asset, there will be people interested in investing in this risky asset instead of throwing all their available funds on the safe asset. That is the diversification in financial investment: Do not put all your eggs in one basket; diversify the available funds between safe assets and risky assets in order to eliminate the internal risk of investment and to acquire the maximum expected utility. Diversification reduces the overall risk the investor faces along with all desirable benefits without any harm.

In the following we use an example to illustrate the concept of diversification. Assume that there are two common stocks A and B issued respectively by two companies A and B, and that an investor needs to select between these two stocks. The probability for an investment in stock A to earn 15% is 0.5, while the probability of earning 5% is also 05. Similarly, the probabilities for an investment in stock B to earn either 15% or 5% are respectively 0.5. So, the expected rate of return from stock A and stock B is the same as 10%. Now, let us analyze the situation in three cases.

> Case #1: The price movements of stock A and stock B are in opposite directions.
> Under such circumstance, when the rate of return of stock A is 5%, the rate of return of stock B is as high as 15%. Conversely, when the rate of return from stock A is 15%, the rate of return from stock B is only 5%. So, by evenly dividing the available funds between these two stocks, the investor can always earn 10% of stable rate of return. That is better than placing all the available funds in one stock, where although the expected rate of return is also 10%, it is not stable.
> Case #2: The prices of stock A and stock B change independently from each other.
> In reality, what is often seen is the case where the prices of these two stocks change independently from each other. Under such circumstance, evenly dividing the available funds between stock A and stock B can not only produce 10% return, but also more possibly produce a 15% return or even higher. So, by diversifying, the consequent investment only becomes better than not diversifying.
> Case # 3: The prices of stock A and stock B move in the same direction.
> In this case, evenly dividing the available funds between stock A and stock B produces the same effect as placing all funds on one stock. In this case, these two methods produce not only the same expected rate of return, but also the same materialized rate of return. So, diversification produces no harm.

In short, for the case of two stocks A and B, no real situation will be beyond the three scenarios discussed above. And no matter which situation is actually happening, diversification will not reduce the return for the investor. Instead, it can more possibly help to stabilize or raise the investment return. So, diversification reduces risk without creating any harm.

Part 3

International financial system: Seen as an ocean of interacting semi-closed systems

Chapter 8

International Monetary System

The International Monetary System (IMS) touches on the most fundamental problem surrounding the international monetary, financial relationships, including the international system of foreign exchanges, international system of payment and settlement, international coordination of payments, international solvency supply, international financial institutions, etc., and the consequent institutional arrangements. The IMS represents an important aspect of the modern international economic relation. It not only plays a major role on the exchange rate system, foreign exchange policy, the management of reserve assets of each nation, but also has a profound impact on the worldwide formation of trade patterns and economic development. Chronologically, the development of the IMS can be divided into three stages: the stage of the international gold standard, the period of Bretton Woods System, and the current time of Jamaica System. This chapter first looks at the concepts and main contents of the International Monetary System, and then along the line of history analyzes the evolution of the IMS, while exploring the improvement of the IMS, the current state of the European monetary system, and introducing the reader to the background on which the IMS is established, and the mission and business activities of the IMS. At the end, we investigate the systemic three-ringed structure of viable, independent economic systems.

This chapter is organized as follows. In Section 8.1, we learn the composites, classification and functions of the international monetary system. In Section 8.2, we look at the development history of the international monetary system by going through the international gold standard, the Bretton Woods and the Jamaica system. Section 8.3 introduces the main suggestions about how to potentially improve the international monetary system and an expected future of the system. Section 8.4 studies the European monetary system by going over the history, the European union of money, and the impacts of the Euro. International organizations of finance, such as the International Monetary Fund and the World Bank, are introduced in Section 8.5, where the proposal of bankruptcy of nations is discussed. After learning the basic composites of the international monetary system, in Section 8.6 we consider the systemic structure that underlies each stable national economic system. Section 8.7 concludes this chapter with a few final words.

8.1 INTRODUCTION OF THE INTERNATIONAL MONETARY SYSTEM

The International Monetary System (IMS) means the totality of the international monetary system, international monetary and financial organizations, and the international monetary order. It stands for such an institutional framework for the relations of money and finance between nations. It is a complex whole consisting of a series of agreements, rules, conventions, agencies, policies, etc. It deals with the monetary standard, the foreign exchange rate system, the adjustment of international payments, the international supply of solvency, and contents of other aspects.

The IMS stands for the standard that normalizes the monetary behaviors between nations, and represents the important basis for various nations to conduct financial activities with foreign entities. There have been two ways with which the IMS is established. One is through the evolution of customs and conventions. For example, the formation of the international system of gold standard has gone through a long and gradual process. When relevant and mutually related customs and procedures were formed, certain ways of activities and behaviors were commonly recognized. When more and more international financial participants follow and comply with certain procedures or customs, a system is established. The other way with which the IMS is established is through formal international treaties or agreements. For example, the Bretton Woods Monetary System was established within a short period of time after World War II through treaties agreed up on in international conventions. However, no matter how the IMS is formed, it represents an objective, historical product of the world economic development.

8.1.1   The composites of the IMS

The content of the IMS includes the determinations of: (1) the key currency to be the international currency, (2) the exchange rate system, (3) international reserve assets, and (4) the mechanism of international payments.

The key currency is such a currency that is used as the base of value for the purpose of exchanges in the international monetary system. It represents a basic element of the IMS. It is because when trading with an external partner, a nation cannot employ its own domestic currency; instead it has to use the currency that is widely accepted by many countries. Only when such a key currency is chosen, the exchange rate and the rate of adjustment of each nation's currency, and the composite structure of international reserves of the nation can in turn be determined. So, determining the key currency, which is the determinations on what material should be currency be made of, the unit of the currency, the status of the currency in the current currency system, becomes an important content of the IMS.

The core of the IMS is the determination of the exchange rate system. International exchanges lead to the need of international payments. When money plays its role as the media of international trades, each nation has to determine a ratio between its own currency and a foreign currency, which is the exchange rate. To this end, various nations generally specify: the basis of how the ratios of currencies are determined, the fluctuation ceilings of the currency ratios, the adjustment of the currency ratios, how the currency ratios are maintained, which is either a fixed or floating exchange system, whether or not a multiple ratio system is employed against a certain currency, whether or not the currency of one nation can be exchanged freely, what method of settlement is utilized when paying off debts between nations, whether or not limitations are used for payments, etc.

In order to meet the need of adjusting international payments and incomes, each nation has to maintain a certain amount of international assets that are commonly accepted by various nations. The supply of international reserve assets consists of an important content if the IMS. The supply of the reserve assets should be under the international control in order to main the appropriate quantity, either too much or little of which will harm the world economy.

A balanced development of international payments and incomes of all nations from around the world is the foundation underneath the smooth operation of the IMS. Under certain circumstances, one nation's imbalance in its international payments can be rebalanced through the nation's adoption and implementation of domestic economic policies and foreign exchange policies. However, under some other circumstances, the imbalance cannot be rebalanced without taking out loans according to international treaties from international financial organizations and/or foreign governments, or through coordinating policies of several national governments by interfering market activities. Such coordination of international payment balances of various nations represents an important content of the IMS.

8.1.2 Classifications of the IMS

The IMS can be divided from the angle of monetary standards and that of exchange rate systems. By using the monetary standards, the IMS can be classified into the gold standard system, gold exchange standard system, and not convertible notes based system. The so-called gold standard system applies gold as the international standard currency. It is a pure commodity standard system. The so-called not convertible notes based system utilizes foreign currency as the international standard currency. It is a purely credit based system. The so-called gold exchange standard system stands for such a system of mixed standards that simultaneously employs gold and free convertibility as the international standard currency.

Based on exchange rate systems, the IMS can be classified into fixed exchange rate system, floating exchange rate system, and systems between these two extremes, such as adjustable peg system, crawling peg system, managed floating system, etc.

Sometimes people also classify the IMS by using the joint criteria of both the monetary standards and the exchange rate systems, such as fixed exchange rate system under the gold standard, adjustable fixed exchange rate system under a mixed standard, floating exchange rate system under paper money standard, etc.

8.1.3 The functions of the IMS

The existence and development of the IMS have brought with the system significant and wide influencing impacts on international trades and international financial activities. It has played an important role in the stability and growth of the world economy and those of various nations. The functions of the IMS have been shown in the following areas:

First of all, it provides a uniform and normalized set of rules of operation for the settlements of international trade and financial activities. The unified IMS not only selects the international currency necessary for the operation of the world economy, but also clearly spells out the regulations regarding the issuance and quantity of the international currency, ways exchanges can be done, and relevant standards. At the same time, it also provides a uniformed computational standard for different nations to account for their perspective national economies. That offers a set of standardized criteria that makes economic exchanges throughout the world possible and stimulates the healthy development of the world economy.

Secondly, the IMS helps to stabilize the exchange rates. One of the most important tasks of the IMS is to promote the stability of exchange rates. The IMS provides a unified set of computational standards for each nation to maintain its stable exchange rate, suggestions and facilities of management regarding the arrangements of foreign exchange systems for various nations, leading to potential stability of world exchange rates. The unified IMS also offers a good external environment for various nations to greatly reduce the impacts of international speculative activities, to stabilize internal values of national currencies and internal circulations of domestic currencies, and to engage in external economic transactions. At the same time, it also provides a solid foundation for developing stable exchange rates throughout the world.

Thirdly, the IMS helps to adjust international payments. One of the basic objectives and main purposes of establish the IMS is to introduce the adjustment mechanism of international payments in order to guarantee the world economy to develop stably and healthily. The introduction of the adjustment mechanism of international payments has to involve the following three aspects: the exchange rate mechanism, the financing mechanisms for nations of trade deficit, and the disciplinary mechanism of the issuing nations of the reserve currencies. Through the effects of these three mechanisms and the impacts of the stable, unified IMS on each nation's trade activities, currency circulations, exchange rates, international reserves, and other areas, the IMS has played an important adjustment role on the international payments of different nations.

Fourthly, the IMS helps to monitor and coordinate international monetary affairs. The establishment and operation of the IMS need to have the relevant coordination authority or organizational management agency. The important function of the management agency of the IMS is to coordinate and monitor the international currencies and financial affairs of different nations in order to ensure the materialization of stable exchange rates and balance of international payments. Along with the current increase in international trades and rapid development of international financial market, how to effectively initiate international collaborations in order to ensure a smooth and effective operation of the international monetary system has become an important topic of discussion in the community of economist and policy makers.

8.2 DEVELOPMENT OF THE IMS

The IMS has been evolving along with the development history of the world economy. Chronologically, the development of the IMS has gone through roughly three stages: the stage of the international gold standard system, the stage of Bretton Woods system, and the current stage of Jamaica system.

8.2.1 The international gold standard

The first IMS in the world is the international gold standard system. It was initially formulated at around the end of 1880s and ended at the break-out of the World War I in 1914. The gold standard means such a system that gold is used as the standard currency. Under the gold standard system, other than gold what is in circulation there are often bank notes that can be converted to gold and some small amounts of other metal coins. However, only the gold money can accomplish all the functions of money, that is, as the measurement of value, the method of circulation, the storage of value, the acceptable way of payments, and the world currency. All the nations that adopt the gold standard system establish the amount of gold to be contained in each unit of their respective currencies, and their currencies can be converted into gold by using the provisioned gold content. Gold has the function of world currency and is commonly accepted for international payments and the final settlements of trades. So, as long as the gold standard system is adopted by most nations, it will naturally play the role of the international currency.

Great Britain, as one of the earliest advanced capitalist nations, implemented the gold standard system in 1816, when gold was used to standardize the value the currency represented. However, with only Great Britain using the gold standard it cannot lead to the formation of an international gold standard system. In the 1870s, after the major capitalist nations in Europe and America adopted and implemented the gold standard system one after another, the international gold standard system was roughly established.

Before the break-out of the World War I, the gold standard system most nations adopted and implemented was the gold specie standard system. Its characteristics include: (i) gold coins of a certain weight and color are used as units of money, where the gold coins are circulated in the society; (ii) the gold coins have unlimited right as legal tender, can be freely molded, freely melted, and the quantity is not limited; (iii) according to its weight and color, a gold coin's office value is determined. The government can either sell or purchase gold without any limitation. And bank notes can be converted to gold freely; and (iv) the reserve currency of each nation is gold; and between nations gold is used to settle trades and gold can be freely imported or exported across national boundaries.

The characteristics of the international gold standard system include: gold plays the role of international currency and functions as the foundation of the international monetary system; the international gold standard system is a strict fixed exchange rate system; and the international gold standard system has a built-in mechanism of adjusting international payments.

During this time period, the international gold standard system was established on the base that domestically each of the major capitalist nations adopted and implemented the gold standard system. Because gold coins could be molded freely, the face value of a gold coin always agreed with its gold content, and the quantity of gold coins could spontaneously satisfy the need of circulation. Because gold coins could be converted freely, various other metal coins and bank notes could be stably circulated while each represented a certain amount of gold so that the value of the local currency was stabilized. And the free international flow of gold kept the foreign exchange markets relatively stable. So, it has been generally believed that the gold standard system is a stable currency system.

Under the international gold standard system, the exchange rate between different national currencies was determined by their respective gold contents, known as the statutory parity. Although the actual exchange rate could deviate away from the statutory parity due to the market supply and demand, the deviation would be limited and was determined by the transportation costs from one location to another. Under the gold standard system, the clearance of foreign trades could be done by using either gold or foreign currency, where value of gold stayed constant while the price of the foreign currency fluctuated with the market demand and supply. When the demand for a foreign currency rose, its exchange rate went up. However, the rise was limited. As soon as the exchange rate was higher than the statutory parity plus the transportation costs of gold, then it would be more advantageous to use gold to settle foreign trades. So, people would directly export gold, leading to a decline in the demand for the foreign currency. So, the statutory parity plus the transportation cost of gold was the ceiling for the exchange rate to increase. Accordingly, the statutory parity minus the transportation cost of gold, known as the point of gold input, represented the floor of how much exchange rate could fall to. When the exchange rate fell below the point of gold input, it would cause more gold to inflow instead of the foreign currency so that the market demand for the foreign currency fell and the exchange rate rebounded.

Under the gold standard, the international payments of each nation were adjusted automatically. Assume that a nation experienced a deficit in its international payments. Then it means that the nation would net export gold. Because the amount of money issuance of the nation was determined by the amount of gold reserves, the outflow of gold would cause the domestic gold reserve to decline, the money supply to dwindle, and the price level to fall. Along with the fall of the price level, the actual exchange rate of the domestic currency would go downward, which lifted the competiveness of the domestic goods in the international market and foreign goods less competitive in the domestic market so that exports would increase while the import decrease. That would lead to the deficit in the international trades to fall until it ultimately disappeared. If there was a surplus in the international payments, then the opposite process of adjustment would take place.

The time when the international gold standard prevailed at its peak level coincided exactly with the heyday of the capitalist free competition. The international and domestic polities were relatively stable, and economies were growing rapidly. The main industrial products of the time came mainly from Great Britain, and the trade deficits of other nations could also be covered with loaned funds from Great Britain; and the international payments were roughly balanced. Within such an advantageous environment, the implementation of the international gold standard had brought forward the fixed exchange rate system, which in turn helped to develop international trades, lending, and investments. So, it could be said rightly that the international gold standard had played an active and powerful role in the economic development and prosperity of the capitalist world of this time period. Even so, the international gold standard systems still suffered from some shortcomings. The automatic adjustment mechanism of the international gold standard system was functioning not as perfectly as it was theoretically claimed. Its functioning had been limited by many factors. In particular,

1   During the time period of over 30 years when the international gold standard was adopted and implemented, the flow of gold between nations was not frequent.

When one nation experienced trade deficit, it was always necessary to export gold. Instead it could use foreign loans to offset the deficit. Similarly, a nation's surplus could also be reduced by exporting capital, such as foreign lending and investment, instead of exporting gold. So, as a consequence, it became difficult for trade imbalances to be rebalanced through opposite directional money supplies and price movements within the involved nations.

2   The automatic adjustment mechanism could eventually cause gold to flow between nations through changes in domestic price levels of different nations, which in turn would cause changes in the import-export transactions. However, in reality the tendencies of changes in price levels in the major capitalist nations were quite uniform, and the changes in price levels did not cause gold to flow. In reality, the prices of goods of international trades were determined by the forces of demand and supply of the international market. They were not completely influenced by domestic production costs; and international trades tended to eliminate the price differences between nations.

3   What is more important is that the smooth operation of the international gold standard was established on the basis that all the nations involved complied with the basic rules of the gold standard without artificially interfering the economic development. However, after this time period ended, the governments of these capitalist nations could no longer allow their national economies to be totally controlled by the market; they naturally applied policy interventions. At the end period of the gold standard system, the central banks or monetary authorities of these nations no longer allowed the gold standard system to function naturally. Instead, they often tried to offset the impacts of gold movements on domestic money supplies. When gold flew into a nation, the money supply should be increased automatically. However, in order to stabilize the prices of goods, the domestic monetary authority would implement measures to control the increase in the supply. Conversely, when gold flew out of a nation, the money supply should be reduced accordingly. However, the monetary authority would take measures to maintain the same level of supply. So, consequently, implementing such monetary policies by each nation's monetary authority destroyed the relationship between money supply and price and gold reserves, making it difficult for the automatic adjustment mechanism of the gold standard to work properly.

Right before the break-out of the World War I, some signs for the international gold standard system to collapse had appeared, which were expressed mainly as follows: (1) The world production of gold could not keep up with the pace of growth of the world economy. (2) The major advanced nations acquired large amounts of gold through maintaining trade surplus and other special rights. In 1913, the gold reserves in Great Britain, the United States, France, Germany, and Russia has reached the level of 2/3 of all the gold reserves of the entire world. That made it difficult for other nations to maintain their gold standard. (3) More and more bank notes in each nation were issued so that converting these notes to gold become more difficult. And (4) exports of gold suffered from more controls. At the time of 1914 when the World War I broke out, each of the participating nations prohibited the export of gold and stopped the conversion of bank notes to gold. That declared the collapse of the international gold standard system.

During World War I, in order to fund the war, each of the participating nations issued large amounts of inconvertible paper notes, which devalued significantly after the war, causing severely inflation. At the same time, the exchange rates of these nations' currencies fluctuated volatilely, that relentlessly affected the stability of the world economy. So, as soon as the war ended, these nations started to recover the gold standard system, and the gold once again could flow freely between nations. However, the recovered international gold standard system was no longer the same as that before the war.

Firstly, the position of gold had been weakened when compared to that before the war. In fact, only the United States was still implementing the complete gold standard system, while both Great Britain and France were employing the gold bullion standard system, where no gold coins were in circulation in the domestic markets. Instead, bank notes or paper currency could be converted into gold only under certain conditions. For example, the Bank of England required that the amount of conversion into gold had to be greater than 1700 pounds. And France required it to be no less than 215000 franc. Each of the other nations that implemented the gold standard, such as Germany, Italy, Austria, Denmark, and other 30 some nations, actually implemented the gold exchange standard system, where the domestic currency of the nation was pegged to the currency of another nation that implemented the gold standard system, (such nations at the time included only the United States, Great Britain, and France), and the nation of gold exchange standard stored in the latter nation a large amount of foreign currencies or gold in order to be able to interfere the foreign currency market when necessary. The international gold standard system recovered after WWI in fact was an international gold exchange standard system centered on U.S. dollar, British pound, and French franc.

Secondly, the role playing of the mechanism that automatically adjusted international payments had been further constrained. After WWI, each nation paid more attention to domestic balances of objectives. When imbalances in international payment caused changes in domestic production and employment, although the automatic adjustment mechanism could rebalance the international payment, it was not beneficial to the stability of the domestic economy. So, fewer and fewer national monetary authorities were willing to sacrifice the objectives of domestic economies so that it made it difficult for the original mechanism of automatic adjustment to play its roles effectively and properly.

During the Great Depression of the world economy from 1929 to 1933, the economies of the Western nations suffered from catastrophic blows. The international society lost its confidence in British pound, forcing Great Britain to give up its gold standard system. After having experienced securities and credit crises, failures of great many banks, outflows of large quantities of gold, the United Stated was also eventually forced to give up its gold standard system. After that, the so-called gold group, consisting of France, Italy, Sweden, and other nations, had also to give up their gold standard, causing the international gold standard system to collapse completely.

Counting from 1816 when Great Britain initially adopted and implemented its gold standard system until 1936 when the last group of nations, such as the Netherlands, Sweden, etc., gave up their gold standard, the system lasted for over one century. After the international gold standard system was officially over, the IMS entered a brand new historical era.

8.2.2 The Bretton Woods System

Before WWII was officially over, the allies had already started to plan for the economic reconstruction after the war, hoping that the chaotic world economic order that existed between the two world wars could be avoided. This reconstruction plan was mainly promoted by Great Britain and the United States. The purpose was to develop international economic collaborations in order to resolve global economic problems.

WWII had caused the relative strengths of the major Western nations changed drastically. Great Britain suffered from great damages and traumas during the war; its economy was severely destroyed. So, the United States, replacing Great Britain, became the center of the world economic system. At the end of the war, American industrial products occupied one half of all the industrial products of the world; and American trades with foreign partners took over 1/3 of the total volumes of world trades. International investments of the United States grew rapidly, making the nation the greatest creditor of the capitalist world. Considering their perspective national interests, the governments of the United States and Great Britain put forward their designs for the new international monetary system. On April 7, 1943, they respectively announced their plans – the Keynes Plan of Great Britain and the White Plan of the United States. The intentions of these two different plans were to compete for the dominance in the international financial field. Due to the political and economic strengths of the United States, Great Britain finally gave in and accepted the American plan after the United Stated made some appropriate concessions. In July 1944, over 730 participants from 44 allied nations attended the United Nations Monetary and Financial Conference, held in Bretton Woods, New Hampshire, in the United States. After 3 weeks of intense discussion, the conference passed agreements on *International Monetary Fund* and the *International Bank for Reconstruction and Development*, known jointly as the Bretton Woods agreement. It was ratified by over 65% of the member nations and became effective on December 27, 1945. So, as of that date, the IMS centered around the United States of America, the Bretton Woods System, was formally established.

The Bretton Woods System was a typical IMS established through international collaborations. The main contents of the system include the following several aspects:

1   It established a permanent international financial organization, that is, the International Monetary Fund (IMF), with the mission of promoting international monetary cooperation. The IMF has been the center of the international monetary system after WWII. Its provisions constitute the basic order of the international finance. It finances its operation from its member nations. To a certain degree, it has maintained the stability of the international financial landscape.

2   It developed the reserve system on the base of parallel gold and the U.S. dollar. In the Bretton Woods system, gold was the foundation while the U.S. dollar was the most important reserve asset. Here, gold played two important roles: On the one hand, it represented an important component of the official reserve assets and the eventual tool for the settlement of international payments. On the other hand, each nation's currency was priced using gold, while the U.S. dollar was directly linked to the gold.

3   All member nations' currencies were pegged with the U.S. dollar. In the Bretton Woods system, the U.S. dollar played a prominent role, because the dollar was

directly linked to gold and it was recognized that the official price of one ounce gold was $35. Correspondingly, the American government promised that it would convert the U.S. dollar using the official price at the request of any national government or central bank at any time. Other nations also respectively determined the gold contents of their currencies without the convenience of being able to convert to gold directly. Instead, through the gold contents, these currencies could be converted into the U.S. dollar at their fixed exchange rates so that these currencies were directly pegged with the U.S. dollar. That was the double linkages of the Bretton Woods system, where the U.S. dollar was linked to gold, and the currencies of all other nations were linked to the U.S. dollar. So, the currencies of all nations were connected indirectly to gold through the U.S. dollar. After these currencies were converted into the U.S. dollar, they could then be converted into gold. From this fact, it can be see that the U.S. dollar occupied a particular position in the Bretton Woods system and became the center of the international monetary system, which naturally made it the most important international reserve asset.

4   An adjustable peg exchange rate system was implemented. Each nation's currency could be converted into the U.S. dollar based on the exchange rate, known as the statutory parity, determined by using the currency's gold content and the price of US$35 per ounce of gold. As soon as the statutory parity was announced by the International Monetary Fund (IMF), it could no longer be changed randomly. If the gold content of a nation's currency had to be varied for more than $\pm 1\%$, then it had to be approved first by the IMF. When participating in the trade of foreign currencies and gold, each member nation could only vary the exchange rates and gold prices within $\pm 1\%$ of the statutory rates and the official price of gold respectively. The government of each nation was obligated to maintain the stability of the exchange rates and gold price.

5   The IMF provided short-term finance to those nations that ran deficits in their international payments in order to help them to overcome the temporary difficulties. The IMF established ordinary loan accounts; and the *Agreements* specified that 25% of the fees each member nation submitted had to be paid in the form of either gold or a currency that could be converted to gold and the rest in the form of its national currency. When a member nation needed international reserves, it could use its currency to purchase a certain amount of foreign currency from the IMF according to the predetermined procedure. And within the set future date, the nation would repay the foreign loan by retrieving the borrowed funds in its currency by using either gold or foreign currency. The greater share a member nation subscribed to in the IMF, the more weight its voting right carried, and at the same time the stronger borrowing power it had. At the time when the WWII was just ended, most nations had very limited international reserves. So, the IMF was basically controlled by the United States. The ordinary capital accounts represented the most basic loans of the IMF. They were only limited to offsetting international trade deficits.

6   It eliminated the exchange control on the transactions of current accounts; however, it limited the international flows of capital. After the international gold standard system collapsed in the 1930s, most nations adopted stringent foreign exchange controls, which seriously damaged the international economic exchanges. In order to improve this situation, the Bretton Woods system demanded

all the member nations to release the controls on the transactions of current accounts as soon as possible. However, because during the previous two wars international capital flows were of strong speculative characteristics and caused tremendous shocks to the stability of the international monetary system, the Bretton Woods system allowed to control the international flows of capital, and required each member nation to limit capital flows across national borders.

During the entire time when the Bretton Woods system was implemented, the world economy grew rapidly, while international trades and investments were also developed great deal. That explains why this period is known as the second golden era of the capitalist world. It has be believed that this time period together with the first golden era, which was the time before WWI when the international gold standard system was in place, had everything to do with the fixed exchange rate system. Of course, the highly prosperous world economy of the 1950s and 1960s were caused by many objective factors and conditions. However, the establishment of the Bretton Woods system created a relatively stable international financial environment, which had indeed played a stimulating role in the development of the world economy.

First of all, the Bretton Woods system, as a product of international monetary collaboration after the war, eliminated the pre-war oppositions of various currency groups, which implemented foreign exchange dumping on each other, and the situation of currency and exchange rate wars. It stabilized the post-war international financial chaos and established a relatively stable international monetary order so that the potential for currency and credit crises were greatly eased while an external environment that was beneficial to economic growth was created. Secondly, the fixed exchange rate system was advantageous for the development of international trades and international investments. Years of fixed exchange rate system that was centered on the U.S. dollar had provided a great convenience for international trades and investments. The post-war growth in these two areas surpassed not only the pre-war growth, but also the speed of growth of the industrial production of the world. Although many factors contributed to this phenomenon, the uniform, fixed exchange rate system had surely played an important role.

Although the Bretton Woods system played an important role in the economic growth of the post-war world, it also suffered from some serious shortcomings. Specifically, Robert Triffin of Yale University proposed the dilemma of the U.S. dollar, pointing out the insurmountable contradiction between the solvency and confidence in the dollar. That was a deficiency of the Bretton Woods system. On the one hand, as the international currency, the supply of the U.S. dollar had to grow constantly in order to satisfy the need of the economic growth of the world and the development of international trades. On the other hand, the constant growth in the supply of the U.S. dollar would make it difficult to maintain the fixed price of gold. The convertibility crisis of the U.S. dollar would become clearer as more U.S. dollars were placed in circulation outside the United States; the convertibility of the dollar into gold would be questioned by people and the credibility of the dollar would be challenged. Such inherent contradiction that was difficult to resolve had determined the instability and the eventual collapse of the Bretton Woods system. Additionally, the adjustable peg exchange rates of the Bretton Woods system turned out difficult to be adjusted often based on practical needs. The Bretton Woods system placed as its top priority the maintenance of fixed exchange rates; only when a member nation suffered from a total imbalance in its international payments, it could apply for changing its exchange rate. However, there was no clear definition for what it was meant to be in total imbalance. As a matter of fact, each nation's exchange rates rarely varied. Although the U.S. dollar was the base currency, when its exchange rate was either too high or too low, it was not convenient for it to make adjustment, either. At the same time, nations with surpluses did not want to raise their exchange rates, and nations with deficits did not like to devalue their currencies. The rigidity of the adjustable exchange rates also placed additional weights on the difficulty of adjusting the international payments. That could be seen as acquired disorders of the Bretton Woods system.

Under the conditions of the Bretton Woods system, in order to maintain international trade balance and stable exchange rates, no nation could afford to pay the price of losing domestic economic objectives. The currencies of nations in trade deficit tend to devalue. However, in order to maintain the fixed exchange rates with the U.S. dollar, the central banks had to throw in dollars to purchase their domestic currencies in the foreign exchange market. That was equivalent to reducing the domestic money supply through operations in the open market. That often led to economic decline and unemployment. On the other hand, the currencies of the nations in surplus tend to strengthen. In order to maintain the parity exchange rates with the U.S. dollar, these nations had to use domestic currencies to purchase U.S. dollars in the foreign exchange market. That in fact was equivalent to adopting an expansionary monetary policy, which often led to inflation. Such situations were completely opposite of the original ideas of the designers of the Bretton Woods system.

In short, the initial design of the Bretton Woods system was based on the fact that the United States had occupied the dominating position in the world economy since after the war. In the early days after the war was over, in order to safeguard its own interests, the United States provided large amounts of foreign aids and made many significant international investments. That, on the one hand, helped the United States stimulate the domestic investments of capital and updates of equipment, and on the other hand, injected livelihood and productivity into the economic developments of the nations that received aids or investments of capital. With time, the United Stated gradually lost its economic and technological advantage. Its growth in productivity eventually fell behind that of Germany, Japan, the Netherlands, and other nations. Such imbalance in development was shown in the international financial field as the dominance of the U.S. dollar was gradually weakened and as a large scale imbalance in international payments appeared, where the main dilemma was the American deficit in its international payments. Facing such severe imbalances, none of the major industrial nations could, considering their own interests, truthfully and holistically adjust their policies. So, the collapse of the Bretton Woods system became inevitable.

Since the late 1950s when the U.S. dollar began to become gradually excessive, large amounts of gold reserves were flowing out of the United States, while short-term foreign debts increased drastically. In the year 1960, the amount of foreign debts of the United States had surpassed its gold reserves and the credit foundation of the U.S. dollar started to become loose. Since October 1960, there had appeared four U.S. dollar crises. It was exactly because of these four U.S. dollar crises that shook the foundation of the Bretton Woods system.

The U.S. dollar crisis that broke out after the war with relatively major scale occurred in October 1960. Before the crisis broke off, the capitalist world had already had a relative excess of U.S. dollars, where selling dollars in large quantities and panic buying of gold had caused American gold reserves to start flow outward. In 1960, short-term debts of the United States surpassed for the first time its amount of gold reserve. In order to maintain the stability of the foreign exchange market, the fixed price of gold, the convertibility of the U.S. dollar into gold, and the fixed exchange rate system, the United States demanded other capitalist nations to work jointly within the framework of the international monetary fund to stabilize the international financial market. To this end, the following measures were implemented:

1   Established a gold pool to sustain the official gold price. After the U.S. dollar crisis broke off, the price of gold started to rise. In October 1961, in order to maintain the gold price and the position of the U.S. dollar, the United States established the gold pool by collaborating with Great Britain, France, Italy, the Netherlands, Belgium, Switzerland, and Federal Republic of Germany. The needed gold for the pool was provided by the United States (50%), Federal Republic of Germany (11%), Great Britain, France, Italy (9.3% each), Switzerland, the Netherlands, and Belgium (3.7% each). The Bank of England was appointed to be the agency of the pool to trade gold on the London market in order to prevent the gold price to go beyond the official price of US$35 an ounce.

2   Developed the general arrangements to borrow. In September 1961, the United States and Great Britain suggested to the IMF to expand the IMF's loan amount to US$6 billion to be used for stabilizing each nation's exchange rate. In November of the same year, Great Britain, the United States, France, Italy, Belgium, the Netherlands, Federal Republic of Germany, Switzerland, Japan, and Canada met in Paris and decided to organize their "Group of Ten", also known as the Paris Club, while reaching one treaty, known as the *General Agreement to Borrow* (GAB), to be effective starting in October 1962. The treaty stipulated that when short-term international flows of large capital could potentially cause fluctuations in exchange rates, the IMF should borrow funds from these ten nations in the amount of US$6 billion to be loaned to the member nations that suffered from currency crisis in order to help the nations to stabilize their currencies.

3   Signed the *Reciprocal Credit Agreement*. In March 1962, the United States signed respectively with each major Western nation, particularly the nations in the Group of Ten, a bilateral *Reciprocal Borrowing Agreement*. It stipulated that the central banks of the two nations would provide reciprocal credit funds to each other to be used to interfere the foreign exchange market in order to stabilize the exchange rate. These measures were mainly aimed at resolving the dilemma problem as caused by the double linkages of the U.S. dollar. Because of the limitations of these measures, the intrinsic defects of the Bretton Woods system could not be corrected fundamentally.

After the Vietnam War started in the mid-1960s, the situation of the international payments of the United States further deteriorated. In March 1968, the gold reserves of the United Stated dropped to the level of about US$12 billion, which were enough for the payment of only 1/3 of all short-term foreign debts. Consequently, there erupted the U.S. dollar crisis of throwing dollars and grabbing gold of an unprecedented scale in the London, Paris, and Zurich gold markets. Within merely half a month, the United States lost US$1.4 billion worth of gold reserves. Just by depending on the gold pool and the gold reserves of the United States, it became impossible to maintain the fixed gold price in U.S. dollars. To this end, emergency measures were implemented. First of all, the London gold market was closed temporarily, starting on March 15, 1968, while announcing that trading gold at the price of US$35 an ounce in that market was halted. Secondly, the gold pool was dissolved and replaced by double-priced gold system, where the United States only traded gold with central banks; although the official market price was still maintained at US$35 an ounce, other than trading gold between central banks, no nation's central bank continued its involvement in gold market with individuals. For individuals, their market gold price would be completely determined by the demand and supply of the market just as other consumer goods. Dissolving the gold pool and implementing the double priced gold system meant that the United Stated could no longer maintain the office gold price, and the U.S. dollar was devalued in disguise. It also implied a partial collapse of the Bretton Woods system that centered on gold and U.S. dollar.

In May 1971, the third U.S. dollar crisis broke off. All major financial markets in Western Europe once again started the wave of throwing U.S. dollars and grabbing gold, or Western German marks, or Swiss francs, or Japanese yen, or other hard currencies. Although all major central banks intervened forcefully, the strong wave of capital movements could not be stopped. Facing the violent crisis, the Nixon administration could help but declared on August 15its implementation of a new economic policy – stop the convertibility of dollar and gold, limit import into the United States while imposing 10% surcharges on all imported goods, and pressuring Federal Republic of Germany, Japan, and other nations to strengthen their currencies – in order to improve the U.S. balance of international payments. Under the extreme chaos of the international financial markets, after several months of consultation, the Group of Ten eventually reached on December 18, 1971 a compromise, the *Smithsonian Agreement*. Its main contents include:

1   Starting on August 15, 1971, the United States would devalue its dollar by 7.89% and the price of gold lifted to US$38 an ounce from US$35 an ounce. The purpose of devaluing the U.S. dollar against gold was to raise the gold reserve of the United States so that the confidence in U.S. dollar could be restored.
2   The currencies of some nations would strengthen against the U.S. dollar. In particular, the Japanese yen would strengthen 16.9%, the Western German mark 13.6%, the Swiss franc 13.9%, the Netherlands' guilder and the Belgian franc respectively 11.6%, the British pound and French franc respectively 8.6%, the Italian lira and the Swedish krona respectively 7.5%.
3   The fluctuation range of exchange rates was enlarged from the original 1% to 2.25%. The purpose of doing so was to increase the flexibility and elasticity of the exchange rate system.
4   The United States removed the 10% surcharge on imports.

The *Smithsonian Agreement* was an emergency product introduced to deal with the crisis of the international monetary system. When seen from the angle of how it re-adjusted the structure of the exchange rates, its basic spirit was still to maintain the fixed rate system. However, because this agreement did not touch on the root problems of the international monetary system, its life span turned out to be very short.

At the start of 1973, another U.S. dollar crisis broke out in the foreign exchange markets. Correspondingly, the U.S. government announced the devaluation of the dollar the second time and the official gold price rose to US$42.22 from US$38 an ounce. In March of the same year, the U.S. dollar was devalued again and the official gold price once jumped to US$96 an ounce. In order to maintain their own interests, each nation one by one announced that they no longer pegged their currencies to the dollar and started to use floating exchange rates.

Since the *Smithsonian Agreement* decided to devalue the U.S. dollar the first time, the dollar stopped its convertibility to gold formally so that the free conversion of the U.S. dollars had been destroyed. After the U.S. dollar crisis of the 1973, each nation started floating exchange rate against the U.S. dollar, including the joint floating implemented by the six nations of the European Communities, so that the fixed exchange rate system no longer existed. Hence, both of the two main supporting pillars of the international monetary system centered on the U.S. dollar failed, leading to the complete collapse of the Bretton Woods system.

8.2.3 The Jamaica system

After the Bretton Woods system collapsed, the state of the international finance became more turbulent, each nation was exploring new plans for reforming the monetary system. In order to establish a new international monetary system, a long period of discussions and consultations ensued. In the process of reforming the international monetary system and instituting a new system, conflicts and struggles were inevitable; and eventually through compromises each party had arrived at a common ground on some of the fundamental issues regarding the international monetary system. In 1976, an agreement was reached in the capital city Kingston of Jamaica, known as the *Jamaica Agreement*. After several rounds of modifications, it was ratified and became officially effective on April 1, 1978. That signaled the entrance of the international monetary system into a new era – the Jamaica system.

Other than maintaining and strengthening the roles of the international monetary fund, there are many major differences between the Jamaica system and the Bretton Woods system. The main contents of the Jamaica system include the following: legalize floating exchange rates, none-monetize gold, increase the fund shares of member nations, expand the financing for developing nations, and raise the importance of the special drawing rights (SDR) as international reserves.

In particular, within the Jamaica system, each member nation could freely select its exchange rate system, where the international monetary fund (IMF) recognized both the fixed and floating rate system. The IMF supervised the exchange rate policies of the member nations, reduced the exchange rate volatility, and did not allow any member nation to manipulate the exchange rate for the purpose of gaining unfair competitive advantage. With 85% approval voting, the IMF could return to the currency parity system that was stable and adjustable. This part of the agreement provided a legal recognition of the managed floating exchange rate system that had been implemented for many years. At the same time, it also emphasized the role of the international monetary fund in terms of stabilizing exchange rates through supervision and coordination.

The clause that abolished gold removed the official price of gold so that the central bank of each member nation could freely trade gold based on the market price, and dropped the member nations' obligation that they had to use gold to clear debts between themselves or with the IMF. The gold held by the IMF would be gradually dealt with, where 1/6 (that was about 25 million ounces) would be sold at the market price so that the gains beyond the official price would be used to establish a credit fund and be utilized to aid developing nations. Another 1/6 would be purchased back by the original contributing member nations using the official price.

The basic share the member nations contributed to the IMF was increased 33.6% to 39 billion special drawing right from the original 29.2 billion. And the weight of each member nation's contribution was also changed, where the weights of major petroleum exporting nations were doubled to 10% from 5%, while those of developing nations stayed invariant. All the major Western nations except Federal Republic of Germany and Japan had their weights decreased.

The credit fund, established by using the monetary gains from selling gold beyond the official price, provided aids to developing nations whose per capita income was less than 300 special drawing rights in order to help them to resolve difficulties experienced in their international payments. The IMF expanded the maximum amounts of unsecured loans from the original 100% of member nations' shares to 145%, widened the loans that compensated export fluctuations from 50% of the shares to 75%, and loosened the lending conditions and prolonged the terms in order to help those nations that suffered from sustained deficits in their international payments to weather the storms.

The agreement stipulated that the member nations could freely trade their special drawing rights without first obtaining approval from the IMF; and the transactions between the IMF and member nations would be carried out using the special drawing rights instead of gold. The purpose of expanding the application range of the special drawing rights was for the SDRs to gradually replace the U.S. dollar and gold as the main international reserve asset.

When compared to the Bretton Woods system, the Jamaica system has the following characteristics: Firstly, gold is no longer used as money. It is not used as the base for currency parities of nations and cannot be used to official means for international clearance. Secondly, the reserve currency becomes more diversified. Although the *Jamaica Agreement* did stipulate that in the future the special drawing rights would be used as the main reserve asset of the international monetary system, in reality the weights of the special drawing rights in various nations' international reserves have been falling instead of rising. Since 1981, the IMF has not issued new special drawing rights. The previous situation of single U.S. dollar has been replaced by that of multiple reserve currencies headed by the U.S. dollar. Thirdly, the exchange rate system has become diversified. According to the statistics of the IMF, as of March 31, 1997, the 181 member nations of the IMF had implemented 9 different exchange rate arrangements, which could be classified into the following three groups:

1   Adjustable peg exchange rate arrangement, including 66 different currencies. They respectively pegged to the U.S. dollar (21 currencies), French franc

    (14 currencies), other currencies (9 currencies), the special drawing rights (2 currencies), and baskets of currencies (20 currencies).

2   Limited floating exchange rate arrangement, including 16 currencies. This group of arrangements consists of two types: floating with a single currency – the U.S. dollar (4 currencies), and joint floating within the European monetary system (12 currencies).

3   More flexible floating exchange rate arrangements (99 currencies). This group includes managed floating arrangements (48 currencies), independent floating arrangements (51 currencies).

    The new international monetary system, after the *Jamaica Agreement* was signed, is in fact such a system that embodies diversified international reserves centered on the U.S. dollar and floating exchange rates. Within this system, the position of gold as an international currency has gradually disappeared, and among the many reserve currencies the U.S. dollar still occupies the dominating position, although this position has been gradually weakened. Within this system, each nation is free to choose its exchange rate system. The exchange rates of the currencies of the major developed nations are either independently floating or jointly floating. For most of the developing nations, they have adopted the pegged exchange rate regime, where the local currencies are pegged to either the U.S. dollar, or French franc, or the special drawing rights, or a basket of currencies. There are also nations that employ other forms of managed floating exchange rate systems. Additionally, with this system, imbalances in international payments are adjusted through multiple channels. Other than exchange rate systems, the international financial markets and international financial organizations have also been playing important roles.

    Since 1973 when the U.S. dollar completely disconnected with gold, the asset composite of international reserves has shown the tendency of diversification. Although the U.S. dollar is still the dominating and the most important international currency, its position has been weakened since 1973. Before 1973, all nations used the U.S. dollar as the standard to value their currencies and tried to maintain a fixed exchange rate against the dollar. After 1973, most nations no longer keep fixed exchange rates with the U.S. dollar, although some developing nations are still doing so. In 1974, there were 61 nations whose currencies were pegged to the U.S. dollar. At the end of 1997, there were only 21 such nations. This drop in number indicates the weakening of importance of the U.S. dollar. Additionally, because of the violent turbulence in exchange rates against the U.S. dollar, since the 1980s, several other hard currencies have been used in the settlement of international trades. Even so, as the most important international currency, there is still not any currency as of this writing that can match the dominance of the U.S. dollar. Some of the very important goods in the international trades, especially raw materials and primary products, are all priced in the U.S. dollar. In the world gold market, gold is priced in U.S. dollar. And the dollar is also the most important media of payments. Within the foreign currency reserves, all central banks still maintain good amounts of U.S. dollars.

    After 1973 when gold was no longer linked to the U.S. dollar, the position of gold as an international reserve has been falling. The second amendment of the organizational agreement of the IMF, which became effective in April 1978, continued the policy of non-monetizing gold. The role of gold as an asset of international reserve continued to deteriorate so that it has become an ordinary consumer good although nations still maintain large amounts of gold reserves.

According to the 1978 amendment of the organizational agreement of the IMF, the member nations could freely arrange their exchange rates. Most of the developed industrial nations adopted independent floating or joint floating, while others chose to peg their self-selected baskets of currencies or some sorts of managed floating exchange rate systems. As for developing nations, most chose to peg the U.S. dollar, or French franc, or the special drawing rights, or self-selected baskets of currencies with few using independent floating system. When choosing its exchange rate arrangement, each economically powerful nation in general is more willing to use a floating exchange rate, be it independent or joint, so that it would have more independence in introducing and implementing domestic policies. As for small nations with limited scales of exchange with the outside world, in order to avoid adverse impacts of exchange rate fluctuations on their price levels, they are more likely to use fixed exchange rates. The nations, whose rates of inflation are very different from those of their trading partners, are likely to use floating exchange rates or crawling peg system. The nations with relatively concentrated trading partners are likely to maintain fixed exchange rates with their trading partners. Conversely, they tend to use such arrangements as floating exchange rates or pegged baskets of currencies.

Within the Jamaica system, imbalances in international payments are adjusted through multiple channels. Imbalances in current accounts are adjusted mainly through the exchange rate mechanism, the international financial market, and the interest rate mechanism, international financial organizations, and by making use of the international reserve assets.

Because the currencies of major nations use floating exchange rates, the mechanism of exchange rates should be the main method used to adjust international payments. The adjustment of exchange rate works as follows: When a nation suffers from deficit in its current account, the exchange rate of the nation tends to drop so that it becomes advantageous for the nation to increase its export and to decrease its import. So, the trade balance and the situation of the current account improve. Conversely, when a nation experiences surplus in its current account, the exchange rate of the nation's currency would rise upward so that the nation's export decreases and import increases, which helps to balance the international payments. However, in reality, the adjustment effect of the exchange rate mechanism is not as big as expected.

The so-called adjustment of the interest rate mechanism means the guidance of the capital flow, either inwardly or outwardly, through the differences between one nation's actual interest rate and those of other nations in order to materialize the objective of adjusting the nation's international payments. The characteristics of the current international monetary system are the existence of a very advanced financial market. So, for a developed nation, the differences between its actual interest rate and those of other nations can easily cause capitals to flow inward or outward. As long as a nation strictly controls its money supply so that its inflation is lower than those of other nations, then lifting the market nominal interest rate can guide capitals to flow inwardly, achieving the effect of improving international payments. Adjusting the imbalance of international payments by using debts and investments through the mechanism of interest rates cannot be accomplished without international financial market. In this regard, the international financial market and private commercial banks have played a very important role since 1973. During the two oil crises of the 1970s, the currents accounts of the petroleum exporting nations enjoyed large surpluses, while the current accounts of non-petroleum production nations, especially non-petroleum production developing nations, suffered huge deficits. By lending the large amounts of deposits of the petroleum exporting nations to non-petroleum production nations, the international financial market and international commercial banks helped to release the deficits of these nations in their international payments, materializing a smooth return of these petroleum dollars.

The IMF has been playing important roles in the adjustment of international payments. It lends to nations of deficits for them to overcome the difficulties in their international payments, guides and supervises nations of deficits and nations of surpluses to adjust their international payments. So, both sides assume an equivalent amount of obligations in the adjustment of international payments. Although the IMF has done a large amount of work in helping with the adjustment of international payment imbalances, comparing to the magnitude of imbalances existing in international payments and the total level of debts developing nations have assumed, its role playing is still very limited.

Additionally, imbalances in international payments can also be adjusted through varying the foreign currency reserves. For nations of surpluses, their foreign currency reserves increase, while the nations of deficits, their foreign currency reserves decrease. However, the effect of the previously mentioned adjustment mechanism is quite limited. Speaking generally, developed nations have more capability and ways to adjust imbalances in their international payments than developing nations.

To summarize, the active roles of the Jamaica system are mainly shown in the following aspects:

1   The reserve currencies are diversified; that greatly eases the shortage of international liquidity. Under the Bretton Woods system, each nation's currency is pegged to the U.S. dollar so that the base currency and the attached currencies are connected to each other. Within the new system, what is implemented are a diversification of international reserve currencies and floating exchange rates so that a devaluation of the U.S. dollar would not cause much impact on other currencies.

2   The flexible exchange rate system is supportive to the development of the world economy. The Jamaica system represents a more flexible, mixed exchange rate system dominated by floating exchange rates. It can sensitively reflect the changes in the world economy. The exchange rates of major currencies are adjusted by the market conditions, and can accurately reflect the need of the economy, while no nation is bound by the obligation to maintain a fixed exchange rate. That avoids large losses of national reserve assets that can be potentially caused by changes in exchange rates. The flexible, mixed exchange rate system can also make a nation's macroeconomic policies more independent and effective without sacrificing domestic economic goals when maintaining a fixed exchange rate.

3   The adjustment of international payment is greatly enhanced. The flexible exchange rate system of the Jamaica system makes the adjustment of international payment adaptable and can acclimate to imbalanced states of economic

development, which makes the system more accord with the laws of market movement. Additionally, within the Jamaica system multiple adjustment mechanisms of international payments work in parallel and complement to each other. With the international monetary fund, international commercial banks, and international financial markets working jointly with each other, the difficulty of effectively making adjustment within the Bretton Woods system is eased to a certain degree.

4   It is beneficial to international financial innovation and development of international trades. Under the Jamaica system, in order to help companies and investors to avoid and reduce the risk associated with floating exchange rates, and to strengthen the competitiveness in the areas of international finance and trade, many financial institutions have introduced new services, new tools, and new products so that much greater spaces are created for international financing, trade, and investment.

To summarize, the Jamaica system is a product of world economic unrests and imbalanced developments. Its operation can roughly adapt to such situation of the world economy so that it exerts an active force for the development of the world economy. However, the malpractices of the floating exchange rate system with multiple reserve currencies centered on the U.S. dollar are also quite obvious, which are mainly shown as follows:

1   The diversified reserve system is not stable. In order to warrant the normal operation of the international monetary system and to avoid the risk of foreign reserves, each central bank has to be very careful in its coordination of various kinds of currencies. When a foreign currency reserve is converted to another currency, it can cause volatility in the foreign exchange market.

2   Large volatility in exchange rates is not good for the development of the world economy. When the exchange rate fluctuates volatilely, it makes it difficult to accurately calculate the costs of foreign trades with increasing of exchange rate risk. Volatile exchange rate leads to increased risk in international credits, and can possibly cause debt crisis. If the exchange rate floats downward, it can easily cause the prices of goods to go up, inducing inflation to occur. Additionally, volatility in exchange rate encourages international speculation, exacerbating the turbulence and chaos of the international financial market.

3   There is a lack of effective mechanism to adjust international payments. Currently, the supervision capability of the international monetary fund is far short from meeting the need of successfully adjusting the international payments for nations with surpluses and nations suffering from deficits. That is why the phenomenon of global imbalance in international payments has been worsening. The international reserves of the nations with surpluses have been surging, making them important capital exporting nations and creditors, while the reserves of the nations that suffer from deficits dropped, making it difficult for these nations to pay off their debts and to eliminate their deficits.

Based on what has been discussed above, although the present international monetary system has displayed its strong adaptability in various aspects, its defects are also very prominent. They have exacerbated the conflicts of the world finance.

So, the discussion on how to reform the present international monetary system has been ongoing and has never been stopped.

8.3   IMPROVEMENTS ON THE IMS

Since the 1970s, the discussion on how to reform the international monetary system has been closely surrounding three major issues: the standard monetary system, the exchange rate system, and the adjustment mechanism for balancing international payments. The eventual resolution of these issues is directly related to the future of the international monetary system. In this section we will mainly look at these issues and related discussions.

8.3.1   Main suggestions for improving the IMS

The main plans and suggestions on how to reform the international monetary system are summarized as follows:

(A)   Establish a reserve system of international commodities

Because many developing nations are affected by the fluctuating prices of primary products and raw materials, these nations often suffer from worsening situations of international payments. That is why some economists propose to introduce an international reserve currency that is based on commodities in order to resolve the instability problem of the international reserve system and price fluctuations of primary products. The main contents of the suggestion include: (1) Create a central bank of the world, issue a new unit of international currency whose value will be determined by a basket of commodities, which will consist of certain basic commodities of the international trade, especially some primary products. (2) The current special drawing rights will be merged into the new international reserve system, and the unit value of the SDRs will be redefined by the basket of commodities, where all other reserve currencies will be replaced by a new international currency that is completed commodity based. (3) The world central bank will use the international currency to purchase the primary commodities that make up of the basket of commodities in order to achieve the desired price stability of the primary products and to reach the objective of stabilizing the international reserve currency of commodities.

Such suggestion for international financial reform is arguably possible in theory. However, in practice, it is difficult to implement. This plan requires a reserve of large quantity of primary products so that the cost will be extremely high. To this end, who is going to pick the tab is a problem not easy to resolve. Additionally, a prerequisite for issuing commodity currency is the establishment of a world central bank. That cannot be materialized in the current situation.

(B)   Establish an international credit reserve system

Robert Triffin (1972) proposed this idea. He believed that the essence of reforming the international monetary system was to establish an international credit reserve system, which was beyond the concept of nations, and to create an international reserve currency on the base of the reserve system. The international reserve currency should not be acted by gold, or any other precious metal, or the currency of any nation. Triffin suggested that each nation should submit its international reserve to the international monetary fund in the form of reserve deposits; and the international monetary fund should function as the clearance organization for all national central banks. When the international monetary fund or some other similar international agency accepts all nations as members, the activities of international payments will be reflected as changes of either increase or decrease of the reserve deposit accounts of the member nations with the international monetary fund. The total amount of the international reserves held by the international monetary fund should be determined by the member nations, and should be adjusted according to the need of world trade and development of production. The creation of the reserves can be done through lending to member nations, purchasing financial assets in the financial market of each member nation, or periodically distributing new special drawing rights. However, it should not be affected by the production of gold or limited by the balance of payments of any nation.

Triffin also pointed out in his writings that the special drawing rights should be employed as the only international reserve asset to gradually replace gold and other reserve currencies. However, his suggestion required that the central banks of all nations comply with an international credit reserve agency that is beyond each and every nation, and an extremely close international monetary collaboration. That currently is not realistic.

(C)   Substitution account of the International Monetary Fund

The substitution account is a particular account the International Monetary Fund establishes and the IMF issues a particular kind of special drawing rights. The central bank of each nation can deposit its special drawing rights that are obtained from converting its excessive U.S. dollar reserves into this account. Then the International Monetary Fund invests in the long-term debts of the Treasury Department of the United States by using the received U.S. dollars. The received interest income will be returned to the depositors of the substitution account. This suggestion was first proposed at the annual meeting of the International Monetary Fund, imagining that through using the special account, all excessive reserves of assets in U.S. dollars could be absorbed so that the SDRs would eventually become the main international reserve assets. This suggestion was never practically implemented, because for a period of time since the year 1980, the U.S. dollar had been extremely strong so that all central banks were more willing to hold U.S. dollars instead of the SDRs. Only when the SDRs or some other reserve asset have become completely functional and a truly better international currency than the U.S. dollar, people would give up their accumulated assets in U.S. dollars.

(D)   Stabilize exchange rates through strengthening economic policy coordination of all nations

Volatility in the currency exchange rates of the major industrial nations has adversely affected the world economy and the stability of the international finance. That has caught the attention of all nations. In October 1985, the meeting of the ministers of the treasury departments and the heads of the central banks of the United States, Japan, Federal Republic of Germany, France, and Great Britain, proposed to coordinate their respective economic policies in order to stimulate the stability of exchange rates. Later in 1986, the ministers meeting of seven nations' treasury departments proposed to materialize their economic policy coordination through controlling the following 10 indicators: GNP rate of increase, inflation rate, interest rate, unemployment rate, fiscal deficit, current account balance, trade balance, increase rate of money supply, foreign currency reserves, and exchange rates. The International Monetary Fund monitors these indicators jointly for all of these nations by classifying these indicators as performance indicators, policy indicators, and intermediate indicators. In the 1987 G7 Summit, specific provisions were made for their coordination of economic policies.

If the proposed coordination of macroeconomic policies could be effectively implemented in practice, it would be very helpful in terms of stabilizing exchange rates. However, truly coordinating the policies was never an easy task, because coordinating macroeconomic policies would greatly weaken the independence of the respective national policies, and erode the national interests of some nations. Additionally, during the time of economic decline, the domestic situations could be so severe that the governments no longer had any spare moment to think of coordination with other nations.

(E)   Establish target zones for exchange rates

By a target zone of exchange rate, it means that each of the relevant national monetary authorities selects a basic adjustable exchange rate as reference, specifies a range around the reference for the actual exchange rate to fluctuate up and down, and reinforce the application. Although there are many different types of the target zones, the main ones can be classified into "hard target zones" and "soft target zones". By a hard target zone, it means that the range of change for the actual exchange rate is narrow and is not often modified, while the specifics of the zone is not openly announced to the public. Generally, it is through using monetary policies to maintain the actual exchange rate within the target zone. By a soft target zone, it means that the range of change for the exchange rate is relatively wide and is modified frequently. The specifics of the target zone are not known to the public; and the target zone is not necessarily maintained by employing monetary policies. The characteristics of target zones of exchange rates are that they combine the flexibility of floating exchange rate system and the stability of fixed exchange rate system, and that they can promote the coordination of various national macroeconomic policies. However, there are many difficulties in implementation, such as the determination of the equilibrium reference rates, effective methods of maintaining the targets, etc. Since the time when the idea of establishing target zones for exchange rates was initially suggested, the comments have been mixed. Developing nations hope to realize the stability of exchange rates through implementing target zones for change rates, while developed nations believe that this idea would not work.

8.3.2 An expected future of the IMS

The afore-mentioned plans are only some of the important ones suggested for reforming the international monetary system. By looking through the literature, it seems that the main focus is on the determination of the international reserve asset and the selection of the exchange rate system. Here, the reform of the international reserve asset is more basic and more fundamental.

As long as what is implemented is neither the gold standard system without any interference nor the complete freely floating exchange rate system, the nation would have to have international reserve assets. Since the end of World War II, the U.S. dollar has been the most important reserve asset; American balance of international payments has been influencing the growth of the international reserves. The amount of U.S. dollar that circulates around the world has been the main factor causing the instability of the international financial situation. However, as of this writing, no better asset has been found to replace the U.S. dollar as the international reserve asset. So, each nation has to hold the U.S. dollar and employ it in the nation's international trades and financial transactions. That is why many plans and suggestions on how to reform the international monetary system are about how to replace the U.S. dollar. Under the current international economic circumstances, it is impossible for us to return to the gold standard system or any other form of gold exchange standard. Starting in 1973, along with the relative weakening of American economic strength and the rise of other economic centers, such as that of Japan, and Europe, a diversification of the international reserve currency has been developing. Even so, as of this writing, the U.S. dollar still occupies the dominating position. In fact, in the diversified international reserve system, no matter which currency's exchange rate goes up or down will affect the stability of the international monetary system. So, looking at the long term, it seems to be reasonable to develop a unified world currency. And this opinion has been reflected in many plans and suggestions of reform. Now the key problem is that to issue such a currency, there is a need to establish a world central bank or a similar agency and to implement unified worldwide monetary policy. That is very difficult to accomplish.

In terms of the reform of exchange rate system, it is very unlikely to materialize the theoretical completely fixed exchange rate system or completely freely floating exchange rate system. From the current situation that the developed nations jointly intervene the foreign exchange market and that many developing nations are implementing peg exchange rate systems, it can be seen that stabilizing exchange rates and reducing the magnitude of exchange rate fluctuation are the common goal of the international society. So, the essence of reforming the exchange rate system is in fact about the problem of how large range the exchange rate should be allowed to fluctuate or the problem of in which form the world could return to the fixed while adjustable parity system. Throughout the evolutionary process of the international exchange rate systems, it can be seen that what appeared first is the strictly fixed change rate system, then appeared floating exchange rate system, following that is the fixed exchange rate system with adjustable rate (the Bretton Woods system), and the last is the current floating exchange rate system. In the past 100 plus years, for the most part what was implemented was the fixed exchange rate system; the floating exchange rate system was practiced during the Great Depression of the 1930s. The current managed floating exchange rate system has been in operation for near 30 years. However, the fact shows that the floating system suffers from many defects.

From the afore-mentioned main plans and suggestions of reforming the exchange rate system, it follows that the majority advocate the establishment of fixed exchange rate system of some sort, while some people suggest maintaining the current mixed exchange rate system where each nation is free to choose its exchange rate arrangement, while requiring the major nations to coordinate their policies in order to materialize the stability of exchange rates. However, implementing fixed exchange rate system requires certain objective conditions, which cannot be met under the [resent circumstances. With the existence of major differences in national inflations, economic growths, balances of international payments, and monetary policies, the foundation for implementing a fixed exchange rate system does not exist. It is possible to materialize the current international effort, at least to a certain degree, of stabilizing exchange rates and reducing the range of fluctuation of the rates through coordinating individual nations' policies and joint interventions. It is because collaborating and coordinating respective national policies, mutually compromising with each other would lead to common goods to all the parties involved. As for the future in the development of the international monetary system, on the one hand, it is dependent on how closely the currencies of the major nations could collaborate with each other; on the other hand, it is also determined by the development of the international reserve currency.

8.4 THE EUROPEAN MONETARY SYSTEM

As a case study in terms of economic development and the systemic yoyo model, this section looks at the history of the European monetary system, how the effort of unionizing the European currencies took off the ground, and what impacts the Euro produces on the regional and international economies.

8.4.1 The history

In order to strengthen the political and economic unity, the six nations of the Western Europe, including France, Federal Republic of Germany, Italy, the Netherlands, Belgium, and Luxembourg, established the European Economic Community (EEC) on January 1, 1958. The common objectives were: Gradually unify economic policies in the economic area, develop a unified market for industrial and agricultural products, realize free movement of capital and labor in the community, coordinate the fiscal, financial, and monetary policies and legislation, and when the time ripe, develop the economic community into a political union. At the early stages of the European Economic Community, because the Bretton Woods system centered on the U.S. dollar was operating well, the community did not clearly propose the objective of unifying its currency; instead, it only paid attention to the problem of coordinating monetary and financial policies.

In the early 1960s, the Bretton Woods system fell into crisis. In October 1960, there appeared the first round of waves of dumping U.S. dollars and panic buying gold since after WWII. That caused the exchange rates of the member nations' currencies of the European Economic Community to fluctuate severely. So, the member nations demanded to coordinate monetary policies in order to gradually materialize the unification of their currencies. On February 9, 1971, the Council of EEC Ministers reached an agreement, setting about to establish their economic and monetary alliance. After the Bretton Woods system collapsed, the financial relationships between nations became more tightly intertwined with each other, and conflicts of national interests became more noticeable. So, it was impossible to develop within a short period of time a unified international monetary system that was acceptable to most nations. Within this background, such nine nations as France, the Netherlands, Belgium, Luxembourg, Denmark, Federal Republic of Germany, and others of the EEC led the establishment of a joint floating group that became effective in March 1973. The currencies of these participating nations maintained fixed exchange rates, while the central banks guaranteed that the fluctuations of the rates would not go beyond ±2.25%. At the same time, these currencies followed floating exchange rates, falling or rising freely, against the U.S. dollar and other currencies, determined by the market supply and demand. In April 1978, the EEC Summit, held in Copenhagen, Denmark, exchanged the views the first time on the possibility of establishing a European monetary system. Respectively in July and December of the same year, the issue of European monetary system was discussed two more times, leading to the decision that on January 1, 1979, the European monetary system would be formally inaugurated with the objective of further eliminating the reliance on the U.S. dollar and the effect of U.S. dollar crises. To materialize the ultimate goal of establishing the joint economic, monetary alliance, the very first issue was to stabilize the exchange rates of the member nations' currencies. To this end, the European monetary system was determined to contain the following main contents:

1   Establish the European currency unit;
2   Implement a stable exchange rate mechanism; and
3   Develop a European monetary fund.

The European currency unit (ECU) was the core of the European monetary system. It was determined by a weighted average of the European Community (EC) member nations' currencies, where the weight of each member nation was calculated based on that nation's economic strength, such as foreign trades and gross national product (GNP). The weights were generally to be adjusted once every five years. However, if the proportion of one currency changed more than 25%, the composite of the weights would be readjusted.

Stable exchange rate mechanism was the core component of the European monetary system. According to the arrangement of this mechanism, each participating nation determined the fixed parity between its currency and the European currency unit, that is, the central exchange rate, and based on this central rate, all parities between the currencies of other participating nations were calculated. Through mandatory interventions of all nations' monetary authorities in the foreign exchange market, the fluctuations of the individual currencies' exchange rates were limited within the allowed ranges so that the desired stable exchange rate mechanism was materialized. In other words, if the exchange rate of two currencies had reached the allowed upper or lower limit of fluctuation, the monetary authority of the weaker currency had to purchase its own currency in order to prevent it from sliding downward further. Correspondingly, the monetary authority of the stronger currency had to sell its currency in order to prevent it from rising further. By employing such symmetric market intervention, the European Community achieved its stable exchange rate mechanism.

The European monetary fund was the foundation of the European monetary system. In April 1973, the Community established the European Monetary Cooperation

Fund for the purpose of stabilizing the exchange rates, and provided credit to the member nations. However, there were only a total of 2.8 billion European monetary units created, that was far from enough to meet the need of mediating the market. In April 1979, 20% of the gold, foreign currency reserves, and the equivalent amount of national currencies of the 9 member nations of the Community of the time were again collected. Fast moving to April 1981, the previous fund had reached such a total value that was equivalent to US$73 billion. At the same time, the fund also stipulated the credit system, where the central banks of the member nations could provide to each other short-term credit without any limit in their local currencies as supplementations to their capabilities of market interference. As a result, a strong common reserve necessary for the purpose of stabilizing the market was established.

8.4.2 The European Union of money

After entering the 1980s, the member nations of the European Community gradually got rid of the internal and external economic crises and demanded to speed up construction of European Union and the unification of the European currency. In December 1985, the European Community Summit in Luxembourg passed the *Single European Act*, deciding that by using 7 years of time to establish a unified European market by the end of 1992, to actualize a true economic and monetary unification, and to realize the free movement of goods, capital, labor, and services between the member nations. In the 39th Summit of the European Community in June 1988, a 17 member committee, headed by the Commission President Jacques Delors with the central bank governors of the member nations and experts as participants, was formed to specifically investigate the feasibility of establishing a European Economic and Monetary Union. In June 1989, the EC Council considered and approved the Report on the EC Economic and Monetary Union, as submitted by the committee headed by Delors. The Delors report suggested materializing the economic and monetary union in three stages, each of which did not have any specified time limits with only the exception that the first stage should be started no later than July 1, 1990, and the main objectives of each of the stages.

During December 9 and 10, 1991, the heads of the 12 member nations of the EC convened in Maastricht, a small two in the Netherlands. On February 7, 1992, the Maastricht Treaty, formally, the Treaty on European Union, was signed. This treaty has been seen as a milestone in the unification of the European currencies. The Maastricht Treaty consists of two parts: one on the political cooperation and the other on the economic and monetary union, and specified a time table for materializing the European economic and monetary union:

Stage #1: July 1, 1992 to end of 1993. During this stage all member nations were required to participate in the exchange rate mechanism of the European monetary system, strengthen the monetary and exchange rate policy coordination between the member nations, and reduce as best as possible the number of adjustments to the central rates.

Stage #2: January 1, 1994 to end of 1996 (and no later than end of 1998). In this stage the work of developing and establishing an independent European monetary institute. This institute would be charged with the responsibility of supervising the

economic and monetary policies of the member nations in order to achieve the
objective of small fluctuation ranges of the exchange rates for the individual national
currencies, and to strengthen the coordination of economic policies so that some
major economic indicators would meet the stipulated goals.

Stage #3: Start to introduce a single currency in 1997 but no later than January 1,
1999. And start to develop the European central bank on January 1, 1997, or
no later than December 31, 1997. This bank would be responsible for the unified
monetary policies. After formally signing the Maastricht Treaty on February 7,
1992, the original plan was to finish the approval procedure by all member nations
by January 1, 1993. However, due to some difficulties that appeared in the process,
the treaty became effective on November 1, 1993, a delay of 10 months from the
original plan. Along with the effect of the Maastricht Treaty, the EC has been known
as the European Union or EU.

Based on the provisions of the Maastricht Treaty, the EU Summit in Madrid,
held on December 15, 1995, formally named the future European currency as euro,
which was to be used to replace the ECU of the then European Monetary System,
and specified the time table for introducing the single currency. According to this time
table, the launch of the euro should consist of four stages. The first stage would last
from December 1991 to December 31, 1998. This time period from the time when the
Maastricht Treaty was passed in 1991 to the end of 1998 was seen as the period of
preparation for launching the euro. The second stage would go from January 1, 1999,
to December 31, 2001. This period would be considered as the transition for each of
the Eurozone nations to convert from its individual currency to the euro. The exchange
rates of the euro were fixed on January 1, 1999, and were not cancellable. Although
the transactions of the financial wholesale market would be in euro, companies and
individual persons could open bank accounts in euro, and payments in euro could be
made between bank accounts, the paper notes and coins of euro were not started in
circulation. The third stage would last from January 1, 2002, to June 30, 2002. During
this time period, the paper notes and coins of the euro would enter into circulation.
Within the Eurozone, the euro and all the original national currencies would be in
circulation simultaneously. The fourth stage would start on July 1, 2002, when all the
individual national currencies of the Eurozone would be taken out of the circulation.
So, Euro would become the only currency circulating in the Eurozone, completing the
formation of the unified single currency in Europe.

8.4.3 Economic impacts of the euro

The introduction of the European single currency, the euro, not only profoundly
changed the appearance of Europe, but also played an important role in the field
of international finance. It generated a far-reaching impact to the world politics and
economic development.

In terms of the European economies, the impacts of Euro can be seen in several
areas: launching and implementing the euro has been beneficial in reducing the risk of
foreign exchange, promoting the production, trade, and investment of the European
Union. It has suppressed inflation and stabilized prices of goods for the Eurozone

nations. It has greatly reduced the trading cost between the member nations. And, it speeds up the formation of the greater Europe center at the EU.

In particular, after the euro was in circulation, the original fluctuations in exchange rates disappeared naturally. That greatly reduced the risk of foreign exchange in the unified market so that the speed of financing and investment between the member nations was increased, and the distribution of various production factors became more logical and reasonable. Suppressing inflation had been for a long time an important policy objective of the nations in the EU. However, due to a lack of effective collaboration, these nations had paid heavy prices, where the effect of the monetary policy of one nation was often cancelled by that of another nation. After euro was launched, monetary policies were established uniformly by the European Central Bank, which has had relatively strong independence and treated controlling inflation and stabilizing prices as its first number objective. That may very likely reduce the likelihood of appearance of hyperinflations institutionally. The direct consequence of using the euro includes that many simplified procedures, leading to enhanced efficiency, accelerated circulation speed of consumer goods, reducing the cost of currency exchanges, reduced costs of clearance, trade, and other nonproduction expenses. Because of the formation of the greater Europe centered at the EU, it has helped the Europe to become one of the poles of the world. That makes it play more important roles and produce more lasting impacts on various international affairs.

In terms of the world economy, the impacts of Euro can be seen from three angles: the role in promoting world trades, the impact on the international monetary system, and influence on the international capital market. In particular, because the Eurozone is one of the growth regions of the world economy, the economic growth of the Eurozone means an expanding magnitude of the import-export market when seen from the angle of other nations and regions. So, it can drive the world growth of foreign trades of other nations and regions. The impact of euro on the international monetary system is reflected in three areas: the impact on the international reserves, the impact on the international exchange rate system, and the impact on the unification of the world currencies.

Soon after the launch of euro, it became the largest international reserve currency, only second to the U.S. dollar. Its appearance has caused a major shock to the existing pattern of the international reserves that has been centered at the U.S. dollar. Currently, the U.S. dollar still occupies the dominating position in the world economy and the international finance. However, when seen from the long term, the cyclic evolution of the American economy and the gradual release of the European economic strength and its size effect will be helpful to lift the competitiveness of euro. The gradual increase of euro in each nation's foreign reserves will make the international single-polar reserve system, which has been mainly centered at the U.S. dollar since the end of WWII, into a system of two poles: the U.S. dollar and the Euro. The euro's position as an international currency has created great impacts on the international exchange rate system. In 1999, over 30 nations outside the Eurozone had linked their exchange rate systems to euro in different degrees. Other than that, euro has also become the lawful currency in some nations and regions. The successful unification of the European currencies possesses a strong modeling effect; it will stimulate other regions and nations from around the world to try to unify their currencies. That of course will strengthen the development of grouping currency tendency in a world scale.

As for the impacts of euro on the international capital market, since January 1, 1999, all stocks listed in the exchanges of Europe have been priced in euros. That has made the stock markets, money markets, and banking business merge into a whole, which has been run using euro since the start. When seen from the mid-term, that will change the existing investment strategies, and will enhance the blue chip market across Europe. When seen from the long term, the European Union will integrate into a unified capital market.

8.5 INTERNATIONAL ORGANIZATIONS OF FINANCE

By an international organization of finance, it means such a supranational institution that participates in the activities of international finance, or the coordination of international financial relations, or the maintenance of the international monetary order, or the normal operation of the international credit system. Such organizations have played important roles in international economic lives, and have become the main body that develops international aids and an important channel of international financing. After WWI, conflicts between Western nations became increasingly intolerable; there appeared serious deficits in international payments; many Western nations experienced difficulties in foreign exchanges, international settlements, and other areas. So, the problem of establishing international financial organizations naturally appeared. The Bank for International Settlements was the earliest international financial organization. Its members were mainly some of the European nations. Its role in the field of international finance was quite limited. After WWII, in order to establish a stable international monetary system and to strengthen the regional financing and economic collaboration, a series of international financial organizations have been established. International financial organizations are grouped into two classes: international and regional. The former includes the International Monetary Fund (IMF), the International Bank for Reconstruction and Development (IBRD, or the World Bank) and its affiliates, the International Development Association (IDA) and the International Finance Corporation (IFC). Currently these four institutions are the international financial organizations that have the most members and far-reaching influences. Among the regional international financial organizations, there are mainly the Asian Development Bank, Inter-American Development Bank, African Development Bank, European Investment Bank, Arab Monetary Fund, and others. This section will mainly discuss the global financial institutions.

8.5.1 The International Monetary Fund

The International Monetary Fund (IMF) started its operation in 1947 and has done a great deal of work in the areas of maintaining the stability of exchange rates, eliminating exchange controls, balancing international payments, and promoting international monetary cooperation. Although it was a product of the Bretton Woods system, it is still playing important roles in many areas in the current international monetary system.

All nations that participated in the Bretton Woods meeting and signed the Bretton Woods Agreement before December 1, 1945, are considered the funding members of the Fund organization. Other nations that joined the organization later are known as other member nations. The IMF organization consists of a board of governors, executive board, a managing director, and many operational agencies. The board of governors is the highest authority of the Fund organization. It consists of a director and deputy director from and named by each member nation, where the deputy director has the voting right only when the director is absent. The main authority of the board of governors includes: approve the admission of new member nations, decide on the shares of the fund, distribute the special drawing rights, determine which member nations to withdraw from the Fund organization, discuss and make decisions on important issues related to the monetary system. The board meets once a year and when necessary it holds special meetings.

The executive board is a permanent establishment of the Fund organization. Its headquarter is located in Washington, D.C. It is responsible for daily affairs of the Fund organization, and exercises all the authorities as either specified in the agreement of the Fund organization or endowed by the board of governors. The managing director represents the chief of the IMF organization. He is elected by the executive board for a term of five years. He is responsible for and manages the operations of the Fund organization. He also fills the chair position of the executive board.

The capital of the IMF organization comes mainly from three sources: the fund shares paid by the member nations, borrowed funds, and the trust fund. Firstly, the fund shares paid by member nations represent the main source of capital of the IMF. The size of a member nation's share is determined with consultation with other member nations by such factors as the national income, gold and foreign reserves, average amount of import, rate of change of export, ratio of export over the national income, etc., of the nation. Other than used as the funds for the IMF to provide short-term loans, the member nations' shares are also employed to determine the amount each member nation could borrow from the IMF, the weight of voting right and size of the special drawing rights of each member nation. Secondly, the borrowed funds are borrowed from member nations based on signed agreements between the IMF and the lending nations. These funds are an important method for the IMF to cover shortages of money. And thirdly, the IMF organization decided in 1976 that it would sell 1/6 of its gold reserve at the market price in four years, where the profit would be placed in a trust fund for the purpose of providing concessional loans to the neediest developing countries.

The business activities of the IMF organization are focused mainly on three aspects: exchange rate surveillance and policy coordination, the creation and management of reserve assets, and loan business. As for exchange rate surveillance and policy coordination, in order to make the IMS operate smoothly and to guarantee the stability of the financial order and the growth of world economy, the member nations promise to collaborate with the Fund organization and other member nations to ensure the orderly exchange rate arrangements to promote the stability of exchange rates. In particular, the Fund organization requires all member nations to:

(i)   Use one's own economic and financial policies to promote the objective of orderly economic growth along with reasonable price stability and appropriate care of own circumstances;

(ii)   Promote stability through creating orderly economic and financial conditions and such a monetary system that will not cause abnormal chaos; and

(iii)   Avoid impeding orderly adjustments of the international payments or attaining unfair competitive advantage against other member nations by manipulating exchange rates or the international monetary system.

As for the creation of reserve assets, the IMF organization formally approved the scheme, as proposed by the group of ten, at the 1969 annual conference, and decided to create the special drawing rights (SDRs) to complement the shortage of international reserves. The SDRs were distributed starting in 1974 by the Fund organization to the member nations according to their respectively contributed shares. The distributed SDRs became the reserve assets of the member nations. As for loan business, lending has been the main business of the IMF organization. It consists of the following several types of loans.

1   Ordinary loans, also known as basic credit loans: They represent the most basic loans provided by the IMF organization. They are used to ease temporary difficulties of the member nations in their international payments. The maximum amount of such loan is equal to 125% of the member nation's contributed share with a term of 3–5 years. The interest rate of the loan increases with the length of the term.
2   Compensation and emergency loans: Any loan of this type is used for resolving the temporary difficulty of a member nation caused either by decreasing export income or by increasing expense of grain imports. The conditions of lending include: the decrease in export income or the increasing expense in grain import is temporary and is caused by reasons beyond the control of the member nation. At the same time, the borrowing nation is obligated to collaborate with the IMF organization to overcome the difficulty jointly.
3   Stock loans: This type of loans was established in 1969 for the purpose of helping exporting nations of primary products to meet the capital need of building buffer stocks in order to stabilize the international market prices of the primary products. Each member nation can borrow up to 45% of its contributed share with 3–5 year terms.
4   Temporary unsecured loans: This type of loans is established temporarily for meeting the rising need by the IMF organization. The source of funds is from temporary loans borrowed from member nations by the IMF organization. For example, the petroleum loans of the time period from June 1974 to May 1976 were provided to the member nations that suffered from difficulties in international payments due to rising prices of the petroleum oil. The system transformation loans established in 1993 were mainly used to help those nations for their difficulties in international payments, as caused by their ongoing economic transitions from clearing account trades of the Council of Mutual Economic Assistance to cash payment trades.
5   Medium-term loans: This type of loans was established by the IMF organization in September 1974. It is also known as expansion loans. These loans are specifically designed to resolve long-term structural deficits in the international payments of member nations. The maximum amount of each loan is equal to 140% of the borrowing nation's contributed share with 4–10 year terms. Generally, such a loan needs to be repaid with 16 installments. The borrowing conditions are: The

IMF organization confirms the existing difficulty in the international payments of the borrowing nation and the need of a loan with a longer term than that of ordinary loans to resolve the difficulty; the borrowing nation has to detail how it is going to improve its international payments within the loan term and to illustrate what policy measures will be implemented within the first year; and the loan will be distributed periodically according to the current policy objectives and the actual situations of implementation of the policies.

6   Trust fund loans: The trust fund was created in May 1976. The source of funds is mainly from the profits of selling the gold held by the IMF organization at the market prices. The trust fund provides loans to developing nations of low income levels. The conditions of acquiring a loan are: Any nation which took a first part of a two-year loan since July 1, 1976, must be such a nation whose per capita income in 1973 was lower than 300 SDRs. There were 61 such nations. Each nation that took the second part of a two-year loan since July 1, 2978, must be such a nation whose per capita income was no more than US$520 in 1975. There were a total of 59 such nations. The annual interest rate of trust fund loans is 0.5% with 10 year terms.

7   Supplemental loans: This type of loans was officially established in August 1977. The capital of these loans was provided by petroleum exporting countries and developed countries with surpluses. They are used for supplementing the shortages of the ordinary loans. When a member nation suffers from severe deficit in its international payments and is in need of a much greater amount of loan of a much longer term than those of an ordinary loan, it can apply for a supplemental loan. The total amount is 8.4 billion SDRs and the maximum loan amount is 140% of the nation's share with 3–7 year terms. The stand-by arrangement term is 1–3 years.

8   Structural adjustment loans: This type of loans was established in 1947 for helping low-income nations to formulate and implement comprehensive macroeconomic adjustments and structural reforms in order to promote economic growth, to improve the difficult situations with their international payments, to strengthen the construction of their national economic infrastructures, and others. The source of funds of these loans comes from the principal and interests of the trust fund. The interest rate of these loans is 1.5% with 5–10 year terms. The maximum loan amount is 250% of a nation's share.

Since 1947 when the IMF organization started its operation, it has done a great deal of work in such areas as strengthening international monetary collaboration, creating multilateral payment systems, stabilizing exchange rates, improving the adjustment process of international clearance, etc. It has played a promoting role in the maintenance of the operation of the international monetary system, and in stimulating international trade and the development of world economy. However, along with the development of the international economic relationships, the IMF organization has also shown some of its defects. For instance, firstly, the amount of funds available for lending is limited. In the area of credit funds, the IFM has provided a large number of short-term loans to developing nations, which has played an important role in buffering the imbalances in these nations' international payments. However, since the 1980s, the balance situations of the current accounts of the developing nations have deteriorated continuously, leading to a surge in the demand for loan funds. Facing this challenge, the available funds of the IMF become obviously too limited and cannot satisfy the demand for loans of the developing nations. Secondly, the allocation of loan funds is no reasonable. The loans of the IMF are allocated according to the member nations' shares. So, the major industrialized nations occupy larger share weights, while the main demanders for these loans are developing nations. Along with each increase in capital, the share of the developing nations in the total amount has been generally going down. And, since 1983, the IMF has lowered the ceiling of borrowing for all nations beyond the ordinary loans, which makes the funds allocation disadvantageous to developing nations. Thirdly, attached to loans are there restrictive conditions, which mean that when IMF member nations use their IMF loans, they must adopt certain economic adjustment measures so that when the IMF loan projects end, these borrowing nations could rebalance their international payments. As the main users of the IMF funds, developing nations are also the bearers of various harsh restrictive conditions of the loans. The stipulated austerity and adjustment conditions attached to the IMF loans have brought forward negative impacts to the economies of the borrowing nations, leading to tensions between the IMF organization and developing nations.

8.5.2 The World Bank Group

The World Bank Group consists of the International Bank for Reconstruction and Development, the International Development Association, and the International Finance Corporation.

First, let us look at the International Bank for Reconstruction and Development, which is known briefly as the World Bank. It was established in December 1945, and officially started its business in 1946. It was an international financial institution created together with the IMF organization, and is also a specialized agency of the United Nations. The mission of the World Bank includes:

(i)   Provide long-term loans for the purpose of productive investments and help the reconstruction and development of member nations through collaboration with other international agencies;
(ii)   Promote private foreign investments in the form of guaranteeing or participating in private lending and investments; and
(iii) Stimulate long-term balanced development of international trade and maintain balanced international payments in the form of developing production resources of member nations through encouraging international investments.

The World Bank is headquartered in Washington DC. Its highest authority is the council, under which the executive board is the body that executes daily operations. The council consists of a director and a deputy director from and named by each member nation with 5-year term and the possibility to serve additional terms. The main authority of the council includes: approve the admission of new member nations, decide on the adjustment of payable shares of member nations, determine the distribution of bet income of the World Bank, and other important matters. Each year, the council meets once jointly with the IMF organization, and might call for special meetings when necessary. The voting right of each member nations is determined by its share, where each nation has 250 basic votes. In general, the share subscribed by a nation is determined by the nation's economic and financial strength in reference to its share in the IMF organization. The United States has always been the largest share owner so that it has the absolute control over the business activities of the World Bank. The executive board is responsible for dealing with the daily affairs of the bank. It consists of 22 members with 5 of them appointed by the top five share owners and the rest 17 produced by regional groups of the other member nations. The executive board elects one person to serve as the executive chair of the board and one person to be the president of the World Bank. The president does not have any voting right except when there is a tie in the executive board, the president can cast his decisive vote. The president serves a 5-year term, and can be re-elected.

The funds of the World Bank come several sources: 1) the share fees member nations paid; 2) issuance of bonds; 3) debt transfers; and 4) net business income. The World Bank requires that each member nation has to subscribe to shares in reference to its shares in the IMF organization. At the initial stage of the World Bank, the statutory base contribution was US$10 billion, which was divided into 100,000 shares making it US$100,000 a share. After then, the share fees had been increased several times, reaching, for example, the price tag of US$188.22 billion for the statutory base contribution. However, not all member nations pay their fees based on the stipulations. According to the original provisions, when a member nation initially joins, it has to pay 20% of the share fee, where 2% needs to be in gold or a free convertible exchange, and 18% in the national currency. The rest 80% of the fee is the to-be-paid-up capital, which is kept by the member nation and needs to be paid when the World Bank experiences difficulties with funds. Issuing bonds is an important source of funds for the World Bank, which issues mid-term and long-term bonds in the international bond market. The funds raised in this way account for about 70% of those used in lending. Since the 1980s, the World Bank has been reselling its debts incurred from its issued loans to commercial banks or private investors. From doing so, it can quickly retrieve a part of the funds to be reused as loans. That has expanded the turnover capability of the funds available for the bank to lend. Since 1984, the World Bank has enjoyed net profit each year. That also becomes a source of supplemental funds. At the end of 1989, the net income of the World Bank reached US$1.1 billion

As for the lending business of the World Bank, it only lends to the government of a member nation, or companies and organized that are guaranteed by the government or central bank of a member nation. Loans from the World Bank are generally used for special purposes and can be considered by the World Bank only when the borrowing nation cannot obtain loans from other sources with reasonable conditions. In order for the World Bank to guarantee that the loans can be repaid on time, it only lend to member nations that have the ability to repay. All loans will be supervised by the World Bank and have to be used as specified.

Each application for a loan from the World Bank must follow a strict procedure along with close examination and supervision of the Bank. Speaking generally, the World Bank first investigates the current state and the future of the economic structure of the applicant nation in order to determine what type of loan to provide. Then a team of experts will be assigned to evaluate the involved project. At the final step, arranged will be negotiations on the loan, signing of the loan agreement, guarantee document, and other legal papers. After the loan is issued, the World Bank also requires the borrower to pay attention to the economic and other consequences of using the loan.

The lending business of the World Bank consists mainly of such loan types as project loans, non-project loans, sector loans, syndicated loans, third window loans, etc. In particular, project loans are provided to member nations for their specific projects in such areas as industry, agriculture, education, transportation, energy, and others. This type of lending represents the most important loans of the World Bank. Non-project loans are the general term of the loans used either for member nations to overcome natural disasters, or for member nations to import raw materials that are in short supply domestically, or for importing advanced equipment. They are also known as emergency loans. Sector loans consist of sector investment and maintenance loans, sector adjustment loans, and intermediary loans, where the loans of the first type are mainly used for improving sector policies and investment focus in order to strengthen the member nation's capability to make and implement investment plans. Sector adjustment loans are used for comprehensive policy and system reforms of a specific sector. Intermediary loans means that the World Bank first lends to an intermediary institution of the borrowing nation, then the intermediary institution lends the received funds to the subprojects of the nation. Syndicated loans stand for such loans giving out jointly by the Works Bank and other financial institutions outside the borrowing nation. The funds are mainly from governmental aids, export credit agencies, and private financial institutions. Third window loans were loans created in 1975 in between the ordinary loans and the concessional loans issued by the International Development Association.

Secondly, let us look at the International Development Association (IDA). This entity was officially established on September 24, 1960, and started its business in November of the same year. It is headquartered in Washington DC. According to the provisions, each member nation of the World Bank can join the IDA, but is not necessary.

The mission of the IDA is to provide concessional loans to low income nations with longer terms than loans of the World Bank in order to help these nations develop their economy and improve their citizens' life quality. Although the organizational structure of the IDA is identical to that of the World Bank, they are completely independent legally and financially. The council is the highest authority of the IDA, and the executive board is responsible for the daily business operation. The chair and associate chair of the council and the executive board, and the heads of other subordinate agencies are all the same people as those of the World Bank. The funds of the IDA consist of the subscribed shares of member nations, supplementary funds provided by the member nations, the appropriation of the World Bank from its net business income, and the net earing of the IDA itself. In particular, the original authorized capital of the IDA was in the amount of US$1 billion. As the number of member nations increased, the total amount also increased accordingly. As of June 1995, the total capital as subscribed by the member nations reached US$92.891. Because the amount of the share fees paid by member nations is limited, while the constitution of the IDA does not allow the IDA to raise funds by issuing bonds in the international financial market, the IDA needs its member nations to provide supplementary funds. Since 1964, the World Bank has appropriated a part of its annual income to the IDA as its funds for lending. Because the loans of the IDA are very favorable, from its business operations, the IDA produces very little income.

The lending business of the IDA focuses on providing long-term concessional loans to low income developing nations. According to the criteria of 1999, only member nations with per capita GNP below US$925 were qualified to obtain loans, which included 81 nations. The lending targets are member nations' governments. The borrowed funds are mainly used for the development of energy, transportation, water conservation, port construction, and other public facilities, and for areas of agriculture, education, and family planning, etc. The term of the loans provided by the IDA is 50 years, where no principal needs to be repaid within the first 10 years. Starting in the second ten years, repay 1% each year; and for the rest 30 years, repay 3% each year. The repayments of the loan can be made by using the member nation's currency either totally or partially. The loans do not bear interest except an annual fee of 0.75% on the disbursed amounts.

Thirdly, let us look at the International Finance Corporation (IFC). The IFC was created in July 1956 and is headquartered in Washington DC. As of 1999, it had 174 member nations. The mission of the IFC includes:

(i) Provide various investments to private enterprises in developing nations without government guarantee in order to promote the economic development of the member nations; and

(ii) Promote foreign private capital to be invested in developing nations in order to stimulate the development of capital markets in developing nations.

The IFC is an affiliate of the World Bank. Its method of management and organizational structure are identical to those of the World Bank. However, it has its own legal and business personnel. The chair positions of the council and the executive board of the IFC are served by those from the World Bank, while the president and vice president of the World Bank serve as the manager and assistant manager of the IFC. According to the articles of agreement of the IFC, only member nations of the World Bank can become members of the IFC. However, not all member nations of the World Bank join the IFC.

The funds of the IFC consist of membership share fees, borrowed funds, net income of business. Specifically, at the time when the IFC was initially established, the initial membership share fees were US$0.1 billion, divided into 10000 shares, making it US$1,000 a share. After several rounds of fee increases, a more recent statutory membership fees have reached US$2.35 billion. The IFC raise its funds mainly through taking out loans from developed nations or the World Bank, and issuing bonds in the international capital markets. Borrowed funds have become the largest source of funds for the IFC. Other than donating to some projects, all of the IFC's net business income is used as its own funds. The main targets of the IFC's lending are private companies in under-developed member nations without requiring the governments of the member nations to provide any guarantee. The subsidized sectors of the IFC's loans are mainly those of manufacturing, processing, and mining. The amount of each loan is between US$2–US$4 million with the term of 7–15 years. All repayments need to be done by using the original loan currency. The interest rate of the loan is determined based on who the borrower is, and is generally higher than the interest rate of loans from the World Bank. For the undrawn portion of the loan, a 1% of commitment fee is charged annually. When the IFC provides a loan, it generally uses the form of joint investment with some private investors, commercial banks or other financial institutions. This kind of investment not only expands the business range of the IFC, but also encourages private capital to be invested in developing nations.

8.5.3 Bankruptcy of nations: An ongoing thought

During the Argentine crisis that started in 2000, some people believed that Argentina, as a nation, had gone bankrupt. Perhaps since that time, the idea for a nation to go bankrupt has been in the making. At the annual meeting of the IMF and the World Bank, which started on September 27, 2002, this idea was proposed openly. After negotiations, the members of the IMF policy committee, who were finance ministers of different nations, demanded the IMF organization to adopt a brand new method to declare bankruptcy for a nation in order to resolve the debt crises that existed from Asia to South America. Some exports of the international community of finance identified this bankruptcy plan for nations as shocking as the China trip President Nixon announced some years ago. If this plan could be implemented, it would lead to revolutionary changes in the global financial system.

The report of the Stand and Poor, one of the well-known credit-rating companies in the world, published on September 25, 2002, pointed out that because of the global economic slump, more and more nations could not make their debt payments on time. This conclusion was made based on a study on the debt situations of over 70 different countries and regions. The report pointed out that in the first three seasons of the year 2002 there were six countries from around the world – Argentina, Gabon, Indonesia, Madagascar, Moldova, and Nauru – that could not afford to repay their maturing debts on time. That made the total number of nations which could not afford to repay their debts reach 28. The Standard and Poor believed that the reason why the global rate of non-performing national debts showed a rising tendency was due to three main reasons:

1 After 9.11, the global economic sump had eroded the ability of debt service of more nations;
2 Some developing nations were experiencing two domestic pressures – political and economic – so that they could not afford to consider the issue of repaying debts; and
3 The international effort of reducing debts and restructuring debts could no longer save the entire deteriorating situation.

It was commonly acknowledged in the world community of economists that the plan of bankruptcy for nations was the greatest achievement of the 2002 annual conference of the IMF and the World Bank. However, at the same time of recognizing the good intention of this idea, people also realized the difficulty of actually implementing this plan. Firstly, a business entity can be liquidated, while a nation cannot. So, in terms of nations, the idea of bankruptcy is only a soft constraint. Secondly, if the IMF plays the role of a regulator for the bankruptcy of a nation, it has to be further involved in sovereignty of the bankrupt nation. Thirdly, doing so will slow and weaken the international capital circulation. Additionally, the implementing the bankruptcy plan will also dependent on whether or not the United States and other Western nations would alter the old international economic order. If the international economic order could not be truly altered, it would be very difficult for developing nations to acquire good space and opportunity of development, and the global debt problem would not be resolved completely. In terms of the IMF, how to design a bankruptcy plan for nations that is readily acceptable, that can well coordinate the relationship between creditor nations and debtor nations, and that can effectively stimulate the reform of the international financial system, would be a far-reaching mission to accomplish.

8.6   THE THREE-RINGED STRUCTURE OF STABLE ECONOMIC SYSTEMS

In this section, we study the systemic relationship between a domestic economy and its international trades with other national and regional economics. In other words, we address such questions as: why is it necessary for an economy to interact with other economies? Can an economy stay closed from other economic systems and still maintain its viability?

First of all, each solvent nation is a relatively independent political system as well as a relatively independent economic system, which interacts with other national and regional economic systems. Based on the systemic yoyo model of the universe, as presented in Chapter 2, the economic totality of the world can be seen as a huge ocean of "fluid", which is made up of commercial goods, information, knowledge, money, etc. In this "ocean", each regional pool of rotational field stands for a temporally stable economy. So, when the communication of information and knowledge is not advanced and even hindered by natural barriers, then an economy could stay relatively closed for a period of time from other economic systems while still maintaining its viability. It is because the intensity of clashes between different economic systems is not as strong and severe as when the communication is more timely and speedy.

As the technology of communication advances, to maintain its sustainability, each rotational pool in the economic "ocean" has no choice but interact with other economic systems (spinning fields) in order to acquire additional strength to stay relative stable. In other words, without constantly learning and strengthening itself by borrowing useful elements, such as new innovations, scientific breakthroughs, etc., from other economic systems, a once prosperous economy will soon become unsustainable and subject to deadly attacks from other vigorous economies. This end explains why it is necessary for any economy to interact with other economic entities. For an expanded and detailed discussion along this line, please consult (Lin and Forrest, 2011).

In the rest of this section, we will address the previous questions from the angle of the internal structure of each and every viable economic system.

8.6.1   Three-ringed systemic structure of each economic system

Because information is not distributed evenly across the land, international trades take place. In the form of importing and exporting commercial goods and exchange of knowledge, the cross-regional trades create new business opportunities and give

their expressions in the existence of both of the relative structural stability and instability to economic development. At the same time, we do see stable existence and
instable evolution in existing economies. More specifically, we see the existence of
three-ringed stabilities. For instance, the outer circulation of international trades, the
second-level circulation of domestic economic activities, and the third-leveled ring of
various regional economic activities can be stably observed with each independent
economic system, where the second-level circulation is concentrated with the main
economic intensity and energy of the particular economy. This observation is analogous to the three-leveled cosmic galaxy systems, star systems, and planetary systems,
where the star systems are the second-level circulations concentrated with the main
materials and energies of the universe. It is also analogous to the microscopic scales of
materials consisting of the molecular systems, atomic systems, and electronic systems,
where the atomic scale systems, as the microscopic second-level circulations, also possess high concentrations of energy with the theory of nuclear energy proven. These
observations suggest that the quasi-stable existence of economic systems needs at least
three levels of circulations with the middle level circulations holding huge amounts of
energy. In particular, the national level of economic systems, the star level of cosmic
systems, and the atomic level of the microscopic system represent high concentrations
of economic or nuclear energy.

If we expand the afore-mentioned observation to the scale of human activities or
the so-called meso-scale systems, we can observe the three-level circulation system of
frigid zones, temperate zones, and torrid zones. Here, the temperate zones contain the
major amounts of energy. Additionally, each typhoon (or hurricane) is also a three-
leveled circulation system, consisting of the typhoon eye, the region of torrential rains,
and high-speed winds, and the outer-ring-shaped region of subtropical high pressures.
Once again, it is also the second-level circulation area that holds high concentrations
of energy.

Based on the afore-mentioned observations, the inspiration of the stable mechanic
device for three-ringed energy transformation, invented during Bei Wei period of China
(386–534 A.C.), and the observation that each natural river is always accompanied by
lakes of various sizes to store floodwater, it is natural for us to see that three-leveled
circulations, shown in each energy transformation, have played the role of coordinating and/or restraining the energy transformation for the underlying stable existence,
the destructive nature of non-three-leveled circulations to instable evolutions, and the
dynamic equilibrium that without eddy motions there will be no kinetic energy transformation and the amount of energy determines the internal heat of the eddy motions
(OuYang, 1998).

8.6.2 Economic energy and three-leveled energy transformation

As we have seen in Chapter 4, in the study of economics, such elementary concepts as
supply and demand really stand for the acts of economic forces. And, from physics it
follows that with forces there are different forms of energy, especially kinetic energy,
which is expressed as the square of speed. Symbolically, we have

$$e = \tfrac{1}{2}mv^2 \text{ (Newton)} \tag{8.1}$$

and

$$E = mc^2 \text{ (Einstein).}$$ (8.2)

where m is the mass of the object of concern, v the speed at which the object is moving, and c the speed of light.

With this concept of kinetic energy in terms of squared linear speeds, known as irrotational kinetic energy, being introduced, the law of conservation of kinetic energy is consequently established in modern science. At this junction, a natural question arises: Since the square of the linear speed constitutes the kinetic energy, can the square of the angular speed of a rotation make up of a different energy? (For more on models of rotation, see (Lin, 2007)). In order to understand how economies work, in this section all the general materials considered are seen as systems with internal structure. That is, we treat materials as non-particles.

Because squared linear speeds represent kinetic energy, squared angular speeds should also stand for kinetic energy. Squared (linear) speed (v^2) and squared angular speed (ω^2) stand for different concepts so that they should play different roles in the evolution of systems. The unit for speed is (m/s), while that for angular speed is (1/s), indicating unevenness in changes and distributions of the movement. In terms of quantities, angular speed measures the rotation of materials, while linear speed is the measurement of straight-line distance traveled by the object within the unit time. That is, we have

$$\vec{\omega} = \vec{i}\omega_x + \vec{j}\omega_y + \vec{k}\omega_z.$$ (8.3)

where the symbol $\rightarrow$ represents vector, $\vec{i}, \vec{j}$, and $\vec{k}$ are respectively the unit directional vectors of the x-, y- and z-axis, and $\vec{\omega}$ a 3-dimensional rotation with ω_m being its m-component along the m-axis, for $m = x, y, z$.

Due to the unevenness of the distribution of the movement, the angular speed $\vec{V}$ can be written as follows:

$$\vec{V} = \begin{vmatrix} \vec{i} & \vec{j} & \vec{k} \\ \dfrac{\partial}{\partial x} & \dfrac{\partial}{\partial y} & \dfrac{\partial}{\partial z} \\ u & v & w \end{vmatrix},$$ (8.4)

where u, v, and w respectively stand for the x-, y-, and z-components of the vector $\vec{V}$. For the horizontal two-dimensional plane, the angular speed in the vertical direction is

$$\omega_z = \frac{\partial v}{\partial x} - \frac{\partial u}{\partial y}.$$ (8.5)

Introducing the flow function ψ gives

$$v = \frac{\partial \psi}{\partial x} \quad \text{and} \quad u = -\frac{\partial \psi}{\partial y}.$$

Similar to the traditional method of quantitative analysis, assume that ψ is twice continuously differentiable and can be represented as simple harmonic disturbances. Looking at the whole, let us introduce the combined disturbance of various scales:

$$\psi = \sum_n \psi_n.$$

where ψ_n stands for the disturbance at the scale level n. Then, we have

$$\nabla^2 \psi_n = -\mu_n^2 \psi_n. \tag{8.6}$$

Let us consider the horizontal problem, and take $V^2 = (\nabla \psi)^2$. From equs. (8.5) and (8.6) and the theorem of orthogonal plane divergences (OuYang, McNeil and Lin, 2002), it follows that

$$\oiint_\sigma V_n^2 \, d\sigma = \oiint_\sigma \left(\sum_n (\nabla \psi_n)^2 \right) d\sigma = \oiint_\sigma \left(\sum_n \nabla \psi_n \cdot \nabla \psi_n \right) d\sigma$$

$$= \oiint_\sigma \left(\sum_n \nabla(\psi_n \nabla \psi_n) \right) d\sigma - \oiint_\sigma \left(\sum_n \psi_n \nabla^2 \psi_n \right) d\sigma$$

$$= \oiint_\sigma \left(\sum_n \mu_n^2 \psi_n^2 \right) d\sigma, \tag{8.7}$$

and

$$\oiint_\sigma \omega_z^2 \, d\sigma = \oiint_\sigma \left(\sum_n (\nabla^2 \psi_n)^2 \right) d\sigma = \oiint_\sigma \left(\sum_n (\mu_n^2 \psi_n)^2 \right) d\sigma$$

$$= \oiint_\sigma \left(\sum_n \mu_n^2 \cdot (\mu_n^2 \psi_n^2) \right) d\sigma.$$

By substituting equ. (8.7) into this last equation produces

$$\oiint_\sigma \omega_z^2 \, d\sigma = \oiint_\sigma \left(\sum_n \mu_n^2 V_n^2 \right) d\sigma. \tag{8.8}$$

By comparing equs. (8.7) and (8.8), it can be seen that the closed line integral of the squared rotational angular speed contains the traditional closed line integral of the squared speed. That implies that even if we use the method of quantitative analysis, the concepts of linear speed and angular speed have different physical meanings. What is very interesting is that equs. (8.7) and (8.8) reveal a major negligence of the traditional conservation of kinetic energy or that of total energy. In particular, a conservation of stirring energy not only contains the conservation of the linear speed kinetic energy, but also shows how the kinetic energy transforms and transfers. For more in depth discussion on stirring energy and its conservation, please consult Chapter 7 of (Lin, 2009).

8.6.3 Non-conservative evolution of economic energy

Because of the rotational nature of economic systems, let us treat the stirring energy of an economy as the system's economic energy. For a given three-ringed economic circulation, if we look at the second level circulation, from the conservation of stirring energy in equ. (8.8) it follows that

$$\mu_1^2 v_1^2 + \mu_2^2 v_2^2 = c_1 = \text{const.} \tag{8.9}$$

From equ. (8.7), it follows that

$$v_1^2 + v_2^2 = c_2 = \text{const.} \tag{8.10}$$

Let Δv^2 stand for the changes in the linear speed kinetic energy from one time moment to the next neighboring moment, then the conservation of kinetic energy implies

$$\begin{cases} \mu_1^2 \Delta v_1^2 + \mu_2^2 \Delta v_2^2 = 0 \\ \Delta v_1^2 + \Delta v_2^2 = 0. \end{cases} \tag{8.11}$$

By eliminating Δv_1^2 and Δv_2^2 respectively in equ. (8.11), we obtain

$$(\mu_2^2 - \mu_1^2)\Delta v_2^2 = 0 \quad \text{and} \quad (\mu_1^2 - \mu_2^2)\Delta v_1^2 = 0. \tag{8.12}$$

Since $\mu_1 \neq \mu_2$, equ. (8.12) implies that it must be $\Delta v_1^2 = \Delta v_2^2 = 0$. This end implies that with second level circulations only (that is, if we assume that there is only national level domestic circulations of economy), no energy transformation and transfer can be carried out, causing blockage or high concentration of energies and consequently instable economic, and in turn political evolution of the nation has to take place. From this explanation, one can see the reason why each viable, prosperous economic system has to have well developed domestic markets and international markets to dredge locally blocked accumulation of energy either inwardly or outwardly.

If we introduce a third leveled circulation, from equs. (8.7) and (8.8) we have

$$\begin{cases} v_1^2 + v_2^2 + v_3^2 = c_1 = \text{const.} \\ \mu_1^2 v_1^2 + \mu_2^2 v_2^2 + \mu_3^2 v_3^2 = c_2 = \text{const.,} \end{cases} \tag{8.13}$$

where c_1 and c_2 are constants. Let Δv^2 represent the changes in the speed kinetic energy from one time moment to the next neighboring moment. Then to satisfy the law of conservation we must have

$$\begin{cases} \mu_1^2 \Delta v_1^2 + \mu_2^2 \Delta v_2^2 + \mu_3^2 \Delta v_3^2 = 0 \\ \Delta v_1^2 + \Delta v_2^2 + \Delta v_3^2 = 0. \end{cases} \tag{8.14}$$

By eliminating Δv_1^2, Δv_2^2, and Δv_3^2 respectively in equ. (8.14), we obtain

$$\begin{cases} (\mu_1^2 - \mu_2^2)\Delta v_2^2 + (\mu_1^2 - \mu_3^2)\Delta v_3^2 = \mu_1^2 c_1 - c_2 = \text{const.} \\ (\mu_2^2 - \mu_1^2)\Delta v_1^2 + (\mu_2^2 - \mu_3^2)\Delta v_3^2 = \mu_2^2 c_1 - c_2 = \text{const.} \\ (\mu_1^2 - \mu_3^2)\Delta v_1^2 + (\mu_2^2 - \mu_3^2)\Delta v_3^2 = c_2 - \mu_3^2 c_1 = \text{const.} \end{cases} \qquad (8.15)$$

Assume $\mu_1 > \mu_2 > \mu_3$ (the same results follow, if we let $\mu_1 < \mu_2 < \mu_3$). Then, the coefficients on the left-hand sides of the first and the third equations in equ. (8.15) are positive and those on the left-hand side of the second equation are negative. So, we have

1   Both equs. (8.11) and (8.14) indicate that the conservation of the linear speed kinetic energy cannot limit the dissemination of energies, because outside the linear speed kinetic energy, there is still the spread of the stirring energy. In other words, artificially keeping an economy closed from the rest of other economic systems will make the economic instable.
2   When both v_1^2 and v_3^2 decrease, v_2^2 will increase. Conversely, when v_1^2 and v_3^2 increase, v_2^2 will decrease. That is, each three-ringed circulation can complete its energy transfer and transformation through the second level circulation with such a process that is clearly shown. In other words, only when a national economic system possesses a well-developed, three-ringed economic circulations, consisting of domestic markets, national coordination, and international trades, its economic robustness and prosperity will permeate all corners within the system.
3   If $\mu_2^2 - \mu_3^2 < 0$, when v_1^2 increases, so does the corresponding v_3^2. Conversely, when v_1^2 decreases, v_3^2 also decreases. In other words, in this case, the scales of economic energies of the first and third levels of the national economy changes synchronously.

This analysis indicates that each transformation or transfer of economic energy is carried out and completed through the second level circulations, and between the first level and third level circulations, there does not exist any direct energy transformation or transfer. That is, the previous discussion explains not only that artificially closing up an viable economy will naturally lead to national instability, but also that the existent instable economic energies constitute the mechanism of transfer of realistic existing economic processes. This end reveals the theory for the evolutionary processes of economies.

8.7 SOME FINAL WORDS

The quasi-three-ringed stability principle of the conservation of economic energies, which can be seen as a realization of the concept of multi-level systems (Lin, 1989; 1990), reveals the procedural aspects that all economic activities have their respective start and end, and the transformation of energies. It not only contains the quantitative laws of economics but also reflects the laws of naturally existing economic processes.

In terms of predictions of economic developments, this principle involves how instable economic energies and the corresponding irregular economic events should be understood: Irregular information comes from irregular events and has to be reflected in instable economic energies. So, in order to make valid economic predictions one has to make use of irregular information (OuYang and Chen, 2006).

Corresponding to the study of economic systems, there should be relevant laws in terms of economic energies. In particular, in terms of the study of problems of economic development, laws on processes have to be established. Because at the same time when economic energies reveal the transformation process of economic prosperities they also describe the effects of economic instabilities, the significance of the concept of economic energies is shown vividly in terms of the underlying physical mechanism.

Chapter 9

International reserves and capital flows

In the name of international reserves and capital flow across national borders, in this chapter we will study how national economies affect and influence each other. Here, international reserves represent one of the central problems facing the reforms of the international monetary system since WWII. On the one hand, it touches on each nation's ability to adjust its international payments and to stabilize its exchange rates; and on the other hand, it also deeply affects the development of the world prices and international trades. As the interdependence of individual national economies on each other gradually strengthens, international repayment behaviors of economic trades become increasingly more important day by day. The quantity, structure, and management of international reserves can directly affect the adjustment of each nation's international payments, stability of exchange rates, and prevention of monetary crises, etc. At the same time, the flow of international capital represents the worldwide movement and transferring process of capitals across national borders. Along with the development of economic globalization, the international flow of capital has becoming the most active factor of the world economy. However, at the same time when the international flow of capital continuously promotes the growth of the world economy, it also provides the ground for wide-ranging financial crises to occur. Since the 1990s, because of the increasing occurrence frequency of major, large-scale financial crises, the international flow of capital has caught people's attention.

This chapter is organized as follows. Section 9.1 looks at the characteristics and components of international reserves and borrowed international reserves. Section 9.2 presents how international reserves are managed in terms of magnitudes and structures. As case studies, we will closely exam the reserves of the USA, Europe, Japan, and Great Britain. Section 9.3 considers such potential problems regarding foreign currency reserves as capital flight, credibility, and currency substitution. Capital flows across national borders are presented in Section 9.4, where what's given includes types of international capital flows, causes, present state and expected future of capital flows, impacts of international capital flows, and the relationship between international capital flows and individual national financial systems. The final Section 9.5 looks at the interaction of individual national economic systems from the angle of the systemic yoyo model by showing that small economies do not have much choice in dealing with large economies; if conditions are right, these small economies could potentially chain up together to form a heavy weight economic player in the international arena.

In terms of references, some parts of this chapter are based on (Jiang, 2008; Johnson and Echeverria, 1999; Wang and Hu, 2005).

9.1 INTRODUCTION TO INTERNATIONAL RESERVES

By international reserves, it means such assets that are commonly accepted in the world for a national government to hold in order to meet the needs of making up the balance of international payment deficit, stabilizing exchange rates, and dealing with emergent costs. As for the definition of international reserves, although there have been disagreements in the community of economists, international reserves are divided into special and general reserves. By special international reserves, it means those freely convertible assets the monetary authority of a nation holds for the purposes of making up the international payment deficit and maintaining the stability of exchange rates. By general international reserves, it means all such liquid assets, including both the official and non-official reserves, that the monetary authority of a nation can use to make up the balance of its international payment deficit and to stabilize its exchange rates. After the mid-1960s, the definition of international reserves has been converging and focused more on special international reserves. For example, in 1965 the Group of Ten provided the following definition of international reserves: International reserves stand for all such assets that are held by the monetary authority of a nation that can be used to either directly support the nation's exchange rates or indirectly by using guaranteed conversions into other assets. Based on this explanation, international reserves must be those reserve assets that are materially, actually, and effectively controlled by the central monetary authority, and include gold, foreign currencies, special drawing rights, the reserve position and the credit line at the international monetary fund, and existing non-residential debts.

9.1.1 Characteristics of international reserves

In general, an international reserve should possess the following characteristics: widespread acceptance, sufficient liquidity, free convertibility, and official holding. In particular, an international reserve asset should be such an asset that is factually recognized, accepted, and employed by all nations. If a financial asset is merely recognized, accepted, and employed within a small region, it cannot be seen as an international reserve asset even though it might enjoy the freedom of convertibility and the sufficiency of liquidity. Each international reserve asset must possess sufficient liquidity so that national governments and monetary authorities can acquire it unconditionally and can mobilize it when necessary. As an international reserve, the asset has to be freely convertible into other forms of financial assets, reflecting its internationality as a reserve asset. Without the needed freedom of convertibility, the reserve asset will not be commonly accepted in the international world so that it cannot play its role of making up the balance of international payment deficit. At the same time, international reserve assets must be directly controlled and can be mobilized by the central monetary authorities. The gold, foreign currencies, and other assets held by non-governmental financial institutions, business entities, and private individuals are not considered as international reserves. This characteristic separates international reserves from international liquidity. It is because of this reason that international reserves are also known as official reserves.

International reserves and international liquidity are two concepts that are different but also related. By international liquidity, it means a nation's capability to repay international debts without employing any adjustment measure. Other than the international reserves in various forms, as held by the nation's monetary authority, it also includes the nation's borrowing ability abroad, which is the ability to borrow from foreign governments or central banks, international financial organizations, and international commercial banks. Hence, the range of measure of international liquidity is wider than that of international reserves, where international reserves are held by an individual nation, which represent the actual ability to make international payments. That is why international reserves are also known as the ability of making unconditional international payments. They are an indicator of a nation's financial strength. On the other hand, the international liquidity stands for the sum of the nation's actual ability of making international payments and the nation's potential and conditional ability to making international payments. It reflects not only the financial strength of the nation, but also, to a great extent, the economic status and financial credit status of the nation.

9.1.2 Components of international reserves

At different historical times, the particular kinds of assets used as international reserves are different. For example, before WWII, the international reserves of all nations were made up of gold and foreign currencies that could be converted into gold. After WWII, the International Monetary Fund (IMF) has provided at different times two kinds of assets to its member nations to supplement their international reserves. Therefore, the international reserves of a member nation of the IMF generally contain the following four kinds: gold reserves, foreign exchange reserves, the reserve position at the IMF, and special drawing rights (SDRs). The composites of these different kinds of reserves are shown in Table 9.1.

By gold reserve, it means the monetary gold held by a nation's government. It does not include the gold to be used beyond the monetary purpose, no matter who owns the gold. Within the international gold standard system and the Bretton Woods system, gold was the most important international reserve asset for each nation, where gold performed the functionality of the world currency and was the final method of clearance of international trades and payments. The second amendment of the Agreement of the

Table 9.1 The composite of international reserve assets (unit: %), based on the annual information issued by the IMF.

Year	1975	1980	1985	1990	1998
Foreign exchange	70.7	82.6	79.4	88.1	91.1
Reserve position at IMF	6.5	4.8	8.9	3.8	4.7
SDRs	4.5	3.3	4.2	3.1	1.6
Gold	18.3	9.3	7.6	5.0	2.6

International Monetary Fund, which became effective on April 1, 1978, proposed the non-monetization policy of gold. That is, all monetary authorities sold their holdings of gold so that all monetary gold was converted to gold to be used for non-monetary purposes. Because gold is, since then, detached from the international monetary system and each national currency, it is no longer the base of any currency system and can no longer be used to clear international trades between national governments.

When national monetary authorities mobilize their gold reserves, they cannot make payment by directly using the actual gold. Instead, they have to trade their gold into a convertible currency by selling the gold in the gold market. That makes the application of gold reserves indirect and inconvenient. Additionally, gold reserves do not bear any interest and carry high storage costs. However, gold possesses the safety and value keeping characteristics that other forms of reserves do not have. That explains why gold still represents an indispensable component in each nation's international reserves. There are three ways for a nation to calculate its gold reserves:

(i)   Use weight in ounces to do the computation;
(ii) Use the price of gold to compute, where one ounce gold = 35 SDRs; and
(iii) Use the market price of gold to calculate.

Since the 1990s, the amounts of monetary gold held by nations from across the world have not changed much. Currently, the gold reserves owned by various central banks and international monetary organizations amount to about 1 billion ounces. That is less than 5% of the total of international reserves. Of the total of gold reserves developing nations own about 17%, while developed nations about 83%.

Foreign exchange reserves are maintained by the monetary authorities in the forms of bank deposits, treasury stocks, long- and short-term government bonds, and other credits that can be used to make up the balance in international payment deficit. In other words, the main forms of foreign exchanges consist of deposits in foreign banks and the bonds of foreign governments. Different from the gold reserves, after WWII, the proportion of foreign exchanges in each nation's international reserves has grown increasingly, becoming the main body of the current international reserves. This end is reflected in two aspects: 1) the weight occupied by foreign exchanges in the international reserves is far more than that of any other form of reserves. For example, in 1999 the proportion of foreign exchanges within the total international reserves of the member nations of the IMF was 91%. 2) When making international payments or intervening exchange rates, foreign exchange reserves are often applied, while gold reserves are rarely touched.

Foreign exchange reserves consist of various currencies that can be used as reserve currencies. Only those currencies that can be exchanged freely and commonly accepted by all nations with relatively stable values can act as reserve currencies. Before WWI, British pound was the most important reserve currency. After WWII, because the U.S. dollar was the only currency that could be converted into gold, occupying an equivalence relation with gold, it became the most important reserve currency in each nation's foreign currency reserves. Since the late part of the 1960s, the economic strength of the United States has been continuously declining with increasingly deteriorating balance of international payments, the gold reserve and domestic capital have been outflowing in large quantities, and U.S. dollar crises occurred frequently. That severely damaged the position of U.S. dollar in international reserves. In 1973, along with the collapse of the Bretton Woods system and the adoption of floating exchange rate systems, fluctuations in the exchange rates of U.S. dollar were further enlarged so that the risk of holding U.S. dollar was greatly increased. At the same time, with the rapid development of German and Japanese economies, the international status of German mark and Japanese yen rose quickly. In order to avoid losses of foreign currency reserves due to fluctuations in the exchange rates of U.S. dollar, some nations adjusted their currency composites of their foreign exchange reserves by holding more German marks, Japanese yens, Swiss and French francs, Dutch guilders, and other currencies, while reducing the amount of U.S. dollars. The IMF's non-monetization of gold in 1976 also introduced SDRs as a form of reserve asset. Since the end of the 1970s, the European Currency Unit (ECU), as a basket of currencies, had become a reserve currency. Starting in January 1, 1999, Euro replaced ECU as a new reserve currency. All these factors led to the appearance of the current diversified international reserve system. The diversification of the reserve currencies is mainly caused by the uneven economic development of different nations from across the world, reflecting the relative position changes of national economies. Additionally, under the floating exchange rate system, changes in the exchanges rates of the major currencies, especially the downward float of the exchange rates of the U.S. dollar, also promoted the diversification of foreign currency reserves.

On April 5, 2005, the IMF published the semiannual Global Financial Stability Report, which revealed that the world foreign exchange reserves had reached the historical level of US$1600 billion, where China owned over US$610 billion, India and Russia respectively owned over US$100 billion. Presently, because the U.S. dollar has devalued considerably, many private entities from across the world have reduced their holdings of U.S. dollars. However, because governmental institutions have not reduced their holdings of assets priced in U.S. dollar, it has significantly helped to stabilize the American capital market and to support the growth of American economy. According to relevant statistics, the proportion of U.S. dollar in the foreign exchange reserves of the world has increased to 64% in 2004 from 56% in 1997. So, the U.S. dollar still occupies an important position in the international monetary system. On the other hand, since the time when euro was born, the central banks of various nations, including those of emerging market economies, have increased their reserves of euros. That played an important role in terms of weakening the strong position of the single U.S. dollar, and enhancing the balance of international financial structure. As of recently, the weight of euro in the foreign exchange reserves of the world has risen year after year reaching over 20%. Although it is still far behind the weight of the U.S. dollar, it has been far ahead of all other currencies. The economic strength and the magnitude of international trades of the Eurozone are roughly the same as those of the United States. The size, depth, and liquidity of the euro bond market are also similar to those of the American bond market. All these comparabilities have provided powerful support to the euro's rising status in international trades, clearance of payments, and reserve currencies.

The reserve position at the IMF is also known as the general drawing rights. That represents the funds a member nation has in its general capital account with the IMF that are available for the member nation to withdraw and use. When a nation joins the IMF, it is required to pay a certain amount of fees, known as the nation's share.

Table 9.2 The reserve status of a member nation in the IMF and possible borrowed reserves.

Reserves status of a member nation in the IMF	Convertible currency balance provided to IMF		The general drawing right of member nations in IMF.
	1st tranche (reserve portion) drawing right:	25% of share	
The possibly borrowed reserves from the IMF by member nations	2nd tranche (credit) drawing right:	25% of share	The conditions of drawing becomes stricter downward
	3rd tranche (credit) drawing right:	25% of share	
	4th tranche (credit) drawing right:	25% of share	
	5th tranche (credit) drawing right:	25% of share	

Twenty-five percent of the subscribed share must be paid using foreign currencies or special drawing rights, while the rest 75% using domestic currency of the nation. When a member nation experiences difficulties in its international payments, it has the right to apply to the IMF for withdrawing convertible currencies by using the domestic currency as the collateral. The withdrawal amounts are divided into five tranches, each of which is equal to 25% of the nation's subscribed share. Because the first tranche is exactly equal to the amount of convertible currencies the member nation paid in the first place, the condition of withdrawing is most loose. As long as an application is submitted, the money is available for use. That is the reason why this tranche of the drawing right is also known as the reserve tranche drawing right. The relevant conditions of all possible reserves to be borrowed from the IMF are listed in Table 9.2.

Since the 1990s, the total of the reserve positions of the whole world in the IMF has been increasing drastically. Currently, the sum of the reserve positions of all member nations of the IMF has surpassed US$60 billion. As for the world distribution of the reserve positions, it has been extremely uneven, where the total of the reserve positions of the industrialized nations has occupied over 86% of the world total, while that of developing nations merely 14%.

The so-called special drawing right stands for a credit asset the IMF provides to its member nations without charge based on the magnitudes of the members' subscribed shares. It represents another right beyond the general drawing right to use funds. The IMF established the special drawing rights (SDRs) in 1969; as long as a nation is a member of the IMF, it can unconditionally participate in the allocation of the SDRs. The current magnitude of the SDRs of the whole world is maintained at the level of about US$22 billion. Because the issuance quantity is limited, the proportion the SDRs occupy in the international reserves is small. Additionally, the allocation of the SDRs throughout the world is extremely uneven with industrialized nations taking more than 3/4 of the total amount of SRDs issued while developing nations only less than 1/4.

The special drawing right has the following characteristics: (i) It is an account unit without any material foundation. It cannot be directly applied to make international payment of international trade. It cannot be directly converted into gold. (ii) It can only be held by the monetary authorities of the member nations of the IMF. It can only be used between the monetary authorities of member nations and the IMF (and the international clearance bank). Non-governmental financial institutions cannot hold and use the SDRs. (iii) Member nations can unconditionally enjoy their allocations of the SDRs without the need of repayment, while the general drawing rights must be repaid back to the IMF within specified time frames.

When the IMF distributes the SDRs to its member nations, it records the allocated quantity each member nation receives onto the nation's account of the SDRs. When a member nation suffers from difficulties in its international payments and needs to mobilize its SDRs, the IMF will coordinate according to the relevant regulations a nation, which generally is such a nation that has a strong surplus in its international trades, to accept the SDRs. Let us look at an example of two nations A and B, assuming that both A and B have received 1 billion SDRs respectively. When nation A experiences a deficit in its international payments and needs to apply 0.2 billion SDRs, and when nation B is assigned to take the SDRs, in the SRDs account of nation A there will be record of debt entry of 0.2 billion SRDs while in the SRDs account of nation B there will be a record of credit of 0.2 billion SRDs. At the same time, the central bank of nation B transfers the convertible currency of the equivalent value into the central bank of nation A so that the central bank of nation A can apply these newly arrived funds in the form of a convertible currency to balance its international payments.

At the start, the method of pricing the SDRs is US$1 = q SDR or 35 SDRs are equal to one ounce of gold. Because of the decoupling of gold and the international monetary system in the 1970s and the recent volatility of the U.S. dollar, the IMF introduced a weighted mean method to determine the value of a SDR. At the current moment, the IMF has chosen four major nations or regions of active international trades (the United States, Japan, Great Britain, and Eurozone) with their individual weights equal to their respective percentages in the overall international trades from across the world. These weights have been adjusted once each five years. As of February 2001, the particular weights were 45% for the U.S. dollar, 29% for the euro, 15% for Japanese yen, and 11% for British pound. To determine the value of one SDR on a specific day, one only needs to multiply the exchange rates of the four currencies against the U.S. dollar of the day or from the day before from the London foreign exchange market. Through the market exchange rate, the value of one SRD can be calculated in terms of any other currency. In particular, the value of one SRD is computed as follows: The value of one SRD in the U.S. dollar = $1 \times 45\%$ + the exchange rate of dollar/euro $\times 29\%$ + the exchange rate of dollar/yen $\times 15\%$ + the exchange rate of dollar/pound $\times 11\%$.

From the afore-mentioned method of calculating the value of one SDR, it follows that the value of the SDR is relatively stable. It is because the effect of the fluctuation in each currency is minimized through the weighted mean smoothing. Additionally, these four currencies represent the most important currencies used from around the world. When the exchange rate of one of these currencies falls, the exchange rate of another one or more of the currencies will likely to go up. These opposite directional movements in the exchange rates can help to offset their combined effect on the value of the SDR, making the value of the SDR relatively stable. Because of the value of the SDR is calculated by weighted averaging of these four major currencies, the interest rates of the corresponding assets in the SDRs are also obtainable by using the weighted averaging of market interest rates of these four currencies.

Within the scope of the IMF, the SDRs can be used for the following purposes: (i) Obtain convertible currencies through using the method of debits, for details see previous discussions; (ii) pay off debts with the IMF; (iii) pay the subscribed shares of member nations; (iv) donate or provide funds to the IMF; (v) use the value of the SDRs as the base of domestic currencies (as of February 1993, five member nations had kept fixed parities of their currencies and the SRDs); (vi) establish reciprocal credit agreements between member nations; (vii) Use it as the account unit of the IMF organization; and (viii) enrich the variety of reserve assets.

9.1.3 Borrowed international reserves

The afore-mentioned four forms of reserves belong to the range of special international reserves. The general international reserves also contain borrowed reserves, which mainly include standby credits, mutual credits and payment agreements, short-term foreign assets in convertible currencies of domestic commercial banks, etc. Currently, the IMF organization categorizes borrowed reserves in the area of international liquidity.

As for standby credit, it stands for a standby loan agreement signed between a member nation that suffers from or expects to experience difficulties in its international payments and the IMF organization. The content of the agreement mainly includes the loan amount, term, interest rate, how the funds are used periodically, kinds of currencies involved, etc. After the agreement is signed, the member nation can use the funds when necessary. The member nation needs to pay an annual maintenance fee for the unused funds, which is recorded into borrowed reserves of the member nation. The signing of a standby credit agreement strengthens the capability of the government of the member nation to intervene the foreign exchange market and influences the psychology of the speculators and investors of the foreign exchange market. Hence, even if the member nation eventually did not use the entire amount or nothing at all, the signing of the agreement itself would have played its role of adjusting international payments and exchange rates.

As for mutual credit and payment agreement, it represents such an agreement between two nations where they could cross-use their domestic currencies. After the agreement is signed, when one of the signing nations experiences difficulties in its international payments, it can spontaneously apply the other nation's currency according to the pre-determined conditions. Such agreements possess the same characteristics of standby credit agreements, belong also to the category of borrowed reserves, and can be deployed at any time. However, these kinds of agreements are also different from each other. For example, both mutual credit agreement and payment agreement are signed by two parties and can only be used to resolve payment imbalances between the signing nations, while standby credit agreements are signed among many nations, and can be used to clear trades involving, say, three nations.

As for short-term foreign assets in convertible currencies owned by domestic commercial banks, although the ownership and use right do not belong to the government, the government can still influence the flow direction of the assets through using various methods, such moral advices, policy advocacy, and others. So, these assets can indirectly play the role of adjusting international payments.

Based on the discussion above, the relationship between international reserves and international liquidity is given in Table 9.3.

International reserves are an indicator of a nation's economic strength. Without any exception, each nation maintains a certain quantity of international reserves, although different nations may have different reasons for maintaining their

Table 9.3 Relationship between international reserves, borrowed reserves, and international liquidity.

Composites of international liquidity	
Self-owned reserves (international reserves)	*Borrowed reserves*
1. Gold reserves	1. Standby credit
2. Foreign exchange reserves	2. Mutual credit and payment agreement
3. Reserve positions at the IMF	3. Short-term assets in convertible currencies
4. SDRs allocation from the IMF	4. Other arrangements

international reserves. Generally speaking, the basic functions of the international reserves are listed as follows:

1   Maintaining the capability of repaying international debts and adjusting temporary imbalances in international payments

Adjustment of international payments is the number 1 reason why a national government maintains international reserves. When a nation suffers from international payment deficit, the government could intervene by using either various domestic economic adjustment policies or international reserves. When the deficit is temporary, the government could simply employ international reserves to make the necessary adjustment without the need of adopting fiscal or monetary policies that might bring forward side effects on the macro-economic level. In other words, either by mobilizing foreign exchange reserves, or by reducing the reserve position and the SDRs at the IMF, or by offsetting the shortage of foreign currencies, as caused by international payment deficit, from selling gold in the international market, the government could prevent the domestic economy from suffering from potential adverse effects as that might be caused by adopting fiscal and monetary policies. That of course would be helpful for the government to materialize its domestic economic objectives. On the other hand, if the international payment deficit is either huge or fundamental and long term, then international reserves can still be used as supplementary measures to provide a necessary buffer for the implementation of fiscal and monetary policies in order to prevent turbulence to appear in the domestic economy as that might be caused by fast application of adjustment measures. That would make it possible for the government to progressively advance the employment of fiscal and monetary policies.

2   Intervening the foreign exchange market and maintaining the stability of domestic currency

After the floating exchange rate system was initially adopted in February 1973, although in theory central banks are no longer responsible for maintaining stable exchange rates and the rates are determined by the market, considering the national interests and the fact that fluctuations of exchange rates bear adverse effects on the domestic economies, and hoping to control the exchange rates of domestic currencies at the desired levels, various nations have more or less interfered the exchange rates by making use of international reserves. When the exchange rate of a foreign currency rises while the domestic currency falls beyond the targeted range of the government, the monetary authority will throw out the foreign currency reserve (increase the supply of the foreign currency) by purchasing the domestic currency for the purpose of either inhibiting the fall of the domestic currency or making the domestic currency rise. Conversely, when the exchange rate of the domestic currency rises, then the rise is constrained by increasing the market supply of the domestic currency through purchasing the foreign currency. So, international reserves are the "intervention assets" used to maintain the exchange rate stability of one nation's domestic currency. The magnitude of a nation's international reserves represents the strength of how much the nation could intervene with the foreign exchange market in order to maintain the stability of its exchange rates; sufficient international reserves are the material foundation for the nation to support and strengthen the reputation of its domestic currency.

There are many reasons behind why the Southeast Asian monetary crisis broke out in 1997. Other than such reasons as economic policy mistakes in general and monetary policy mistakes in particular, large international payment deficits, defective economic structures, and foreign exchange speculations, one important cause is that the foreign currency reserves of the Southeastern nations were not sufficient. When the monetary crisis emerged, these nations did not have enough capability to defend their domestic currencies so that the crisis eventually broke out inevitably.

3   Guaranteeing the credit for the nation to take foreign loans and its ability to repay the debts

Under the normal conditions, sufficient international reserves are helpful in terms of strengthening the nation's international credit, attracting foreign investments, and promoting economic development. When an international financial institution or commercial bank gives out a loan, it first investigates the solvency of the nation within which the borrowing company is located. Therefore, the state of a nation's international reserves constitutes one of the important indicators specialized credit agencies and international financial institutions use to make their credit investigations and to evaluate the risks of the nations involved. At the present, there are numerous specialized agencies and financial magazines from around the world that make periodic credit assessments for all nations in order to determine each nation's safety factor for loans. This kind of evaluation generally includes the payment pattern of the target nation's current account, the ratio of maturity debt service over the nation's export income of the current year, the state of the nation's international reserves. So, maintaining a certain quantity of international reserves represents one of the important safeguards for the government of a nation or a commercial firm to conduct international financing and to ensure the payment of debts and services.

4   Maintaining the internal balance of the nation

The so-called internal balance represents the realization for a nation to enjoy sufficient employment and stable price level. Having sufficient international reserves is extremely important for a nation to maintain and to materialize the internal balance of the nation's domestic economy. For example, when a nation experiences inflation, the nation can draw on its international reserves to increase the quantity of goods in circulation by expanding import and appropriately reducing its reserves, and to ease the pressure of inflation by reducing the quantity of its domestic currency in circulation, without employing any contractionary fiscal and monetary policy.

5   Gaining competitive advantage for the nation

When a nation holds sufficient international reserves, it means that the government of the nation has a relative strong capability to interfere the foreign exchange market. That is, the government has the strength to raise or lower the exchange rate of its currency so that it gains competitive advantage in international trades. If the domestic currency is a major reserve currency, holding sufficient international reserves is very important in maintaining the currency's international status.

9.2 MANAGEMENT OF INTERNATIONAL RESERVES

The so-called management of international reserves means the process a national government or the nation's monetary authority uses to optimize the magnitude, structure, and application of the reserve assets while meeting the needs of the nation's international payments and the nation's domestic economic development by planning, adjusting, and regulating the magnitude, structure, and application of the reserve assets.

When seen from the angle of a nation, the management of international reserves involves two aspects: One is about the quantitative management, which decides and adjusts the magnitude of the international reserves, while resolving the problem of how to maintain an appropriate level of international reserves; the other involves how to determine and how to adjust the structure of international reserves, which is about the problem of how to optimize the structure of international reserves. The former is often known as the scale (or magnitude) management of international reserves, while the latter the structure management of international reserves. The management of international reserves, on the one hand, maintains the normal operation of a nation's international payments, and on the other hand, promotes the use efficiency of the nation's international reserves.

9.2.1 The management of magnitudes

How much international reserve should a nation maintain? The answer to this question not only affects this nation's economic development and its economic exchanges with the outside world, but also bears a degree of influence on the development of the world economy. The practice indicates that it is not true that the more international reserve a nation has the better. Within a specific time period, if a nation's international reserve surpasses the need to service the nation's international trades of the spot period and this accumulation has reached a certain scale, then the reserve is considered excessive. The first side effect of excessive reserve is that valuable resources are idling. For nations that are short of capital, they should avoid such situation. Excess reserve also means that the nation has given up the use of part of its foreign assets, which makes the cost of holding international reserves rise and represents a waste of the reserve assets. It is because the excessive international reserve could have been employed either to develop production by importing capital goods, proprietary technology, etc., or to satisfy the national consumer needs by adjusting the domestic market of through importing consumer goods. Additionally, the scale of international reserves, especially the scale of foreign currency reserves, is closely related to the quantity of domestic currency in circulation. The greater the scale of foreign currency reserves is, the greater quantity of the domestic currency there is in circulation, where increasing quantity of domestic currency in circulation can very possibly cause the severity of inflation.

Then, what will happen if the scale of a nation's international reserve is too low? In this case, the low scale of international reserve cannot satisfy the need of the nation's foreign trades, which bonds to reduce the level of economic exchange with the outside world. Also, too little reserve can weaken the monetary authority's capability to balance the nation's international payments, which further affects the national government's ability to deal with various financial dealings with other countries. Too little reserve makes it more difficult to handle emergent situations. If a nation maintains its international reserve at a low level for a long period of time, an international payment crisis may occur adversely. So, the international reserve of a nation can be neither too much nor too little. It should be maintained at an appropriate level, which is known as the moderate scale. The moderate scale of a nation's international reserve is mainly determined by the supply and demand of the nation's international reserve assets.

In terms of the demand of a nation's international reserve, it stands for a central problem studied in the management theory of international reserves. The demand of a nation for international reserve comes mainly from the following aspects: (i) the need to offset the international payment deficit. If the monetary authority of the nation holds sufficient international reserve, then when international payment deficit appears, as caused either by falling export or rising costs of import, the monetary authority can mobilize its international reserve while avoid implementing fiscal and monetary policies that might produce adverse effects on the domestic economy. (ii) The need to intervene with the foreign exchange market. When the exchange rate of a nation's currency drops such a large magnitude that starts to either affect the international reputation of the currency or produce adverse effects on the domestic economy, the nation can apply its foreign currency reserves to support its domestic currency by purchasing the currency. (iii) The need to maintain international reputation. And (iv) the need to cover international payments due to emergencies.

In terms of the supply of a nation's international reserve, it depends on the rise and fall of the four components of the international reserve. As we have discussed earlier, the general drawing rights and special drawing rights of the international reserve cannot be actively changed by any nation. It is because they are based on the nation's subscribed share in the IMF, while this share is determined by considering the economic strength of the nation. When the nation's economic strength did not change significantly or when the IMF organization did not adjust the subscribed share, the available general drawing rights and the special drawing rights of the nation will not change. So, the scale of a nation's international reserve is determined mainly by the nation's gold reserve and foreign currency reserves.

As for the gold reserve, its increase is materialized through purchase in either the international or domestic market or both. As for a nation that issues a reserve currency, it can increase its international reserve by applying its domestic currency to purchase gold in the international market. However, for most other nations that do not issue reserve currencies, their domestic currencies are not accepted in international payments. So, if any of these nations wants to purchase gold in the international market, it has to apply a freely convertible currency (a reserve currency). That is, what the nation changes is only the structure of its international reserve, where the gold reserve is increased while the foreign currency reserves are decreased, without increasing the total magnitude of the international reserve. However, when seen from the angle of the domestic gold market, no matter whether the nation issues a reserve currency or not, when the central bank uses the domestic currency to purchase gold in the domestic market, it can increase the gold reserve so that the scale of the international reserve is increased. This method is known as the monetization of gold. That is, non-monetary gold is transformed into the gold that will be used as money. Of course, the method of increasing the international reserve through monetizing gold is limited, because this method is constrained by such conditions as the production of gold and others.

As for the reserve of foreign currencies, the following methods can be employed to increase this reserve: (i) Surplus of international payments is the fundamental way for a nation to increase its reserve of foreign currencies, where surplus of the current account represents the most reliable and most stable source. At the same time, surplus of capital and surplus of financial accounts are of the characteristics of borrowed reserves, which are instable. (ii) Obtaining foreign currencies through interfering with the foreign exchange market. When the monetary authority of a nation sells its domestic currency for foreign currencies in the foreign exchange market, the newly acquired foreign currencies will be listed in the reserve of foreign currencies. And (iii) International credit is acquired when the nation obtains government loans or financial institutional loans from the international market. And the mutual credit between central banks can also be employed as international reserve.

The factors that influence the determination of the appropriate scale of a nation's international reserve include: (i) The size and stability of international payment flow; (ii) applications and effectiveness of other measures designed to adjust international payments; (iii) selection of the exchange rate system and exchange rate policy; (iv) the financing ability of the nation on the international financial market; (v) the level of development of the nation's domestic financial market; (vi) the international status of the nation's currency; (vii) the opportunity cost of holding international reserve; and (viii) the state of cooperation of international currencies. In particular, the number 1 role of international reserve is to adjust international payments. So, the size and stability of international payment flow are the main factor that determines the magnitude of a nation's international reserve. As for the stability of trade payment and income, if the demand elasticity and the supply elasticity of the exported goods of a nation are all greater than 1, while the demand elasticity of imported goods is less than 1, then the trade income and payment are relatively stable. Under such conditions, if the import and export are roughly balanced or show a slight surplus, then there is no need for a large scale of international reserve. Conversely, if the supply elasticity of exported goods is less than 1 while the demand elasticity of imported goods is greater than 1, then it means that the nation's condition of foreign trade is relatively poor so that the nation needs to maintain a relatively sufficient international reserve. And as for the balance and stability of international payments, speaking generally, if a nation maintains a surplus, then the nation needs a small scale of international reserve. Conversely, if the nation continuously holds a deficit in its international payments, then the nation has a relatively large demand for international reserve and needs to have a relatively large scale of international reserve.

Other than making use of international reserves, macro fiscal and monetary policies, exchange rate policies, foreign currency and foreign trade controls, etc., can also be employed as measures to adjust international payments. If a nation deals with its international payment deficit effectively by using fiscal and monetary policy or exchange rate policy, or the nation can effectively control its import and import and the flow of foreign capital by implementing a strict foreign currency and trade control, then the nation can maintain a relatively low level of international reserve. Otherwise, it has to maintain the large scale of international reserve. International reserve is closely related to the exchange rate system, because one of the roles of international reserve is the intervention in the foreign exchange market in order to stabilize the exchange rate. If a nation employs a fixed exchange rate system or a pegged exchange rate regime, and the government does not like to frequently change the level of exchange rate, then the nation needs to hold a relatively large scale of international reserve in order to deal with exchange rate fluctuations as may be caused by emergent factors of the foreign exchange market. Conversely, under the floating exchange rate system or the flexible exchange rate regime, the nation would need to maintain a relatively small scale of international reserve. Additionally, frequently interfering with the foreign exchange market requires more international reserves than interfering with the market only occasionally; and target-range interventions would need more international reserve than corrective interventions. As for the ability for a nation to finance in the international market, if the nation has a relatively good reputation in the international financial market, then the nation can quickly and conveniently obtain loans from foreign governments and international financial institutions with stable sources of funds. In this case, the nation does not need to maintain a large scale of international reserve. Conversely, if a nation's credit rating is low, the nation will have difficulty to raise funds. In this case, the nation will need more sufficient international reserve. What needs to be pointed out is that if a nation's scale of international reserve is too low, the nation's credit rating will not be high so that its ability to borrow foreign funds will also be lowered.

Developed financial markets can, on the one hand, attract injections of international capital, and, on the other hand, react to interest rate and exchange rate policies more sensitively so that the adjustment efficiency of interest rate and exchange rate policies is enhanced. In this case, the scale of international reserve can be accordingly reduced. Conversely, the more backward a financial market is, the more the adjustment of international payments will be dependent on the scale of international reserve. If a nation's currency is in the position of a reserve currency, then the nation can offset its international payment deficit by increasing the external liabilities of its currency so that the nation does not need to maintain much international reserve. The international reserve of one nation is often left with foreign banks in the form of deposits. That means that the price for a nation to maintain international reserve is to sacrifice parts of its investment or consumption. So, the opportunity cost for a nation to maintain international reserve is the difference between the return of the imagined investment and the interest income. The greater this difference is, the higher opportunity cost the international reserve carries; and the smaller this difference is, the lower the opportunity cost the international reserve bears. As constrained by its economic interests, the magnitude of international reserve a nation needs is inversely proportional to the opportunity cost of the international reserve the nation holds. The higher the opportunity cost is for a nation to maintain its international reserve, the less needed the international reserve becomes; and vice versa. If the monetary authority of a nation is in good partner relationships with other foreign governments and international financial institutions, and has signed a good number of mutual credit and standby credit agreements, then when the nation suffers from international payment deficit, the monetary authorities of other nations would be able to jointly intervene with the foreign exchange market. In this case, the international reserve of the nation does not have to be large. Conversely, the nation will need a large scale of international reserve.

From the discussions above, it follows that determining the appropriate scale of a nation's international reserve is a complicated matter. Because there are many interacting factors that influence the appropriate scale of international reserve, and there are also a large number of stochastic variables existing in the actual economy, it makes the accurate determination of the scale of a nation's international reserve difficult or even impossible. At the present, a widely employed method that can be conveniently applied to determine the appropriate scale of a nation's international reserve is known as ratio approach. This approach is first introduced by Robert Triffin (1966). By analyzing the historical records of the international reserves of several tens of nations for the time period between the two world wars and the early period from 1950–1957 after WWII against the exchange rate control situations of the nations, Triffin discovers that the demand for the reserve increases with trade growth so that he suggests employing the ratio of the scale of international reserve and the amount of imports to measure the sufficiency of the international reserve. His basic conclusion is that for nations that implement strict foreign currency control can afford to hold relatively small reserves because the governments can effectively control imports; however, the bottom line is not less than 20%. As for most nations, maintaining an international reserve that amounts to 20%–40% of the annual amount of imports would be appropriate. The characteristic of this ratio approach is that the two variables – the scale of reserve and the amount of imports – are related. That is why it is often known as the reserve – import (R/M) ratio scheme.

The advantage of this ratio approach is: Firstly, it creates the precedent of systematically studying the appropriate scale of international reserves. After then, the relevant investigations have been developed great deal with various approaches introduced. Secondly, because the most basic use of international reserves is to offset the international payment deficit, while the trade income and cost represent the most important, and most fundamental items in international payments, under the normal circumstances, it agrees with the reality that the ratio of the scale of reserve and the annual amount of imports is used to determine the level of international reserve. Thirdly, this method is straightforward and applied readily. Because of this reason, as of this writing, the IMF organization and international banks are still using this method to evaluate each nation's risk, and each nation is using this method to roughly estimate the scale of its international reserve.

Summarizing what has been discussed above, it follows that the most appropriate scale of a nation's international reserve is a variable that is difficult to determine accurately. As of this writing there is still not any method to correctly measure this variable. So, in terms of the overall management of international reserve, what is more practically significant is perhaps to consider comprehensively a group of factors that affect the scale of a nation's international reserve; and to determine the appropriate region of the scale by using the objectives of macroeconomic development as the fundamental premise and by following the principles of minimizing the marginal costs and maximizing the return of holding the international reserve. In short, the rule that should be used to determine the appropriate region of the scale of a nation's international reserve is: Determine the upper and lower limits by using the actual circumstances of the nation's economic development. Here, the upper limit stands for the maximum scale of reserve that is needed to satisfy the high speed annual economic development with drastically increased imports. It is known as the insurance scale of reserve. And the lower limit represents the minimum scale of reserve needed for slow economic development with greatly reduced imports. It is known as the constant scale of reserve. The upper limit represents when the nation possesses the sufficient international liquidity, while the lower limit the critical constraint of the national economic development. The range bounded by the upper and lower limits constitutes the appropriate region for the scale of the nation's international reserve.

9.2.2 The structural management

The objective of the structural management of international reserves is to constantly adjust the quantities of the four reserve assets in order to optimize the composite and effectiveness of the reserve by considering the respective liquidity, safety, and profitability of the assets. In the international reserve of a member nation of the IMF, the total of the reserve position at the IMF and the SDRs amount to less than 10%, while the gold reserve and foreign currency reserves take up more than 90%. The reserve position and the quantity of SDRs are determined by the IMF. They can hardly be changed quantitatively and structurally by the member nation. So, the structural management of a nation's international reserve focuses on the gold reserve and foreign currency reserves.

Gold possesses a relatively good degree of safety. However, it lacks liquidity and profitability. Due to this safety, in international reserves, there must be the appropriate component of gold. However, because of the poor liquidity and forever changing price, the amount of gold reserve cannot be too large and should be kept stable. The forms of gold reserve consist mainly of the following:

(i)   Direct operation: Raise the rate of return of the gold reserve through the international gold market by trading spot contracts, futures contracts, and options.

(ii) Indirect turnover: The monetary authority of a nation materializes indirectly the goal of raising the value and maintaining the value of the gold reserve by making and selling gold coins and by engaging in the business of leasing gold, handling gold loans, etc.

(iii) Form asset portfolios: Based on the principles of liquidity and profitability, a certain portion of the gold reserve is promptly converted into reserves of foreign currencies that bear high returns and liquidity. The portfolio is appropriately adjusted according to the changing environment of the market exchange rates.

Under the floating exchange rate system, because the exchange rates and interest rates of the reserve currencies change constantly, the reserve currencies are somehow less safe. Even so, each reserve currency, as a world currency, can be directly employed to make international payments and to intervene with the foreign exchange market. Additionally, foreign currency reserves can increase value through earning interests. So, Foreign exchange reserves possess a good degree of liquidity and profitability, which make these reserves occupy high proportions in the entire international reserve. At the same time, the management of foreign exchange reserves becomes the focus of the management of international reserve. The structural management of foreign exchange reserve is mainly about the composite management of different reserve currencies and the investment management of the reserve assets.

In terms of the management of different currencies, it is about how to select and adjust promptly the weights of various reserve currencies in the overall foreign exchange reserve. Differences in the exchange rates and interest rates of the reserve currencies and changes in these rates lead to uncertainties and differences in the return of the reserve assets held in these currencies. Hence, when the monetary authority of a nation arranges the composite of reserve currencies, it mainly considers the exchange rate risk and interest rate risk as may be caused by the fluctuations in the exchange rates and interest rates. Speaking more specifically, in order to reduce the exchange rate risk, the currency selection in the management of reserve currencies is done mainly by considering the following factors: (i) the currencies to be used in foreign trades; (ii) the currencies needed to repay foreign debts; (iii) the currencies the government will use when it needs to intervene with the foreign exchange market in order to stabilize the exchange rates; (iv) the comparison of the changes in the interest rates of various reserve currencies and the changes in the relevant exchange rates; and (v) the diversification of reserve currencies.

In terms of the asset structure of foreign exchange reserve, it means the combinatorial proportions of such foreign currency assets as cash, deposits, priced short-term and long-term bonds, etc., in the foreign exchange reserve. In order to make the employment of foreign exchange reserve satisfy collectively the requirements of liquidity, safety, and profitability, the management of foreign exchange reserve, in general, is done hierarchically based on how conveniently each asset class can be converted into cash, while determining the appropriate proportions of the hierarchies. By using the liquidity to make the division, the foreign exchange reserves are classified into three hierarchies as follows:

1   Primary reserves: This hierarchy includes cash and quasi-cash, such as demand deposits, short-term treasuries, commercial papers, etc. This hierarchy of foreign exchange reserves has the highest degree of liquidity with the lowest profitability without much risk. These reserve assets are mainly used to clear a nation's regular and temporary foreign payments.
2   Secondary reserves: This hierarchy contains mainly mid- to long-term bonds. These reserve assets produce higher levels of profits with lesser liquidity and greater risk than those assets of the primary hierarchy. The management of the secondary reserves focuses on the profit along with adequate liquidity and risk. These reserves are mainly employed as supplementary liquid assets as guarantees to cope with temporary and emergent foreign payments.
3   Tertiary reserves: These are various tools of long-term investments. These reserve assets have the highest level of profitability, the worst liquidity, and the greatest risk. Only after the monetary authority has decided the scales of the primary and secondary reserves, it will generally consider making the rest reserve as long-term investment.

9.2.3   Reserves of the USA, Europe, Japan, and Great Britain

Firstly, let us look at how the United States manages its international reserve. According to the statistics published by the Treasury Department, since the year 1996, the foreign exchange reserve of the United States has been maintained at around US$30–45 billion. And at the end of April 2005, the foreign exchange reserve of the American government was about US$41.7 billion, where the government held foreign bonds of worth US$26.9 billion and US$14.8 billion in foreign currency deposits. The special position of the U.S. dollar in the international monetary system determines that the United States does not need to maintain a large magnitude of reserve assets. That is something unmatchable by any other nation.

This international reserve of the United States is jointly managed by the Treasury Department and the Federal Reserve, where the Treasury Department is actually responsible for introducing international financial policies, while the Fed is responsible for making and implementing domestic monetary policies. In terms of the management of foreign exchange reserves, these two agencies work jointly to guarantee the continuity of the American international monetary and financial policies. Since 1962, the Treasury Department and the Fed have been coordinating their interferences in the foreign exchange market, while the specific intervention operations are carried out by the Federal Reserve Bank of New York. This New York Bank is not only an important part of the Fed, but also an agent of the Treasury Department. Starting in the late 1970s, the Treasury Department owns about half of the foreign exchange reserve of the United States, while the other half owned by the Fed.

The Treasury Department manages the foreign exchange reserve mainly through the Exchange Stabilization Fund (ESF). In as early as 1934, the Gold Reserve Act has stipulated that the Treasury Department has the total control of the assets of the ESF. Currently, the ESF is made up of three kinds of assets, including assets in U.S. dollars, assets in foreign currencies, and SDRs, where the part involving foreign currencies is delegated through the Federal Reserve Bank of New York, which mainly invests in foreign government bonds and deposits in foreign central banks through the New York foreign exchange market. Under particular circumstances, the ESF can exchange currencies with the Fed in order to acquire additional usable assets in U.S. dollar. In such cases, the ESF sells spot contracts of foreign currencies to the Fed and buys back more distant futures contracts using the market price. All the operations of the ESF need first to obtain approvals from the Treasury Department, because the Treasury Department is responsible for the making and perfecting American international monetary and international financial policies, including the policies on how to intervene with the foreign exchange market. Additionally, the Exchange Stabilization Fund Act requires the Treasury Department to report annually to the President and the Congress on the operations of the ESF, while the Treasury Department is also required to arrange the audit report for ESF.

On the other hand, the Fed manages the foreign exchange reserve mainly through the Federal Open Market Committee (FOMC) while keeping a close cooperation with the Treasury Department. The Fed mainly trades its foreign exchange reserve on the New York Foreign Exchange Market through the manager, acting as the agent of the Treasury Department and the FOMC, of the Federal Open Market Account of the Federal Reserve Bank of New York. The scope and method of the Fed's interference operations of the foreign exchange market have changed with the evolution of the international monetary system, which can be divided into three stages. The first stage is the time period when the Bretton Woods system was in use. During that stage, what the Fed was most interested in was whether or not the price of the U.S. dollar could be maintained constant in the gold market instead of the foreign exchange market. The second stage started in 1972 when the floating exchange rate system began to form. During this stage, the Fed started to actively intervene with the foreign exchange market through using the main method of exchanging currencies with other nations' central banks. The third stage began after the signing of the Plaza Agreement in 1985. Since then, the Fed has interfered with the foreign exchange market rarely by exchanging currencies. Instead, it directly purchased U.S. dollars or foreign currencies from the open market.

Secondly, let us look at the management of international reserve of the euro zone. According to the statistics of the European Central Bank (ECB), in 2001 the foreign exchange reserve of the euro zone reached the peak of €235.4 billion. Since then the scale drastically dropped to €136.3 billion in 2004. As of April 2005, the reserves of foreign currencies from outside the euro zone, as held by the ECB, were €135.7 billion, where foreign bonds were worth €100.3 billion and foreign currencies in the amount of €35.4 billion. Additionally, the scales of the reserves held by the central banks of the member nations of the euro zone were about €322.0 billion.

The management of the reserves of the euro zone belongs to the responsibilities of the European System of Central Banks (ESCB), which was established in 1998 and consists of the ECB (established in 1998) and the central banks of the member nations of the European Union. The ESCB also constitutes the euro system with the ECB playing the role of decision maker. The ECB and the member nations' central banks all hold and manage foreign exchange reserves.

The ECB manages the reserves mainly through introducing and making strategic investment decisions. The ECB's objective of managing the foreign exchange reserves is to maintain the liquidity and safety of the foreign currencies in the reserve in order to satisfy the need of intervening with the foreign exchange market and to maximize the value of the reserve assets. According to the articles of the European Central Bank System Act, the total amount of international reserve assets transferred from a member nation's central bank to the ECB is determined by the share of capital the nation occupies in the ECB, where 15% of the transfer needs to be in the form of gold and 85% in the foreign currency form that combines U.S. dollars and Japanese yens. The ECB can require member nations to transfer additional foreign exchange reserve. However, these additional transfers can only be used to offset the reduced amount of the international reserve instead of be used to increase the original international reserve. The management system of the ECB's foreign exchange reserve consists of two layers. One is that the decision-making body of the ECB formulates strategic investment decisions, which mainly involve the currency composite structure of the foreign exchange reserve, while meeting the needs of balancing the interest rate risk, return, credit risk, and liquidity. The other is that corresponding to the strategic investment decisions, all member nations' central banks take coordinated actions to manage the foreign exchange reserves of the ESCB. In particular, the ECB's management committee decides the investment strategy for ECB's foreign exchange reserve by considering the needs of future operations. After the ECB informs the member nations' central banks about its decisions, these nations' central banks take joint operations through relevant organizations. Then the ECB manages the transaction information of the individual nations' central banks received through the communication network of the ESCB.

For the purpose of managing the reserve, the ECB defines four key parameters. One is the introduction of the two-leveled investment benchmarks – strategic and tactical – for each reserve currency. The strategic benchmark is defined by the ECB management committee. It mainly reflects the preference of the ECB for policies regarding long-term risk and return. The tactical benchmark is introduced by the executive board of the ECB and mainly reflects the preference of the ECB regarding the short- and mid-term risk and return. The second key parameter is the allowed degree of deviation between risk/return and the investment benchmark and relevant measures of correction. The third key parameter is on the operating mechanism that handles the reserve transactions and the investable bonds. And the fourth key parameter is about the limitation on the exposure of credit risk. The ECB does not reveal the details of the parameters in order not to create unnecessary consequences to the financial market.

The member nations' central banks of the euro zone manage reserves mainly through implementing tactical investments that are consistent with the reserve strategy of the ECB and through implementing their own independent decisions for their respective reserves. According to article 31 of the European Central Bank System Act, if an investment operation in the international financial market of a member nation's central bank might create such an effect on exchange rate or on the nation's domestic liquidity that might go beyond the range as specified by the ECB's guiding principle, then this investment transaction has to be approved by the ECB in order to guarantee the continuity of the ECB's exchange rate and monetary policies. Additionally, without advanced ECB approvals, all member nations' central banks are allowed to make investment operations in foreign currencies in the international financial market or to perform other operations to fulfill their obligations in such international organizations as the IMF. Each nation's central bank holds and self manages its international reserve that was not transferred to the ECB. Since the time when the ECB first started to intervene with the foreign exchange market, member nations' central banks no longer need to formulate their respective objectives of foreign exchange intervention. Instead, they only need to design and implement their tactics.

Thirdly, let us look at how Japan manages its international reserve. According to the statistics of Japan's Ministry of Finance, from 1996 to 2002, Japan's foreign exchange reserve rose from US$217.9 billion to US$451.4 billion; after that time period, due to the weakening of the U.S. dollar, the reserve rose quickly to US$824.2 billion at the end of 2004. Since then, Japan's foreign exchange reserve has been lowered some to US$818.6 billion as of March 31, 2005, where foreign bonds amounted to US$696.1 billion and deposits in foreign currencies to US$122.4 billion. The management system of Japan's foreign exchange reserve belongs to the responsibility of Ministry of Finance. According to the provisions of Japan's Foreign Exchange and Foreign Trade Law, in order to maintain the exchange rate stability of Japanese yen the Finance Minister can employ whatever intervention method necessary in the foreign exchange market. And according to the provisions of the Bank of Japan Law, as the bank of the government, when the finance minister believes that there is a need to interfere with the foreign exchange market, he will carry out the specific operation as instructed by the Ministry of Finance. The Finance Ministry manages its foreign exchange reserve mainly through the special account of foreign exchange assets that are deposited at the Bank of Japan.

Based on the guidance announced by the Ministry of Finance on April 4, 2005, the Bank of Japan's management of the foreign exchange reserve is carried out by mainly following the decisions of the Ministry of Finance within the following frameworks:

(i) The Objective: Maintain the stability of exchange rates of Japanese yen by keeping a foreign exchange reserve of sufficient liquidity so that necessary trades of foreign currencies can be made;

(ii) The principle: On top of the primary goal of maintaining the safety and liquidity of the foreign exchange assets, pursue after possible profit in order to eliminate the disadvantageous fluctuations of the foreign exchange market. When necessary, closely collaborate with relevant monetary authorities of foreign countries;

(iii) The composite of the reserve: Foreign exchange assets consist mainly of national debts and governmental agency bonds of strong liquidity, bonds of international financial organizations, asset-backed securities, deposits at various nations' central banks, and deposits at financial institutions that are highly rated with strong repayment ability; and

(iv) The management of risk: The credit risk, liquidity risk, and interest rate risk of reserve assets are strictly managed and controlled by using internal models.

The Bank of Japan materializes its intervention in the foreign exchange market mainly through the Forex Balance Operation of the Financial Market Department and the Backup Operation of the International Department. The Forex Balance Operation is responsible for analyzing the market and providing policy recommendations. With approvals from the Ministry of Finance, the Back Operation is then responsible to make the actual foreign exchange trades. The intervention operations of the Bank of Japan generally take place in the Tokyo foreign exchange market. However, if that market is closed, the intervention operations are carried out in a European market or the New York market with the necessary funds from the foreign exchange special account. When there is a need to sell foreign currencies, the Bank of Japan does it by selling the assets in its foreign currency assets special account on the foreign exchange market. When there is a need to buy in foreign currencies, the necessary funds in Japanese yen are raised by issuing short-term government bonds. If necessary, the Bank of Japan can also request foreign monetary authorities to intervene in the market with funds or other relevant means, which need first be approved by the Ministry of Finance. The Forex Balance Operation maintains close ties with market participants, such as Forex brokers, foreign branches of the Bank of Japan, and central banks of other nations. Additionally, it is also responsible for monitoring the exchanges rates, and studying the changes in foreign debts, foreign stock markets, commodity markets, etc. The Forex Balance Operation summarizes the financial, economic situations and reports to the Policy Board of the Bank of Japan, and at the same time, provides the daily report of the foreign exchange market to the International Department of the Ministry of Finance. On the basis of this report, the Ministry of Finance decides whether or not to intervene in the foreign exchange market. As soon as the decision of intervening in the market is made, the actual operation is carried out by the Backup Operation of the International Department. Additionally, the actual investment of the foreign exchange reserves is also done by the Backup Operation.

Fourthly, let us look at the management of the international reserve of Great Britain. Since 1999, the foreign exchange reserve of Great Britain had been maintained in the range of US$30–40 billion. As of March 5, 2005, the foreign exchange reserve of British government was about US$34.9 billion, where foreign bonds amounted to US$347 billion and foreign currencies to US$0.2 billion. This magnitude is much smaller than to those of the reserves of the German and French central banks. That reflects the tradition of allowing British pound to fluctuate more freely.

The management system of the reserve of Great Britain belongs to the responsibility of the Ministry of Finance, where the Bank of England is only responsible for the daily managerial details. The Ministry of Finance implements the management strategies of the reserve mainly through setting the balance for the Exchange Equalisation Account (EEA). In as early as the year 1931 when the gold standard system collapsed, Great Britain's foreign exchange reserve and gold reserve were transferred to the Ministry of Finance. In 1932, established was the EEA, the reserves of which constituted the foreign exchange reserve of Great Britain. Any intervention activity in the foreign exchange market of the British government has to be carried out through this account. And, this account also provides foreign exchange services for the government and its agencies. Because the Exchange Equalisation Account Act does not allow the EEA to borrow from foreign sources, the British government issues foreign debts through the National Loan Fund in order to supplement the foreign exchange reserve for the EEA. The Ministry of Finance is responsible for the strategic management of the foreign exchange reserve. It decides whether or not there is a need to intervene in the foreign exchange market. However, it does not operationally participate in the actual market operation. On the contrast, the reserve management of the Bank of England is tactical. It attends to the actual market operations and the daily management, while playing the agent role of the Ministry of Finance. Each year, the Ministry of Finance provides guiding recommendations for the management of foreign exchange reserve. The content of the guiding recommendations mainly includes:

(i) The benchmark return of the investment of the reserve assets and allowed deviation, the composite of the reserve assets, the composite of foreign currencies, the rate of return of investment, etc.
(ii) The framework for controlling credit risk and market risk; and
(iii) The framework for loan projects of the National Loan Fund, etc.

Based on the reserve strategies of the Ministry of Finance, the Bank of England carries out the detailed management of the reserve with the objective of maintaining the liquidity and safety of the reserve assets while maximizing the return. Each year, the Bank of England and the Ministry of Finance jointly decide on the benchmark return of investment as outlined in the guiding recommendations of the Ministry. This benchmark return is determined based on the past risk, return, trade, the possible kinds of foreign currencies needed for the intervention of the foreign exchange market, and other factors. It is announced in the annual report of the EEA. Once each 6 months, the Bank of England reports its most recent investment performance to and discuss the relevant reserve tactics with such a special group that consists of a commissioner of the EEA, the manager of the macroeconomic policies and International Finance of the Ministry of Finance, the executive market director of the Bank of England, and some other government officers. In each quarter, the Bank of England provides its independent feedbacks on the effectiveness and sufficiency of the reserve management through the internal audit department; and the person in charge of the audit department reports to the executive director, who in turn reports to the commissioner of the EEA. Additionally, British national audit office conducts an external audit of the EEA once a year. The Bank of England reports its investment performance at the monthly meeting organized by Debt and Reserve Management Department of the Ministry of Finance. Additionally, it conducts periodic pressure tests on the market risk of the EEA in order to measure the risk resistance of the assets of the EEA when experiencing market fluctuations and related potential losses. The control of credit risk belongs to the responsibility of the Credit Risk Advisory Committee that is internal to the Bank of England. Other than playing the agent role of the Ministry of Finance in the management of foreign exchange reserve, the Bank of England also holds its own foreign exchange reserve, which is not a part of the foreign exchange reserve of the British government. Instead, this reserve is solely used by the Bank of England in its potential intervention of the foreign exchange market in order to support its independent monetary policies.

9.3 POTENTIAL PROBLEMS REGARDING FOREIGN CURRENCY RESERVES

In the government's management process of foreign currencies through the means of systems or the method of policies, there might appear some problems that could affect the stability of the economy due to the changing levels of control, the stages of economic development, and external conditions. This section studies some of the selected often-seen problems

9.3.1 The problem of capital flight

The so-called capital flight means such abnormal outflows of capital that are caused by panics, doubts, or for the purpose of risk avoidance or for the purpose of escaping certain market controls. Beyond the outflow of funds, capital flight stands for abnormal movement of capital that is caused by safety considerations or consequent to some other purposes. There are different understandings regarding the concept of capital flights. Some analysts, such as the World Bank, treat all flows of capital from developing countries to developed countries as capital flight. This comprehension has expanded the connotation of the concept.

Because of the invisibility of capital flights and the imperfection in the statistics of international payments, it is often the case that capital movements cannot be fully reflected in the account of capital exports. It has been difficult to quantify how much capital has been gone. To this end, there are two general methods to make the estimate: one is the direct method. That is, it estimates the capital movement by using the sum of the errors in the international payments or missing account balances and the private (non-banking) short-term capital movements. This method is simple and straightforward. However, it suffers from some problems. For example, the errors or missing account balances contain not only the capital flows that are not recorded, but also some real statistical errors. Additionally, long-term capital movements may also include the problem of capital flight. The second method is indirect, which subtracts the current account deficit (ΔCAD), as applications of capital, and the increase (ΔR) in the reserve assets from the sum of the increase in foreign debts (ΔD), as a source of capital, and net inflow of foreign direct investments (ΔFDI). Symbolically, we have

$$\text{Capital flight} = (\Delta D + \Delta FDI) - (\Delta CAD + \Delta R) \tag{9.1}$$

Capital flight appears when the owner of the capital reconfigures his asset portfolio. So, differences in potential foreign and domestic returns and risks constitute the main reasons for the appearance of capital flights. From the angle of the return factor, a relatively low rate of return on domestic assets can be possibly caused by the following reasons: the exchange rate of the domestic currency is estimated high (because the potential devaluation of the domestic currency could lead to losses), the low interest rate as caused by the domestic implementation of financial repression policies, decreasing domestic, actual interest rate as caused by relatively high inflation, etc. When seen from the angle of the risk factor, asset losses could be caused by such reasons as instable domestic political situation, implementation of new control policies or frequently policy changes, imperfect legal system, etc. Additionally, if the capital came from illegal sources, then flight to a foreign is obviously a safer option.

Capital flight can be done through various legal and illegal means. Capital flight of large scales often appears after the conversions of funds and financial accounts, because in this case, capital flight can be readily done through a lawful method. When the conversion between funds and financial accounts is controlled, capital flight is generally materialized through various unlawful methods in order to avoid the control. One often seen method is to forge prices in international trades by using internal and external collusions, where over-invoiced imports and under-invoiced exports are employed. In other words, prices higher than those of actual trades are employed to apply for import usage of foreign exchange, while prices lower than those of actual trades are applied for export currency settlements. From doing so, the part of price difference is transferred abroad. Another often seen method is to establish false investments. In particular, through internal and external collusions, non-existing projects are used to apply for foreign currencies of investments; then the desired amount of capital is transferred abroad. A third often seen method is to transfer funds abroad through underground banks.

Capital flight is extremely unfavorable to the economic development of a nation. Within a short period of time, capital flight of large scales can cause economic chaos and turbulence. In the long run, capital flight reduces the quantity of money available for domestic use, lowers the tax income of the government from domestic assets, and increases the domestic burden of foreign debts, leading to a whole series of serious economic consequences. So, the national government has to create a long lasting macro environment that is stable for economic growth; and before that is materialized, a relatively strict control of capital and financial accounts needs to be adopted in order to prevent capital flight from appearing.

9.3.2 The problem of credibility

This is really the problem about the credit of the government itself and the trade-off between flexibility and stability when the government implements its policies. In the foreign exchange management, the problem of credibility is mainly shown in the process of maintaining the fixed rate system (or the exchange rate interval) and interventions in the foreign exchange market. In particular, the common objective of intervention in the foreign exchange market is to maintain a specific exchange rate level. So, for our purpose here, we will use the scenario of a government trying to maintain its fixed exchange rate system as our example to discuss the problem of credibility in the management of foreign currencies.

When there appears a large amount of unidirectional currency trades in the foreign exchange market, and the market believes that there is a good chance for the exchange rate of the domestic currency to change, the government signifies its determination to safeguard the fixed exchange rate system and the exchange rate level by using public statements, warnings, and other means. Those can be seen as signals the government sends to the market, indicating that if the exchange rate continues to fluctuate, the government will intervene in the foreign exchange market or implement some other measures of control. Such government signals of promise are generally extremely effective when the market expects chaos and speculation is rampant. When the market opinions about the future changes of the exchange rate are very different, a promise from the government will be possibly used as commonly acceptable evidence that will direct the expectations of the market participants and contain the development of speculative activities. So, without any actually expense, the excessive fluctuations in the exchange rate are resolved. However, for this method of sending signals to the public to work in terms of maintaining the fixed exchange rate system, the following conditions have to be satisfied:

1   The signal the government sends has to be clear that the purpose of potential intervention is for maintaining the exchange rate. Otherwise, the market participants might not be able to tell whether this signal is about the exchange rate or something else.
2   The future policy, as signaled, must be able to move the exchange rate in the desired direction. To this end, let us look at the monetary model as an example. The money supply of the government has the most direct effect on the exchange rate. If what the intervention signal indicates is not a future adjustment of the money supply, or if what the signal implies is an incorrect adjustment of money supply, then the exchange rate will not be adjusted as what is wished for.
3   The government has to establish the reputation that it practices what it says so that its signals have credibility. If in the past the government did not maintain the exchange rate as promised, then the market will not believe in what the government signals. In this case, all intervention signals the government sends will not be considered credible, and will not alter the expectations of the market participants.
4   Any promise the government makes cannot be severely inconsistent with the basic situation of the economic reality. For example, when the government promises to intervene in the market by using foreign exchange reserve, it should have sufficient scale of the reserve; when the government promises to tighten the money supply, it should not have a large fiscal deficit that needs financing of the central bank; etc.

Otherwise, even if the government made a relevant promise, the market participants would still question the promise based on their rational expectations so that the government suddenly loses its credibility. In other words, when the basic economic situation did not change or only changed slightly, the government should pursue after stability in its policy and economic environment; through making public promises, the government should resolve fluctuations with economic intervention as the backup so that the economy would return to the desired level of stability. When the basic economic situation has changed to such a degree that the original objective of the government could no longer be maintained, the government should alter the objective flexibly according to the new situation, while attempting to eliminate the adverse side effects of the lowered objective instead of rigidly insisting on the impossible objective. That is also a principle the government should comply with in its management of foreign currencies.

9.3.3 The problem of currency substitution

By currency substitution, it means such a phenomenon that a foreign currency can replace the domestic currency either wholly or partially in terms of the functions of money. Speaking simply, currency substitution can be defined as an excessive demand of the domestic residents for a foreign currency. There are two kinds of currency substitutions. One is that both domestic residents and foreign residents simultaneously hold the domestic currency and a foreign currency. The other kind of currency substitution is that the domestic residents unilaterally hold and use a foreign currency. The former is known as a symmetric currency substitution, which generally appears between advanced nations. The latter can be seen as an asymmetric currency substitution, which often appears in a developing country where its domestic currency is replaced wholly or partially.

There are many reasons for currency substitution to occur. Firstly, when the value of a nation's currency is highly unstable, as might be caused by high levels of inflation or various other factors, the residents will lose their confidence in the purchasing power of the domestic currency so that they no longer use the domestic currency as a means to store value. Next, if the national government cannot effectively control what currency is used in trades and transactions, then sellers of goods will demand some other more reliable currency as the media of business transactions instead of the domestic currency. That makes the domestic currency lose its function as the media of circulation. When the domestic currency is no longer an effective media of circulation, people will naturally give up the domestic currency as the unit of accounting and method of payments so that the functions of the domestic currency are entirely replaced. At the same time when the domestic currency loses all of its functions, because some currencies are commonly accepted, such as the U.S. dollar whose purchasing power has been successfully maintained for a long period of time, they will pass across the national borders and enter into the storage field, the payment field, and even the circulation field of another nation. A good example is the American dollarization of the Latin American countries. What is worth noticing is that different of the Gresham's Law that bad currency drives out good currency, the currency substitution what is just described is the phenomenon of how a good currency drives out a bad currency.

As for the estimate of a currency substitution, it is difficult to obtain the exact statistics of how much of a foreign currency is in the circulation of a domestic economy. Hence, it is difficult to quantify the depth and width of a currency substitution. Because of this reason, we can only comprehend currency substitution through some rough numerical estimates. Due to a lack of actual data, people commonly use the proportion of foreign currency deposits in the total financial assets as an indicator for the degree of foreign currency substitution. In many economic analyses, the ratio F/M, foreign currency deposits over domestic currency, is often used to represent currency substitution, where M is the cash (C) held by the domestic public, defined as the sum of demand deposits (D), savings deposits (S), and time deposits (T), and F the foreign currency deposits within the domestic financial system. Additionally, there are also some other ratios, such as $F/(M + F)$, M/F, and F/D, that are used to represent currency substitution. Within all these estimates, the demand of foreign residents for the domestic currency is not considered. So, these ratios are often used to estimate asymmetric currency substitutions. At the same time, these ratios ignore the foreign currency in the domestic circulation and the foreign deposits of the domestic residents.

The most serious economic consequence of currency substitution is the instability of the domestic financial order and the weakened ability for the government to apply its monetary policies. Under the condition that there is an existing currency substitution, inflows and outflows of the foreign currency disturb the supply-and-demand mechanism of the domestic currency so that the determination of interest rate becomes more complicated. At the same time, they also weaken the ability of control of the central bank over the total quantities of credit and money in circulation so that it makes it difficult for the central bank to design and implement monetary policies. Additionally, because a large quantity of foreign currency is used, the inflation tax revenue of the government is greatly reduced.

What needs to be pointed out is that the afore-mentioned negative effects of currency substitution can only appear when the said currencies can be exchanged freely. So, after its domestic currency can be exchanged freely, the government has to adopt effective measures to prevent currency substitution from happening. To this end, improving the stability of value and the actual rate of return of the domestic currency, and the confidence in the domestic currency is the most essential method of resolving the problem of currency substitution. That requires the government to effectively control the inflation and any other macro, instable situations. When this end cannot be materialized, it is necessary to limit the exchange of the domestic currency.

This section analyzes the pros and cons of various systems and policy adjustment methods used in the management of foreign currencies. In the following two sections, we will expand our perspective to the entire world, while discussing the impacts of financial globalization on internal and external equilibria, adjustments of these equilibria, and international coordination.

9.4 CAPITAL FLOWS ACROSS NATIONAL BORDERS

Flows of international capital represent processes of movement and transfer of capital across national borders throughout the world. Along with the globalization development of the world economy, flows of international capital have become the most active factor in the world economy. However, at the same time when the international capital continuously helps to stimulate the world economic growth, it also provides the soil for financial crises of the international scale to occur. Since the 1990s, along with the frequent occurrence and increasing severity of international financial crises, flows of international capital have caught the major attention of the world.

9.4.1 Types of international capital flows

By flow of international capital, it means such a movement of capital from one country to another country or from one region into another region in pursuit of some economic or political purposes. That is, it stands for transfer of capital in the international arena. By capital inflow, it means that capital moves into a country, indicating that foreign debts of the country are increased or that the foreign assets held by the country has been reduced. By capital outflow, it means that capital moves outward to another country and that more domestic asset is held abroad or that foreign debts held by the country are reduced. The state of capital flow of a country or region within a certain time period is reflected comprehensively on the balance sheet of the country's international payments and incomes and its financial account.

International capital flows consist of short-term flows and long-term flows. In particular, by short-term capital flow it means such movements of capital that have duration of one year or less. Such type of capital flows takes mainly the form of short-term financial assets, such as demand deposits, treasury bonds, transferrable large CDs, commercial papers, banks promissory notes, and other financial tools of the money market. Short-term capital flows are facilitated through telephones, telegrams, faxes, and other modern methods of communication.

According to their different purposes, capital flows can be classified into the following four groups: (i) Capital flow of trading. It stands for transfers of capital as caused by international trades. In order to clear the credits and debts involved in international trades, capitals have to move from one country or region to another, forming capital flows of trading. For this kind of capital movement, the funds generally flow from a goods importing country into the goods exporting country. (ii) Capital flow of financing. It stands for such international movements of capital as caused by activities of financing between banks that manage foreign currencies and other financial institutions. Such flows of capital are mainly provided to serve banks and financial institutions that either have surpluses of funds or suffer from shortages of capital. These flows of capital take the forms of exchange rate arbitrage, interest rate arbitrage, swap, position allocation, interbank lending, etc. Because these flows involve large amounts of capital, high frequencies of movement, and the business of foreign exchange, they produce certain impacts on the short-term interest rates and exchange rates. (iii) Capital flow of value keeping, which is also known as capital flight. It represents international movement of capital as caused for the purpose of safeguarding the short-term safety and profitability of the capital by employing measures to avoid or prevent losses from happening. The main reasons that might cause capital flows of value keeping include: fluctuations in the domestic political situation no longer provide the necessary safe environment for investment; variations in exchange rates become dangerously high so that the value of capital might drop; the control on foreign currencies is tight or the imposed taxes are high so that the liquidity of capital is threatened, etc. (iv) Capital flow of speculation. It means such international capital transfers that are caused by various speculative activities in order to take advantage of changes and differences in interest rates, exchange rates, gold prices, bond prices, and prices of financial products. Such flows of capital aim at making profit from price differences; and whether or not the speculators could be profitable depends entirely on whether or not their predictions and judgments are correct. For example, a nation's temporary deficit in its international payment can pressure the exchange rate to float downward. If speculators believe that this downward float were temporary, they would then buy the nation's currency at a relatively low price and sell soon after for a higher price. From doing so, they earn a speculative profit from the changes in the exchange rate. The specific forms of this kind of capital flows include: speculation on temporary changes in exchange rates; speculation on permanent changes in exchange rate; speculative capital flows related to trades; reactive capital flows to interest rate differences between countries; etc.

Short-term flows of international capital represent a form of international capital movement that is most complicated and involves large sums of money. Such flows possess the following four characteristics: (i) Complexity. Short-term capital flows have not only variations of complicated forms, such as the afore-mentioned flows of trade, bank, value keeping, and speculation, etc., but also involve various sophisticated tools, such as cash, bank demand deposits, and other credit tools available on the money market, such as short-term bonds, papers, etc. (ii) Policy. That is, economic policies of different countries, such as interest rate policies, exchange rate policies, etc., bear significant effects on the short-term flows of international capital. When the interest rate of a country is lifted higher than other countries, international capital will flow into that country. Conversely, international capital will flow out of that country. Additionally, if a country does not have any control on the exchange rate or has a relatively loose control on the exchange rate, then short-term international capital can also easily flow into that country. (iii) Speculation. Within the floating exchange rate system, short-term international capital, especially the "hot money" within the short-term capital, has very strong characteristics of speculation. Speculativeness constitutes a clear characteristic of short-term flow of international capital. And (iv) marketability. That means that the hot money truly "complies with" the laws of market: the hot money will flow to wherever there is higher profitability. And, some people even fabricate news about where there might be more chance for profit and where there would be fewer opportunities. By driving up or suppressing the currency (or currencies) of a nation or region, they artificially establish capital flows from around the world.

In terms of long-term capital flow, it means the movement of capital that lasts 3 or more years or has no definite framework of time. It represents the most important mode of international capital movements. Its basic forms include direct investment, purchase of bonds, international lending, and international economic aids. The root cause behind international capital flows is the continuous development of the world productive forces and increasing refined division of international labor. The fundamental conditions for long-term international capital flows to form are the comparative advantage of different countries, such as ownership advantages, internalization advantages, geographic location advantages, etc. However, the motivation of long-term international capital flows is very diverse, including those driven by profit, by factors of production, by markets, by political speculations, etc.

Flows of long-term international capital appear in the following main forms:

1   Direct investment. It means such business activities that investors of one country invest directly in the industry, mining, commercial ventures, financial services, and other areas of another country, while acquiring partial or whole management controls of the invested companies. When seen from the angle of investment methods, direct investments can be divided into the following four forms: (i) The creation of new enterprises. It means that investors directly establish in another country wholly owned new enterprises, branches or joint venture subsidiaries of multinational companies. The investments for new enterprises, especially those involving joint investments, do not have to be limited to monetary capital, where machinery and equipment, even inventories, can be used as investment capital. (ii) Direct purchases. It means that the investors directly purchase established businesses in another country. When compared to the investment form of creating new enterprises, direct purchase has the following advantages: Firstly, it can save more time and money than those spent in creating new enterprises. It simplifies unnecessary steps and procedures; it can inherit the technology, management skill, and the market share of the original company, while pushing the products quickly into the international market. Secondly, it can lower the management cost, while lifting the economic efficiency. (iii) Purchase of company stocks while reaching a certain proportion. As what the United States government rules, when a foreign entity owns more than 10% of the stock of an American company, the foreign entity is considered making direct investment. (iv) Reinvestment. It means that the foreign investors reinvest some or all of their profit made in another country in the original company or other companies of the same country. This kind of investment also belongs to the form of direct investment. However, in practice, reinvestment does not truly involve inflows or outflows of international capital. The reinvestment of profits represents an important form of direct foreign investment of multinational companies. It occupies about 40% of all international direct investment.

2   Bond investment. It means such a form of investment that investors purchase priced bonds in foreign currencies in the international bond market. When seen in the angle of a country, when bonds are purchased inward in the international bond market, it is known as an investment, where what it means is an outflow of international capital. Conversely, selling bonds in the international bond market is known as raising funds. It implies an inflow of international capital. Investors and fund raisers in the international bond market can be governments, or business entities, private persons, or international financial institutions. Bond investment represents an important form of international capital flow. Especially since the 1980s, as affected by various factors, the securitization of international investments has shown the tendency of non-stopping increase and played its important roles.

The investments in securities take mainly two forms: One is the investments in stocks. In this case, the investors purchase stocks issued by foreign companies in the international stock market. Stocks represent the certificate of ownership of the investors in publicly traded companies. They do not carry terms. When the investors need cash, they can sell their stocks in the stock market in order to retrieve the cash. So, investment in stocks belongs to the category of long-term capital investments. The other form is the investments in bonds. Such investments mean that investors purchase the bonds of foreign governments or foreign companies in the international bond market. The holders of bonds can retrieve their principals and interests on the maturity dates from the bond issuers. The terms of bonds can be divided into short term, midterm, and long term. Purchases of mid-term and long-term bonds are considered as long-term investment. In terms of the investors, because bonds can be either converted back into cash or held until their maturities to retrieve the principals and interests, investments in bonds have relatively low risk and guarantee for return. When a bond is freely traded in the market, its market price is inversely proportional to the bank interest rate.

3   International credit. It means such a credit event that a nation's financial institution, the government, and a business company, or an international financial organization lends funds to the government, enterprise, or financial institution of another nation, or an international organization. International credit includes mainly the following three forms: government loans, international financial organizations' loans, and international bank loans. More specific forms also include export credits, leasing credits, compensation trade credits, etc. International credit possesses the following main characteristics: it represents the international movements of purely borrowed monetary assets. Different of either direct investments, which involve the creation of the physical entities of enterprises or the purchase of the stocks of existing companies in another country, or bond investments, which touch on the issuance and trade of bonds, the returns of international loans are seen as interests and relevant fees.

4   International economic aids. It stands for both gifts and concessional loans international economic organizations provide to developing nations. It is also a component of the long-term international capital flow.

9.4.2 Causes, present state and expected future of capital flows

The root reason underlying the international flows of capital is the ultimate pursuit of profits, especially economic profits. The globalization process of production and capital after WWII has caused the scale of international movement of capital increased tremendously and helped to accelerate the flows. Studies of the flows of international capital that pass across national borders reveal that there are many different reasons behind these movements of capital.

First of all, the national differences in the rates of return motivate capitals move from locations of low rates of return to those of high rates of return. If there were no obstacles that hindered the free movement in the world, factors of production would always gather in places where they could maximize their efficiency, which implies that the maximum return of the factors would be achieved. In the practice of international economics, capitals always flow from places of relative abundance (while the return is relatively low) to places of relative scarce with relatively high returns. In real life, although factors that hinter the free movement of capital always exist and it is very difficult for capital to freely move to places that provide the maximum return, differences in returns and interest rates always motivate capitals to find their innovative ways to flow from regions of lower return to regions of higher return. So, the rate of return of the countries with outflows of capital goes upward, while that of the countries with inflows of capital goes down. When the rates of return of these two places are roughly equal, the capital flows will gradually slow down until the eventual halt.

Secondly, changes in exchange rates and imbalances in international incomes and payments cause international flows of capital. In the international financial market, people always sell the currencies whose exchange rates are weakening and pursue after the currencies whose exchange rates are strengthening in order to avoid exchange rate risk or to make profits in the process of changing exchange rates. In the process of switching currencies to hold, flows of capital in different currencies will appear. And large surpluses and deficits in international payments can also cause capitals to flow across national borders in the settlements of international credits and debts. At the same time, imbalances in international incomes and payments can weaken the currencies of the nations that suffer from deficits and strengthen the currencies of the nations that enjoy surpluses. Within such international flows of capital, large scale, across national border movements of short-term capital can impact the exchange rates severely, while turbulent fluctuations in exchange rates in turn can cause short-term capitals to move from one country to another in large magnitudes. That end had been shown vividly in the financial crisis of the Southeast Asia in the late 1990s.

Thirdly, exchange rate risk, market risk, and other risk factors can cause capitals to flow internationally. The relationship between risk and return is extremely important for investors. Speaking generally, each movement of capital needs to collectively consider safety, liquidity, and profitability. When the levels of risk are equivalent, investors pursue after higher rate of returns. When the rates of return are the same, investors then go after the opportunities with better safety. However, different people have different understandings on the same risk, such as economic risk, political risk, financial risk, etc. So, in the practical operations of international economy, when people who are risk averse pursue after the objective of safety, they move their capitals from places of high risk to places of low risk. And at the same time, risk lovers earn their returns by moving their capitals into areas of high risk. That can also cause international flows of capital.

Fourthly, speculation, circumventing trade protection, international division of labor, and other factors also cause international flows of capital. In the international economy, because of the constraints of trade protections, there are trade barriers. In order to bypass these trade barriers, people often make direct international investments. That is, companies will invest directly in the places of market by erecting factories to produce and to sell their products at the same places. Such processes will cause international movements of capital. Additionally, the multinational companies, which stated to appear rapidly after WWII, have greatly encouraged international movements of capital in their across-border business operations no matter whether they invest in their vertical division or horizontal division of international labor. On the other hands, the economic globalization that began to develop after WWII has greatly eliminated the barriers of trades and capital flows between the member nations, which has accelerated the speed of international flows of capital. For example, the process of European economic unification not only pushed for free movements of capital within the territory of the European Union, but also encouraged movements of capital within the areas that have close ties with the European Union.

Since the 1980s, the scale and direction of international capital flows have been constantly changed and adjusted, showing the characteristics of pattern changes and development tends. (i) Private capital instead of governmental capital becomes the main player in the international capital movements; (ii) the time structure of international capital flows become increasingly blurred; (iii) the structure of the international capital market is further adjusted; (iv) the strategy of "flight to quality" has dominated the basic movement of international capital; (v) the regional structure of investment becomes increasingly more imbalanced; and (vi) it is becoming clear that foreign investments of the advanced nations are retreating back to their home countries. In particular, based on investment entities, international capital flows are divided into two groups: official development finance and foreign private capital. Before the 1980s, the main carriers of international capital flows were governments and international financial institutions. However, according to the statistics of the World Bank, since the 1990s, the amount of private investments that flew into developing countries has been four times of the combined total of international financial organizations and governments. Currently, the movement of private funds has occupied 3/4 of the total capital that flow around the world. According to the 2005 report on the global development of finance of the World Bank, the amount of capital that flew into developing nations in the year 2004 had reached the unprecedented historical high level since the financial crisis of the 1990s. The movement of net private capital, including debts and equity capital, into developing countries increased US$51 billion in 2004, reaching US$301.3 billion, where the net amount of foreign direct investments (FDI) reached US$165.5 billion in 2004, an increase of US$13.7 billion from the previous year. Among private investments, the investment behaviors of multinational corporations had become forever more important, and the weight of private investments had been further strengthened.

International capital flows are often studied as movements of short-term capitals and long-term capitals. As for the flow of long-term capitals, it mainly includes international direct investments, indirect investments (investments in bonds), international credits, and other forms. The flow of short-term capitals on the other hand mainly includes such capital flows of trading, bank funds, value keeping, and speculation. The term standard commonly used to divide long-term and short-term capital flows is 1 year. Evidently, these two classes of capital flows have different motivations, objectives, and impacts on individual nations' balances of international payments, and even on the stability and development of the world economy. And demands for regulating the flows and degrees of recognition of the need to regulate the flows are different. However, along with loosening of the world financial and trade control, and appearance of financial innovations one after another, especially the introduction of new financial products and securitization of assets, it has made it very convenient to frequently convert back and forth the long-term and mid-term assets of the international capital flows. For example, the appearance of large certificates of deposits, swaps of currency and interest rate, discounting and extending bills, and various fund operations has made the term structure of international capital flows increasingly blurred. In the real economic life, clearly distinguishing long-term and short-term capital flows has become very difficult. At the same time, large amounts of short-term capitals are often mixed in with international trades and flow alongside of long-term capitals, making regulating capital movements more difficult and more costly.

Along with the further strengthening of financial innovations and the securitization of assets, investments in international bonds has become a new trend of development and another important method for international capital flow. According to the 2005 report on the global development of finance of the World Bank, the issuance of bonds in 2004 reached another record level, where the movement of private debt capital experienced strong growth. At the same time when the net amount of bank loans continued to drop, the movement of net bond capitals rebounded strongly, reaching a new high of historical record. Although more and more countries could obtain bank loans, the total amount of bond issuance had surpassed the total sum of bank loans the first time in history.

Theoretically speaking, the regional (nations) structure of international capital flows is determined by individual countries' economic growths, differences of interest rates, states of market structures, changes in psychological expectations, and other factors. Under the normal circumstances, the rate levels of return dominate the basic movement of international capitals. However, under particular circumstances, such as wars, financial storms, economic recessions, credit crises, etc., the owners of capital will pay more attention to the safety of their investments and the quality of the assets they hold. The basic flows of capital in the current and future time periods will be dominated by the flight-to-quality strategy and by first selecting the assets of the lowest risk.

In the regional structure of international capital flows, for a long period of time, most of the flows have appeared between advanced nations and regions. Although the proportion of flows that go into developing countries, when compared to the total amount of capital in the world market, has shown an uptrend, it is still at a relatively low level, and in some years, the proportion had been dropped. According to statistics, the amount of direct foreign investments in developing countries increased in 2004, which partially offset the falls in the previous two years, reaching 11%, the highest level in the past 15 years. Even so, the proportion of capital movements that flew into low-income countries was still very low. Additionally, the outflow of direct foreign investments, as published by developing countries, had shown dramatic increase. It was estimated to have reached US$40 billion in 2004 (it was only US$3 billion in 1991). The outflow of direct foreign investments mainly appeared in the countries that received most of the direct foreign investments in the recent years.

According to the Commerce Department statistics of the United States, by using the historical cost, as of the end of 2003 the direct investment abroad made by the United States had reached an accumulated total of US$1788.9 billion with a market capitalization of US$2730.3 billion. Within this accumulated total, in order to dodge the domestic income taxes and to make it convenient to apply capital anywhere in the world stage, multinational corporations of the U.S. capital have left surprising amounts of profits abroad.

Since the economic recovery seemed to have gained momentum in 2004, there have appeared signs that the Japanese manufacturing sector, which has been moving abroad massively since the 1980s, started to retreat back to its homeland. Starting in late 2004, many multinational corporations of the Japanese manufacturing sector, especially the electrical and electronic sector, have strengthened their domestic investments while

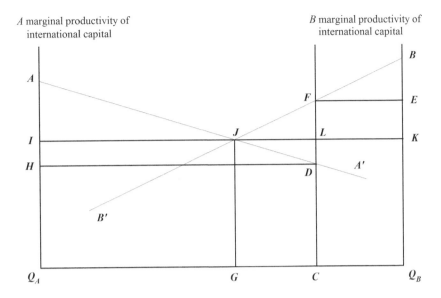

Figure 9.1 General model for the flow of international capital.

moving parts of the production of high tech and high value-added products back to Japan.

9.4.3   Impacts of international capital flows

Along with the expansion in the scale of international capital flows, the acceleration of the movement speed, and the increase in the innovations of financing and trading tools, the impact of international capital flows has become more widely felt. This subsection will discuss the pros and cons of international capital flows by using theories and models.

First, let us look at what effects capital flows have on the total economy and welfare by using MacDougall's welfare effect model (MacDougall, 1960). MacDougall established the welfare effect model for two countries by using the neoclassical general equilibrium analysis so that he could compute the optimal scale and the corresponding effects of capital inflows. He believed that the reason for international capital to flow is that there are differences between individual countries' interest rates and their expected interest rates. When the markets of all nations are perfectly competitive, funds can move freely from nations with abundant capital into nations with shortages of capital. The international movement of capital will make the individual marginal productivities of the nations to converge so that the total output of the world and the welfares of all nations will improve.

This model is developed on the basis of the following assumptions: (1) The world economy consists of two nations A and B, where A represents the nation with abundant capital and exports it, while B is short of and imports capital. (2) The marginal productivity of capital decreases. That is, when all other factors are held constant, additional capital will lead to less productivity per unit capital. (3) The domestic economies of these nations are in perfectly competitive state, where the marginal productivity of capital is equal to the marginal yield of capital. (4) The rate of savings represents a fixed proportion of the per capita income. That is, the savings propensity is an exogenous variable. (5) The inflowing capital is entirely used for investment without being spent in consumption.

In Figure 9.1, the vertical axis stands for the marginal yield of capital, AA' and BB' are respectively the curves of marginal productivity of the nations A and B; they are also the demand-for-capital curves of the nations. The horizontal axis represents the quantity of capital, where $Q_A C$ and $Q_B C$ are respectively the quantities of capital of the nations A and B, and $Q_A Q_B$ the total quantity of capital of the world. Because the initial marginal productivity CD of nation A is lower than that CL of nation B, it causes the capital of nation A to move into nation B. When the amount of capital moving from nation A to nation B is equal to GC, the respective marginal productivities of these two nations are equal, that is, $GJ = Q_A I + Q_B K$. The total output of nation A changes to $Q_A AJG$ from $Q_A ADC$ before the capital occurs, the total output of nation B changes to $Q_B BJG$ from $Q_B BFC$, and the world total changes to $Q_A AJG + Q_B BJG$ from $Q_A ADC + Q_B BFC$. That is, the output of the area of the triangle JFD, the welfare triangle, is increased. So, it follows that the free movement of international capital can help improve the total output of the world, although the effects on the inflow and outflow nations are different.

As for the outflow nation, the output decreases ($GJDC$), the return of labor also drops (from HAD to IAJ). However, the return of the capital increases from $Q_A HDC$ to $Q_A IJG + GJLC$ with an addition of $HILD$. As long as the amount of investment abroad $GCLJ$ is greater than the amount of decrease in production $GJDC$, then the overall national income of the nation with capital outflow will increase, JLD in Figure 9.1. Conversely, for the nation with capital inflow, the output, labor income, and the total national income (JLF in Figure 9.1) increase, while the return of capital decreases. Additionally, this model implies that taxing the profit of foreign investment might make foreign investors to reduce their input of capital, although the welfare of the nation with inflow of capital will continue to rise. By benefiting from the effect of size, elimination of monopoly, advance in technology, removal of production bottlenecks, etc., the external economy can make the marginal productivity curve of capital move and increase the output so that the overall welfare level of the nation with capital inflow will rise. The degree of rise is affected by various factors, such as laborers' wages, and the changes in the taxation as caused by the changes in the returns of the domestic and foreign capital.

MacDougall's neoclassical model also suffers from some weaknesses. For example, when analyzing the effect of capital inflows on the receiving nation's economy, the model is more appropriate to explain how direct foreign investments affect the increase of the receiving economy. And, the capital accumulation, as caused by capital inflow, does not guarantee the welfare of the receiving nation to definitely rise.

Next, let us look at how the two-shortage model explains the relationship between capital inflows and economic development. The two-shortage model is initially proposed by Chenery and Brune (1962). It explains why theoretically it is necessary for developing countries to apply foreign capital to offset their domestic shortages of funds. In order to maintain the economic growth at a certain speed, the difference between investments and savings, known as the savings shortage, must be kept the same as the difference between imports and exports, known as the foreign exchange shortage. Because investment, savings, import, and export are all independently variable, these two differences in general are not equal. In order to achieve the equality, one of the two methods of adjustment can be applied. One is the method of not employing foreign capital. When the domestic savings shortage is greater than the foreign exchange shortage, either the investment needs to be compressed or savings increased. When the foreign exchange shortage is greater than the savings shortage, either the import needs to be reduced or the export increased. By using this method, unless it is very possible to increase savings or export, otherwise the speed of economic growth will be slowed down. Another method of adjustment is to look for capital beyond the differences. That is, use foreign resources. For example, import equipment and machinery by using foreign capital. On the one hand, such import does not need to be balanced by export; and on the other hand, this investment project does not need to be offset by domestic savings. So, making use of foreign capital can offset two shortages. That is, it can satisfy the need of investment, and it can reduce the pressure of paying for the import so that economic growth is guaranteed.

In macroeconomics, $S = I$, which means that the savings can be smoothly transformed into investments, is a basic condition for economic growth. However, when the domestic source of capital is not sufficient to support the desired speed of growth, there appears a shortage in such areas, say, as savings, foreign exchange, government income, technology, etc. So, introducing foreign resources to offset the shortage becomes necessary. To this end, Chenery and Brune's two-shortage model mainly considers the savings shortage and foreign exchange shortage. From the fundamental identity of the national economy that the total income is equal to the total supply, we have

$$\text{total supply } Y = C + S + T + M \tag{9.2}$$

where C stands for consumption, S the savings, T the tax income, and M the import.

$$\text{total demand } Y = C + I + G + X \tag{9.3}$$

where I represents the investment, G the government spending, and X the export. If the tax income is equal to the government spending, that is, $T = G$, then we have

$$S + M = I + X \quad \text{or} \quad I - S = M - X, \tag{9.4}$$

where the left hand side $(I - S)$ is the difference between investment and savings, the savings shortage, while the right hand side $(M - X)$ represents the difference between import and export, the foreign exchange shortage. Because investment, savings, import, and export are four independent variables, the purpose of making adjustment is to make the previous equation hold true.

In terms of the policy implications, the model emphasizes the necessity for a developing nation to utilize foreign resources, which can strengthen the nation's capability of export, and help to support the appearance of the virtuous circle of higher income and higher levels of savings, making the nation's resource allocation more reasonable. Therefore, developing nations need to adjust their domestic economic structures to meet the need of attracting foreign resources by sufficiently making use of the roles of their governments in adjusting economic activities. However, the foreign capital has to be repaid eventually. So, there is a need to improve the use efficiency of foreign capital so that the foreign resources can either directly or indirectly stimulate the growth in export, in savings, and in the ability of repaying the debts.

In the rest of this subsection, let us look at how international capital flows affect such economic variables as exchange rates and money supply.

First of all, let us look at the effect of international capital flows on exchange rates. There is no doubt that exchange rate represents one extremely important economic variable in an open economic system. Along with the increasing importance of international capital flows in the world economic trades, changes in their scales and structures have constituted the important economic structure behind the determination mechanism of exchange rates. The effect of capital flows on exchange rates is shown in two aspects. One is their effect on the nominal exchange rate, and the other on the actual exchange rate.

The effect of the net inflow of capital on the nominal exchange rate varies from one nation's form of exchange rate arrangement to another nation's form. Specifically, for the floating exchange rate system, the net inflow of capital shows up as an increase in the supply of foreign currency in the foreign exchange market. If the demand for the particular foreign currency is fixed, that will surely cause the value of the domestic currency of the receiving nation to rise. The rise in the nominal exchange rate in turn affects the trade surplus. In this case, using trade deficit to attract a net inflow of capital can help reduce the appreciation pressure of the domestic currency while materializing an automatic adjustment of the international payments. For the fixed exchange rate system, each nation's monetary authority is responsible and has the obligation to maintain the exchange rate of its domestic currency at a fixed level, or to limit the fluctuation of the exchange rate within a specified range. So, although foreign capital inflows continuously, the nominal exchange rate will still be maintained at the pre-determined level. That will lead to an increased supply of foreign currency, which will eventually become part of the foreign exchange reserve held by the central bank.

The appreciation pressure of the domestic currency, as caused by capital inflows, can cause changes to the actual exchange rate through the path of increasing the domestic inflation. When the net inflow of capital is absorbed by the central bank as a part of the foreign exchange reserve, if this additional amount is not balanced off somehow, the increased reserve will cause the domestic currency supply to rise. If the demand for the domestic currency did not increase accordingly, it will cause the domestic price level of goods to go up. If the foreign price level and nominal exchange rate are maintained constant, then the actual exchange rate of the domestic currency rises. And when the magnitude of the domestic price increase is lower than that of the foreign inflation, the invariant nominal exchange rate means that the actual exchange rate of the domestic currency has been lowered. From some of the changes in the actual exchange rates as experienced by capital receiving countries, it can be seen that the actual exchange rates of Latin American countries, such as Argentina, Mexico, and others, jumped in large magnitudes after foreign capital was introduced, while the scales of rise in the actual exchange rates of Asian countries and regions, such as South Korea, Singapore, Taiwan, etc., after they received foreign capitals, were relatively small. Additionally, there are also a few countries, such as Brazil and Indonesia, that their actual exchange rates fell after importing foreign capital.

Secondly, let us look at the effect of international capital flows on the money supply. The effect of capital flows on the money supply is mainly shown in the movement and stock of international payments. Changes in the movement and stock of international payments will surely be reflected in the changes of the foreign exchange reserve. That is, inflow of international capital → surplus in international payments → increase in the foreign exchange reserve, and vice versa.

If analyzing such nominal variables as foreign exchange reserve, money supply, and price level from the angle of creating the domestic currency, the following relationship holds true:

$$M = K \cdot B \tag{9.5}$$

where K stands for the money multiplier, and B the base currency. The base currency $=$ the net foreign assets F of the central bank (the foreign exchange reserve of the nation) $+$ net claims on the government $D +$ other money supply projects, such as re-lending to financial institutions, bond investments, and other loans $=$ the net amount of foreign assets $R +$ the quantity of domestic credit W. That is, $B = R + W$. If the increment of change is written as $\Delta B = \Delta R + \Delta W$, then

$$M = K \cdot B = K(R + W) = K \cdot W + K \cdot e \cdot u \tag{9.6}$$

where e stands for the exchange rate and u the amount of foreign currency held in hand.

Now, let us assume that the money multiplier is a constant. And for the convenience of our discussion, we also assume that the exchange rate is also constant. Let ΔM be the increment in the money supply, ΔW the increment in the quantity of the domestic credit, and Δu the increment in the foreign exchange reserve held in hand. Then we have

$$\begin{aligned} M + \Delta M &= K(W + \Delta W) + K \cdot e \cdot (u + \Delta u) \\ &= K \cdot W + K \cdot e \cdot u + K \cdot \Delta W + K \cdot e \cdot \Delta u \end{aligned} \tag{9.7}$$

This equation implies that in an open economy, the foreign exchange reserve and the domestic money supply of one nation are related in the following ways:

1   Changes in the foreign exchange reserve stand for an important channel of money supply so that short-term capital flow influences money supply through affecting the foreign exchange reserve.
2   The three scenarios of capital flow have different effects on the money supply. If the inflow of capital is greater than the outflow, the foreign exchange reserve increase the amount of Δu, and the money supply is expanded $(K \cdot e)$ times. That is, M is increased as much as $(K \cdot e \cdot \Delta u)$. If the capital inflow is equal to the outflow, the change in the foreign exchange reserve will be 0 so that the money supply will not change. If the capital inflow is smaller than the outflow, the foreign exchange reserve will be reduced by the amount Δu so that the money supply is decreased by the amount of $(K \cdot e \cdot \Delta u)$.

3   The mechanism of how long-term capital flow affects the money supply is the same as that of short-term capital flow, except that other than affecting the money supply, long-term capital flow can also lead to indirect consequences on the money supply through capital inflow's combined effect, multiplier effect, and induced effect, which affect the demand for currency of the receiving nation so that a pressure on money supply is created.

9.4.4   International capital flows and national financial systems

International capital flows not only bring forward huge effects on the macro economy and policies of the receiving nation, but also create not ignorable risk to the financial system of the receiving nation. After the Asian financial crisis of the 1990s, financial risks that might be triggered by international capital flows have caught unprecedented attention. For example, Krugman (1997), who was the first to have predicted the appearance of this crisis, proposed the "eternal triangle" theory, believing that the independence of monetary policies, the stability of exchange rate of currencies, and the free flow of capital constitute a triangle. However, in economic development, not all nations could materialize these three objectives simultaneously. So, each nation chooses two of three that are most beneficial to the nation.

Speaking generally, the financial risks brought along by international capital flows are mainly reflected as currency crises, bubble economies, bank crises, etc. However, the financial crises brought forward by capital flows do not necessarily evolve into currency crises or bank crises. Whether or not a financial crisis can be effectively contained depends on the adjustment capability of the monetary authority and the international economic environment of the time.

In terms of currency crises, the exchange rate, as a ratio of two currencies, should be mainly determined by the force of market supply and demand. Along with the relative ups and downs of the demand and supply forces, there naturally appear certain fluctuations in the exchange rate. However, when the fluctuations of the exchange rate are too wild, it can easily lead to instabilities of the foreign exchange market and even currency crises. Depending on the exchange rate system installed, currency crises can be divided into two groups: For the fixed exchange rate system, a change in the exchange rate means the domestic currency is devalued (or appreciated) when compared to the foreign currency, making the pre-determined level of exchange rate difficult to maintain. For the floating exchange rate system, when the fluctuation in the market exchange rate goes beyond the range that can be explained by using the actual economic factors, it is referred to as a currency crisis.

Speaking generally, the development process of capital flow and currency crisis can be summarized into three stages. For the first stage, before the crisis breaks out, the nation in danger experiences inflows of large amounts of capital so that the stock prices and real estate property prices rise rapidly, and the value of the domestic currency goes up. For the second stage, various factors, such as the appreciation of the domestic currency, cause the economic growth to slow down, the imbalance in the current account to become more severe, and the essential economic situation to gradually reverse. On top of all these happenings, the expectation for the domestic currency to depreciate starts to appear in the financial market. For the third stage, the panicky mood of the market gradually accumulates to such a point that a crisis could be triggered readily. Then speculators launch their attack on the exchange rate of the domestic currency by dumping the currency. That makes the capital flow reverse its direction, and causes the domestic currency to depreciate drastically; a currency crisis breaks out. It is often the prelude of bank system crises.

In terms of bubble economy, The New Palgrave: A Dictionary of Economics (1987) defines bubble economy as the price of an asset or the prices of a series of assets suddenly jumped higher in a continuous process; and at the start, the rising prices make people expect further price increase so that new buyers appear. These people generally only care about making quick profits from their trades without any interest in the use of these assets or in the earning potential these assets might have. Accompanying the rising prices are there often reversed expectations, following which prices slump and end eventually in a financial crisis. Bubble economies are instigated by many factors, such as virtualization of economies, financial liberalization, etc., where international capital flows surely play the fueling role of creating and busting bubble economies.

The financial globalization of the developing countries and regions has provided new playgrounds for speculations of international capitals. And because of lack of domestic capital and relatively high interest rates, at the starting period of the financial globalization these countries and regions almost always experience large amounts of inflows of foreign capital, creating a necessary condition for the appearance and expansion of financial bubbles.

Firstly, in the money market, if the receiving country implements the floating exchange rate system, the domestic currency has to appreciate and will be overestimated. The bubble part of the domestic exchange rate weakens the competitiveness of the department of export and trades, causing deficit in the current account and deterioration in the income and payment of trades. If the receiving country implements the fixed exchange rate system, even only for the stability of nominal exchange rate, the inflation that is caused by the inflow of capital can also make the actual exchange rate rise. In the late 1990s, because of large sums of foreign capital inflows, there appeared exchange rate bubbles in most of the Eastern Asian nations as a consequence of large jumps in the actual exchange rates of the domestic currencies. That laid the crucial ground and provided the opportunity for speculative attacks on the domestic currencies of the region.

Secondly, the stock and real estate property bubbles. If the receiving country implements the fixed exchange rate system, while the central bank does not apply sterilization policy, then the supply of the domestic currency will have to increase. The abundant quantity of currency will naturally flow to stock market and the housing market, forming bubbles. In the late 1990s, forty-six percent of international bond capital and 16% of the international stock capital of the Asia-Pacific Economic Cooperation (APEC) went to the finance and real estate sector. In Thailand, the percentages were even higher, reaching 60% and 18% respectively. Inflows of large amounts of capital have to cause the prices of the financial assets with relatively small supply-demand elasticity to suddenly rise. Even when the floating exchange rate system is implemented, if the domestic interest rate stays continuously higher than the rate abroad, the inflows of large quantities of foreign capital will also create huge bubbles in the stock and real estate markets. In his analysis of the causes behind the Asian financial crisis, Krugman

(1999) believed that there existed serious moral hazard problem in the financial structure of Southeast Asian countries. That caused bubble economies to appear; and when the bubble economies popped, that became the most important reason of the Southeast Asian financial crisis.

In terms of bank crisis, according to the definition established by the IMF organization, it means real or potential run that causes either banks to stop cashing their debts until they are forced to close down or the government is forced to intervene or rescuer the banks. Because of the banks' characteristics of high debts, banks suffer from their inherent fragility. At the same time, as a place that gathers borrowers and lenders, the prerequisite for banks to operate normally is, on the one hand, that the released loans can be repaid on schedule, and on the other hand, that the depositors do not show up to withdraw their funds at the same time, which makes the law of large numbers hold true. However, international capital flows constitute shocks to both the asset side and liability side of banks.

Along with the relaxation of control and the inflow of large quantities of foreign capital, banks' liquidity surges rapidly and banks' capability of credit expands. Faced with these two situations, the management of banks, when seen from the angle of assets owners, likes to expand asset scales, pursue after the extraordinary high risk profit, and lend funds to high risk businesses and ventures, such as real estates, securities, etc., in order to make the asset prices of these entities rise quickly. On the other hand, the rising prices of these assets stimulate for loans to gather in these business ventures, forming a bubble and causing the debt structure of banks inappropriate. As soon as the bubble busts, what is left with the banks will be large amounts of bad and doubtful accounts. When seen from the angle of the liability side, when international capital outflows quickly, it creates withdrawal pressure on banks, making banks trapped in the difficulty of liquidity. That difficulty in turn generates a herd effect in the domestic depositors, causing large scale bank runs. That adds "frost" on top of the "snow" of the banks' liquidity problem.

Another reason that helps to make bank crises more severe is the trading of financial derivatives. These derivatives represent an important component of international short-term capital. They were originally introduced for the purpose of hedging and discovering prices. However, as the virtual economy departs further from the real economy, these derivatives have been excessively virtualized. Together with their ability of leverage, they are widely used by international speculators, creating more risk for the financial market and helping to expand bubble economies. Such accidents of risk as how Barings Bank, a British veteran investment bank, ended its business forever as a consequence of losing billions of U.S. dollars from speculating financial derivatives, how Japan's Sumitomo Corporation lost over US$2.8 billion, and others, further confirm the fact that international capital flows have greatly create the risk of the financial market.

If international capital suddenly leaves from the banking system, it can not only make individual banks collapse, but also cause potential problems for the entire banking system. It is because the central bank involved can only assist the commercial banks with its limited foreign currency reserve and reserve capital. It cannot print and issue foreign currencies, and cannot play the role of the ultimate lender of foreign currencies. The severity of the problems can easily go beyond the boundary the central bank can stand, leading to system risks.

The afore-given analysis indicates that under the current condition that the flows of international capital are becoming increasingly more frequent than before, these flows may very well cause damages to an individual nation, making the nation suffer from risky financial and economic turbulences. Studies in Chapter 11 will show that when the macro economy of a nation experiences certain problems, that nation may very well become the target of financial attacks of international speculators, forcing the nation into financial crisis. To this end, please go to Chapter 11 for more details.

9.5 INTERACTIONS BETWEEN ECONOMIC YOYO FIELDS

When more than two economic entities are concerned with, the interaction between these economies, a multi-body problem, becomes impossible to describe analytically. This end is witnessed by the difficulties scholars have faced in the investigation of the three-body problem since about three hundred years ago when Newton introduced his laws of motion regarding two-body systems. So, to comprehend meaningfully how international money flows affect individual national economies, in this section, we will focus on figurative expressions developed for the interactions between economic entities.

9.5.1 Classification of economic yoyo fields

Just as electric or magnetic fields, where like fields repel each other and opposite fields attract, for the yoyo field structures underneath economic systems, the same principle holds true, where the like-kind ends (either the north poles or south poles) repel and the opposite attract. In particular, when the south poles of two economic yoyo fields face each other, Figure 9.2(a), where S stands for the black hole side and N the big bang side, although the two south poles have the tendency to attract each other, the meridian fields X and Y actually repel each other apart, where m_{X_1} repels against m_{Y_1} and m_{X_2} against m_{Y_2}. So, in this case, there does not exist any attraction between the two black hole sides, since these S sides cannot feed anything to each other directly, and both have to get input materials from their meridian fields. When the north poles of yoyos X and Y face off (head on), Figure 9.2(b), the yoyos are pushed apart, where in comparison the meridian fields of X and Y only come into place with attraction when the edges of the north pole sides of X and Y directly face each other. When the south poles of an economic yoyo structure faces the north pole of another economic yoyo structure (Figure 9.3(a)), where what is spurted out of the economic yoyo structure Y can go and do go directly into the economic yoyo structure X. At the same time, the meridian fields of both economic yoyo structures X and Y have the tendency to combine into one field too (Figure 9.3(b)).

In terms of the relative positioning of two economic yoyo structures, assume that two spinning yoyo fields X and Y are given as shown in Figure 9.4. Then the meridian field A of X fights against C of Y so that both X and Y have the tendency to realign themselves in order to reduce the conflicts along the meridian directions. Similarly in Figure 9.4(b), the meridian field A_1 of the economic field X fights against B_1 of Y. So, the economies X and Y also have the tendency to realign themselves as in the previous case.

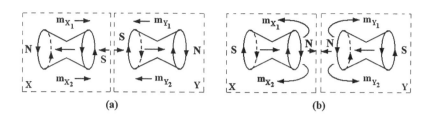

Figure 9.2 Repulsion of like economic yoyo fields.

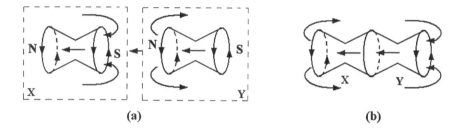

Figure 9.3 How two economic yoyos can potentially become one economy.

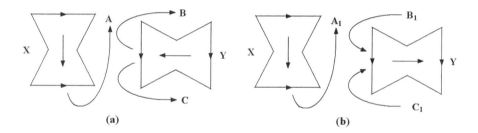

Figure 9.4 The tendency for economies to line up.

Of the two economic structures X and Y above, let us assume that one of them is mighty and huge and the other so small that it can be seen as a tiny particle. Then, the tiny particle economy m will be forced to line up with the much mightier and larger spinning economy M in such a way that the axis of spin of the tiny economy m is parallel to that of M and that the polarities of m and M face the same direction. For example, Figure 9.5 shows how the particle economy m has to rotate and reposition itself under the powerful influence of the meridian field of the much mightier and larger economic structure M. In particular, if the two economic yoyo structures M and m are positioned as in Figure 9.5(a), then the meridian field A of M fights against C of m so that m is forced to realign itself by rotating clockwise in order to reduce the conflicts with the meridian direction A of M. If the economic structures M and m are positioned as in Figure 9.5(b), the meridian field A_1 of the economy M fights against B_1 of m so that the particle economy m is inclined to readjust itself by rotating clockwise once

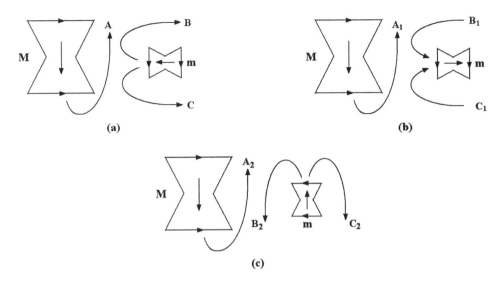

Figure 9.5 How a mighty economic field M bullies a particle economic entity m.

again. If the economic structures M and m are positioned as in Figure 9.5(c), then the meridian field A_2 of yoyo M fights against B_2 of m so that the tiny particle economy m has no choice but to reorient itself clockwise to the position as in Figure 9.5(b). As what has been just analyzed, in this case, the economy m will further be rotated until its axis of spin is parallel to that of M and its polarities face the same directions as M.

Now, let us imagine a tiny particle economy m that plays the role of a unit in the supply chain of an international business activity, where the supply chain is theoretically seen as the circular loop in Figure 9.6(a), where the X's stand for an overriding economic yoyo field of the environment going into the page. Then, the axis of spin and its polarities of the tiny economy m have to live up with those of an abstract, but realistically existing mighty spinning economic yoyo, part of whose eddy field is symbolized by the X's in Figure 9.6(a).

Based on the Left Hand Rule 1, for details, see Chapter 2, and the north direction of the economic yoyo field outside the supply chain, (the reason why we use the direction outside the circuit instead of the inside is because in the eddy field of the mighty economic yoyo, the circuit would be only the size of a dot), we can see how the particle economies, such as different units in the supply chain, line up inside the circuit as shown in Figure 9.7. This line-up of particle economies inside the circuit also determines the north direction of the superpositioned eddy field along the circuit, which fights against the increase in the intensity of the original eddy field. This end in fact provides a systems theory support for the validity of the following generalized Lenz's Law, where Lenz's Law is well-known in the theory of electromagnetism.

Generalized Lenz's Law: *There is an induced yoyo current in any closed, conducting loop if and only if the yoyo flux that is a yoyo field through the loop is changing. The*

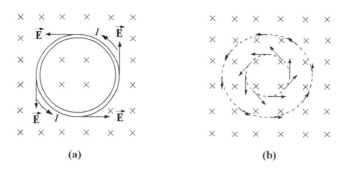

Figure 9.6 The manifestation of an induced economic yoyo field.

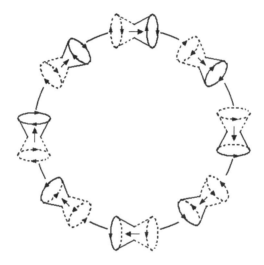

Figure 9.7 How particle economies line up inside the circuit.

direction of the induced current is such that its induced yoyo field opposes the change in the yoyo flux.

If we ponder over the discussion above in more details, it is not hard to see that the imaginary return circuit is not necessary. As long as the economic yoyo field (an eddy or meridian field), as indicated by the X's, changes, there will be a manifested yoyo field in the space (Figure 9.6(b)). This kind of manifested economic yoyo field is a special kind systemic field.

To make sense out of the generalized Lenz's Law beyond the traditional understanding in the field of electromagnetism, let us imagine an underdeveloped economy. It exists in an established international business order. If one day, the economy suddenly plans to change its business conducts completely, what will happen to the economy is that it will find that every other economic entity in the established international environment instantly become united in the fight against its planned changes. The more

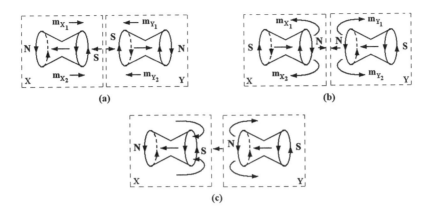

Figure 9.8 Systemic yoyo structure of economies combine through their meridian fields.

the underdeveloped economy is determined to carry out its planned changes, the more resistance it will experience from the united environmental front. This example is one case of application of the generalized Lenz's Law in the area of international business.

9.5.2 Combinations of economic yoyos through meridian fields

When systemic economic structures combine through their meridian fields (Figure 9.8), because these meridian fields have different properties in different directions such that the same poles repel each other and opposite attract, between the opposite poles of two meridian fields, there exists a field potential pit (trap) over which the economies attract to each other. At the same time, between the same poles of the meridian fields, a field potential rampart exists to repel the poles from each other. No matter whether it is the field potential pit or rampart, the field intensity is different from one economic yoyo structure to another. For economic yoyos of relatively stronger field strengths, their field potential pits are relatively deeper and their potential ramparts are relatively higher, meaning further away from the economic centers of the yoyo structures; and for economic yoyos of relatively weaker field strengths, their field potential pits are relatively shallower and their potential ramparts relatively lower. When systemic structures of economies link to each other through their meridian fields so that the black hole sides (the south poles) face with their big bang sides (the north poles), their eddy fields also link and interact with each other. For example, in Figure 9.9 the field structure of a ^{4}He nucleus is shown. This nucleus consists of two protons and two neutrons. Its yoyo structure looks like a large ring along which the protons and neutrons are connected through their quarks. In Figure 9.9(a), the oval arrows stand for the eddy fields' directions of the protons and neutrons, and the line arrows the meridian fields' directions. Because the protons and neutrons are connected through their quarks and their eddy and meridian fields' directions satisfy the Left-Hand Rule 1, the ^{4}He nucleus can be depicted well in Figure 9.9(b). In this figure,

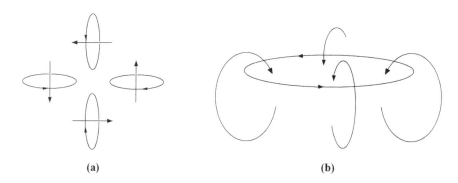

Figure 9.9 The yoyo structure of a ^{4}He nucleus.

the horizontal ellipse stands for the loop made up of the protons and neutrons connected through meridian fields and the vertical ellipses the eddy fields of the protons and neutrons.

As indicated by Figure 9.9, each field bond involves such parameters as bond strength, bond angle, bond force, bond length, bond energy, etc. The so-called bond strength is exactly the field intensity. The stronger the bond strength is, the greater the bond force and the higher the bond energy. Conversely, the weaker the bond strength is the smaller the bond force and the lower the bond energy. That is, the bond strength is directly proportional to the bond force (and the bond energy). Within the effective range of a yoyo field its field bond possesses elasticity. That is, both bond length and angle can be changed. When the bond length increases, the bond strength, force, and energy decrease.

When the strength of attraction between two opposite meridian fields increase, that is, when the bond length shrinks, the strengths of repulsion between the same poles of the meridian fields also increase accordingly. For example, when a proton attracts a neutron through the opposite poles of their meridian fields (Figure 9.9(c)), their same poles repel each other. When the nuclear bond shrinks due to the attraction of the opposite meridian poles, the strength of repulsion between their like meridian poles becomes stronger correspondingly. Both of the attraction and repulsion keep the active economic systems apart.

In short, what is discussed in this section indicates how small economies have not much choice in dealing with those economies that are many magnitude levels larger. However, if conditions are right, these small economies could potentially chain up together to form a heavy weight economic player in the international arena.

For more in depth discussion about interactions of systemic field structures, please consult (Lin, 2010).

Chapter 10

Financial globalization and its consequences

Because of the development of communication technology, there has appeared the tendency of globalization in the international economy. And the magnitude of capital flows around the world has also gradually surpassed the scale of international trades, forming a huge international financial market. This financial market, which goes beyond the limitation of geography and the constraint of time, makes it easy for the demanders and suppliers of capital, who are located in different countries, to conveniently complete their trades. Such convenience makes the depth and width of international business trades greatly strengthened and has helped the world economy to develop unprecedentedly. However, at the same time, speedy flows of large capital amounts have also created huge impacts on the international payments and national economies of all countries directly or indirectly through capital and financial accounts, and caused monetary, financial crises that went across national borders.

In this chapter, we will study the structure of the international financial market, causes and effects of international capital flow, impacts of financial globalization, and why financial attacks are always possible. In particular, Section 10.1 will look at the international currency market, capital market, and foreign exchanges. Section 10.2 investigates the reasons, characteristics, and effects of international capital flows. Section 10.3 considers debt and bank crises caused by international capital flows. In Section 10.4, such topics as the concept and causes currency crises, the Asian currency crisis of the 1990s and its resolution are discussed. The final Section 10.5 considers theoretically the question of whether or not financial attacks are always possible by looking at how environmental conditions determine the natures of individual economies, why there will always be financial centers under different sets of rules and regulations, and why emerging economies will always exist, providing easy targets for financial attacks.

In terms of references, this chapter is based on (Jiang, 2008; Lin, 1999; Lin and Forrest, 2011). For relevant, in depth discussions, please consult these works and the references listed there.

10.1 THE STRUCTURE OF THE INTERNATIONAL FINANCIAL MARKET

By international financial market, it means such a place where either capitals flow internationally or financial products are traded internationally. The international financial market mainly includes the international money market, international capital market, Euro currency market, foreign exchange markets, and some other important submarkets.

Since the 1970s, the direct connection between flows of capital and actual production and trades in the international financial market has been diminishing, showing obviously more of the financial characteristics of currencies. So, the international financial market has shown its own laws of independent movement, and produced important effects on the operation of open economies and the maintenance of internal and external balances.

10.1.1 The international money market

By international money market, it means the market that discounts short-term (less than one year) bank credits, short-term bonds and bills, where intermediaries include commercial banks, bills accepters, discounters, security dealers, and security brokers. Because of the differences in national traditions and customs, the main businesses of the money market and the positions of the intermediaries are also different. For example, in the United States, the short-term money market mainly deals with short-term bank credits and short-term bonds, where commercial banks occupy the most important position in this market. On the other hand, in London, the short-term money market is mainly involved in the discount business, where discounters take the center stage.

The international money market consists of the following submarkets: short-term credit market, short-term securities market, and discount market. In particular, the short-term credit market is a market that aims at serving banks. This market provides loans with such short-terms as one year or less. The purpose of these short-term loans is for meeting temporary needs of capital and adjusting positions. The shortest term of these loans is one day, while the longest term is one year, with such other terms as 3 days, one week, one month, 3 months, half a year, etc. The interest rates are often based on the rates of interbank loans of London Bank. The transactions are often done in the form of whole sales with as small amount as several thousands of thousands of U.S. dollars, and as large amount as millions and tens of millions of U.S. dollars. The transactions are straightforward and carried out completely on reputations through telephones and faxes without any need for guarantees and collaterals. The short-term securities market is the place where short-term securities with terms not longer than one year are traded internationally. The transaction objects include short-term Treasury bills, transferrable banks' certificates of deposits (CDs), banks promissory notes bills, and commercial promissory notes. In terms of the discount market, the so-called discount means that credit notes, which will mature on future dates, are converted into cash at discounters by subtracting the interest computed using discount rates from the date of conversion to the maturity date. The discount market is the place where funds are raised through converting not matured notes into cash by using the method of discount. The notes to be discounted include mainly Treasury notes, bank bonds, company bonds, bank promissory notes, and commercial promissory notes. The discount rates are generally higher than banks' interest rates.

The international money market is the place where international short-term financial assets are traded. In this market, companies with temporary surplus of capital and companies with temporary deficit of funds can satisfy each other's needs simultaneously. On the one hand, this market provides the companies that experience temporary deficits of funds various short-term capitals from overnight to one year. On the other hand, the companies that wish to earn some extra returns by making use of their temporarily idling funds acquire a channel of investment. Because this market transcends national boundaries, it can align short-term capitals within the entire world and increase the efficiency of monetary funds. However, because the funds in this market are large in quantity and have strong liquidity, they can easily cause violent impacts on the international order, and affect the external equilibria of individual nations' economies, leading to the break-out of monetary, financial crises.

10.1.2 The international capital market

The international capital market represents the mid-term and long-term (over one year terms) financing market, where the participants are banks, companies, security dealers, and government agencies. The main business of the capital market consists of two large classes: (i) banks loans and securities trading, and (ii) mortgage loans and leasing loans. Other businesses that involve the function of long-term financing can also be classified into this market. What needs to be pointed out is that the international money market and international capital market are divided by using the terms of the financial products that are traded. They may overlap with the Euro currency market, foreign exchange market, gold market, financial derivative markets, etc., that will be studied in the following.

The international capital market consists of the credit market and securities market. In particular, the credit market stands for such a place that government agencies, including international economic organizations, and multinational banks provide long-term financings to their clients. Here, the basic characteristics of government loans are that the terms are long, interest rates are low, and there are attached conditions. The terms of government loans can reach as long as 30 years, the interest rate could be as low as 0%, and the attached conditions in general limit the range of how the funds could be used. For example, they might strict that the funds can only be used to purchase the goods produced by the lending nation, or that borrowing nation has to make certain promises or adjustments to its economic policies or foreign policies. So, government loans belong to the class of binding loans. On the other hand, bank loans in general do not carry constraints; the interest rates are determined by the market conditions and the credit worthiness of the borrowers. For loans of large amounts, banks generally provide syndicated loans, where a group of banks jointly lend to the same borrower with one bank acting as the leader, several banks as managers, and the rest as participants. The leading bank in general is also a managing bank. It collects the finder's fees and management fees, and shares the management works of the loans with other managing banks. Scale of syndicated loans once dropped during the debt crisis of the early 1980s. And then starting in 1986, these loans have been out of the low and the scale of this business activity has developed rapidly in the following years, for details see Tables 10.1.

The securities market is mainly made up of the bond business. The bond issuers can be government agencies, international organizations, as well as business firms,

Table 10.1 International syndicated loans (unit: US$0.1 billion) (from Global Financial Stability Report, published by the IMF in September 2007).

	2002	2003	2004	2005	2006
International banks	12,969	12,414	18,069	22,323	21,212
Syndicated loans	11,996	11,304	16,354	19,909	18,223

Table 10.2 Amounts Outstanding and Net Issues of International Debt Securities (unit: US$0.1 billion) (from Global Financial Stability Report, published by the IMF in September 2007).

	2002	2003	2004	2005	2006
Amounts outstanding	88,309	111,350	132,758	139,600	175,739
Net issues	9,938	13,525	15,413	18,000	261,777

Table 10.3 Foreign financing of major emerging market countries: Issuance of stocks (Unit: US$0.1 billion) (from Global Financial Stability Report, published by the IMF in September 2007).

	2002	2003	2004	2005	2006
China	24.8	64.2	141.9	257.2	418.1
India	3.5	13.0	43.5	67.1	91.4
Brazil	11.5	2.9	16.5	34.3	87.1
Russia	13.0	3.7	24.8	68.1	176.0

companies and banks. The issuance of most bonds is underwritten by banks or securities dealers, acting as intermediaries. Other than the bond business, since the 1990s, international financing activities, which mainly focus on equities, have also been quite active. For example, many companies from around the world have gone public in the stock market of the United States so that these companies have acquired their desired international financing. All businesses of obtaining international financing through using equities as collaterals or directly selling equities can be classified into the category of international capital market. For relevant information, see Table 10.2.

In terms of the roles of the international capital market, it is the place for different countries to conduct international long-term financing. A domestic company can acquire its necessary capital needed for the long-term management through borrowing long-term loans, issuing bonds, and selling stocks in the international capital market. Because long-term funds are mainly needed in the area of production, through the international capital market long-term capitals are allocated throughout the world, flowing to the country and economic entities that have the relatively high efficiency of productivity. So, the international capital market is beneficial to the improvement of the global total output. For relevant information, see Table 10.3.

10.1.3 The Euro currency market

The predecessor of the Euro currency Market was initiated in the Eurodollar market in the 1950s. At that time, the Soviet Union and the Eastern European countries were afraid that their U.S. dollar assets located in the United States could be frozen. So, they transferred these capitals into British banks. In order to acquire foreign funds, to restore the status of the British pounds, and to support the development of the domestic economy, all major commercial banks of London were allowed to receive offshore U.S. dollar deposits and to handle U.S. dollar lending business. Consequently, the Eurodollar market appeared. After 1958, the international payment deficit of the United States provided huge amounts of capital to the Eurodollar market. The capital outflow controls of the United States forced American citizens abroad to conduct their borrowing transactions in the Eurodollar market. After the 1970s, two substantial price increases in the world oil market created large export surpluses for the petroleum oil exporting countries. Because these surpluses were expressed in the U.S. dollar, they have been known as petrodollars. The oil exporting countries put large amounts of the petrodollars into the Eurodollar market, while the non-oil production developing countries took out loans from the Eurodollar market to offset their deficit incurred in their petroleum oil trades. Consequently, the capital supply and demand of the Eurodollar market increased continuously. By comparing the financial markets of the United States and Europe of the same time period, it can be seen that because trades on the Eurodollar market are done using non-domestic currency, they do not cause any effect on the stability of the domestic financial system. So, the country that hosted the market only regulated the Eurodollar market loosely without any requirement for reserve capitals and any limitation on interest rates. That of course greatly stimulated the development of the Eurodollar market.

Since the late 1960s, along with the frequent occurrence of U.S. dollar crises, the currencies traded on the Eurodollar market were no longer limited to the U.S. dollars. Instead, they were expanded gradually to include German marks, Swiss francs, and other currencies. At the same time, the geographic location of this market was also expanded to Singapore, Hong Kong, and other places, where the business of lending in U.S. dollars, German marks, and other currencies was also started. Consequently, the original Eurodollar market was evolved into Euro currency market, where Euro is no longer a word used to represent a specific geographic location. Instead, it means "offshore". The so-called Euro currency stands for a currency that is in circulation abroad of the country that issues the currency. For example, Eurodollar, Euromark, etc., represent the U.S. dollars that are in circulation outside the United States, the German marks that are used outside Germany, etc. The so-called Euro currency market stands for such a market where foreign currencies are traded. It includes not only the money market of terms one year or less, but also the capital market of terms longer than one year. After the 1980s, the meaning of "abroad" involved in the Euro currency market has be changed again. In 1981, the U.S. Federal Reserve Bank approved to establish a facility in New York to handle international banking services, to accept foreign deposits in U.S. dollars and other currencies, and to provide credits to foreigners while waiving the constraints of reserves and interest rate. Evidently, this international banking facility possesses the special right of conducting non-residential business operations in foreign currencies and being not constrained by the laws of the countries that issued the currencies. It is really a Euro currency market within the territory of the United States. The offshore financial market, which was established in 1986 in Japan, also manages Japanese yen. It stands for a Euro currency market located within the territory of Japan. In terms of the Euro currency markets that can manage domestic currencies, the domestic currencies the markets handle are still coming from nonresidents.

The transactions completed within the modern Euro currency markets have two different forms. In one form, one of the two involved parties is resident while the other party is non-resident. This form of transaction is known as an onshore transaction. In the second form, both parties involved in a transaction are non-residents. Such a transaction is known as an offshore transaction. And, offshore transactions also contain two different types. In one type, the locations of the actual transaction and the recording of the transaction are the same, while in other type, these locations are different. For example, a European currency trading takes place in London, while the transaction is recorded on the account of a company located on an island of the Pacific or Caribbean Sea.

By considering the relationship between onshore businesses and offshore businesses, Euro currency markets can be grouped into three classes. The first class involves the system type. That is, there is no strict separation between onshore transactions with residents' participations and offshore transactions with non-residents' participations. Domestic funds and foreign capitals can be converted at any time. Both London and Hong Kong belong to this class of markets. On such markets, the boundary between onshore and offshore financial services is no longer clearly defined. The second class contains such Euro currency markets where onshore businesses and offshore business are registered individually; they are supervised separately. The markets of the separation type help insulate the effects of the flows of international capital on the deposit quantity of the domestic currency and on the macro domestic economy. The international banking facility of the offshore financial market in New York, the overseas special account of the offshore financial market in Japan, and the Asian currency account of the offshore financial market in Singapore all belong to this class of Euro currency markets. The third class includes such Euro currency markets as that almost no actual offshore business transactions are taken place. These markets exist for the sole purpose of escaping from certain legal constraints, such as income taxes, by providing book-keeping and accounting. Some of the island nations or territories of the Pacific Ocean and in the Caribbean Sea represent the centers of such offshore book-keeping financial centers.

Euro currency markets entertain mainly such transactions as interbank lending and borrowing, European bank loans and European bonds. The European bank loans can be short-term, mid-term, and long-term and bear either fixed interest rates or variable interest rates. The organization form of the lending business consists mainly of syndicated loans. European bonds represent such bonds a national government or a business company issues on the bond market of another country with face values priced in a foreign currency. For example, such a bond that is issued by a British organization in the American bond market with face valued priced in Japan yen represents a European bond. Because the issuance of European bonds is done by using a foreign currency as the basis of the face value, it is hardly constrained by the monetary authority of the issuance country.

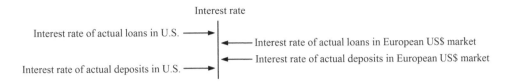

Figure 10.1 Interest differences of European market of U.S. dollars.

The Euro currency market is totally internationalized. It represents the main body of the international financial market. Because it is operated with foreign currencies, it possesses many particular characteristics of operation. First of all, the coverage of the market is wide without any geographic limitation. Because of the modern communication system, it has become a truly global market with some geographical centers. These geographical centers are developed in the cities and on the basis of the traditional financial centers, such as London, New York, Tokyo, etc. These cities are located in such countries that have developed advanced economies with sufficient supply of capital and have been historically places where money is transacted. These financial centers are located within stable economies, political environments, and have good communication and financial infrastructures, skilled managerial talents in the financial sector, and incentives and conditions of free operation provided by the governments. Secondly, the market enjoys a huge transaction size with a large variety of products while involving many different kinds of currencies. The relevant financial innovations are extremely active. Most of the individual transactions on the Euro currency market involve amounts of capitals beyond US$1 million with over billion U.S. dollar deals often seen. Other than the U.S. dollar, Japanese yen, and the Euro, such other currencies as Swiss franc, British pound, Canadian dollar, etc., are also involved frequently in the Euro currency market. Additionally, the currencies of developing countries are also seen occasionally, and even transactions priced in the SDRs have appeared. All these transactions have made the Euro currency market very closely connected to the foreign exchange market. Thirdly, the Euro currency market has its particular interest rate structure. Specifically, the interest rate structure of the Euro currency market is based on LIBO, which is somehow related to the interest rates of individual countries. But at the same time, it is also influenced by the supply and demand relation of the Euro currency market. The key reason why depositors and borrowers are attracted to the Euro currency market is that the difference between the interest rate of loans and that of deposits in the Euro currency market is smaller than the difference of the American market, Figure 10.1.

There are many reasons for why the Euro currency market enjoys the interest rate advantage. Firstly, commercial banks in the domestic financial market are affected by the requirement of maintaining deposits reserves and the constraint of upper limit of interest rates. All of these increase the cost of doing business. On the other hand, in the Euro currency market, none of these constraints exists. Banks can autonomously provide more advantageous interest rates and new and innovative products. Secondly, the Euro currency market to a great degree is a whole sale market for interbank businesses, where the amount involved in each transaction is very large so that the average fees and other service costs are relatively low. Thirdly, the Euro currency market generally serves large companies or government agencies with good credit ratings and relatively low risk. Fourthly, the competition at the Euro currency market is extremely intense, which helps to lower the transaction costs. For example, there were 114 foreign banks that established branch offices in London. And in 1990, this number had been increased to 451. When so many institutions conduct international financial operations at the same location, they naturally result in increased competition so that the service fees are consequently lowered.

Because it is generally involved in the non-residential lending in foreign currencies, the Euro currency market is much less regulated. That not only is beneficial to the innovation and development of the market, but also exacerbates the potential market risk of itself and its potential threat to the stability and equilibrium of the domestic economy.

10.1.4   International foreign exchange

The foreign exchange market is the place where trading of foreign currencies takes place. Its participants consist of all the institutions and individual persons who buy or sell currencies, such as central banks, commercial banks, foreign exchange brokers, companies that deal with foreign currencies, etc. The transactions of the foreign exchange market include spot transactions, forward transactions, futures transactions, and options trading. As of April 2007, the average daily trading volume of the foreign exchange market from around the world had reached US$3200 billion. The U.S. dollar is the most traded currency. Other than that, what were traded also included Euro, Japanese yen, and British pound. Here, the trading of the U.S. dollar occupies such a heavy weight of 86.3% among all the completed trades with Euro being the second (37.0%), Japanese yen (16.5%), and British pound (15.0%).

London is the largest center for foreign currency trading. It amounts to about 1/3 of all the trades in the foreign exchange market. New York, Zurich, Frankfurt, Tokyo, Hong Kong, Singapore, etc., are also important locations of the foreign exchange market. Along with the development of electronic communication technology, more and more trades of foreign currencies are completed through the internet, fax, and telephone. Because the foreign exchange transactions are mainly done between banks, the foreign exchange market in fact stands for a place for buying and selling currencies between banks. For the most part, trades in foreign exchange are speculative in nature, where profitable trades are executed by making use of the slight differences in exchange rates that exist at different times and different places. The foreign exchange market operates 24 hours a day continuously by tightly connecting together all the trading centers located at different time zones through the electronic media.

The foreign exchange market has two trading systems: the market maker system and the auction system. Under the market maker system, each investor provides the market maker the quantity and direction of his trade. When the market maker collects the quantitative information of sellers and buyers, he quotes the price. And he is obligated to complete the transaction with the investor at the quoted price by using his own funds or securities (that is, he is obligated to maintain the market liquidity). The market make plays the roles of a buyer and a seller at the same time and makes his profit through the difference of buying and selling prices. Under the auction system, the buyer and seller respectively quote the directions and quantities of their willing trades. They trade either directly or through their respective brokers who submit the trading orders to the market, where deals are made by matching the trading orders. Based on the continuity of trades, the auction system can in turn be divided into call auction system and continuous auction system.

Other than the international money market, the international capital market, the Euro currency market, and the foreign exchange market, there are also two other important markets that are worth mentioning. One is the international gold market, and the other the international financial derivatives market. Although gold is no longer a world currency, it is still used as a part of the world reserve; and the gold price is still utilized as the barometer of the world economy. On the other hand, the international financial derivatives market has been developed a great deal in the past thirty some years. Because of the trading leverage system of the derivatives, the impact of the capital flows, especially the flows of speculative capitals, of the derivatives market, has been magnificent, constituting a main force of the speculative capitals on the world economy.

10.2   CAUSES AND EFFECTS OF INTERNATIONAL CAPITAL FLOW

Capital flows throughout the world can be classified into two kinds: One contains such flows that are closely related to the actual production and business trades and can be referred to as international fund flows. This kind of flows belongs to the research of international investments. The other kind includes such capital flows that do not have any close relationship to the actual production, business trades, and investment activities. These flows stand for the international money movement of the pure "financial" characteristics and exist mainly for the purpose of capturing the differences in assets' prices and for financial returns. They are referred to as international capitals, and belong to the research category of international finance. What this section discusses is these international capital flows.

10.2.1   Characteristics of capital flow in the international market

Since the 1980s, the capital flows that existed in the international financial market reflect the following characteristics:

1   The transaction amounts are huge, and grow independently from the real economy;
2   The international financial market has shown the characters of a whole sale market, where organizational investors represent the main carriers of the international capital flows; and
3   The proportion of the trades on derivatives has been increasingly rising, making the degree of trade virtualization much higher than before.

In particular, the size and growth speed of the international financial market are much higher than those of the real world economy. As of the end of April 2007, the daily average trading volume of the foreign exchange market from around the global has reached US$3200 billion. In comparison, the total world trade of goods in

2005 was US$10200 billion and the total trade of service was US$2400 billion, which represents the trading volume of the foreign exchange market of merely a few days. So, it can be seen that no matter whether or not it is the growth speed of the total, the actual trades are far behind the capital flows of the international financial market. The world GDP of 2006 was US$48140 billion, while according to the incomplete statistics of the World Federation of Exchanges, the total capitalization of the world stock markets in 2006 was about US$50640 billion, which was more than the total world GDP. In many industrialized countries, savings diversification of residential families and the openness of the financial sector have made the amount of investment capital held by organizational investors surge rapidly. In 1990, the total amount of capital under the management of organizational investors of the member nations of the Organization for Economic Cooperation and Development had reached US$13800 billion, which represents 77.6% of the total GDP of these member nations. And in 2003, the total amount of capital under the management of organizational investors reached US$46800 billion, which represents 157.2% of the total GDP of the member nations. If it is seen from the angle of how international capitals flow traditionally and how the derivatives are traded, then the quantity of international capital flows that is created out of these derivative trades has been in an absolute dominant position. And the growth speed of derivative trading also increases very rapidly. As of the end of June 2006, the nominal balance of the world over-the-counter derivatives market was US$3699.1 billion, which represents a 288.6% growth when compared to the value of 2000. And at the end of 2005, the total of the capitalizations of the world bond market, the world stock market, and banks' assets was US$165120 billion, which is not even 1/2 of the nominal balance of the derivatives market.

10.2.2 Reasons for international capital flows

Since the 1980s, other than the rapid growth in foreign trades and investment activities in the world, other main reasons that have stimulated the large rise in the scale and speed of international capital flows include:

1   there is sufficient supply of funds in the international financial market;
2   the globalization of the financial market has promoted the worldwide configuration and arbitrage of capitals;
3   capitals move for the purpose of circumventing controls or risks;
4   financial innovations have provided the convenience; and
5   there have been deregulations on the movement of international capital.

In particular, in terms of the supply of capital in the international financial market, the main source of capital in the international financial market is the worldwide accumulation of the huge amount of financial assets that are no longer related to the actual production. These capitals are originated in different countries on the bases of inflation of those main countries that issue convertible currencies, of the huge amount of petrodollars obtained by the oil producing countries from substantially increasing the oil prices, and the large quantity of the U.S. dollars spilled into the world market through their huge amounts of international payment deficits of the world-currency issuance countries, such as the United States.

The development of the financial market and financial intermediary organizations has also derived many financial assets, making more and more social assets unrelated to the actual production. Through the creation of credits, international capital flows also have the tendency of growing automatically. For example, the Euro currency market to a certain degree possesses a deposit creation process. That is, when a deposit is made into the Euro currency market, if all the savings and loans of this initial deposit stay within the Euro currency market, then this initial deposit can generate a whole series of deposits. When we analyzed in one of the earlier chapters the deposit creation process in the domestic financial market, we used the following expression to indicate the deposits derived multiplier:

$$m = \frac{1}{1 - (1 - r)(1 - c)} \tag{10.1}$$

where m stands for the money multiplier, r the rate of the statutory reserve, and c the proportion of cash leakage. Because the Euro currency market does not require any statutory reserve, there is not in theory any cash leakage. So, the deposit creation on this market is infinite. However, on the other hand, the proportion c of leakage in the Euro currency market is quite high. It is shown as capitals moving into domestic markets or other international financial markets. Theoretically, the magnitude of this proportion is yet to be determined. However, for now as this book is written, this value is commonly taken to be somewhere between 2.5 and 5.5.

In terms of the worldwide configuration and arbitrage of capitals, along with the development of communication technology, the degree of integration of the world financial markets has been constantly heightened. Owners of capital can allocate his funds with small fees into various financial markets throughout the world in order to materialize their desired combination of risk and return. Each process of allocating financial assets in various markets throughout the world and in various asset classes represents the process of international capital movement. The increasing frequencies in investment diversification and allocation adjustment also stimulate the movements of international capitals. On the other hand, the strengthening connections between financial markets make market participants more aware of price spreads of the same financial assets traded in different financial markets in a timely fashion so that they can mobilize large quantities of capital within short moments of time to participate in arbitrage activities. At the same time when these activities eliminate the price spreads of different financial markets, they also form large scales of capital movements.

Since the 1980s, the economies of some emerging market countries have grown rapidly. However, these countries implemented strict financial market controls without adequate property protections. After accumulating certain magnitudes of wealth, the residents of these countries often wish to move their funds to foreign countries in order to circumvent the strict controls or to avoid non-economic risks. Such migration of funds is often done through informal channels, constituting very large flows of international capital. At the same time, since the 1980s, there has appeared a wave of financial innovation throughout the world. The general reasons behind financial innovations include avoidance of risks, advancement of technology, the inverse effect of government control, high inflation, etc. International financial innovations motivate capitals to move internationally, especially the movements of financial assets that are detached from actual production and investment. First of all, the international financial innovations provide effective supports for capitals to avoid the risks experienced in their international movements. Considerable part of the exchange rate risk can be avoided by using such tools as foreign exchange forwards, futures, options, etc. The interest rate risk can be avoided by utilizing such tools as forward rate agreements, interest rate futures, interest rate collars, etc. The credit risk can be reduced by employing such methods as exchanging equities and debts and others. Secondly, international financial innovations enhance the efficiency of capital flows. By employing new financial tools and methods of financing, the cost for international capital movements has been greatly reduced. Thirdly, international financial innovations and the introduction of derivatives provide new playgrounds for speculators. In the derivatives market, the price fluctuations are more dramatic with high levels of trading leverage. Speculators can control large quantities of contracts with little amounts of earnest money. That makes the potential profits and losses much greater than those achievable in the more traditional markets. And, most trades of derivatives belong to the sheet business, where the investors acquire a new way of making profits without affecting the proportion of funds in the underlying investments. In the current international financial market, a good amount of capital moves for the purpose of reaping in speculative profits. The effect of international financial innovations on this type of capital movement can be seen readily.

Within a quite long period of time after WWII, individual countries had implemented strict controls on the movement of international capitals. After the 1970s, a wave of relaxing foreign exchange controls, capital controls, and even financial controls appeared in various countries, where the domestic banks' credit markets and securities markets were also gradually liberated: foreign financial institutions were allowed to enter the domestic financial markets, non-citizen residents were allowed to raise funds in the domestic financial markets, and the controls on financial institutions were relaxed. As of 1992, most advanced countered had relaxed their controls over the flow of international capital. Instantaneously, the relaxation on asset controls in the emerging markets also became very obvious.

10.2.3   Positive effects of international capital flows

International capital flows possess the following positive effects.

1   They improve the optimal allocation of assets within the world.
2   They stimulate the conduction of wealth effect.
3   They strengthen the liquidity of capital. And
4   They help integrate the international financial market.

In terms of the return of capital, the financial securities issued in the international money and capital markets on the one hand provide a channel for the public to earn good returns on their savings, and on the other hand, deliver a source of funds for the business entities that need capital so that idling funds flow to where they are most needed through the financial market. That helps lift the marginal output of capital and improve the quality of life for many people by stimulating the production of goods and services. In terms of trading risk, under the condition that the domestic financial market is improperly developed with unsophisticated facilities, all the microeconomic entities have to bear the risk of change in economic variables, such as exchange rate, interest rate, etc. If an economic entity can enter the developed international financial market, it can then employ such financial tools as derivatives to eliminate or reduce its trade risks. So, the development of the international financial market can help optimize the capital allocation in the world in two aspects: making good returns and avoiding or reducing risks.

When international capital flows from the financial market of one country to another, it helps spread the wealth effect of the securities market of one country. That is why the economic growth of the country that hosts an important financial market can often promote the economic prosperity of a much greater area of the world. For example, the American Dow Jones Industrial Average had been reaching historical highs one after another on the back of the network economy. The swelling wealth not only filled up the pockets of the citizens of the United States, but also looked for other investment opportunities elsewhere in the international financial market. It, of course, brought forward some bright hopes to the exhausted Asian economy that just came out of the 1997 financial crisis.

In the past, because the individual national markets had different structures under different sets of rules, the degree of difficulty in converting assets into cash varied from one country to another. When the returns on investment in two countries were very different, if an investor wanted to make profit by holding foreign assets, he would then have to bear the risk of unable to convert back into cash. The expansion in the magnitude of international capital movement, the richness in the variety of consumer products, and the increase in the number of participants, have improved the depth and width of the international financial market. The holders of financial products can convert their assets into cash readily with low costs so that the liquidity of capital has been increased, which in turn stimulates the further expansion in the magnitude of international capital movement.

The most direct benefit of the speedy growth in the international capital flow is the integration of the international financial market. The so-called integration means that all parts are connected together in a certain fashion. One indicator for the degree of integration of all the national financial markets is the price of capital, known as interest rate. In the following, we will look at the performance of the interest rate index, and investigate the effect of international capital flow on the integration of the international financial market.

Firstly, let us look at the difference between the interest rate of a financial asset in the domestic financial market and that in the offshore financial market. In recent years, the domestic interest rate of the U.S. dollar and the offshore interest rate have always experienced synchronized changes with slight difference. That indicates that the degree of integration of the domestic market and the offshore market has been deepened. Here, for the domestic interest rate of the U.S. dollar we considered the overnight Federal Funds rate, and for the offshore interest rate we used LIBOR overnight rate of the U.S. dollar.

Secondly, let us look at the tendency of change in the interest rates of various national financial markets. The development of the international financial market makes it possible for capitals to be allocated within a much greater range. When the demand for capital in one domestic financial market is greater than the supply, capitals will flow in. When the demand for capital is smaller than the supply, the

unit: %

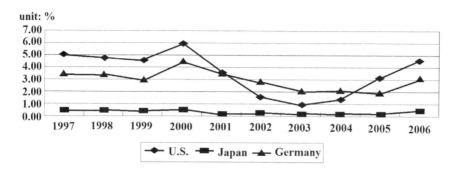

Figure 10.2 Trends in short-term bonds of U.S., Japan, and Germany (Note: For Japan, the 60-day discount rate is taken as the short-term rate, while for U.S. and Germany, the interest rates of 3-month Treasuries are taken as the short-term rates) (Data from 2006 December issue of International Financial Statistics, published by IMF) .

unit: %

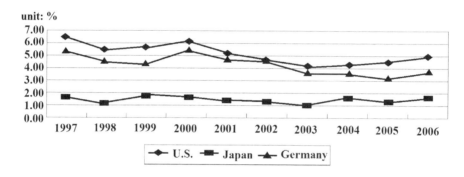

Figure 10.3 Trends in long-term bonds of U.S., Japan, and Germany (Note: The interest rates of 10-year Treasuries of these countries are taken as long-term interest rates) (Data from 2006 December issue of International Financial Statistics, published by IMF).

abundant capital will flow outward. So, the levels and changes of interest rates on the individual nations' financial markets tend to gradually converge. When measuring the degree of integration of various national financial markets, the synchronization in the tendencies of changes in interest rates is more important than the actual levels of the rates, because each nation has different inflation and tax rate so that there will naturally be differences in the nominal interest rates. However, as long as capitals can flow freely, the tendencies of change in interest rates should converge. Figures 10.2 and 10.3 show the levels and tendencies of the short-term and long-term interest rates of several major industrial countries, where the short-term interest rate is based on that of short-term treasury and long-term interest rate is based on the 10-year treasury. For the short-term interest rates, the national tendencies of change in the cited interest rates are fairly consistent, reflecting the fact that the tendencies of change in the supply-and-demand relationship of the short-term capital are roughly the same from the money market of one country to another. Contrary to the money market, it is more important to compare the supply-and-demand relationships of mid-term and long-term capitals in the capital market. Short-term capitals have high levels of liquidity, while mid-term and long-term capitals are relatively more stable. Their interest rates can more adequately reflect the basic situation that underlies the supply and demand of the country, and the depth of the integration of various national financial markets. When looking at the long-term interest rates, changes in the levels of the long-term interest rates of various nations maintain a quite consistent tendency. This situation illustrates that the flows of international capital have closely connected various national domestic financial markets.

10.2.4 Impacts of capital flows on domestic and international equilibria

At the same time when the flows of international capital promote the integration of the international finance and improve the global welfare, they also cause the following difficulties for the maintenance of the internal and external balances of various nations:

1   They increase the difficulty for a nation to maintain its external equilibrium;
2   They makes it difficult to manage the micro economies;
3   They magnify the trading risk;
4   They affect the realization of internal balance; and
5   They spread the fluctuation in one country's economy into other countries.

In particular, the integration of the international financial market makes the transfer of funds from one country to another exceptionally fast and convenient. The international payments of the nation are mainly affected by the frequent and large scale inflows and outflows of capital. So, the decision makers have to balance between maintaining the stability of the actual economy and upholding the solidity of the international payments. Under the floating exchange rate system, flows of huge amounts of capital alter the supply and demand relationship of the foreign exchange market and amplify the fluctuation of exchange rates. Under the fixed exchange rate system, flows of international capital can cause severe impacts on the foreign exchange reserves of a nation. And at extreme circumstances, some flows of short-term capital with speculative purposes can lead to the break-out of currency crises. Under the condition of sufficient flows of international capital, large fluctuations in the foreign exchange market can increase the uncertainty for the businesses that are involved in import and export trades. The floating relationship between individual nations' exchange rates and interest rates can make the decisions of microeconomic entities of one nation on investment and financing difficult to control as these decisions might very well be affected by the economic policies of other nations. Enterprises either employ such tools as financial derivatives or other methods to reduce risks while paying additional expenses or place hopes on their correct predictions on the future so that precautious measures are used to digest risks. Both of these methods increase the operational difficulty of the microeconomic entities.

Financial derivatives are not only methods that can be employed to manage risks, but also themselves trades involving the most risks. The derivatives market represents the most actively participated place by international short-term speculative capitals. The security deposit system of the derivatives market makes the trading of derivatives highly leveraged. The change in the prices of the underlying contracts can be many times more than the security deposits paid so that the potential for profit and loss can be enormous. That is a characteristic any other financial market does not have. And, the transactions of most financial derivatives belong to the surface business, which is not limited by the ratio requirement of the available funds in bank accounts. It has been objectively difficult to regulate the transactions of derivatives. The price fluctuations in the derivatives are short-term, virtual, and highly reversible. The severity in the price fluctuation can affect the prices of the underlying products so that the fluctuation can be spread into the markets of other financial products, creating threats to the stability of the international financial system.

In terms of the effect of international capital flows on the internal equilibrium, for the short-term, the effect is mainly expressed as that in order to intervene in the foreign exchange market the government has to give up the independence of its monetary policies. If the moving amount of the short-term capital can be maintained at an appropriate dynamic equilibrium for the long term, that is, with a period of time the inflow of capital is equal to the outflow, then the movement of the short-term capital will not impact the monetary policy much. However, if the net inflow of short-term capital is maintained at the same direction while the quantity is huge due to predictions of currency appreciation or devaluation or the outlook on the nation's economy, then it will cause greater impact on the domestic money supply and the interest rate so that the real economy and the internal equilibrium will be affected. For the long term, if the movement of international capital causes currency crises or other fluctuations in the financial field, it will create far-reaching effect on the real economy. The investment and financing capabilities of enterprises and the consumption of individual families will be hit. Those countries that went through either Latin American debt crisis or Asian financial crisis all experienced slowed economic growth, reduced productivity, contracted demand, and other difficulties in the ensuing years after the crisis.

Under the condition that international interest rates fluctuate synchronically and regional capital markets move jointly, the fluctuations of one country's financial market, as impacted by domestic policies or international speculations, will spread to another country through the interest rate effect or other means. Such spread can be in the same direction or the opposite direction. For example, when the exchange rate of two countries' currencies is kept relatively stable, then the fluctuations in the interest rate of one country, as caused by the country's policies, will cause the interest rate of the other country to move in the opposite direction. As an another example, if a tightening of one country's money causes a slump in the country's capital market, then capital will tend to leave the country for another, leading to a prosperity in the other country's capital market. Additionally, speculators can acquire further financial support in the international financial market after they achieved the initial advances through using the leverage of derivatives. With additional speculators joining in the advances, the fluctuations that appeared initially in one country will eventually be spread to many other countries, causing a large scale monetary and financial crisis. To this end, the Asian financial crisis of 1997 is a good example.

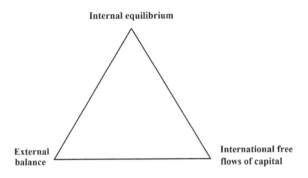

Figure 10.4 A generalization of the impossible trinity.

Based on what has been discussed above, it follows that when the frequency of international capital flows is high, it will make it difficult to maintain the international and external balances. If we change the independent monetary policy of the Mundellian Trilemma (also known as the impossible trinity), as studied in one of the earlier chapters, to internal equilibrium, and change the stability of exchange rate to external equilibrium, then we obtain a generalized trilemma as shown in Figure 10.4. This new ternary structure tells us the following: Once the international capital flow becomes instable, it will make it difficult for a country to simultaneously achieve its internal and external balances. In the following, we will look at the shock effects of international capital flows.

10.3   DEBT AND BANK CRISES: IMPACTS OF FINANCIAL GLOBALIZATION

Since the 1970s and 1980s, the steps of international financial liberalization have been accelerated. Capital flows and financial globalization have impacted the internal and external balances of each country. Typical examples include the Latin American debt crisis of the 1980s, the whole series of bank crises of the 1970s, and the currency crisis of the 1990s.

10.3.1   Debt crises of developing countries

In the early 1980s, a worldwide debt crisis broke out. A so-called debt crisis means that a series of non-oil producing developing countries were unable to repay their maturing debts, which were mostly owned to the international banking industry; that caused not only these developing countries' credit ratings to drop drastically, nut also the international banking industry trapped in a funding crisis. That severely affected the stability of the international banking industry in particular and the stability of the international monetary system in general.

There are various reasons that cause debt crises. First of all, the old colonial rule had made the economic structures of the developing countries obsolete. Under such

Table 10.4 Trade payments and incomes of oil-producing and non-oil-producing developing countries (unit: US$0.1 billion) (from 1992 Yearbook of International Financial Statistics, pp. 127).

	Oil-producing developing country	Non-oil-producing developing country		Oil-producing developing country	Non-oil-producing developing country
1973	+67.0	−113.0	1978	+449.9	−311.1
1974	+683.0	−371.0	1979	+1150.0	−426.5
1975	+557.5	−375.0	1980	+1712.5	−609.5
1976	+657.4	−237.4	1981	+1227.1	−701.7
1977	+608.9	−215.2	1982	+615.5	−405.5

conditions, these developing countries attempted to employ foreign debts to offset the shortage of their domestic savings and funds. However, due to the backwardness of their domestic economic structures and lack of the basic infrastructures, these countries could not obtain the corresponding outputs. Secondly, price system on the international market has long been swayed by the developed countries. Consequently to their backward economic structures, the developing countries export mainly produces, raw materials, and primary products. The prices of these goods in the international market are very low, placing the developing countries in disadvantageous positions in the international trade. That surely affected the export incomes of these developing countries. Thirdly, because of the low levels of education, the developing countries do not have sufficient well trained professionals. Along with political unrests, the management of domestic economies and foreign debts is often inappropriate, leading to inefficient usage of foreign debts. Fourthly, other than previous three basic reasons, the Organization of Petroleum Exporting Countries (OPEC) had raised oil prices two times, once during 1973 and the other during 1979–1981. That caused the non-oil producing developing countries to suffer from huge deficits in their international payments. And that was the direct trigger for the break-out of the debt crisis, see Table 10.4 for more details. The oil price increased about three folds from U.S.$3.01 a barrel in October 1973 to U.S.$11.65 a barrel in January 1974. And, the price of U.S.$14.327 a barrel in January 1979 was lifted up to U.S.$34 a barrel in October 1981. So, during the eight years from October 1973 to October 1981 the oil price was increased 11 times, leading to the huge international payment surpluses of the OPEC, see Table 10.4. These surpluses were deposited into the international financial market, constituting the so-called petrodollar, because petroleum oil is generally priced and traded in U.S. dollars. In June 2008, the price of oil reached as high as U.S.$140 a barrel.

Other than the four reasons as listed above, the blindness and commercial standards of the lending practice of the international banking industry also helped to cause the debt crisis to break out. After the first price increase of the oil, large amounts of petrodollars flew into the international financial market, filling the market with abundant capital. So, all the multinational banks were looking for ways to lend their available funds. As a consequence, the non-oil-producing developing countries became their lending targets. The major characteristics of the international bank loans of the 1970s were that they targeted at sovereign nations. Because government guarantees were often in place, a large number of loans rushed into non-oil-producing developing

Table 10.5 Foreign debts and capabilities of repayment of non-oil-producing developing member nations of the IMF organization (unit: US$0.1 billion) (cited from IMF Occasional Papers, vol. 40).

Items	Year											
	1973	1974	1975	1976	1977	1978	1979	1980	1981	1982	1983	1984
Total foreign debts	1,300	1,610	1,910	2,280	2,910	3,430	4,060	4,900	5,780	6,550	6,942	7,310
Long-term debts	1,120	1,380	1,640	1,950	2,380	2,830	3,400	3,960	4,640	4,230	5,800	6,720
From Governments	510	600	700	820	1,000	1,190	1,360	1,580	1,770	1,980	2,210	2,410
From private banks	610	780	950	1,150	1,380	1,640	2,040	2,390	2,870	3,250	3,890	3,850
Rate of debt repayment (%)	16	14	16	15	16	19	20	18	22	25	22	22

countries, forming the debts of these countries owed to foreign individuals and banks. From Table 10.5, where the rate of debt repayment means the annual repaid amount of interest and principle over the income from exporting labor and goods (the higher this rate is, the lower the repayment capability), and the amount of short-term debts is equal to the amount of total foreign debts minus that of long-term foreign debts, it follows that during the time period from 1973 to 1984, the amount of foreign debts of the non-oil-producing countries was increased about 5 times, where the debts owed to foreign governments rose 4 times, while the debts owed to foreign individuals rose about 6 times. According to the statistics of the World Bank, from 1973 to 1982, and the amount of lending of private banks was increased at the average rate of 19.4% a year, while that of foreign governments at the average rate of 14.5% a year.

The blindness in banks' lending was shown not only in the quantity of loans issued, but also in the regions to which the number of loans was concentrated and in the distributions of terms when the loans need to be repaid. The speeds of obtaining bank loans for countries like Brazil, Venezuela, Mexico, and others were much faster than the average speeds listed in Table 10.5. And, because the arrangement of repayment terms was not done appropriately, the maturity dates of many loans were too concentrated at around the same times, causing many countries unable to repay several loans within a short window of time.

To measure a country's foreign debt affordability and solvency, there is a whole set of debt metrics available. The set of metrics includes:

(i)   The debt ratio, which is the ratio of the foreign debt balance and the gross domestic production (GDP). This ratio should normally be less than 10%.
(ii)   The debt-export ratio, which is defined as the foreign debt balance over the export of goods and labor of the current year. This ratio should normally be lower than 100%–150%.
(iii) Debt service – GDP ratio, which stands for the quantity of the annual amount of payments toward the interest and principle over the GDP. This ratio generally should not be higher than 5%.

(iv) Foreign debt service ratio (also known as debt service ratio), which is equal to the ratio of the annual total toward the payments of interest and principle over the annual income earned from exporting goods and labor. This ration generally should not be higher than 20%.

Among these four metrics, the first and the second indicators represent the foreign debt affordability of a country, while the third and the fourth the solvency or the repayment capability of the country. Here, the foreign debt service ratio is considered the most direct and most important indicator that measures a country's credit and ability to repay the debts. This indicator has a upper threshold of 20%, beyond which the country's credit and solvency are in serious doubts. However, in the 1970s, in order to make profits, the international bank industry often employed government guarantees in the place of objective analyses. Despite of the declines in the credit ratings of some of the non-oil producing developing countries, the international banking industry continued to lend to these countries with additional harsh business conditions attached. According to the publications of Morgan Guaranty Trust Corporation, as of 1983, the foreign debt service ratios of the non-oil-producing developing countries had commonly broke beyond the threshold, see Table 10.5, with some of these countries breaking 100%, where Brazil reached 117%, Mexico 126%, Argentina 153%, Venezuela 101%, Poland 94%, Zaire 83%, and Peru 79%.

Other than the excessive number of loans issued, the over concentration of geographic regions, and inappropriate time zones of maturity, the international banking industry also employed some of the new financial tools and techniques when it issued commercial loans. To this end, there were two outstand performances. The first is that the proportion of loans with floating interest rates rose from 10% in the early 1970s to 40% in the 1980s, which pushed up the service costs of the loans for the developing countries. The second is that in order to reduce risk, the growth in short-term loans was faster than that of long-term loans, making the debtor countries have to use new loans to pay off the old ones. That caused the use of debts as a short-term behavior.

Although the international banking industry was also responsible to the onset of the debt crisis, because the debt crisis involved the reputations and stabilities of the individual banks themselves, after the crisis broke out, the international banking industry worked jointly with the international financial institutions and relevant national governments to actively come up with methods to ease the crisis in order to maintain the stability of the international finance. In the process of easing the debt crisis, the international banking industry took the following main measures:

1   Increased lending to heavily indebted countries, while combining the new loans with the conditionality of the international monetary fund organization so that the loans were associated with the adjustment policies of international payments and the economic stability policies of the indebted countries
2   Rearranged the debts. That is, rescheduled the repayments of the maturing debts by expanding the terms. The debt rearrangements were not simply matters that involved the two sides of the loans, the lending banks and the borrowing countries. Instead, they involved multi-party negotiations. For example, in 1981 and 1983, the numbers of parties involved in the multilateral negotiations of rearranging the payments of debts respectively reached 13 and 25 indebted countries; and the

amounts of bank debts to be rearranged respectively totaled US$4.5 billion and US$52.1 billion.

3   Applied new financial tools to ease the debt crisis, including debt paybacks, debt swaps, and debt exchanges. By debt buyback, it means that some indebted countries were allowed to purchase back with some discounts its debts by using cash. Debt paybacks were generally conducted together with debt rearrangements. By debt swap, it means that a debt was converted into bonds and equities which then were sold in the marketplace. By debt exchange, it means the exchanges of two or more different debt tools. For example, at the end of 1987, with a joint arrangement of the American government and Morgan Bank, Mexican government spent US$2 billion cash and purchased US$10 billion of 20-year U.S. treasuries from the American Treasury Department with interest rate 0%. It then deposited the treasuries with the New York Federal Reserve Bank as the guarantee for Mexican government to issue US$10 billion worth of new bonds. At the same time, the creditor bank converted the original US$20 billion Mexican debt into a new loan using 50% discount. By doing so, Mexican government applied merely US$2 billion cash to exchange for its US$20 billion debt.

Since the time when the international debt crisis broke out in the 1980s, nearly thirty years have passed. During passing years, the international financial market and the flows of international capital have been further developed. However, the important lessons this debt crisis had taught us are that on the one hand, each country should pay attention to the economic policies that are applied to strengthen the stability of the internal and external balances; on the other hand, the characteristic of the international capital flows is that they are driven by profits, which can weaken the constraints of international payments of each country, causing the balance gap of payments to expand until the payments could no longer be sustained. So, each country should pay special attention on how to stick to its discipline of international payments

10.3.2   Bank crises

Since the 1960s, the internationalization of the banks of the Western countries has been developed rapidly so that these banks' business range and methods have been changed majorly. Many once regional banks are now multinational, and have established branch offices and facilities abroad. In order to escape the supervision and control of any national monetary authority, many overseas markets have emerged. The dependence of each multinational bank on foreign currencies is growing day by day; and the international competition of banks becomes increasingly fierce. All these circumstances together with the collapse of the fixed exchange rate system of the Bretton Woods system in the 1973 make the risk of international management of banks greatly increased. In 1973 and 1974, there appeared a series of bankruptcy events of banks in the international banking industry. After the bankruptcy of San Diego National Bank in the United States in 1973, Franklin Bank, one of the twenty largest banks in the United States, and Herstatt Bank of the Federal Republic of Germany also went bankrupt one after another in 1974. Additionally, the Union Bank of Switzerland, the Westdeutsche Landesbank of the Federal Republic of Germany, and the Lloyds Bank of England, respectively experienced losses of about US$100 million.

All these vents together constituted the first major bank crisis of the Western world since WWII. The notable characteristic of this bank crisis is its internationality, because the closures and losses of these banks were all related to their respective international transactions.

After entering the 1990s, there appeared a wave of financial integration and financial innovation, which makes, on the one hand, in terms of quantity and types new financial tools, methods and techniques of financing for banks emerge endlessly, and on the other hand, various risks for banks much, much greater. Within merely the year 1995, there happened two serious events where Barings Bank of England went bankrupt and the Daiwa Bank of Japan suffered from heavy losses. In particular, Barings Bank was initially established in 1793. As of the end of 1993, the amount of assets of the bank totaled at £5.9 billion; the bank's core capital ranked at the 489th among 1000 largest banks in the world. However, the manager of the Futures Department of Barings Bank in Singapore, who also oversaw concurrently the Liquidation Department, established accounts against the regulations, longed Nikkei futures and shorted Japanese bonds using the funds owned by Barings Bank. When Japan suffered from the great Kobe earthquake, the Nikkei index crashed, causing Barings Bank to lose £860 million. At that particular moment, the amount of all assets Barings Bank owned only totaled at £470 million. That forced Barings Bank to go bankrupt. On the other hand, Daiwa Bank was initially established in 1918. In the year 1995, the bank owned a total amount of assets valued at ¥24300 billion, positioning itself at the 13th position among all Japanese commercial banks. On September 26, 1995, Daiwa Bank uncovered that because since 1984 one employee at the New York branch office started off-the-book trading of U.S. bonds, costing the bank such huge losses as ¥110 billion, which was equal to roughly US$1.1 billion. During the 11 years since 1984, the person made over 30 thousand off-the-book trades, and lost an average of US$400,000 a day. In order to conceal the losses, at the New York branch office he forged the bond account, which showed a balance of US$4.6 billion for July 1995. However, in fact, the balance was only US$3.5 billion. On July 13, 1995, after this employee took the initiative and reported the matter, this major loss was uncovered.

The stories of Barings Bank and Daiwa Bank imply that after the banking business and banking institutions were internationalized there were still loopholes within the internal control system. The work of the Barings Bank employee should be trading derivatives on behalf of Barings Bank's customers, and pursuing arbitrage opportunities for the bank. These operations did not bear too much risk. At the same time, the Liquidation Department could monitor the risk positions of each trader through its daily settlements. However, the particular employee, who caused Barings Bank to go under, happened to be responsible in two roles: trading and settling, so that his works were not properly supervised. The situation with Daiwa Bank was similar. The particular bank employee not only traded bonds, but also managed the bonds so that his bond trading records and bank's records of outstanding bonds were held in his own hands. That created a convenient opportunity for him to trade speculatively and to fabricate the records.

Although the direct causes underlying these events with Barings Bank and Daiwa Bank are the loopholes in the internal management system, they were still directly related to the internationalization of the banking industry. Banks' internationalized trades have increased the risk potential. Past bank crises have caused huge impacts on the international financial market, as well as severe negative effects on the relevant national economies and financial stabilities. That has caused severe anxieties in the community of international finance. In February 1975, under the auspices of the Bank for International Settlements, Basel Committee, a coordinating agency, was established to supervise the international activities of banks. This Basel Committee consists of such member nations as the United States, United Kingdom, France, Japan, Belgium, Luxemburg, Canada, the Federal Republic of Germany, Italy, Netherlands, and Switzerland. The committee was made up of the bank supervisors of 12 national financial authorities and represented a permanent organization, which met three times a year regularly to discuss matters relevant to the international supervision of banks. On September 25, 1975, Basel Committee reached its first contract, known as Basel Concordat. In May 1983, the concordat was revised. The widely referred to Basel Concordat is mainly this revised version of 1983.

Basel Concordat confirms that the overseas institutions of any bank have to be under supervision, where the authorities of the motherland and the host country are jointly responsible for the supervision. The host country is obligated to supervise foreign banks located within its territory; the supervision authorities of both sides communicate with their information to each other, and exams the overseas institutions of the other side. Although Basel Concordat did not have any mandatory legal binding on the member nations, the committee wished it could become the supervision objective and effective behavioral standard of all nations. Additionally, Basel Concordat further clarified the distribution of the supervision power:

(i) The supervision on the solvency of a branch is within the responsibility of the mother country's authority, and the supervision on the solvency of a subsidiary is jointly conducted by the mother country and the host country, while the supervision on the solvency of a jointly funded bank belongs to the nations that are part of the joint business venture.

(ii) The supervision on liquidity of branches and subsidiaries falls within the authority of the host country, while the management authority of mother country is responsible for the liquidity of the entire bank group.

(iii) The supervision agencies of the host country and the mother country share the responsibility of supervising foreign exchange activities and positions, where the mother country is responsible for the bank's global foreign exchange positions while the host country for the foreign exchange transactions and foreign exchange exposures within its territory. The supervision of the aforementioned aspects includes audition, assessment, and direct evaluation and other procedures on the safety of the bank's business.

The 1983 version of Basel Concordat and its revisions only proposed abstract management principles and allocation of responsibilities of supervision without mentioning any concrete, operational management criteria of supervision regarding the liquidity, settlement capability, foreign exchange activities and positions, and others. So, speaking wither legally or practically, each country's management on the international banking industry has formed a self-contained system, which in practice means that there is really not any sufficient supervision. Because of this problem, in July 1988, the Basel Committee passed the so-called Basel Accords, which established the method for calculating the ratio of a bank's assets and risky assets, known as capital adequacy ratio, and the required minimum. The Accords represents an important document of the Basel Committee toward the development of operationality. In particular, when computing the quantity of risky assets, Basel Accords connected the line items that were either listed within or outside the bank's balance sheet with different risk ratings. The Accords classifies capitals into two classes. The first class consists of the core capitals, also known as tier 1 capitals or the first tranche of capitals. And the other class is made up of supplementary capitals, also known as tier 2 capitals or the second tranche of capitals. There is a certain boundary or limitation between these two classes of capitals. Basel Committee decided that the targeted standard ratio of capitals and risky assets should be 8%, where the rate of the core capitals should be at least 4%. That represented the minimum standard for the international banks of the member nations of the Basel Committee to comply with starting in 1992.

The connotation of the Basel Accords is that it evolves continuously along with the objective needs of the international supervision. From the 1995 accidents of Baring Bank and Daiwa Bank, regulatory bodies recognized that defects in the business procedures and the system of the banking industry are also sources of risks. On June 25, 2004, the central bankers of the Group of Ten unanimously adopted the "International Convergence of Capital Measurement and Capital Standards: A Revised Framework", that is, the final draft of the New Basel Capital Accord, and decided that the G* group will start to implement it at the end of 2006. The New Capital Accord clearly included the market risk and operational risk into the calculation of risky assets, and proposed its innovative three pillars: the minimum capital requirements, the supervisory review of regulatory authority, and market discipline. It was more sensitive to risks, and included more pervasive range of supervision.

Responding to the deficiencies in financial regulation as by the late 2000s financial crisis, the third installment of the Basel Accords, known as Basel III or the Third Basel Accord, was agreed upon by the member nations of the Basel Committee in 2010–2011. It was scheduled to be introduced from 2013 until 2015. This Accord was supposed to strengthen the capital requirements of banks by increasing liquidity and by decreasing leverage. It is a global, voluntary regulatory standard addressing the capital adequacy, stress testing, and market liquidity risk of banks. The changes made on April 1, 2013, extended the implementation until March 31, 2018.

10.4 CURRENCY CRISES: IMPACTS OF FINANCIAL GLOBALIZATION

Currency crises, which might be caused by flows of international capital, bring forward much greater impacts on the internal and external balances of each country. Many countries and international economic organizations have been actively looking for countermeasures, hoping that future crises can be prevented and the damages of such crises can be reduced. The investigation of this problem has become in recent years one of the important research topics in the field of international finance.

10.4.1 The concept of currency crises

Speaking generally, a currency crisis means that the fluctuation in a nation's currency within a short window of time has gone beyond certain amplitude. And speaking specifically, it means such an event that market participants have caused through manipulating the foreign exchange market the nation's foreign exchange rate system to collapse, while leaving the foreign exchange market fluctuating continuously. One clear characteristic of a currency crisis is that the domestic currency depreciates substantially. There are distinctions and connections between currency crises and financial crises. For financial crises, there are fluctuations not only in the exchange rates, but also in the prices of such financial markets as stock market, banking systems, etc., as well as operational difficulties and bankruptcies of financial institutions. Currency crises can trigger financial crises; and financial crises, which are caused by domestic reasons, can also lead to the break-out of currency crises.

From the specific definition of currency crises, it follows that currency crises can be divided into the following types based on their causes:

1   The economic fundamentals deteriorated due to the government's expansionary policies, which triggers attacks by international speculative capitals, and led to the eventual currency crisis;
2   Although the economic fundamentals are relatively health, a currency crisis breaks out by the attacks from international speculative capitals that are triggered by either political event(s) or the effects of psychological expectations; and
3   A currency crisis breaks out by the spillover of an ongoing currency crisis in another country. This kind of currency crisis is known as a contagious currency crisis.

To explore the mechanism underneath the occurrence of currency crises, let us assume that there are two currencies A and B, where 1 unit of currency A = 25 units of currency B. At the start of our analysis, the exchange rate of currency B has been overestimated so that there is strong expectation for currency B to depreciate. Speculators generally employ one of the following three methods to launch attacks on currency B:

Method 1: Speculators short sell a distant currency B, assume two weeks or one month, using the exchange rate of 1 unit of currency A equal to 26 units of currency B or higher. That causes other market participants to imitate one after another (that is the herd effect), making the exchange rate of currency B to gradually fall to that of 1 unit of currency A = 30 units of currency B. Now, the speculators first convert currency A into currency B using the exchange rate A1/B30 in the spot market and at the same time, they cover their future currency contracts using the exchange rate of A1/B26 by buying A with payments in currency B so that each a unit of A earns 4 units of B.

Method 2: The speculators first borrow currency B in the interbank market of currency B and then buy into currency A in phases at the exchange rate A1/B25 or higher. When the exchange rate of B gradually falls to A1/B30, the speculators convert A back into B and then repay their debts in the interbank market. In this way, each unit of A earns 4 units of B (minus the interest costs of borrowing B). So, if the central bank of currency B drastically raises the interest rate of interbank borrowing, it will then increase the speculation costs so that the speculation using this method will lead to less profit or even losses to the speculators.

Method 3: Because speculative attacks generally cause the interest rate of interbank lending to rise, if the central bank attempts to defend its currency by raising interest rate to increase of cost of speculation in the foreign exchange market, then it will cause the stock market to fall sharply. When this is the case, before launching their attacks, speculators will first make short sellings in the stock market or the stock index futures market (or take long positions after launching their speculative attacks in the stock market or the stock index futures market). Even if the exchange rate stays invariant because of the intervention of the central bank, the speculators can still make their profits in the stock market or the stock index futures market.

The specific steps of applying method 3 can be expressed as follows:

1   Short sell in stock market → launch speculative attack → interest rate rises → stock market falls, → cover short positions in the stock market.
2   Launch speculative attack → interest rate rises → stock market falls → go long in stock market → withdraw from speculation → interest rate is adjusted downward → stock prices go up → cover long positions in stock market.
3   Short sell stock index futures in the futures market → launch speculative attack → interest rate goes up → stock market falls → stock index futures falls → cover positions in stock index futures.

No matter it is to the economy of the involved country or the entire world, each currency crisis can cause major effect. Its damages are shown in the following areas:

(i)   The process of each currency crisis causes harm to economic activities;
(ii) The currency crisis causes economic conditions to from before to after of occurrence of the crisis; and
(iii) When the government is forced to adopt remedial measures after the break-out of a currency crisis, contractionary fiscal and monetary policies are the most commonly employed.

For example, during the ongoing process of a currency crisis, in order to prevent outflows of capital, the government might adopt the measure of raising interest rate and the control over the foreign exchange market might be maintained for a long period of time. That would consequently cause economic contraction, and economic activities with the outside world would be suppressed. During the time of a currency crisis, large quantities of capitals would be frequently flow in and out of the country, causing severe fluctuations in the country's financial market. Additionally, the instability and panics that appear during the time of crisis could bring forward very large disturbances to the normal lives of the public. At the same time, currency crisis can easily trigger financial crises, economic crises, and even political crises, and societal crises. Next, foreign capitals often leave the country after the break-out of a currency crisis, which deals a heavy blow to the country's economic development. And, the depreciation of the domestic currency, as caused by the currency crisis, would cause the foreign debts, as measured by using the domestic currency, increase substantially. If the government were forced to adopt the floating exchange rate system after the break out of a currency crisis, the fluctuation in the exchange rate would be extraordinarily large due to the government's lack of experience in managing the system effectively. That of course would create inconveniences for the normal production and trades. If after a currency crisis broke out the government were forced to adopt contractionary fiscal and monetary policies to remedy the consequent losses, if the crisis were not caused by any expansionary policy, then this measure could very possibly lead to major disasters to the society. Additionally, in order to receive foreign financial aids, the government is often pressured to implement various conditions attached to the aids, such as opening the domestic commodity market, financial market, etc., which generally brings more risks to the operation of the domestic economic.

Of course, each currency crisis can expose many of the hidden problems of the domestic economy. At the same time, substantial depreciation of the domestic currency can also correct the existing phenomenon of overestimate of the currency, which is beneficial to the improvement of the balance sheet of international payments. However, all these advantageous factors play their roles under extreme levels of pains. So, looking at the issue holistically, it can be seen that the harm of a currency crisis is much greater than the benefit.

10.4.2   Causes of currency crises

Since the 1970s, crises in the international financial market have broken out frequently, leading to a major development of the theory of financial crises, where the study on how currency crises occur has especially caught people's attention. At the end of the 1970s, the theory of currency crises began to take shape with its relatively independent and complete system. The 1997 Asian financial crisis revealed many brand new characteristics that are different from those observed in the previous crises. That made the following investigations much richer and diversified than ever before.

Summarizing the available literature, the causes of currency crises can be categorized into two main groups: macroeconomic fundamentals and the continued expansion of the domestic credit. In terms of the former, the deterioration of the fundamentals includes external imbalances, such as current account deficit, actual appreciation of the domestic currency, etc., and internal imbalances, such as unhealthy operation of the economic system, huge fiscal spending, high level of inflation, relatively low level of foreign currency reserves, etc. It can be quantified by using a system of multiple macroeconomic and financial indices. This system can be employed to warn the potential break out of a crisis by monitoring the changes in the numerical readings of the indices.

The traditional early warning indicators of crises generally include the following indicators:

1   The current account deficit. When this deficit stays 5% or more of the GDP, it is often seen as a sign of not being sustainable for long term. So, it has been used as a warning signal for a forthcoming crisis.
2   The debt indicator. It includes: (i) total foreign debts/GDP; (ii) short-term debts/GDP; (iii) (current account deficit – foreign direct investment)/GDP, that is, (CA – FDI)/GDP; (iv) short-term debts /international reserve; (v) maturing

debts/international reserve; (vi) maturing debts/export income; (vii) the number of months the international reserve is able to pay for imports, etc. After the Asian financial crisis, people specifically include the problems of the financial sectors in the fundamentals.

Fundamentals represent the necessary conditions, instead of sufficient conditions, for crises to occur and for crises to worsen. However, they cannot necessarily predict accurately when a financial crisis will break out in a country. And, even for the time of normal economic conditions, there is no guarantee that the fundamentals are perfect. All that can be said is that when the fundamentals start to deteriorate, speculative attacks, political crises, and other factors can easily become the catalyst for currency crises to occur.

As for continued expansion of the domestic credit, which represents the explanation of the first generation theory of crises, Paul Krugman (1979) developed the first relatively plausible theory on currency crisis. This theory believes that under the condition that a country's demand for money is stable, the domestic credit expansion can cause losses in the foreign currency reserves and deterioration in the fundamentals so that under speculative attacks the original fixed exchange rate system will collapse, leading to the occurrence of a crisis. In particular, assume that a country's demand for money is very stable, where the money supply M consists of two parts: the domestic credit and foreign currency reserves. If all other conditions are held fixed and the domestic credit is continuously expanded in order to intermediate the fiscal deficit, then the money supply will be consequently increased. When the money supply is in excess of the residents' demand for money, then the residents will purchase goods, labor, and financial assets from abroad. The deficit in international payments will reduce the foreign currency reserves, which will eventually bring the money supply back to the level of the demand for money. By using the method of currency analysis, we have

$$M = D(t) + R(t) = M^d = M_0 \tag{10.2}$$

where the demand for money M^d is a constant, denoted as M_0, the domestic credit is $D(t)$, the foreign exchange reserve is $R(t)$. That is to say, when the domestic demand for money is stable, when the domestic credit (D) expands, the foreign exchange reserve (R) will have to reduce. Such relation is depicted in Figure 10.5.

Because the foreign exchange reserve is limited, continued expansion of the domestic credit will eventually exhaust the foreign exchange reserve of the monetary authority. Foreign exchange reserve is the important tool for the government to apply in its attempt to maintain the fixed exchange rate system. If there is a presumed minimum level of the foreign exchange reserve, then when the reserve falls below that minimum level, the government will have to give up the fixed exchange rate system, and the exchange rate level determined by the floating mechanism will be much lower than the originally fixed exchange rate. In Figure 10.5, when time reaches the moment t_0, the foreign exchange reserve is exhausted so that the fixed exchange rate system naturally collapses.

If there were no speculator in the marketplace, then the collapse of the fixed exchange rate system would merely be the consequence of the gradual exhaustion of

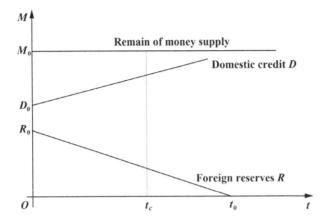

Figure 10.5 Loss of foreign reserves caused by expanded domestic credit.

the foreign exchange reserve as the credit gradually expands. However, if there were speculators who launched an attack on the domestic currency by selling the domestic currency and buying foreign currencies in the foreign exchange market, then the government would try to maintain the stability of the exchange rate by making use of the foreign exchange reserve. That would speed up the exhaustion of the foreign exchange reserve and the collapse of the fixed exchange rate system. In order to illustrate the problem of how to select the time of speculative attack, let us introduce a new concept, known as the shadow floating exchange rate. It means the free floating exchange rate of the foreign exchange market determined by the economic fundamentals when the government does not intervene in the market. Under the floating exchange rate system, the shadow floating exchange rate is the same as the actual exchange rate. Under the fixed exchange rate system, the shadow exchange rate is estimated by the market participants through using the fundamentals, where credit expansion can make the shadow exchange rate to rise if the method of direct pricing is used.

Let us now use Figure 10.6 to depict how speculators' expectation causes the fixed exchange rate to collapse ahead of time. In this figure, the vertical axis stands for the exchange rate given by using the method of direct pricing, e_0 the exchange rate under the fixed exchange rate system, the straight line ZAC the level of the shadow exchange rate. As the time t moves forward, the domestic credit expands continuously and the foreign exchange reserve is being exhausted so that the shadow exchange rate rises. If the marketplace did not have any speculator, then at time moment t_0, the foreign exchange reserve were exhausted completely (or the reserve fell below the accepted level of the government), the government would give up the fixed exchange rate system by allowing the rate to freely float. At this moment, the exchange rate jumps from point B to point C, which means the domestic currency is depreciated majorly.

Considering the psychological expectations of speculators, if speculators can accurately predict based on the fundamentals, such as the continued domestic credit expansion, the collapse of the fixed exchange rate system, then speculators will go long the foreign currency by using the current fixed exchange rate and cover the long

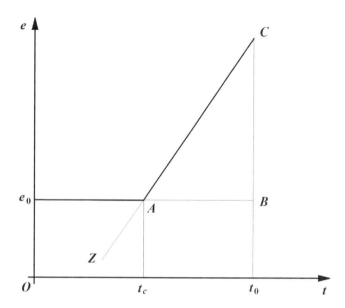

Figure 10.6 Time when a monetary crisis will occur as expected by speculators.

position to make a profit as soon as the domestic currency is depreciated after the collapse of the fixed exchange rate system. If the speculators sell the domestic currency and buy the foreign currency at the same time moment, it will then constitute a speculative attack on the fixed exchange rate system of the country. This attack will instantly exhaust the foreign exchange reserve, making the fixed exchange rate system collapse ahead of its time. Speaking more specifically, when the shadow exchange rate rises to the level of the fixed exchange rate e_0 at time moment t_c, speculators will launch their attack, causing the foreign exchange reserve to exhaust instantly and the fixed exchange rate system to collapse right away, as shown in Figure 10.6. After then, the exchange rate begins to float freely. During the process when the fixed exchange rate system is transformed to the floating exchange rate system, the exchange rate will experience jumps, which develop as what is shown as the sharp turning line e_0AC in Figure 10.6. This analysis shows that the slower the credit expansion is, the smaller the slope of the straight line ZAC is, and the later the currency crisis will break out.

Krugman's model is of the following characteristics when it is used to analyze currency crises:

1 In terms of the causes of currency crises, it believes that each currency crisis is caused by the conflicts between government's macroeconomic policy and the policy of maintaining the fixed exchange rate system (to this end, please recall the Mundellian Trilemma). This analysis treats the problem of international payments as the process of self-adjustment of the currency supply. It bears a heavy characteristic of the monetarism. Domestic credit expansion is the root reason for losing reserves. So, government's expansionary policies push the economy into a currency crisis.

2   On top of the mechanism of how currency crises occur, the model emphasizes on the general process of how speculative attacks cause the reserves to fall below the minimum allowed level so that a currency crisis eventually break out. In this process, the central bank is basically in a passive position. Expectations only move the time forward for the currency crisis to occur, while the scale of the foreign exchange reserve is the core variable that determines when to give up the fixed exchange rate.

3   In terms of the policy significance of the model, the main conclusion is that tightening fiscal policies are the key for the prevention of currency crises. Considering the fact that the root reason for a currency crisis to occur is the economic fundamentals, speculative attacks only represent an exogenous factor. Hence, such measures as taking foreign loans, limiting capital flows, etc., can only temporarily stabilize the exchange rate. If there is no adjustment at the level of the basic economic policies, the fixed exchange rate system will eventually collapse.

Within a quite long period of time, most of the currency crises that occurred from around the world have been caused by domestic credit expansions. Since the end of WWII, relatively influential are the three currency crises in Mexico during the 1970s, the Chilean currency crisis in 1982, etc. However, after entering the 1990s, flows of speculative capital have caused greater and greater impacts on macro economies; and when the economic fundamentals were still relatively healthy, these capital flows could also cause currency crises. To explain this phenomenon, the second generation model for currency crises was born.

The second generation theory of currency crises was proposed by (Maurice Obstfeld, 1986) and others with a different logic of thinking. It believes that the reason why speculators would launch an attack on a currency is not because of the deterioration of the economic fundamentals instead it is because of the self-realization of the depreciation expectation. In particular, this theory believes that with the particular laws of motion regarding international short-term flows, although a country did not implement any expansionary policy and has sufficient foreign exchange reserve, it can still experience a sudden speculative attack on its currency, which appears mainly because of the psychological expectations. The steps speculators take to attack a country's currency often start with borrowing the specific currency within the domestic money market, and then sell short the domestic currency in the foreign exchange market. If this attack is successful, the speculators will purchase the domestic currency after it has depreciated and repay the borrowed funds in the domestic currency. In doing so, the cost of the speculative attack is the interest paid to serve the loan from the domestic money market, while the expected return is the interest income from holding the foreign currency assets and the income from the depreciation of the domestic currency. As long as the depreciation magnitude of the domestic currency, if the expected attack is successful, is greater than the difference of the afore-mentioned interest rates, speculators will launch the attack. On the opposite side, if the government raises the interest rate of the domestic currency, it will increase the cost for speculators to launch their attacks. Theoretically speaking, the government can always raise the interest rate to a certain level to maintain its fixed exchange rate system. However, the problem is that raising interest rate is not costless. The costs might include:

(i)   If the government has a lot of debts, raising interest rate can make the budget deficit worse; and

(ii)   High interest rate is not good for financial stability. High interest rate implies economic contraction, which brings along with economic decline and high unemployment. Especially in the modern economy, stock market and real estate market are closely connected to interest rate. If a high interest rate causes a crash in the stock market and makes the prices of the real estate market fall, then it will cause the entire economy to fall into a recession or even a crisis.

The benefits for the government to maintain its fixed exchange rate system generally include:

(i)   Eliminate the adverse effects of freely floating exchange rate on the international trades and international investments. That will create a relatively stable external environment of the national economy.
(ii)   Make the fixed exchange rate to play the role of a nominal anchor that can suppress inflation.

And

(iv)   Acquire the reputation from maintaining the exchange rate that the government is consistent with its policies. That is extremely important when the government policy objects are the public with rational expectations; that will make it easier for the government to achieve its desired effects of future economic policies.

When the government experiences a speculative attack, whether or not to raise the interest rate for the purpose of maintaining the fixed interest rate is in fact about the balance between cost and benefit. That is mainly expressed as the balance between maintaining the reasonable openness of the economy and realizing stable economic development. So, in a certain sense, it is about the contradiction between the external balance and international balance, which can be illustrated by using Figure 10.7, where the vertical axis stands for the cost of maintaining the fixed exchange rate, the horizontal axis the domestic interest rate, and the CC curve the change in the cost of maintaining the fixed exchange rate. It varies with the interest rate. Assume that in the economy there is an optimal level of interest rate $\hat{i}$, where the cost of maintaining

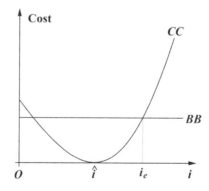

Figure 10.7 The cost and benefit of maintaining fixed exchange rate.

the fixed exchange rate is the lowest. When the interest rate gradually moves away from this level, the cost will increase so that the CC curve can be seen as U shaped. The BB curve stands for the return from maintaining the fixed exchange rate system. According to the previous discussion, this return has nothing to do with the level of interest rate. That is why when it is plotted, it is a horizontal line. When the interest rate is at i_e, the cost and the return of maintaining the fixed exchange rate are the same. From the figure, if follows that when the interest rate needed to maintain the fixed exchange rate is higher than i_e, the cost is higher than the return. So at this moment, the government will give up the fixed exchange rate system and when a currency crisis will break out.

From the discussions above, we can see that the currency crises the second generation model describes have the following characteristics:

1 A hidden condition for a currency crisis to occur is the existence of two equilibria in the macro economy: one is the equilibrium of different public's expectations on whether or not the fixed exchange rate system can be maintained, and the self-realization of each of these different expectations. Here, a "good equilibrium" is that the expectation of depreciation is zero so that the exchange rate is kept stable. The other equilibrium is the expectation of depreciation. When this expectation reaches a certain degree, the government will continuously raise the interest rate in order to maintain the exchange rate until eventually it gives up the fixed exchange rate system. The consequence of this equilibrium is a currency crisis. The jump from the "good equilibrium" to the currency crisis is often irrelevant to the economic fundamentals. Instead, it is often caused by the so-called sunspot phenomenon. These so-called sunspot events are often closely related to the laws of motion of the international short-term capital flows. When these events lead to expectations of depreciation, speculations will be triggered, leading to the realization of these expectations. So, this kind of currency crises is also known as expectation self-fulfilling crises.

2 Let us look at the general process of occurrence of currency crises where in order to fight against speculative attacks the government continuously raise the interest rate until it has to give up the fixed exchange rate system. The expectation factor determines whether or not a currency crisis will break out, and to what degree of severity the crisis will reach, while the level of interest rate is the central variable that dictates the decision of whether or not to give up the fixed exchange rate system. Speaking specifically, when the public expects that the currency will depreciate, the mechanism for a currency crisis to occur represents a vicious circle: The government raises the interest rate to maintain the fixed exchange rate → the cost for the government to maintain the fixed exchange rate goes higher → the market expectation for the currency to depreciate is strengthened → the interest rate is stimulated to further increase. So, it can be seen that whether or not for a currency crisis to break out stands for the dynamic game between the government and speculators. Under the condition of information asymmetry, the market can only produce an interval estimate $[\underline{f}, \bar{f}]$ for the true cost f for the government to give up the fixed exchange rate. Within this interval, speculators continuously launch their attacks. If during the time when the government tries to defend the

fixed exchange rate there is such a piece of sufficiently good news that can alter the speculators' expectations, then the occurrence of the forthcoming currency crisis will be stopped. Otherwise, the government will be forced to depreciate the domestic currency. If during this time the fundamentals, including the economic growth, level of employment, inflation, etc., suffer from substantial deterioration, then the occurrence of a currency crisis will be unavoidable. So, although occurrence of speculation may not be related to the fundamentals, the process for a crisis to break out can often cause the fundamentals to change.

3   One of the main measures the government can use to prevent currency crises from happening is to increase its policy credibility. The higher this credibility is, the smaller chance for a currency crisis to occur.

The second generation model of currency crises introduces the effect of expectations. It greatly expands the explanation capability of the theory of financial crises. For instance, the September crisis of the British pound in 1992 was caused by the speculators' expectation that the British government would prioritize domestic economic growth, stability in employment rate, and other goals above the stability of exchange rates. However, this understanding did not provide an explanation for how the attack was practically initiated, what factors led to the expectation, and any constructive policy suggestions. Obstfeld and others believe that because expectations are self-fulfilling, to prevent the crisis to occur one only needs to change the expectations. At the same time, they also explain that expectations are determined by factors outside the model. So, how to affect expectations is not among the problems the model attempts to address.

The diffusion mechanisms of currency crises can be classified into the following three classes:

1   The effects of the fluctuations of other countries' exchange rates on the domestic macro economy;

2   The model actions of how speculative activities have achieved magnificent returns; and

3   The alternative arbitrages or alternative divestments in the international financial market.

For example, the depreciation of the currency of a major commodity competitor country can lower the competitiveness of the domestic goods so that the domestic economy and employment are affected. If there is such an expectation in the marketplace, speculators will take trading actions before the macro economy experiences any real changes. That of course constitutes a depreciation pressure on the domestic currency. The initial success of the speculative attack on one country will encourage other market participants to imitate and to launch attacks another country or to participate in the speculative activities of the financial market. In these two situations, the country whose currency is overestimated is generally the first one to be attacked. If the assets of one country do not have any advanced market of hedging instruments, then the speculators may very well make use of relevant substitute assets of another country as their hedging instruments. Because of this reason, the currencies issued by countries with advanced money markets and foreign exchange markets could be impacted by external non-economic factors. This diffusion model additionally reveals why the currencies of such countries that enjoy health economic fundamentals can also be attacked.

With today's integration of the financial market, when one country suffers from a currency crisis, the crisis can most easily spread to the following three types of countries:

1   The major trade partner countries and major competitor countries. When a currency crisis breaks out in one country, it is either the case that the imports of goods from the major trade partner countries of the country in crisis drops or the case that the exports of the competitor countries are under much increased pressure. So, the trade incomes and payments of both the partner and competitor countries suffer from major changes, which induce speculative attacks.
2   The countries that share similar economic structure and development models, especially similar economic problems, such as overestimated exchange rates, with the country in crisis. The speculative capitals will attack these countries one after another without any discrimination.
3   The countries that excessively depend on the inflow of foreign capital. After the relatively influential currency crisis broke out, the speculative capital in the international financial markets will generally be adjusted and reduce its holding of foreign assets, especially the assets from countries of relatively high risk. Consequently, many countries will inevitably experience the phenomenon of capital outflows. If such outflows affect a country's international payments, then that country may also suffer from its own currency crisis.

10.4.3   Asian currency crises: A real-life case

In the 1990s, there had appeared three severe currency crises in the world: the September crisis in Europe (1992), Mexican crisis (1994), and Asian crisis (1997). The European crisis forced Great Britain and Italy to withdraw from the European Monetary Union. Both the Mexican and Asian crises respectively brought the entire Latin America and entire Asia into an economic and financial chaos for over a year. Among these tree crises, the Asian crisis had created the most long-lasting and severe effects. In this subsection, we will use this crisis as a case study to analyze the occurrence, evolution, and resolution of a currency crisis.

Firstly, let us look at the background on which the Asian crisis occurred from four different angles:

1   Long-term fixed exchange rate system and appreciation of the domestic currency,
2   Current account deficit and inflows of capital,
3   Continued deterioration of the operating conditions of the financial system, and
4   Fall in the government creditability.

In terms of long-term fixed exchange rate system and appreciation of the domestic currency, Asian countries are mostly export-oriented economies. Since the late 1970s, they had implemented a fixed exchange rate system where the value of each domestic currency was pegged with a basket of foreign currencies that was dominated by the U.S. dollar. The fixed exchange rate system was useful in the stabilization of the export and import prices, and convenient for manufacturers of exported goods to arrange productions. However, the foundation of the fixed exchange rate system was that the economic developments of all the involved countries were similar; when the domestic economic development deviated from that of the country, whose currency was pegged by the domestic currency, the domestic currency would be either overestimated or underestimated. In the 1990s, the fast growth of the U.S. economy made its dollar appreciate against all other major currencies. So, the Asian currencies that were pegged with the U.S. dollar were also floated higher and were increasingly overestimated. That made these Asian countries situated in disadvantageous positions in their trades with other advanced countries.

In terms of current account deficit and inflows of capital, since the early 1990s, the Asian countries had started to experience their current accounts' deficits. And the magnitudes of the deficits continued to enlarge. During 1985–1989, the average current account deficits of Thailand, Indonesia, South Korea, Malaysia, Philippines, were respectively only about 0.3% of their GDPs. However, after entering the 1990s, the average current account deficits of these five countries had respectively reached 4% of their GDPs with the tendency of continuously rising. Thailand and Malaysia were the worst cases. There were two reasons for this phenomenon to occur: Firstly, each of the Asian currencies that were pegged with the U.S. dollar was depreciated with the dollar and appreciated with all other currencies; secondly, the export of each country consisted of mainly semiconductors and other primary processed products. Along with the economic stagnation of Japan that was the main country of export and the competition from other developing countries, the international market share of the Asian countries had been shrinking. In order to offset the current account deficits, and because the fixed exchange rate system weakened the independence of their monetary policies, the differences between the foreign and domestic profits of these Asian countries were enlarged. The high interest rates induced large scales of foreign capital to flow into these countries. To encourage foreign capitals, Asian countries adopted the policy of combining their fixed exchange rate system with free conversion of capital assets. That created a beneficial opportunity for the speculation of short-term capital.

In the early 1990s, the economies of these Asian countries reached their peak of fast growth. The weight of bank lending to private entities in each of these countries compared to its GDP had been increasing rapidly. However, the investment behaviors of the financial institutions suffered from major abuses. Firstly, there was the problem of mismatching currency structure that existed in the lending practice, where borrowing was in foreign currencies while lending was in the domestic currency; and there was the problem of mismatching term structure, where borrowing was in short-term foreign capital while lending was in long-term foreign capital. These two mismatches created huge interest rate risk and repayment risk. Secondly, large quantities of funds entered into such high risk long-term investment projects as real estate through financial institutions, which helped to push the assets' prices higher, creating the false prosperity of the national economies. Financial institutions not only provided loans to real estate companies, but also entered into the real estate market directly by using their own capitals and borrowed funds. That practice had surely increased the operating risks of these institutions. Thirdly, the financial laws and regulations of these Asian countries were not complete. None of these countries had either laws or regulations regarding deposit insurance or the requirement for financial institutions to comply with the accounting and auditing standards that were internationally practiced. And worst of all, none of these countries had a warning mechanism in place for financial crisis.

Any fixed exchange rate system should be implemented on the bases of political stability and good government credibility. However, when the crisis first emerged, these Asian governments demonstrated their policy incapability and poor credibility. Before these countries started to intervene in the market and to defend their currencies, they made policy mistakes and showed their indecisiveness. The public's doubt on the credibility of the governments directly eroded the foundation of the fixed exchange rate system. At the same time, the ruling of Indonesian president Suharto had entered its final stretch. People predicted that Indonesian political situation would become turbulent. All of these factors had strengthened the expectation for the fixed exchange rate system to collapse.

Secondly, let us look at the evolution process of the Asian currency crisis. This crisis that started in 1997 can be divided into three important periods: The first period: Thailand started to implement the floating exchange rate system, which triggered the break out of the crisis. The second period: The slide of the Korean currency led to the second round of storms. The third period: Indonesian crisis became of the fuse for the third wave of the Asian currency crisis.

Thailand was the first country that was affected by the Asian currency crisis. The tactics speculators used to snipe Thai Baht (and other currencies) consisted of the following: use foreign exchange swaps to sell Thai Baht in the marketplace; at the same time, apply forward foreign currencies, foreign currency futures, currency swaps, and other methods to hedge risk (this tactics were once used effectively in the Mexican crisis and the September European crisis), where the so-called forward means either that at the same time when spot contracts of foreign currencies are bought distant future contracts are sold or that at the same time when spot contracts of foreign currencies are sold distant future contracts are bought. In early 1997, international hedge funds purchased in the foreign exchange market large amounts of forward contracts of foreign currencies in U.S. dollars, which promised to make payments at maturity in Thai Baht. As of the second and the third month of the year, the trading volume of the forward contracts had reached US$15 billion. However, at the same time, the Thai government still continuously provided the market with forward foreign currencies without implementing any measure of intervention and control. That cost the government over US$5 billion of its foreign exchange reserve.

On July 2, 1997, Thai Baht could no longer hold on to the speculative attacks, and slumped 10%. On the same day, the central bank of Thailand issued a statement with the endorsement of the Ministry of Finance that the exchange rate of Thai Baht would be determined jointly by the market mechanism and supply and demand situations in the domestic foreign exchange market. In other words, a managed floating system was implemented to replace the fixed exchange rate system with Thai Baht against a basket of foreign currencies that had been in place since 1984. After that date, other Asian countries also gave up their fixed exchange rate systems. For example, on July 11, 1997, the central bank of Philippines decided to widen the foreign exchange trading range for its peso; Indonesia also expanded the amplitude of fluctuation of its rupiah to 12% from 8%, and then decided to simply implement the floating exchange rate system on August 14. Soon after that date, Malaysia, South Korea, and Taiwan also followed the suit one after another.

In November 1997, after two months of continued fall, Korean won was dero-gated with immeasurable weight, leading to Korean won crisis. As of November 20, the exchange rate had fallen 20% to 1,138 won against 1 U.S. dollar. Soon after that day, it was uncovered that Yamaichi Securities of Japan had suffered from huge losses in foreign exchange and in securities transactions. That caused Tokyo stock market crash. On December 8, the exchange rate of Japanese yen on the Tokyo foreign exchange market fell to 130 yen against 1 U.S. dollar, a historical low since May 1992. Under the pressure of these two shocks, the currencies of the Southeast Asia were fully depreciated, stock markets plunged, and the Asian currency crisis reached another elevation. During this time period, political factors started to affect the progress of the financial crisis. In May 1998, there appeared massive riots within Indonesian territory. The uncertain political and economic situations once again caused the currencies of Southeast Asia to fluctuate severely. In fact, along with the proliferation of the currency crisis, it had been transformed into such an economic crisis that fully affected the national economies and the international financial order, making the Asian area and even the financial situation of the world economy more chaotic, unpredictable.

10.4.4 Solution to currency crises as analyzed using Asian scenario

Limited by the length of this book, we will only discuss the measures of revolution of currency crises, as caused by changes in the exchange rate system, strengthening management of capital flows, and reforms of the financial system, from the angle of finance by using the Asian currency crisis as our example.

First of all, let us look at changes in the exchange rate system. During the Asian currency crisis, the substantial depreciations in the various countries' currencies and the adoptions of the floating exchange rate system were the helpless consequences of the speculative attacks. So, we refer to these alterations as changes in the exchange rate system instead of reforms of the exchange rate system in order to separate the passiveness and activeness. The essence of giving up the fixed exchange rate system and adopting the floating exchange rate is to place the focus of attention to the internal balances, while letting the external balances be achieved by the fluctuations of the market exchange rate. After having implemented the floating exchange rate system, all the Asian countries and regions went through the painful process of currency devaluation. However, by the 20-20 hindsight, it can be seen that the devaluation had played its roles very well:

1   The currency devaluations had stimulated exports, limited imports, while the trade deficits continued to shrink. Because of the damaging aftermath of the crisis, the effect of currency devaluation was not clear during the first year (1998) after the crisis. However, starting in 1999, the Asian economies generally recovered with their falling speed of export not only slowed but also starting to grow positively.

2   The devaluation eased the pressure of the currencies being overestimated and reduced the losses of foreign exchange reserves. In fact, in the first quarter of 1999, the foreign exchange reserve of each country had started to rise.

3   The devaluation strengthened the confidence of international economic organiza-tions on the prevention of crises and helped the organizations to increase their aid

efforts. And, the IMF organization also included currency devaluation as one of the main conditions for receiving international aids.

4   After large magnitudes of drops in the levels of exchange rates, starting in June 1998, they began to rebound. And the capability of resource allocation and balance of the international financial market started to play its roles. In early 1999, all the major currencies of the Asian area started to strengthen. And

5   The floating exchange rate system recovered the independence of each country's monetary policies so that each of the monetary authorities could once again promote its country's economic growth by flexibly using the lever of interest rate.

In the process of changing the exchange rate system, each of the Asian countries learned the following lessons: Firstly, at the same time of implementing the floating exchange rate system, the exchange rate should still be regulated. That could make not only the exchange rate system more elastic, but also the monetary authority in full control of the level of the exchange rate. For example, the central bank of Thailand positioned its exchange rate system as a managed freely floating exchange rate system of the market, instead of a mere floating exchange rate system. Secondly, as soon as hints of a crisis are discovered, the earlier the level of the exchange rate is adjusted, the more beneficial it will be in terms of controlling the appearance of the crisis. By doing so, it could not only reduce the amount of losses in the foreign exchange reserve, but also support the domestic currency to return to its equilibrium state quickly under the effect of the market forces. That of course is beneficial for the easement of the crisis.

What should be noted is that in the Asian currency crisis, only mainland China, Singapore, and Hong Kong did not change their exchange rate systems. However, that was caused by their particular reasons. For example, the mainland China implemented capital control; Singapore had always been implementing its managed floating exchange rate system, where the monetary authority had relatively large roles in the arrangement of foreign exchanges; and the situation in Hong Kong was quite specific. It was the only region in Asia that had maintained its fixed exchange rate system. Because the economic structures and societal states of these three regions and countries were relatively stable and healthy, they had their strong abilities to withstand the attacks.

Secondly, let us look at the appropriate control on capital flows. The combination of the fixed exchange rate system and free currency conversions under capital accounts provides a convenient door for international hot money to go in and to go out. After the currency crisis, all the Asian countries recognized the serious harm of international hot money so that they to different degrees strengthened their controls on capital flows. There are the following main types of measures of control:

1   Limit the forward and non-trade exchange currency transactions.
2   Ban non-residents from speculating using the domestic currency or financing activities that could potentially be transformed into speculation activities.
3   Restrict the foreign exchange exposure of domestic companies and banks, or establish time limits for foreign exchange exposures. And
4   Tax all inflows of short-term capital.

Thirdly, let us look at reforms of financial systems. Imperfection in the financial system of each of the Asian countries was another important reason for the success the international hot money could achieve in the Asian currency crisis. So, reforming the existing financial system had become one of the important tasks of defusing crises. The reforms of the financial systems of the Asian countries can be divided into two types: The first type is short-term restructures of the existing financial systems, and the second type is the strengthening of long-term monitoring of the existing financial systems.

In terms of restructures of the existing financial systems, after the break out of the crisis, in order to reduce the shock of large number bankruptcies of financial institutions on the economy, each national government restructured its financial system, such as the introduction of deposit insurance mechanism, providing capital supports directly to financial institutions, directly taking over or closing down financial institutions while creating special agencies to implement the restructuring, creating financial restructuring authority, asset management companies, and other organizations to specifically deal with banks' bad debts, to help those banks that suffered from serious difficulties so that they could recover from their trauma and start to grow again. In particular, the purpose for the Asian countries to adopt the deposit insurance mechanism is for the depositors to have confidence in the market during the process of restructuring. The form of the adopted deposit insurance mechanism was not necessarily the creation of a specialty company; it could also include oral and written promises from the government. For example, Indonesia in early 1998 declared that the government would guarantee the legitimate interests of all domestic depositors and creditors of state-owned, private, and joint venture banks; however, the warranty period was two years. South Korea on the other hand expanded the original upper limit of the deposit insurance to the full amount from 20,000,000 for a period of three years, while at the same time the central bank would purchase ₩5,000 billion worth of the bond issued by the deposit insurance company in order to guarantee the payments to depositors. In order to make some of the financial institutions that bear significant effects on the national economy stay in business, the government provided capital supports to these firms either by directly injecting funds or providing concessional loans. After the crisis broke out, there appeared huge amounts of bad debts, leaving many financial institutions unable to meet the operating capital requirements so that they had no choice but declaring bankruptcy or closure. In order to reduce the impact on the economy, the government in general got directly involved in the mergers, purchases, and sells of financial institutions.

As for strengthening financial monitoring, in order to completely eliminate many persistent ailments left over through the history, each of the Asian countries reformed its financial supervision mechanism. Speaking specifically, each country's reform of its financial supervision was started mainly in the following two aspects: First, improve the setup of the financial regulatory body. In terms of the institutional settings, the regulatory bodies of the Asian countries and regions can be roughly classified into the following two classes: One class is that the central bank is responsible for the unified regulation of the financial industry, including banks, securities, insurance, etc. This setup is seen in Thailand and Indonesia. The other class is to establish special regulatory agencies, which conduct separate supervision of the financial industry. This setup is employed in Singapore, South Korea, and Hong Kong. Overall, all countries tend to establish a unified regulatory agency. That not only saves money, but also provides fast and flexible reactions to the occurrence of emergencies. Secondly, establish clear regulatory principles and improve supervision quality.

After the crisis, the supervision objective of the Asian countries generally aims at establishing a stable and competitive financial system in order to prevent potential financial crises. To this end, most of these countries and regions have followed the following supervision principles:

(i)   The market, that is, market forces will occupy the dominating position in the process of supervision;

(ii)   The openness, that is, at the same time when supervision is strengthened, the market will continued to be open in order to sufficiently make use of international capital to accelerate the national economic growth;

(iii) Internationality. It is mainly expressed in introducing internationally adopted supervisory standards and methods and in collaborations with authoritative international institutions;

(iv)   Legalization. Supervision principles and special issues of financial reforms are finalized in the form of law; and

(v)   Transparency, that is, through collaboration with other agencies, the transparency of the supervision policy will be improved.

10.5   WILL FINANCIAL ATTACKS BE ALWAYS POSSIBLE?

Based on what is discussed above, a natural question is: Would a well-established financial system with adequate monitoring system not suffer from major financial crisis? To this end, what is studied in Section 3.7 in Chapter 3 implies that the answer to this question is: No, even with an adequate monitoring system in place, no matter how well established a financial system is, it will (quasi-) periodically experience crises of different scales. In particular, when an economic system is considered well-established with an adequate monitoring mechanism in place, it can be theoretically identified as a realistic copy of the dishpan used in Hide's dishpan experiment, where the fluid (money in circulation) is flowing uniformly along with the dish. Hence, both Hide's dishpan experiment and Theorem 3.1 (of Never Perfect Value Systems) imply that chaos will be forthcoming next. In terms of economics, it means that financial crises of various sorts are inevitable. On the other hand, as argued in Section 3.7 in Chapter 3 innovation is the livelihood of financial infrastructures. That implies that no realistically existing financial system can be forever considered well-established, because when a financial product is seen as innovative, it means that the product fills a niche in the market place. So, the recent worldwide heat wave of financial innovations will continue to make each and every established financial market inadequate or incomplete, the implemented monitoring system obsolete. That is, both financial markets and their monitoring systems are always in the process of improvement. Therefore, whichever system could lead the international effort of financial innovation, this particular financial infrastructure will have an edge over other financial infrastructures and capital flows from this leading infrastructure would have the advantage to launch attacks on other weaker markets that are following the pace of innovation.

10.5.1 How environments determine the nature of economies

Corresponding to the development of the recent internationalism and the ideal world culture, it is natural to think about that if a uniform, synchronized world financial system could be established, then there would be no weaker markets that were following the leading markets in terms of innovation. Therefore, potential financial attacks would no longer be possible. However, the following mathematical reasoning shows that in terms of economics, there will always be different financial centers working under different sets of rules and regulations. Therefore, some of the centers will inevitably emerge as leaders in one way or another. In other words, as long as the technology of communication continues to improve, financial attacks will be inevitable.

To reason using our systemic yoyo model, let us place the totality of all different existing economies in the historical flow of time and see each economic system as a spinning yoyo field coexisting with many other economic yoyo fields. By doing so, the methodology of thinking of the systemic yoyo analysis indicates that as long as an economy has uneven internal organization, there must be uneven moments of economic forces. Combining with the naturally existing uneven gradient forces produced by the uneven internal organization, the abstract but realistic yoyo structure of the economy will have to spin. That is, the uneven internal organization of an entity produces eddy motion for the entity. And, the more uneven internal organization the entity possesses, the greater the naturally existing uneven gradient forces will be, and the faster the entity will spin. In a fast spinning economy, where the high speed of rotation represents a high degree of unity of the underlying economy and a high level of fragility, the number of economic centers changes over time just as what is shown in the dishpan experiment with the number of eddy leaves. This analysis also indicates that it is also possible for an economy to have many financial centers as long as the overall internal organization of the economy is relatively even. This end combined with what is obtained in (Lin and Forrest, 2011) leads to the following conclusions:

1   Economies that are situated in such lands that have severely, unevenly distributed natural resources tend to have higher degrees of unity and higher levels of fragility than those that occupy lands with relatively even distribution of necessities for the existence of ordinary lives;
2   Economies existing on rich lands with relatively even distributions of natural resources tend to have multiple financial centers, leading to coordinated competitions for innovations.

10.5.2 Existence of financial centers under different regulations

The phenomenon of financial units in an economy can be modeled by using the concept of centralized and centralizable systems. In particular, (Hall and Fagen, 1956) introduced the concept of centralized systems, where a system is centralized if one object or a subsystem of the system plays a dominant role in the system's operation. The leading part can be thought of as the center of the system, since a small change in it would affect the entire system, causing considerable changes across the entire spectrum. (Lin

1989) applied the concept of centralized systems to the study of some phenomena in sociology, obtaining several interesting results, including an argument on why there must be a few people in each community who dominate over others. In particular, a system (Lin 1987) is an ordered pair of sets, $S = (M, R)$, such that M is the set of all objects and R a set of some relations defined on M. The sets M and R are known as the object and relation set of S, respectively. Here, each $r \in R$ is defined as follows: There is an ordinal number $n = n(r)$, as a function of r, known as the length of the relation r, such that $r \subseteq M^n$, where

$$M^n = \underbrace{M \times M \times \cdots \times M}_{n \ times} = \{f : n \to M \ is \ a \ mapping\}$$

is the Cartesian product of n copies of the object set M. A system $S = (M, R)$ is trivial, if $M = R = \emptyset$. Given two systems $S_i = (M_i, R_i)$, $i = 1, 2$, S_1 is a partial system of S_2 if either (1) $M_1 = M_2$ and $R_1 \subseteq R_2$ or (2) $M_1 \subset M_2$ and there exists a subset $R' \subseteq R_2$ such that $R_1 = R'|M_1 = \{f : f$ is a relation on M_1 and there is $g \in R'$ such that f is the restriction of g on $M_1\}$. In symbols, we have

$$R'|M_1 = \{f : g \in R' \ (f = g|M_1)\},$$

where $g|M_1 \equiv g \cap M_1^{n(g)}$. A system $S = (M, R)$ is called a centralized system, if each object in S is a system and there is a nontrivial system $C = (M_C, R_C)$ such that for any distinct elements x and $y \in M$, say, $x = (M_x, R_x)$ and $y = (M_y, R_y)$, then $M_C = M_x \cap M_y$ and $R_C \subseteq R_x|M_C \cap R_y|M_C$. The system C is known as a center of S. The following has been shown:

Theorem 10.1. (Lin 1988b). Assume ZFC. Suppose that $S = (M, R)$ is a system such that $|M|$, the cardinality of M, $\geq c$, where c is the cardinality of the set of all real numbers, and that each object in S is a system with finite object set. If there exists such an element that belongs to at least c objects in M, there then exists a partial system B of S with an object set of cardinality $\geq c$ and B forms a centralized system.

One interpretation of this result is that as long as an economy forms a system, as defined here, where different parts of the economy are closely connected by various relationships, and some special elements exist such that each of these elements transcends through a great number of the parts of the economy, then in this economy, at least one financial center (or financial unit) will appear. And, when dictated by nature, such as severely uneven distribution of resources, various existing centers (financial units) will naturally fight for control of greater power and influence. In the process of power struggle, this theorem indicates that many financial units along with their spin fields could be eliminated without damaging the underlying structure of the economy. This theoretical result might be the explanation for the phenomenon of China's highly centralized political and economic power structure and the historical fact that after each political and economic unification of China, all oppositions were mercilessly eliminated, which, according to our yoyo model analysis, created a uniform motion in the political and economic spin field.

10.5.3   Centralizable economies

Let $S_i = (M_i, R_i)$, $i = 1, 2$, be two systems and $h\colon M_1 \to M_2$ a mapping. Define two classes $\hat{M}_i$, $i = 1, 2$, and a class mapping $\hat{h}\colon \hat{M}_1 \to \hat{M}_2$ by using the transfinite induction as follows:

$$\hat{M}_i = \bigcup_{n \in \mathrm{Ord}} M_i^n, \quad i = 1, 2,$$

and for each $x = (x_0, x_1, \ldots, x_\alpha, \ldots) \in \hat{M}_1$,

$$\hat{h}(x) = (h(x_0), h(x_1), \ldots, h(x_\alpha), \ldots) \in \hat{M}_2,$$

where Ord is the class of all ordinals. For each relation $r \in R_1$, $\hat{h}(r) = \{h(x) : x \in r\}$ is a relation on M_2 with length $n(r)$. Without confusion, h will be used to indicate the class mapping $\hat{h}$ and is seen as a mapping from the system S_1 into the system S_2, denoted $h\colon S_1 \to S_2$. When $h\colon M_1 \to M_2$ is surjective, injective, or bijective, the mapping $h\colon S_1 \to S_2$ is also seen as surjective, injective, or bijective, respectively.

The systems $S_i, i = 1, 2$, are similar if there is a bijection $h\colon S_1 \to S_2$ such that $h(R_1) = \{h(r) : r \in R_1\} = R_2$. The mapping h is known as a similarity mapping from S_1 onto S_2. A mapping $h\colon S_1 \to S_2$ is termed to as a homomorphism from S_1 into S_2 if $h(R_1) \subseteq R_2$.

A system $S = (M, R)$ has n levels (Lin 1989), where n is a fixed whole number, if (1) each object $S_1 = (M_1, R_1)$ in M is a system, called the first-level object system of S; (2) if $S_{n-1} = (M_{n-1}, R_{n-1})$ is an $(n-1)$th-level object system of S, then each object $S_n = (M_n, R_n) \in M_{n-1}$ is a system, called the nth-level object system of S. For a graphic representation of this concept, see Figure 10.8.

A system S_0 is n-level homomorphic to a system A, where n is a fixed natural number, if there exists a mapping $h_{S0}\colon S_0 \to A$, known as an n-level homomorphism, satisfying the following:

1   The systems S_0 and A have no non-system kth-level objects, for each $k < n$.
2   For each object S_1 in S_0, there exists a homomorphism h_{S1} from the object system S_1 into the object system $h_{S0}(S_1)$.
3   For each $i < n$ and each ith-level object S_i of S_0, there exist level object systems S_k, for $k = 0, 1, \ldots, i - 1$, and homomorphisms $h_{Sk}, k = 1, 2, \ldots, i$, such that S_k is an object of the object system S_{k-1} and h_{Sk} is a homomorphism from S_k into $h_{S_{k-1}}(S_k)$, for $k = 1, 2, \ldots, i$.

A system S is centralizable (Lin 1999), if it is 1-level homomorphic to a centralized system S_C under a homomorphism $h\colon S \to S_C$ such that for each object m in S the object systems m and $h(m)$ are similar. Each center of S_C is also known as a center of S.

Theorem 10.2. (Lin 1999) A system $S = (M, R)$ with two levels is centralizable, if and only if there exists a nontrivial system $C = (M_C, R_C)$ such that C is embeddable in each object S_1 of S; that is, C is similar to a partial system of S_1.

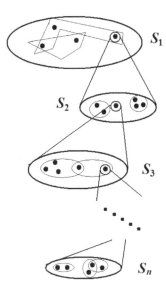

Figure 10.8 A graphical representation of an n-level system.

This result provides the theoretical foundation for checking whether or not an economy that is in a chaotic state would soon become organized. When this theorem is mapped over to the current economic state of Chinese affairs, one should be able to see where Chinese economy is at in its potential economic revival for the ultimate appearance of its powerful economic spin field. For our purpose here, all the details are omitted.

10.5.4   Forever existence of emerging economic centers

Americans consumed millions of Japanese cars, cameras, television sets, and electronic gadgets during the 1970s and 80s, which created considerable antagonism towards Japan, and lived off Chinese goods starting in the late 1980s, which similarly led to resentment towards the Chinese. In 1993, eighty-eight of the hundred films most attended throughout the world were American, and two American and two European companies dominated the collection and dissemination of news on a global basis (Havel 1995), which also seemed to have produced alienation from across the world against America. These economic realities in fact indicate how new economic activities appear and how new economic centers are established. In particular, Figure 10.9 shows the various ways for sub-eddies (new economic activities and financial centers) to appear, when there are at least two visible centers N and M. And if the parental fields N and M do not keep up with their innovations, the sub-eddies will eventually take over their dominance because these sub-eddies are born and grow on the back of the combined strengths of the parental fields. In fact, this analysis explains why historically economic centers travel from places to places.

At this junction while talking about emergence of new economic centers and considering the emergence of the Western civilization in the 11th and 13th centuries, let us

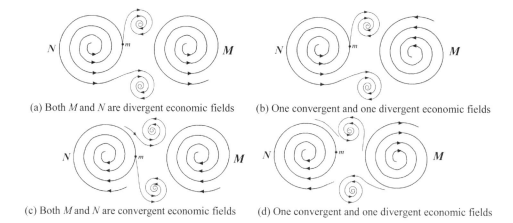

(a) Both M and N are divergent economic fields   (b) One convergent and one divergent economic fields

(c) Both M and N are convergent economic fields   (d) One convergent and one divergent economic fields

Figure 10.9 Object m might be thrown into a sub-eddy created by the spin fields of N and M jointly.

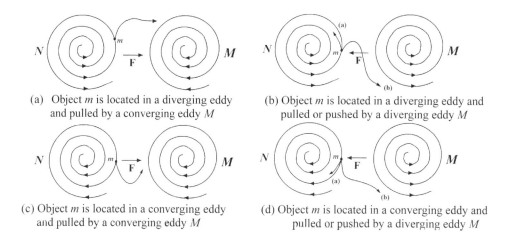

(a) Object m is located in a diverging eddy   (b) Object m is located in a diverging eddy and
     and pulled by a converging eddy M pulled or pushed by a diverging eddy M

(c) Object m is located in a converging eddy   (d) Object m is located in a converging eddy and
    and pulled by a converging eddy M pulled or pushed by a diverging eddy M

Figure 10.10 Acting and reacting models with yoyo structures of harmonic spinning patterns.

see how a new emerging economic center adopts useful elements of different existing centers to build its own organizational structure.

For a new emerging economic center to adopt useful elements from different existing centers, the new center has to be created with such objects m out of scenarios of Figures 10.10(b) and (c) and Figures 10.11(a) and (d). For scenarios in Figures 10.10(a) and (d) and Figures 10.11(b) and (c) no viable sub-eddies can be fruitfully produced. If a sub-eddy is indeed created in cracks between existing spinning fields, elements of these centers useful to the newborn might not be naturally adopted by the sub-eddy. They could also be forced on to the offspring. For example, Japan's forcible opening to the West by Commodore Perry in 1854 vividly shows that after tasting the greatness of the Western civilization by external force or pressure, Japan voluntarily made its

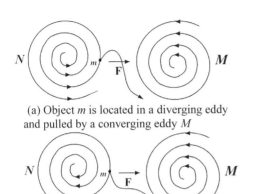

(a) Object *m* is located in a diverging eddy
and pulled by a converging eddy *M*

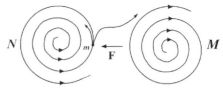

(b) Object *m* is located in a diverging eddy
and pushed or pulled by a diverging eddy *M*

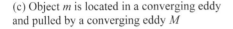

(c) Object *m* is located in a converging eddy
and pulled by a converging eddy *M*

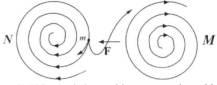

(d) Object *m* is located in a converging eddy
and pushed or pulled by a diverging eddy *M*

Figure 10.11 Acting and reacting models with yoyo structures of inharmonic spinning patterns.

dramatic decision to learn from the West following the Meiji Restoration in 1868. Figure 10.9 provides different possibilities for a new economic center to appear.

What is discussed in this subsection once again supports the claim that as long as the technology of communication continues to improve, financial attacks are inevitable, because newly emerging markets and centers in general are immature and their monitoring systems inadequate.

Chapter 11

Financial crises and currency wars

As we have accustomed to in this book, when seen from systems research, each economy can be righteously treated as a pool of rotational fluids of information, knowledge, money, etc. That is, the international economy is an ocean of spinning fields, some of which exist peacefully alongside of each other, while others fight against each other, destroy each other, and takeover each other. Analogous to the phenomenon of tsunamis that appear in the earthly ocean, there should be crises in the international financial world. That is exactly what we are going to study in what follows next.

This chapter is organized as follows. Section 11.1 introduces the concepts of financial and currency crises and glances over several representative theories. Section 11.2 focuses on speculative attacks across national borders and relevant currency crises, their characteristics, and then analyzes the success story of Chilean capital account liberation. In Section 11.3, we investigate currency wars and a possible strategy of self-defense against currency wars based on recent advances in the research of feedback systems.

In terms of references, this chapter is based on (Johnson, Darbar and Echeverria, 1997; Wang and Hu, 2005; Lin, 1994; Lin and Hopkins, to appear).

11.1 INTERNATIONAL FINANCIAL CRISES

Financial crises include all those crises with respect to either respectively or jointly currency, credit, bank, debt, and the markets of stocks, bond, and financial derivatives, etc. They generally mean severe fluctuations and chaos that appear in the financial area of a nation, interfering very negatively with the operation of the real economy. Each of these crises is accompanied by sudden deterioration of all or most financial indexes, stock market crash, capital flight, credit destruction, extremely tight money supply, rising interest rate, bank runs, bankruptcy of a large number of financial institutions, major decrease in the official reserves, inability to repay the interest and principal of maturing debt, currency devaluation both internally and externally, etc.

So-called international financial crises stand for rapid spreads of severe unrests and chaos, which first appear in the international financial market, into the relevant nations and regions through payment and/or financial manipulations, or through financial panics, causing upheaval and disorder in these relatively regional financial fields.

Because the modern-day national economies have been closely intertwined with each other to a high degree of globalization, when a nation or geographic region suffers from economic and financial crises, in terms of their damaging effects and adverse impacts, the crises tend to possess international and global characteristics. Since the 1920s, there had appeared many large scale financial crises. The most noteworthy of these crises include

1   The global stock market crisis that started on October 28, 1929, in the New York Stock Exchange and spread to other nations quickly, leading to a worldwide financial and economic crisis until 1933;
2   Six US dollar crises one after another during the time period from 1960 to 1973;
3   A bank bankruptcy wave that started with the failure of the United States National Bank in San Diego and quickly spread to many other western countries during 1973–1975;
4   The debt crisis that broke out in August 1982 and quickly involved over 50 developing countries from around the world;
5   The international stock market crisis that started with the plummet of the Dow Jones Industrial Average on October 19, 1987; the consequent chaos of the Wall Street momentarily spread throughout all the major stock exchanges of the western world;
6   The British pound crisis of 1992;
7   Mexican financial crisis of 1994;
8   The southeast Asian financial crisis of 1997;
9   The Russian currency, financial crisis of 1998;
10 The Brazilian currency, financial crisis of 1999;
11 The Argentinean currency, financial crisis of 2001;

Among these financial crises, the global stock market crash during 1929–1933, the repeated U.S. dollar crises during 1960–1973, the bank bankruptcy wave of 1973–1975, the debt crisis that broke out in 1982, and the 1997 southeast Asian financial crisis had led to the most widely felt impacts and pains.

11.1.1 Financial crises and currency crises

The general concept of financial crises includes the following four classes of crises: currency, bank, foreign debt, and systems. Here, by currency crisis it represents such a situation that due to purposeful and targeted speculative activities against a nation's currency, the nation's currency suffers from devaluation, or the nation's government is forced to drastically increase its interest rate or spend a large amount of foreign reserves to defend its currency. By bank crisis, it means actual and/or potential bank runs or such a scenario that a number of banks stop repaying their debts because of their bankruptcy or the government is forced to interfere by providing large amounts of support. By foreign debt crisis it implies such a case that a nation can no longer repay its foreign debts on time, no matter whether the debtors are governments or private individuals. By systematic financial crisis, it stands for the destructive effects on the real economy due to severe damages of the financial markets so that the efficiency of the markets is greatly affected. The connotation of systematic financial crises might overlap with those of other kinds of crises, while currency and bank crises might not lead to severe damages to a nation's payment system. So, neither currency crises nor bank crises can be identified with systematic financial crises.

In many circumstances, the specific definition of financial crises simply means currency crises. The early investigation of currency crises can be traced back to at least (Krugman, 1979a). Paul Krugman treats a currency crisis as an international balance-of-payments crisis. He believes that in order to prevent their currencies to devaluate, countries with either a fixed exchange rate system or pegged exchange rate system would pay the price of either spending their international reserves or increasing inflation due to their raising domestic interest rates; when the governments give up on the fixed exchange rate system or pegged exchange rate system, their currencies experience drastic drops in value, leading to currency crises. Currency crises are generally indicated by the collapse of the fixed exchange rate system or forced adjustment to the system, such as official devaluation of the local currency, enlarged floating range of the exchange rate, drastic decrease in international reserves, noticeable rise in the interest rate of local currency, etc.

According to the literature, there are four main criteria for judging currency crises: (1) Sudden and large scale changes in exchange rate; (2) the weighted average of exchange rate and foreign reserves fluctuates widely; (3) the weighted average of exchange rate, foreign reserves, and interest rate vibrates wildly; (4) Import drops drastically. The second criterion was established by (Kaminsky, Lizondo and Reinhard, 1998) who believe that a currency crisis represents such a scenario that is caused by either devaluation of the nation's currency or drastic drop in the nation's international reserve or both as a consequence of an attack on a nation's currency. Because of this reason, currency crises can be verified afterward by using the index named exchange market pressure (EMP). This index stands for a weighted average between the monthly percentage change in the exchange rate of the local currency and that of the international reserves. Because this index increases with the devaluation of the local currency and the loss of international reserve, major increases in this index indicate a strong pressure to sell off the local currency. Because this second criterion possesses very practical operationality, it has been employed most widely among these four criteria.

Most of the financial crises that occurred since the 1990s have brought along with them clear characteristics of currency crises; and another outstanding feature of recent currency crises is they have been accompanied by bank crises. Such scenarios are referred to as twin crises.

11.1.2   Theories of international financial crises

After the 1997 Asian financial crisis broke out, a large amount of theoretical explorations have been done in order to pinpoint why it happened. In the following, we will introduce a few of the major theories, jointly known as new theories on financial crises.

Theory of moral hazard

The so-called moral hazard represents the possibility of loss in terms of assets, rights, and/or interests to one of the parties involved due to their mismatched rights and responsibilities. In financial crises, moral hazards are expressed as highly risky investments made by financial institutions, as a consequence of various kinds of promises governments made for depositors, creating huge amounts of bad and doubtful debts. That in turn leads to crises of public confidence and payment crises of financial institutions, and ultimately causes financial meltdowns. Ronald Mckinnon is a major representative of this opinion, which believes (Mckinnon, 1989; Mckinnon and Pill, 1999) that moral hazards is the mechanism underlying the break-outs of financial crises.

In the 1980s, there appeared the phenomenon of bankruptcy of a large number of savings and loans associations in the United States. By careful post-event reflections of this wave of bankruptcies, economists and policy makers realized that within a loosely managed environment, the deposit insurance system could initiate austere moral hazard problem. Because Federal Savings and Loan Insurance Corporation provides deposit guarantee for the Savings and Loan Associations, these associations did not need to worry about excessive risk taking (credit or investment) could potentially cause them the trust of the depositors. At the same time, because the insurance premium was the same for all the Savings and Loan Associations, these institutions did not need to worry about the potential of paying increasing amounts of insurance premiums as a consequence of excessive risk taking. Within such an environment of loosening control, it was natural for the Savings and Loan Associations to pursue after risky activities in order to achieve higher returns. That is the so-called financial excess, which inevitably exacerbates the vulnerability of the financial system. So, when the external conditions change, the financial bubble would rupture, triggering a financial crisis.

In terms of the East Asian financial crisis, the moral hazard played a more complex role. Among these relevant countries of East Asia, most of them did not really have any agency that was equivalent to a deposit-insurance corporation, and the governments did not clearly make any promise to insure financial institutions, either. However, due to the intricate interweaving monetary, business, and human relationships between the governments and financial institutions, and the constant indications of the news media, the public had strongly believed that as soon as financial institutions experience difficulties, the governments would definitely provide protection. Of course, this belief was only an informal impression. Because such so-called protection lacked legal documentation and backup, we call it as implicit guarantee. Because of the existence of the implicit guarantee, the impact of moral hazards becomes more damaging.

According to the theory of moral hazards, the development of each financial crisis consists of the following three stages:

Stage #1: Financial institutions make investment decisions under the condition of implicit guarantee. Assume that a financial institution raised $1 million from the market place that were considered to be government insured. Then the institution has two investment choices: risk-free investment or risky investment. Assume further that the return on the risk-free investment is $1.07 million, while the rate of return on the risky investment is a probability distribution. In particular, if successful, the return on the risky investment is $1.2 million, and if not successful, the return is merely $0.8 million, where the probability is 50% for each scenario to occur. Then the expected return is $1 million ($120 \times 50\% + 80 \times 50\%$). Therefore, any investor, who is risk neutral, would prefer the risk-free investment based on the expected return. However, the financial institution, which obtains an implicit guarantee, would choose the risky investment. It is because if it made the risky investment, under the optimal circumstances it would obtain the additional return in the amount of $200,000; and if it suffered from losses under the worst scenario, the financial institution would not be responsible for any debt so that in such a case, its return would simply be zero. In particular, the expected return of the financial institution from taking the risky investment is $100,000 $(= 20 \times \frac{1}{2} + 0 \times \frac{1}{2})$, which is greater than $70,000, the return from the risk-free investment.

Stage #2: Although the financial institution made its distorted investment decision, because of the existence of the implicit guarantee, the public still rests assured with its capital stored at the institution. That consequently stimulates financial intermediaries to make excessive loans, leading to rapid rise in asset prices and investment booms of the entire economy. Hence, a financial bubble is created.

At the same time, capital market liberation exacerbates excessive investment. If the capital market were not open, the capital available to financial institutions would be constrained by the limited amount of domestic savings. Therefore, the unsatisfied monetary hunger of the financial institutions would only make the interest rate to go up. The rise in financing cost would to a certain degree impede the desire to invest. On the other hand, capital-market liberation implies that one could obtain funds from the international capital market so that the demand for investment or speculation swells continuously, while spreading the relevant risk into the international market. If, at the same time when the capital market is open, a certain kind of fixed exchange rate is maintained, the high interest rate would attract non-stoppable influx of foreign capital.

Stage #3: After the bubble economy endures for a period of time, the "good" fiscal states of the financial institutions, formed with inflated asset prices, start to be noticed and start to cause concern. Such concern gradually evolves into wide-spread financial panic, when the risky investments appear to falter, causing the economic bubble to burst. Throughout this process, what bear the brunt are the skyrocketed asset prices. As soon as the asset prices start to drop, the fiscal states of the financial institutions rapidly deteriorate, leading to payment crises. At this juncture, although the business of the financial institutions is already in a precarious state, the long-awaited government help does not arrive. The shattered hope soon causes unrest in the financial market, further depressing the asset prices. The payment problem of the financial institutions spreads quickly; the financial system collapses; and a financial crisis break out.

The conclusion of the moral hazard theory is that the true cause of financial crises is the implicit guarantee of the government that leads to moral hazards; the fluctuation in the value of money is merely an expression of the crises. However, if the political and economic systems evolve in a beneficial direction, the damage of moral hazards will be weakened. For example, if the incumbent administration is expected to hold the office for another term and is determined to reform, then the investment decision making of financial institutions will likely converge to the direction of risk-neutrality from risk preference. So, the key for the prevention of crises is to cut off the nepotism and close contacts between the government and financial institutions as much as possible, while strengthening the regulatory management of the capital markets and financial institutions.

As for the evaluation of moral hazard theory, the theory can be well employed to explain the characteristics of the East Asian financial crisis. In particular,

Firstly, the good financial situations of the East Asian countries, which suffered from the crisis, were not true reflections of the real economic states. Behind the displayed balance sheets were huge amounts of hidden debts, which were the large amounts of implicit guarantees provided to the financial institutions.

Secondly, before the crisis broke out, these nations experienced slumping drop in assets prices. For example, after Thailand experienced a sharp rise in its stock prices during the early 1990s, the market began to decline in 1995, followed by sharp drops in 1996. It was the sharp drops in the asset values that triggered the financial crisis.

Thirdly, the reason why the financial crisis of the East Asia could spread to other nations and regions with different geographic locations and different export structures of products is mainly because the economies of these nations and regions were in metastatic equilibrium with extremely weak financial systems. So, as soon as a minor triggering catalyst appeared, self-realizing financial crises occurred.

Even though the theory of moral hazards possesses such a capability that explains the East Asian financial crisis, it also suffers from major weaknesses.

Firstly, this theory is developed on three main propositions: (1) when there is an implicit government guarantee, financial institutions should definitely invest excessively; (2) Risky investment behaviors of financial institutions and other financial agencies would replace all other kinds of "normal" investment behaviors of their entities; and (3) Foreign capital would surely first consider those enterprises and financial institutions that are under the protection of implicit government guarantees. However, before the crisis broke out, there was a rise in all kinds of investment behaviors in the East Asian countries, including direct investments from foreigners. And more than half of the loans provided by international banks and almost all direct investments and security investments, which were equivalent to 3/5 of the total amount of foreign capital inflow, went into non-banking enterprises that did not have the protection of implicit governments' guarantee. This fact implies that the actual situation faced by the East Asian countries was not in total agreement with the assumptions of the moral hazard theory.

Secondly, the theory does not provide an explanation on what factors triggered the break-out of the crisis, while it ignored potentially malicious attacks of international hot money.

Theory of fundamental determinants

This theory is one of the most classic and most ancient theories on financial crises. After the East Asian financial crisis, the community of theoretical economists enriched the conventional theory of fundamental determinants by supplementing additional determinants so that the explanatory power of the theory is enhanced. Dominick Salvatore (1998; 1999) and Giancarlo Corsetti (1998) are some of the main representatives of this important work.

The basic arguments of the theory of fundamental determinants include: Firstly, the fundamentals of a nation are the most critical factor that determines whether or not crises would break out. They also represent the root cause for crises to spread and worsen. Secondly, deterioration of the fundamentals consists of external imbalances, such as current account deficits and appreciation of the real exchange rates, and internal imbalances, such as unhealthy runs of domestic financial systems, huge governments' implicit fiscal costs (created by the attempt of helping with non-performing loans), and relatively low levels of foreign reserves. It can be measured by using a system of macroeconomic and financial indices, and can be employed to warn the break-out of a crisis based on numerical changes of the index system. Thirdly, although the fundamentals could not be adequately utilized to predict when a nation would suffer from a financial crisis, it could indicate the trend of the developing crisis. Fourthly, under many circumstances, speculative attacks and political problems are catalysts that trigger the break-out of financial crises.

The conventional early warning indicators of financial crises generally include the following stems: (1) current account deficits; the ratio of current account deficits over GDP; (2) debt indicators: (i) total foreign debts/GDP; (ii) short-term debts/GDP; (iii) (current account deficits − direct oversea investment)/GDP; (iv) short-term debts/foreign reserve; (v) maturing debts/foreign reserve; (vi) maturing debts/income from exports; and (vii) number of months the foreign reserve is able to cover all imports.

After the break-out of the East Asian financial crisis, problems the financial sector experiences are also included in the consideration of the fundamentals, believing that the financial fragility is one of the most important factors that trigger crises. Here, financial fragility is a concept with a very wide range implication. It covers the totality of all the operational risks of the financial system. The following is one typical example: If the total number of loans increases rapidly and the quality of the loans deteriorates at the same time, then a good portion of the loans will become non-performing when the overall economic situation deteriorates.

If the new theory of fundamental determinants is employed, one can see that before the break-out of the East Asian financial crisis, each of the indicators out of the involved nations implied that the risks were growing. These indicators clearly revealed the facts that the exchange rates were overestimated, that the current account deficits were growing, and that the conditions of the financial systems were deteriorating. In particular,

1   Since 1990, the currencies of the East Asian countries were strengthened continuously. It was because the high-speed development of American economy had made its dollar strengthening against all other major currencies. So, the strength of the East Asian currencies, which were pegged with the U.S. dollar, consequently rose, leading to continuous overestimations.
2   Starting in the early 1980s, current account deficits began to appear in these East Asian countries and had been growing. Two reasons had caused such a phenomenon. Firstly, each nation's real exchange rate had been rising; and secondly, these nations' exports were dominated by semiconductors and other primarily processed products. Along with the economic stagnation of Japan, which was the main importer of these products, and export competitions from other developing countries, their market shares continued to shrink and their earnings ability of foreign currencies was greatly affected. During the time period from 1985 to 1989, the current account deficits of Thailand, Indonesia, South Korea, Malaysia, and Philippines, amounted to only 0.3% of their respective GDPs. However, after

entering the 1990s, the current account deficits of these five nations had raisin to 4% of their GDPs, while still showing the tendency of continuous growing. Especially, Thailand and Malaysia experienced the most difficulty.

3   The operating conditions of the financial systems continuously deteriorated. At the start of the 1990s, the economic development of these East Asian countries reached their peak. However, their financial systems became increasingly fragile. The underlying reasons for that to appear included: (i) Along with the rapid economic development and the rising prices of assets, such as real estate properties, the demand for money of the private sector had increased drastically; and (ii) high interest rates and open capital markets attracted huge amounts of foreign capital. As the intermediary between the demand and supply of capital, the financial institutions on the one hand attracted foreign investments from offshore markets, and on the other hand made huge amounts of loans to the private sector. From 1990 to the end of 1996, the proportions of the loans made to the private sectors within the national GDPs in these countries grew quickly. For example, in 1990, these proportions of the bank loans in Malaysia, Thailand, and South Korea were respectively 111.4%, 83.1%, and 102.5% in terms of their GDPs. However, at the end of 1996, they reached respectively 144.6%, 141.9%, and 140.9%.

However, there are major drawbacks in the behaviors of the financial institutions. These drawbacks can be summarized as follows: (i) borrow externally and lend internally. That is, borrow foreign currencies while making loans in the local currencies. That created a huge risk in terms of exchange rates; (ii) Borrow short-term while lending for long short. That is, the influx of foreign capital is short term, while the local lending is for long term. Such investment behavior created term mismatches on the balance sheets of the banks and increased the domestic reliance on foreign debts. Such mismatches easily increased the risk of debt services. At the mid-year of 1997, the amounts of short-term foreign debts in Indonesia, Thailand, and South Korea all exceeded their respective foreign reserves; and (iii) invest in risky assets. Financial institutions invested heavily in real estates, while real estates possessed the effect of enlarged feedback on the economy: When the economy grew, the speeds of increase in the prices of real estates and investments in real estates all surpassed that of the actual economy, creating the illusion of rapid economic growth through wealth effects. That is, an economic bubble was produced.

Although the glitches of the financial systems had gradually taken shape in as early as 1996, the entire world cheered for the "Asian miracle". So, instead of attempting to improve/reform their individual financial systems, all of these Asian countries rode on the momentum of economic growth and intensified their efforts to attract foreign capital. For example, Thailand established the "Bangkok International Banking" facility to streamline business transactions, and to greatly make the dealing of foreign capital more convenient. On the other hand, as counterexamples, some of the nations and regions with relatively better fundamentals survived the East Asian financial crisis, such as Hong Kong and Taiwan.

The theory of fundamental determinants possesses a good degree of explanatory power. Especially because it introduces various performances of the financial system into the conventional system of indices, the theory provides more realistic descriptions of financial crises. However, the weakness of the theory is also very obvious. Specifically, (i) the fundamentals are surely important factors that trigger financial crises. However, even during periods of normal economic development, there is no guarantee that the fundamentals are perfectly inline. Hence, the fundamentals can only be necessary conditions, instead of sufficient conditions, for the break-out and worsening of financial crises. If we use the Asian crisis as our example again, we can see that: Firstly, before the break-out of the crisis, the deterioration of each economic index was still far less severe than that of the corresponding index of the Latin American countries during the Mexican crisis in 1994. Secondly, the degrees of deterioration were different from one country to another so that it seemed impossible to derive at a uniformly useful warning for crises. For example, both Thailand and Malaysia suffered from relatively high levels of deficits in their current accounts, while both Indonesia and South Korea experienced much lower levels of deficits. Thirdly, some of the insightful scholars in the theoretical and practical communities had recognized many of the existing problems in East Asia before the break-out of the crisis. However, most of the international investors shrugged off the warnings and treated the deterioration in the fundamentals as an expression of imbalanced economic developments. They believed that as long as appropriate regulations were introduced or reinforced, the chance for a major crisis to break out was very small. This example illustrates that although the theory of fundamental determinants points out that emergencies, such as political changes, are triggers of financial crises, it does not explain the mechanism behind the break-out of crises. (ii) The illustrated connection between the deterioration of fundamentals and the break-outs of financial crises is not sufficient to demonstrate their causal relationship. Many scholars point out the fact that such problems of the financial institutions as non-performing and bad debts, increasing fiscal deficits, etc., all appeared after the break-outs of financial crises. So, to a certain degree, these problems could be seen as consequences of the crises instead of the causes. If the crises did not break out, these problems could stay dormant and hidden forever, or be manifested in some milder fashion. Of course, we could not deny the fact that it was these "dormant" problems that attracted the international speculators to pay particular attention to East Asia. Therefore, the fundamentals and financial crises represent reciprocal causations.

Theory of financial panics

This theory provides a relatively complete description for the crises with deteriorating capital flows. It believes that during the eve of the crisis, all the East Asian countries went through the process of rapid inflow of capital. However, the inflow of foreign capital was very fragile, and could easily be affected and reversed by financial panics. As soon as the reverse occurred massively, a financial crisis would follow.

In as early as 1983, Diamond and Dybvig (1983) and other economists had pointed out that financial panics could cause and worsen economic crises. The so-called financial panic stands for the sudden and large-scale withdraw by the creditors of the short-term capital from the debtors who still have the ability to repay the debts because of some unrelated reasons. That is a collective action. In particular, there are multiple equilibria in the financial market; financial panics are a consequence of the financial market evolving away from the equilibria. There are three conditions that could cause a financial panic: (i) The amount of short-term debts of one nation or a financial institution exceeds that of short-term assets; (ii) A nation or a financial institution does not have sufficient liquid funds to serve the short-term debts; (iii) There is not any single economic unit that could play the role of the final lender. If the aforementioned conditions hold true, when a creditor, including a depositor, discovers that other creditors have withdrawn their capital, he would do the same. Conversely, if a creditor realizes that other creditors continue to lend out their funds, he would likely follow the suite. Speaking economically, these two kinds of behaviors are considered rational. However, the first kind of behavior could cause large scale financial panic, leading to major economic losses. This theory was developed on the basis of reflection of the American crisis of savings and loan associations, while introducing the concept of multiple equilibria. It became the origin of the second generation of theories on financial crises.

Steven Radelet and Jeffrey Sachs (1998) systematically surveyed and carded the theory of financial panics. Based on their work, the highlights of the theory can be summarized as follows:

1   The foreign capital that suddenly flowed into a country during a short period of time right before the break-out of a financial crisis brings along a tremendous risk. After the East Asian countries opened up their capital markets in the early 1980s, these nations attracted a large amount of foreign capital at a very high speed. Initially, the capital entered into various fields of production of the actual economy instead of the speculative real estates and financial areas. It greatly enhanced the economic development. This state of affair continued until the 1990s. After then, major changes occurred in the economic and financial landscape of these countries. The financial markets matured and the need for speculation grew continuously. At the same time, along with the high expectation of further economic prosperity, the inflow of foreign capital rose rapidly. However, the inflow was mostly for short-term purposes and was mostly directed to economic areas with strong characteristics of speculation. Hence, the inflow could be easily reversed.

2   Right before and right after the break-out of the crisis, a series of events occurred in the financial markets that helped to trigger the onset of financial panics. Bankruptcy of financial institutions and enterprises, renege of governments' commitments, and malicious speculation of the speculators were some examples of the triggering events. For example, in January 1997, Hanbo Iron & Steel Co. of South Korea filed for bankruptcy under the pressure of over $6 billion debt. In February of the same year, Kia, South Korea's third-largest automaker, also suffered from the same fate. The fall of these major corporations had created a huge pressure on the commercial banks that provided finance for the corporations. In the same year, Samprasong Land, a financial company of Thailand, missed its February deadline for payment on its foreign debt. However, Thai government reneged its own words, and did not follow through its earlier promise to help. So along with it, Samprasong brought a good number of financial institutions, including Financial One, Thailand's largest financial institution, into nightmares. This event created a crisis of confidence, causing the foreign capitals to withdraw from East Asia, while leaving behind a major financial crisis.

3   After the crisis broke out, various factors helped to magnify the financial panics, which made the crisis much worse. Blind and short-sighted prevention and control

measures and many economic phenomena, which appeared after the break-out of the crisis, magnified the financial panics and helped to accelerate the capital flight. The combined effect was the complete collapse of the financial markets of the relevant countries within a very short period of time. In particular, the following factors helped to magnify the financial panics:

(i) Failures of the policies of the governments and the international society. Each of the national governments employed its foreign reserves as its stake to defend its fixed exchange rate system along with some other radical measures, which were far beyond its ability to pay. Consequently, the market started to suspect the sustainability of the fixed exchange rate system. For example, Indonesian government demanded state-owned enterprises apply their savings to purchase notes of the central government. That of course worsened the pressure for bank run, directly helped to lift the interest rates, and affected adversely the ongoing investment projects. During the time of upheaval in the stock market, Malaysia decided to establish a large scale foundation to push stock prices higher. However, due to lack of funds and determination, this plan was abandoned soon. Both Thailand and South Korea injected large amounts of cash into those financial institutions that were on the brink of bankruptcy. However, that only helped to enlarge the fiscal deficits. Furthermore, inflammatory remarks of government officials and the news media stimulated and accelerated the capital flight. These immature behaviors of the governments caused the investors to lose confidence on their abilities to maintain the markets' orders.

(ii) Political turmoil appeared. After East Asian financial crisis broke out, South Korea, Thailand, Philippines, Indonesia and other countries faced the term changes in the administration. In particular, new administrations took office in both South Korea and Thailand at the start of the crisis, while Pilipino president took office in May 1998. The Indonesian election took place hastily in late 1997 under the manipulation of President Suharto, followed by his stepping down as pressured by massive protests. The effect of the political turmoil quickly spread into other neighboring countries.

(iii) Credit rating agencies lowered the credit ratings of the Asian countries.

(iv) Capital flight itself exacerbated the severity of the financial panics, creating a vicious circle. Currency devaluations, stock market volatility, and other economic conditions motivated large amounts of capital flight, which in turn led to new rounds of currency devaluations, bank runs, and rises in interest rates. Exchange rate risk and elevated financing costs deteriorated the operating conditions of some of the companies and banks that once performed relatively well. So, the offshore creditors lost their confidence in their investment projects; they were no longer willing to expand their short-term investments. At the same time, in order to recover the sufficiency of capital, some of the multinational and domestic banks with strong liquidity also started to limit their proposal of loans. In order to repay the short-term loans in U.S. dollars, liquid capitals were preferred so that the problem of bank runs was intensified so that the asset quality of banks and the financial situations of business firms deteriorated quickly, which furthered the capital flight.

These factors continued to intensify the panics of the market participants day by day; and capital flight continued, which eventually made the crisis much worse.

From the previous discussions, it follows that the theory of financial panics has provided the following policy suggestions: Firstly, multiple equilibria could exist in capital markets. So, financial systems need to be improved constantly in order to sustain their healthy operation and to prevent unexpected crises. Secondly, the international financial market could be easily affected by financial panics. Hence, a fair and effective organization should be selected to play the role of the final lender in order to promptly prevent the break-out and consequent spread of financial panics. Thirdly, policy makers should comprehensively and carefully introduce and take up measures; when a crisis first emerges, they would apply fine-tuning means to interfere in order to prevent adverse effects of short-term behaviors on the market mood.

The theory of financial panics maintains that panics are originated in the deterioration of the fundamentals of a nation, triggered by a series of sudden financial and economic events, and inflated by inappropriate reactions of the national government and international organizations. When large scale capital flight started, an eventual crisis would be caused and worsened. This conclusion is theoretically valuable. It introduces the new concept of market mood, while combining the strengths of the theories of fundamental determinants, moral hazards, and others; and it emphasizes the triggering effect of outflows of international capital, and provides a band new explanation for the deterioration of crises.

11.2   SPECULATIVE ATTACKS AND CURRENCY CRISES

Since the 1980s, along with the gradual strengthening of the global capital market, international capital flows have been unprecedentedly developed. Their magnitude has been increasing; their speed of movement has been accelerating; and their implicit danger and risk have been also ballooned. Large amounts of private short-term capitals, which are not under the surveillance of any national government or international financial organization, have been flowing freely and seeking profitable opportunities in the international financial market by making use of various newly invented financial tools and trading platforms. These new characteristics of international capitals have often caused extreme volatilities in the international financial market; speculative attacks with increasing strengths and time duration have happened with growing frequency. The break-out of the 1997 Asian currency crisis further indicates that currency crises, initially caused by speculative attacks could further evolve into a full-range financial crisis and a profound social turmoil.

11.2.1   Short-term capital flows and speculative attacks

The so-called international speculative capital or hot money represents such capital that is frequently moved within and between various markets in pursuit of short-term, high levels of profits without any particular fields of investment focus. Speculative capital tends to be short term even though there are exceptions to this rule of thumb. One of the modern characteristics of international speculative capitals is their camouflage. At the same time, these capitals can also go along with the market cycles by pursuing mid- and long-term investments. Additionally, not all short-term capitals are speculative. For example, the short-term capitals involved in the financial interme-diation and settlement of international trades, short-term interbank funds, banks' short-term positions for allocation, etc., are not speculative in nature.

Along with the expansion in the size, circulation speed, and coverage of the inter-national capital markets have international speculative capitals grown. Based on their predictions on the changes of exchange rate, interest rate, security prices, gold price, or the prices of certain commodities, speculative capitals could be suddenly involved in large scale both long and short trades in a short period of time. By substantially alter-ing the composites of their portfolios and by affecting the confidence of the holders of other assets, these speculative capitals cause severe instability in the market prices so that short-term high profit opportunities could be created. Such market behaviors that disturb the market prices and appear suddenly are known as speculative attacks. As limited by the constraint of pursuing after quick profits, international speculative capitals generally choose to attack such economic sectors and geographic areas that can hold large amount of capitals and allow fast movements of funds with expected high returns, few financial regulations. These economic fields include (foreign) currencies, futures, options, precious metals, real estates, etc.

From what has happened empirically in the past, it seems that international specu-lative capitals have been fond of attacking the currency of one nation or the currencies of several nations at the same time. Most often seen speculative attacks are those that assault either fixed exchange rate systems or regulated exchange rate systems. Speaking generally, when a nation either employs a fixed exchange rate system or pegs a target exchange rate, the speculators often make their judgment that as long as the official parity or the target exchange rate does not conflict with the fundamental conditions and states of the economy, the official exchange rate will be maintained. However, if the speculators believe that the fundamental states of the current economy could not sustain the prevalent level of exchange rate for long, they would launch a speculative attack in order to speed up the dissolution of the fixed exchange rate system.

Under the conditions of either fixed or pegged exchange rate system, as long as there appears either a domestic inflation or recession accompanied with sustained current account deficits, the governmental promise on the fixed exchange rate would lose its reliability. It is because in these situations there is a heavy pressure to devaluate the local currency; in order to maintain the promised exchange rate, the government will be forced to mobilize and spend its international reserve. Even with the support of the international financial markets, the fundamental imbalances in the economic states still cannot be corrected, which can most likely delay the occurrence of the devaluation of the local currency, although the devaluation will sooner or later happen inevitably. If speculators predicted this forthcoming event, they would mobilize their capitals ahead of time and launch their speculative attack. By making use of the spot and forward transactions, futures contracts, options trades, and swaps of various financial tools, speculators carry out their multi-dimensional speculative strategy by positioning their capitals at the same time on the markets of foreign exchanges, securities, and all different forms of financial derivatives. Because of the fixed exchange rate or the promise that the government would maintain the rate fixed, the risk to the speculators is actually quite low, because the direction along which the exchange rate would move is clear. To say the least, even if the prediction is incorrect, the worst is that the exchange rate parity did not change so that the most the speculators would lose is their minimal amounts of trade costs. Hence, once a speculative wave is started, the magnitude in general is large, leading to the expected consequences, as a self-fulfilling prophecy of the humongous scale.

11.2.2 Recent speculative attacks and currency crises

During the time period of Bretton Woods system after World War II, the strength and power of the private capitals grew drastically; their speculative activities evolved with increasing levels of energy. Their attacks on currencies were mostly successful and amplified with ever-growing vigor and intensity. The most typical are the British pound crisis of the late 1967, French franc crisis of August 1969, and the U.S. dollar crisis of 1971–1973. If we say that the root problem for Bretton Woods system to eventually collapse were the defects of the system itself, then the direct triggering factor for the system's collapse would be the speculative attack of the international short-term hot money on the U.S. dollar – the base currency. When Bretton Woods system was over, the world was in a wave of deregulation, strengthening the market mechanism, promoting economic and financial liberalization. Correspondingly, the international financial markets become further liberalized and global. Along with the application of modern technology of communication and computer networks, financial derivatives and methods of trading are developed in abundance. All these political, societal and technological advances provided the space for international capitals to grow and mobilize unprecedentedly. With their greatly increased speed of mobility, international capitals have launched frequent speculative attacks. Among the most typical are the attacks on the pegged exchange rate system employed by some countries of Latin America in the early 1980s, the Mexican peso crisis of 1994, and financial crises of Eastern Asia during 1997–1998. The following case studies are based on (Wang and Hu, 2005).

Case 1: The attacks on the exchange rate system of Latin American countries in the early 1980s

In 1978, Chile, Uruguay, and Argentina decided to employ a crawling peg exchange rate system. Each of these national governments established its plan to gradually depreciate its local currency against U.S. dollar. However, in their implementations, their rates of inflation were much higher than that in the U.S.A., while their degrees of depreciation were much smaller than the difference of the U.S. inflation rates. Therefore, the over-evaluations of their local currencies made the deficits of their current accounts rise. During 1981–1982, the interest rate in the international financial markets reached an historical high, making the burdens of foreign debts and the deficits of the current accounts of these three countries difficult to sustain, so that it became inevitable for the local currencies to devaluate while departing from the targeted exchange rates. Under this background, these countries respectively experienced speculative attacks, corresponding currency devaluations and the consequent capital flights, and crises of domestic financial institutions' runs.

Case 2: Mexican peso crisis of 1994

In 1982 after having suffered from its debt crisis, under the supervision of the IMF Mexico implemented a comprehensive policy for economic adjustment and reform, while tightening its economy and dramatically reducing its fiscal deficit. In 1987, Mexico re-fixed its exchange rate between Mexican peso and U.S. dollar. In January 1989, Mexico started to employ a crawling peg exchange rate system, which was changed to a moving target regional exchange rate system in December 1992, while gradually expanding the floating range for peso. This series of measures of economic reform achieved a certain degree of success; the national economy steadily recovered. However, in 1994, Mexican economy once again stalled while accompanied by political instability. Therefore, the expectation and rumor for peso to depreciate grew; capitals fled one after another. Interventions of the central bank made the market interest rate rise drastically, while the national foreign reserves were depleted quickly. On December 30, Mexican government eventually had to allow peso to depreciate. However, the new exchange rate established after the depreciation immediately suffered from speculative attacks so that Mexican government had to implement a floating exchange rate system. After then, the domestic economic conditions and the political situation made foreign investors extremely nervous, causing continued capital flight, banks subjected to runs, and the economy falling into crises. In the newly adopted floating exchange rate system, peso continued to depreciate; until the end of 1995, peso had reached consecutive historical lows one after another.

Case 3: The speculative attack on the joint floating mechanism of European monetary system (1992–1993)

The national currencies within the European monetary system of the nations that were members of the European Community had followed a joint floating exchange rate mechanism. This mechanism had led to the creation of the European currency unit (ECU) and established the statutory central exchange rates between the individual national currencies and the European currency unit. Therefore, between these member nations a system of fixed exchange rates was employed, while externally a joint floating rate system was implemented. In the early 1990s, the member nations of the European Community experienced varied degrees of economic turmoil. Uniformities in their respective macroeconomic states, such as inflation rates, unemployment rates, fiscal deficits, economic growth rates, etc., started to be broken. Over time, it became clear that some member nations could no longer maintain their statutory exchange rates with ECU, providing opportunities for the international speculative capitals.

In the late 1992, an initial round of speculative attacks appeared. Among the first group of victims of the attacks were Finnish marks and Swedish krona. At the time, neither Finland nor Sweden was a member of the European Community. However, they all hoped to join so that they voluntarily pegged their own currencies with ECU. Under the speculative attacks, Finland quickly gave up its fixed exchange rate and drastically depreciated its currency on September 8. On the contrary, Swedish government decided to protect its krona by raising its short-term interest rate to 500% annually. That eventually defeated the speculative attacks. At the same time, British pound and Italian lira also suffered from continued attacks. On September 11, European monetary system agreed for lira to depreciate 7%. Although German central bank spent around 24 billion marks to support lira, three days later lira was still forced out of the European monetary system. By this time, the Bank of England had lost several billions of U.S. dollars to protect its pound. Even so, on September 16, the British pound was still forced to float freely. Although French franc also suffered from speculative attacks, through joint interventions with Germany and by greatly increasing the interest rate, the value of franc recovered.

This crisis of European monetary system started in 1992 and lasted until 1993, during which speculative attacks often occurred. At the end of 1992, Portuguese currency escudo depreciated; Spanish currency pesetas was devalued once again, while Swedish krona and Norwegian krone started to float. In the early part of 1993, Ireland's pound depreciated, Portuguese escudo depreciated another time, and Spanish pesetas experienced its third round of depreciation. On the other side, French franc and Danish kroner successfully stood against the sporadic speculative attacks.

Case 4: 1997 currency crises of East Asia (1997–1998)

In the 1980s and early 1990s, nations in Southeast Asia sped up their steps toward financial liberalization by completely opening up their domestic financial markets to attract maximal scales of foreign investments. Such fast-speed liberalization led to drastic economic growth, which was known as "Southeast Asia miracle". However, after entering the mid-1990s, the rising labor costs were translated into the decreased international competitiveness in their products so that deficits began to appear in the current accounts of some Southeast Asian countries. Because these countries did not in a timely basis upgrade their industrial structures in order to keep pace with the increasing competitiveness of their products, the continued influx of the foreign capitals together with domestic investments led to the formation of economic bubbles and overheated sector of real estates. For example, in 1996, Thailand's balance of foreign debts had reached over 90 billion U.S. dollars with more than 40 billion dollars of short- and mid-term foreign debts, both of which surpassed the corresponding levels of foreign reserves at the start of 1997. Additionally, because of the overheated investments, particularly the overheated investments in real estates, the bad debts of Thai financial institutions had amounted to more than 30 billion U.S. dollars in early 1997. Therefore, the public and foreign investors started to worry about the economic conditions and financial order in Thailand, which inevitably helped to consolidate the expectation for baht to depreciate. At the same time, international speculators were also building up their monetary energy and preparing to launch their large scale attacks. On February 14, Thai currency baht depreciated 5% against U.S. dollars, making the covered speculative attacks public. After then, baht suffered from ever increasing pressure to depreciation further; and interventions of Bank of Thailand were quickly exhausting the national foreign reserves. After mid-May, speculative capitals launched their new rounds of attacks, creating an eleven-year new low for the exchange rate of baht. Several nations in Southeast Asia jointly intervened in the foreign exchange markets by buying in baht, while Bank of Thailand sacrificed 5 billion U.S. dollars of foreign reserves and once again raised its short-term interest rate. However, all of these still could not rebuild the public confidence and drive off the speculative attacks. Eventually on July 2, Bank of Thailand was forced to allow baht to float freely in the exchange markets, causing baht to depreciate 20% against U.S. dollar on that single day. Subsequently, the speculative attacks quickly spread over to the neighboring nations and regions so that Philippines, Malaysia, Indonesia, Singapore, Hong Kong, and Taiwan were all affected. Other than Hong Kong, the local currencies of these countries and regions all depreciated against U.S. dollar with different scales. At the same time, all these nations and regions except Singapore and

Taiwan fell into deep financial and economic crises. After October of the same year, the crises spread over to South Korea, causing South Korean currency won to depreciate deeply against U.S. dollar and making the economy of South Korea fall deeply into an economic crisis.

Case 5: Russian currency crisis (1998)

At the early stage of Russian economic transition, large amounts of international capitals entered Russia. As of July 1, 1997, the accumulated foreign investments totaled 18 billion U.S. dollars, about 10 billion dollars of which were short-term capitals and invested in the securities markets. Although the economic growth of the real economy was nearly zero, the stock market rallied rapidly in the first half of 1997; bubble expansion appeared in the prices of financial assets. When the currency crises of Southeast Asia broke out and started to spread toward the regions of Northeast Asia, the probability for Russia to experience a similar currency crisis was very big, considering the market long-term expectation of instability for Russian economy. Starting in November 1997, speculators launched their attack on ruble with many others followed. During the end of 1997 and early part of 1998, Russian government and central bank sacrificed foreign reserves to purchase ruble, while expanding ruble's floating range and increasing the interest rate from 21% to 35%. Although the situation was temporarily stabilized, large amounts of foreign reserves were lost. However, after May, another wave of speculation started, causing ruble's exchange rate to fluctuate severely again. Though Russian government raised the interest rate to 150% and sought for international assistance in order to counter the attack, the situation was never successfully under control. On August 17, Russian government had to expand ruble floating exchange corridor; that is, allowing ruble to depreciate. However, unexpectedly, the situation continued to worsen. Eventually on September 2 without a choice the exchange corridor was abandoned. That meant that Russian government gave up its over-two-year old managed target floating mechanism of ruble.

Case 6: Brazilian currency crisis (1999)

The currency crises that appeared in East Asia and Russia increased the market expectation for Brazilian currency to depreciate, causing large amounts of capital leaving Brazil since September 1998. At the end of 1998, Brazilian Congress did not pass the bills to increase the benefit taxes of civil servants and those of retired civil servants as contained in the fiscal adjustment plan, while the 1998 deficits of the fiscal balance, trades, and current accounts exceeded the expected levels. So, along with the decreasing market confidence, speculative attacks were triggered, forcing Brazilian central bank to allow its real to float freely against U.S. dollar.

11.2.3 New characteristics of recent speculative attacks

All the speculative attacks that were launched since the 1980s have shown a good number of new characteristics, including amplifying scales and forces, enhancing dimensionality in terms of their attack strategies, covering larger geographic areas, increasing publicity of planned attacks, etc.

The main reasons for the amplifying scales and forces of the speculative attacks include

1   The size of the international speculative capital has been increasing constantly. Since the 1980s, the economic growth of the main industrialized nations has been slowed, while the reform for financial liberalization of various countries has been strengthened. That caused large amounts of capitals to enter into the international financial markets to look for new opportunities. Such continuous accumulation of international speculative capitals, together with the multiplying effect of international credit, makes the available speculative capitals increase multiple times. Based on the relevant estimates of the International Monetary Fund, there were at least 7,200 billion U.S. dollars of speculative capitals currently floating around the international financial markets. That amount is equivalent to 20% of the world GDP; and each day, over 1,200 billion U.S. dollars of speculative capitals are looking for various profit opportunities. That is nearly one hundred times more than the amount of trades of real consumable goods.

2   The capitals available for speculation purposes have shown the tendency to collaborate and to take actions jointly. Along with the rapid development of communication technology and the expanding availability of internet, a 24/7 operational system for the globalized exchange markets has appeared so that capitals can be transferred from one exchange market to another instantly. That makes the once stranglers and disbanded international speculators develop into powerful speculation assemblages. As a third force different from various national currency authorities and international financial organizations, they have constituted a major thread to the stability of each nation's exchange system and the normal operation of the international currency system.

3   The introduction of financial derivatives provided a leveraged platform of trading for the speculators. Since the 1970s, financial innovations have led to the profligate development of new financial derivatives and their trades. Because of the characteristic of high leverages of the financial derivatives, speculators can trade such financial products that are valued tens or even hundreds of times of the little amounts of capital they mobilized. That enables a hedge fund to embark on trades worth several hundreds of billions of U.S. dollars with a small amount of capital, affecting the entire international financial markets.

Since the 1990s, the strategies of speculative attacks have been further developed. The traditional speculator would simply use the price differences between spot and future trades, while the current strategies of speculative attacks have been quite complicated. By utilizing all kinds of available financial tools, the speculator gets involved in various markets of the traditional tools and derivatives. It can be said that current speculative attacks make use of the inherent linkages among the prices of various financial products, traded in different markets, to make comprehensive and profitable arrangement of capitals.

Traditional speculative attacks have been isolated and scattered in different geographic regions. However, along with the recent deepening globalization of the international financial markets, and along with the further unification of regional economies, speculative attacks have also showed clear regional characteristics. No matter whether it is the Latin American currency crisis of the early 1980s, the Mexican crisis of 1994, the European monetary system crisis of 1992, or the Southeast Sian crisis of 1997, each started with the initial shocks in such a market that contained the most concentrated imbalances. Then, the initial shocks spread over to other neighboring markets, displaying a dynamic process of mutual influences.

Traditional speculations used to be hidden or semi-public arbitrage activities. However, since the 1990s, along with the deregulation and financial liberalization of the international financial markets and the widening use of information and network technologies, originally speculative activities have been gradually evolved into open and purposeful attacks on specified currencies. Such openness and publicity could be and have been strengthening the market expectation of depreciation of the target currency. For example, at the end of 1997, George Soros published an article to openly declare that he and associates would launch another attack on Hong Kong dollars. After that, he announced in newspapers that the next currency crisis would appear in Russia, which was later shown to be accurate. Soon after then, Soros commented that Brazilian currency was evaluated too high so that Brazilian real would be his next target of attack, which was indeed what happened next.

11.2.4   Chilean capital account liberation: A case of success

Among the emerging markets is Chile one of those early nations which opened their capital accounts. Its opening process of over twenty years could be roughly divided into two periods. The first period lasted from 1974 to 1984. Starting in the mid-1970s, within a very short period of time, Chile introduced many substantive measures to liberate current and capital accounts. However, this process of economic liberation was interrupted by the 1982 financial crisis, and was forced to adjust for three years. The second period started in 1984 and is continued until the present day. Compared to the first period, the overall design and specific implementation of this second period have seemed to be more gradual and cautious. It might be because of the failure of liberation of the early period that forced Chile to give up the radical way of opening up its market and to adopt the current way of progressive liberation. The relatively successful design and implementation of liberation have made Chilean economy evolve smoothly since the early 1990s and enjoyed an ever increasing growth without suffering from any severe financial volatility. A good number of renowned economists have treated the Chilean liberation of capital accounts during this period of time as a model of success.

Starting in 1985, Chile walked out of the shadow of the earlier financial crisis; its economic conditions improved gradually. It was at this time that Chile began to reverse the orientation of capital control and trade protectionism developed during the crisis, and renewed its impetus to liberate trades and to open up capital accounts. Because it had suffered through the financial crisis caused by the immature and fast-paced liberation reform and opening up, Chile adopted in the new round of opening up capital accounts a more progressive policy stance. And the coordination among the implementation of opening up capital accounts, other relevant reforms, and the introduction of macroeconomic policies has been more carefully thought of. It might be because of this reason that Chile did not get involved in the financial volatility caused by either the 1995 Mexican financial crisis or the 1997 East Asian financial crisis, making Chile one of the few emerging markets that have sustained a steady economic growth.

Aspect 1: Preparation for opening: Domestic financial reform and trade liberation

The financial crisis that occurred soon after opening up its capital accounts made Chilean government clearly realize the importance of a sound financial system and strengthened financial regulation. That prompted Chilean government proceeds to focus on this area of key reforms. In 1986, Chile introduced many new regulations and arrangements regarding financial monitoring, business range of banks, and deposit insurance system. According to the new banking law, financial supervisory authorities strengthened their monitoring over the financial conditions of banks, and made the management information of banks more transparent. Additionally, the new law also placed each banks' business activity under rigorous scrutiny. For example, the maximum amount of money each lender could lend out has been stipulated; restrictions on banks' foreign currency positions have been imposed; and no loans of preferred conditions are allowed to lend to either companies or individuals that have close connections with banks; etc. The modified law allows banks through subsidiary entities to engage in securities business, such as stock broking, investment fund business, financial advising, etc. However, it requires at the same time the strict separation between the traditional banking business and securities business in the areas of money, personnel, facilities, etc. Additionally, the law makes clear arrangement in term of insuring deposits. To guarantee the normal operation of the payment system, all checking accounts in banks are insured to their total values. If the phenomenon of bank run appears, the central bank is obligated to provide loan assistance as early as possible in order to meet the demand of deposits withdrawals. To suppress the problem of moral hazards, only partial insurance is provided for time and savings deposits. For the central bank to effectively carry out the monetary policy and fulfill the responsibility of supervising financial institutions, in October 1989, Chile established the independent status of the central bank through legislation. The drastic reform of the financial system, alongside with the strengthened financial supervision, made Chilean financial system increasingly sound and the operation of financial institutions increasingly robust. Under these prerequisites, Chilean central bank started to gradually relax the interest rate control adopted when dealing with the financial crisis, while stopping the practice of announcing the guiding interest rate in 1987. Since then the interest rate has been adjusted through open market auctions of the bonds of the central bank. That represented an important step for Chile to transit from direct monetary control to indirect regulation. Through such a series of financial reforms, Chile established the basic framework of a thorough financial system. After 1991, the emphasis of financial reforms moved to strengthening the competitiveness of the financial market, enriching tradable products, and improving the market efficiency.

In order to develop the competitiveness of trading sectors, making them capable of meeting the challenges that would come along with the liberation of capital accounts, starting in 1985, Chile gradually lowered the tariffs, reversed the protectionist trade policies adopted during the crisis, and pushed for trade liberation. In the area of exchange rate arrangements, the exchange rate of peso also evolved in the more elastic direction of market orientation. The maximum daily fluctuation of the exchange rate against U.S. dollars also widened from 1% to 4%; and the value of peso was majorly depreciated two times in 1985. During 1987 and 1988, Chile further eased the control over the transactions of current accounts, adjusted tariff rates, and increased the policy financial assistance for exports. Additionally, the elasticity of peso's exchange rate was also gradually increased; the maximum daily fluctuation of the exchange rate against U.S. dollar was increased to 6% in 1988, to 10% in 1989. Because of the success of the 1985–1990 exchange rate reform, the actual effective exchange rate of peso had shown a depreciation trend.

In the following years, adjusting to the evolving conditions, Chile continued to upturn the elasticity of exchange rate. For example, in 1992, the maximum daily fluctuation of the exchange rate, allowed by the law, rose from the original 10% to 20%. In short, Chilean market-oriented reform that started in the late 1980s in the areas of trading sectors and exchange rate arrangements has effectively strengthened the international competitiveness of its exported products and the national capability to resist various adverse impacts experienced in the process of liberating capital accounts, creating a beneficial environment for the gradual and eventual liberation of capital accounts.

Aspect 2. Cautious process of opening up

At the same time when or slightly after the reforms in the areas of its domestic financial system and trading sector liberation took place, Chile underwent a relatively comprehensive and considerably cautious process of liberating capital accounts. During this period of time, other than selectively and stepwisely allowing inflows of capital, Chile also relaxed the limitation on the outflow of capital and formulated the institutional control on the inflow of short-term capitals.

(I) Cautious opening for capital inflow and limitation on short-term capital inflow

In 1985, Chile permitted foreign direct investments to enter the country in the form of exchanging debts for equity, while requiring that the capital could not leave the country within 10 years and the produced profits could not repatriated within 4 years. As for investments in securities, although foreign currencies were allowed to be used for the purchase of certain bonds by citizens and non-citizens, the foreign currencies applied could not come from the official foreign exchange market. In 1987, a special foundation was permitted to be established to assist foreign investors to directly invest in Chile in the form of exchanging debts to equities. In 1990, Chile issued the first time American Depositary Receipts (ADRs) for the purpose of attracting and directing foreign capital to flow into Chilean capital market. Additionally, Chile lowered the credit rating requirements for its domestic firms to issue bonds overseas, permitted companies to issue negotiable bonds abroad, and reduced the tax burdens for foreign corporations that invested directly in Chile.

At the same time when capital inflow was liberated, Chile also introduced a series of measures to limit the influx of certain kinds of capitals, including short-term capital. In June 1991, the government regulated that other than export credit, for all newly borrowed foreign debts 20% of the total values had to be deposited in Chilean central bank to satisfy the unremunerated reserve requirement (URR) for a period between 90 days to one year depending on the terms of the debts. Additionally, foreign currency loans also needed to pay 1.2% annual stamp duty just as loans of local currency. In July 1991, it was determined that borrowers could also apply cash whose amount was equivalent to the finance cost of the required unremunerated reserve to substitute for the unremunerated reserve deposited in the central bank. In May 1992, the requirement for unremunerated reserve was expanded to cover foreign currency deposits, while the proportion of the reserve was raised to 30%. And no matter what the terms were, the reserves were required to be kept for one year. After then, the requirement of unremunerated reserves was constantly adjusted with the changing circumstances. In 1994, all borrowings in foreign currencies were required to deposit the reserves in U.S. dollars. In 1995, the upper limits of loans in Chilean peso with U.S. dollar reserves were lowered, while the requirement for unremunerated reserves was expanded to the trades of American Depositary Receipts on the secondary markets and foreign capital inflows that did not increase the bank capital. Because the required reserves were the same for all capitals of different terms, the shorter the terms of the inflowing capitals were, the higher the financial costs of the reserves became. Hence, at the same time when the requirement of reserves did not cause major effect on long-term capital inflows, it did create an institutional exclusion of short-term capital inflows, which effectively prevented excessive inflows of short-term capitals.

(II) Opening of capital outflows

After 1985, due to the introduction of a series of measures that encouraged foreign investments, capital inflows steadily increased. In 1990, the surplus of capital accounts reached as high as 9.9% of the year's GDP. In order to reduce the pressure of rising exchange rate, Chile started to carefully open up the outflow of capitals. In 1991, the government allowed the first time its citizens to invest abroad their foreign currencies they acquired from unofficial foreign exchange markets; and the minimum time requirement for direct foreign investments to withdraw was shortened from earlier 10 years to 3 years. In 1992, retirement pension funds were allowed to invest proportionally their capitals overseas.

During 1993–1996, the speed of liberation for capital outflows was greatly sped up. The focus of the government was to shorten the time limits on how soon foreign investors could send their profits back home, to allow insurance companies, banks and mutual funds to invest greater proportions of their capitals abroad, and to reduce the constraints placed on investment tools and geographic locations of Chilean investments overseas. These measures made Chilean citizens invest abroad significantly more convenient and provided greater freedom for foreign investors to withdraw their capital from Chile.

Because capital outflows were opened timely and certain kinds of capital inflows were effectively limited, the surplus of Chilean capital accounts has been resting at a rational level since 1991. That helped effectively reduce the pressure of rising peso value and sustain the export competitiveness and the soundness of current accounts.

(III) Experiences and lessons learned

Chilean failed attempt to liberate capital accounts before 1985 indicates that liberating capital accounts is a highly risky reform experiment. The gradual opening that was restarted after 1985 has been highly regarded by many economists. John Williamson (1996), of the Institute for International Economics, Washington, D.C., believes that the Chilean liberation of capital accounts, which took place after 1985, has been extremely successful; its experience is worth other nations to emulate. By inspecting the practical process of Chilean liberation of capital accounts in the past few decades, the following experiences and lessons should be able to turn everyone's head.

Firstly, the surplus (which sometimes was excessively large) in the capital accounts, which appeared during the Chilean liberation of capital accounts, indicates that under the current situation, where private money instead of governmental funds has become the main body of international capital flows, keeping capital accounts open is an advantageous condition for countries of emerging markets to attract foreign investments.

Secondly, opening up capital accounts will almost inevitably produce large inflows of foreign capital, causing real appreciation of the local currency and very possibly leading to deterioration of trade balance and current accounts. So, to strengthen external competitiveness of the trade sector, to reduce or eliminate the possible adverse effects on the trade balance and current accounts, caused by opening up the capital accounts, the current accounts and trade sector should be liberated ahead of time or at least at the same time. The huge deficit that appeared in Chilean current accounts during the late 1970s and early 1980s proved this point in the opposite direction; and the fact that the state of payments of current accounts could still improve even when the actual Chilean exchange rate rose supported this point positively.

Thirdly, establishing a sound domestic financial system and strengthening financial supervision should take place before the liberation of capital accounts. Before 1985, Chile attempted to eliminate financial repression and develop a market-oriented financial system within a very short time period, while quickly liberating capital accounts. However, the financial crisis that occurred afterward indicates that a huge price would have to be paid if an immature domestic financial system was employed to accommodate the liberation of capital accounts. Regarding nations of emerging markets, the maturity of the domestic financial market includes not only removing controls on interest rates and credits, implementing indirect monetary regulatory mechanism, and other contents of liberation, but also cultivating modern financial institutions and infrastructure of financial markets, developing various regulatory facilities and conditions, redefining the relationship between banks and governments, restructuring the credit culture, etc. In short, establishing an effective domestic financial system takes time and cannot be rushed. So, sufficient time and adequate estimate of difficulty to successfully implement domestic financial reforms should never be overemphasized for nations of emerging markets. These nations should never have excessive confidence on the soundness of their domestic financial systems. That is, the liberation of capital accounts could not be accomplished with undue haste.

Fourthly, at the end of the 1970s, because the actual Chilean interest rate was clearly higher than those from abroad, the opening of capital accounts quickly attracted large inflows of capital. That caused the actual exchange rate rose drastically and eventually endangered the international trade and current accounts. That indicates that in the process of opening up capital accounts, the nation should prevent as much as possible the actual interest rate to rise. To this end, the government should effectively lower the risk premium for investors through various measures, which include:

(i)   Endeavor to control fiscal deficits and inflation, and maintain macroeconomic stability;

(ii) Strive to maintain coherence between exchange rate reform and other macroeconomic policy arrangements in order to lower the effective interest rate later; and

(iii) Effectively reform the domestic financial system, avoid the appearance of banks' non-performing and bad debts in order to reduce financial risk.

Fifthly, before 1977, although Chile relaxed the control on individuals and companies, except banks, to take foreign loans, it did not help clearly increase the inflow of foreign capital. However, after 1977, when the control on banks' intermediation of foreign capital was relaxed, the inflow of foreign capitals increased multiple times. This fact indicates that loosing up the control on banks' capital transactions is the essential contents of capital account liberation. So, it is reasonable to believe that in the process of opening up capital accounts, great importance should be given to relaxing the constraints that limit banks from overseas financing. This realization might be particularly important for countries of emerging markets with immature capital markets.

Sixthly, one of the important reasons why Chile has been relatively successful with its liberation of capital accounts that started after the late 1980s is because it developed and maintained the system of unremunerated reserves to keep short-term foreign capitals from entering Chilean financial market. Studies show that although this system of unremunerated reserves did not seem to bear much clear impact on the inflow of foreign capitals in terms of statistics, it did have visible effect on the term structure of the entering capitals. The investigation on the economic situations of 20 emerging market countries after they experienced the Mexican financial crisis shows that there is no systematic correlation between the amounts of inflows of foreign capitals and the nations' financial fragility and currency stability. However, the amounts of inflows of foreign capitals are closed related to the proportions of short-term capitals. This end explains that although Chilean system of unremunerated reserves did not effectively reduce the magnitude of inflows of foreign capitals, it could and had altered the term structure of the foreign capitals. Hence, it was effective to a degree in terms of maintaining financial stability and preventing currency crises.

After entering the 1990s, the problem of opening up the capital accounts of emerging market countries once again became the focus of attention of economists and policy makers. As for such practical problems as those related to the particular order, steps, and relevant policy assistance to be employed in the process of liberation of capital accounts have there been a good number of different opinions. The past experience has shown that as a matter of fact there is no unique set of methods to deal with these problems; each nation should time its liberation, plan the order, develop steps, and introduce relevant policies according to its particular economic conditions. When facing with adverse attacks, the nation should also adjust its original plan or even alter the strategic thinking based on the particular circumstances. Therefore, it should be recognized that carefully investigating the nations that undergoes or just completed their liberations would be extremely beneficial.

11.3 CURRENCY WARS AND POSSIBLE SELF-DEFENSE

What has been discussed so far in this chapter focused on speculative attacks whose purpose was to make enormous amounts of money. In the rest of this chapter, we see from a different angle why such attacks could be realistically treated as active currency wars against the economy of a region or nation. Following this analysis, we will see how a self-defense could be established.

11.3.1 One possible form of currency wars

According to (Bernanke and Gertler, 1999), the fundamental value of a particular capital is equal to the present value of the dividends the capital is expected to generate throughout the indefinite future. Symbolically, the fundamental value Q_t of a depreciable capital in period t is given by

$$Q_t = E_t \left(\sum_{i=0}^{\infty} \left[\frac{(1-\delta)^i D_{t+1+i}}{\prod_{j=0}^{i} R_{t+1+j}^q} \right] \right) = E_t \left(\frac{D_{t+1}}{R_{t+1}^q} + \frac{(1-\delta)^1 D_{t+2}}{R_{t+1}^q R_{t+2}^q} + \frac{(1-\delta)^2 D_{t+3}}{R_{t+1}^q R_{t+2}^q R_{t+3}^q} + \cdots \right)$$

(11.1)

where E_t stands for the mathematical expectation as of period t, δ the rate of physical depreciation of the capital, D_{t+i} the dividends and R_{t+1}^q the relevant stochastic gross discount rate at t for dividends received in period $(t+1)$. Then, we can rewrite equ. (11.1) as follows:

$$Q_t = E_t \left(\frac{D_{t+1} + (1-\delta) Q_{t+1}}{R_{t+1}^q} \right)$$

(11.2)

Because of various reasons, such as fads, the market price S_t of the capital differs persistently from the capital's fundamental value Q_t. When $S_t \neq Q_t$, we say, as in (Bernanke and Gertler, 1999), there is a bubble. However, to be more specific, when $S_t > Q_t$, we say that there is a positive bubble in the market place; and a negative bubble, when $S_t < Q_t$. In the realistic market place, asset prices mostly like deviate from the fundamental values due to various reasons, such as liquidity trading or to waves of alternating optimism or pessimism.

If a bubble exists at period t with probability p to persist into the next period, then by using the mathematical expectations the difference between the market price and the fundamental value of the capital in period $(t+1)$ satisfies the following:

$$p(S_{t+1} - Q_{t+1}) + (1-p) \cdot 0 = a[(S_t - Q_t)R_{t+1}^q] + (1-a) \cdot 0$$

(11.3)

It means that the mathematically expected $(S_{t+1} - Q_{t+1})$ value with probability p for $S_{t+1} \neq Q_{t+1}$ to happen is equal to the expected growth of the t-period difference $(S_t - Q_t)R_{t+1}^q$ with probability a ($>p$) for $S_t - Q_t \neq 0$. That is, what is expected is a more severe "bubble", since $a/p > 1$. So, if we assume $a/p < 1$, it means that the bubble in period $(t+1)$ is expected to be less severe than in period t.

From equ. (11.3), it follows that we know the following expression is true:

$$p \cdot \frac{S_{t+1} - Q_{t+1}}{R_{t+1}^q} = a(S_t - Q_t)$$

(11.4)

Now by taking the mathematical expected value for $(S_{t+1} - Q_{t+1})/R_{t+1}^q$ in period t, we have the following:

$$E_t \left(\frac{S_{t+1} - Q_{t+1}}{R_{t+1}^q} \right) = p \cdot \left(\frac{S_{t+1} - Q_{t+1}}{R_{t+1}^q} \right) + (1-p) \cdot 0 = a(S_t - Q_t)$$

(11.5)

Equ. (11.2) implies that

$$Q_t = E_t \left\{ \frac{D_{t+1} + (1-\delta)\,S_{t+1} - (1-\delta)\,S_{t+1} + (1-\delta)\,Q_t}{R_{t+1}^q} \right\}$$

$$= E_t \left\{ \frac{D_{t+1} + (1-\delta)\,S_{t+1}}{R_{t+1}^q} - (1-\delta)\,\frac{S_{t+1} - Q_{t+1}}{R_{t+1}^q} \right\}$$

$$= E_t \left\{ \frac{D_{t+1} + (1-\delta)\,S_{t+1}}{R_{t+1}^q} \right\} - (1-\delta) E_t \left\{ \frac{S_{t+1} - Q_{t+1}}{R_{t+1}^q} \right\}$$

$$= E_t \left\{ \frac{D_{t+1} + (1-\delta)\,S_{t+1}}{R_{t+1}^q} \right\} - (1-\delta)a(S_t - Q_t), \quad \text{(from equ. (11.5))}.$$

Therefore, we have

$$Q_t + (1-\delta)a(S_t - Q_t) = E_t \left\{ \frac{D_{t+1} + (1-\delta)\,S_{t+1}}{R_{t+1}^q} \right\}$$

which is the equivalent to:

$$\frac{S_t[Q_t + (1-\delta)a(S_t - Q_t)]}{S_t} = E_t \left\{ \frac{D_{t+1} + (1-\delta)S_{t+1}}{R_{t+1}^q} \right\}$$

By isolating the factor S_t in the numerator and cross multiplying the rest from the left hand side onto the right hand side produce the following:

$$S_t = E_t \left\{ \frac{[D_{t+1} + (1-\delta)\,S_{t+1}]S_t}{R_{t+1}^q[Q_t + (1-\delta)a(S_t - Q_t)]} \right\}$$

$$= E_t \left\{ \frac{[D_{t+1} + (1-\delta)\,S_{t+1}]S_t}{R_{t+1}^q\{[a(1-\delta)S_t - a(1-\delta)Q_t] + Q_t\}} \right\}$$

$$= E_t \left\{ \frac{[D_{t+1} + (1-\delta)\,S_{t+1}]S_t}{R_{t+1}^q[bS_t + (1-b)\,Q_t]} \right\}, \quad \text{where } b = a(1-\delta)$$

$$= E_t \left\{ \frac{D_{t+1} + (1-\delta)\,S_{t+1}}{R_{t+1}^q\left[b + (1-b)\frac{Q_t}{S_t}\right]} \right\}$$

$$= E_t \left\{ \frac{D_{t+1} + (1-\delta)\,S_{t+1}}{R_{t+1}^s} \right\}$$

Therefore, we have derived at

$$S_t = E_t \left\{ \frac{D_{t+1} + (1-\delta)\,S_{t+1}}{R_{t+1}^s} \right\} \tag{11.6}$$

where the return on stocks R_{t+1}^s is related to the fundamental return of the capital R_{t+1}^q as follows:

$$R_{t+1}^s = R_{t+1}^q \left[b + (1-b)\frac{Q_t}{S_t} \right] \tag{11.7}$$

If $S_t > Q_t$, then $R_{t+1}^s < R_{t+1}^q$, meaning that the expected stock return $S_t = E_t \left\{ \frac{D_{t+1}+(1-\delta)S_{t+1}}{R_{t+1}^s} \right\}$ is less than the fundamental return of $S_t = E_t \left\{ \frac{D_{t+1}+(1-\delta)S_{t+1}}{R_{t+1}^q} \right\}$.

From equ. (113), it follows that

$$(S_t - Q_t) = \frac{p}{a} \times \frac{(S_{t+1} - Q_{t+1})}{R_{t+1}^q} \tag{11.8}$$

which implies that at period $(t+1)$, there is a positive bubble if and only if there is a positive bubble at period t. Because $0 < p < a < 1$, if $S_{t+1} > Q_{t+1}$ (assuming a positive bubble) then from $0 < \frac{p}{a} < 1$, equ. (11.8) implies that the current bubble $(S_t - Q_t)$ is smaller than the fundamental return of the bubble in period $(t+1)$. In other words, the bubble in period $(t+1)$ gets more severe than when in period t.

Now, if $S_{t+1} < Q_{t+1}$ with $0 < p < a < 1$, then equ. (11.8) implies that the underpricing S_{t+1} of the asset Q_{t+1} in period $(t+1)$ is more sever than that in period t. Moreover, this equation also indicates that at period $(t+1)$ there is a negative bubble if and only if there is a negative bubble at period t.

These two conclusions evidently contradict the efficient market hypothesis, because these analogies indicate that if in period t the market over prices the asset, then the overpricing will continue forever; and if in period t the market underprices the asset, then the underpricing will also continue forever. That is, neither positive nor negative bubble will ever crash. So there are two possibilities: (1) The model in equ. (11.8) does not hold true in general, or (2) The efficient market hypothesis is not ever true.

Evidence appears to show that the efficient market hypothesis holds true at least occasionally and also bubbles, both positive and negative, do burst (Beechey, Gruen and Vickrey, 2000; Smith, Suchanek and Williams, 1988). So, the model in equ. (11.8) needs to be modified in order to describe the more realistic market situation better.

If in equ. (11.8) we assume $0 < a < p < 1$, then a similar analysis as above indicates that the phenomenon of overpricing or under pricing disappears over time. And, if $0 < a = p < 1$ is assumed, then the model in equ. (11.8) implies that the existing underpricing or overpricing stays fundamentally stable.

If $0 < a < p < 1$ and $a \approx 0$ are assumed, then equ. (11.8) implies that $\frac{p}{a} \approx +\infty$ so that the fundamental return $\frac{(S_{t+1}-Q_{t+1})}{R_{t+1}^q}$ of the asset approaches 0, meaning that the existing bubble gradually disappears with time.

Next, let us focus on the analysis of equs. (11.1) and (11.2). In particular, assume that in the $(t+i)$-th period foreign investments are suddenly increased drastically due to expected appearance of activities from the current weak economic state and weak local currency. So for this period, δ (= the physical depreciation rate of capital) would increase due to the increased amount of money supply, which in turn pushes up the inflation. So $(1 - \delta)^i$ would decrease drastically if the influx of foreign investments is large. At the same time, R^q_{t+i} (= the stochastic gross discount rate of the $(t + i)$-th period) would also increase, because of the increased inflation while the dividends D_{t+i} would generally decrease due to the reason that everybody would like to reinvest much of the available capital back into the market in order to capture the rising book values, including stocks, real estate, and others with of impressive increasing prices. That is, the present value of the return of the $(t + i)$-th period

$$\frac{(1 - \delta)^i D_{t+i}}{R^q_{t+1} \cdots R^q_{t+i}}$$

would drop from the level of expectation. In reality, the investor would hold onto the increased book value by receiving less tangible returns, hoping that the book value would continue to rise drastically.

Due to the wide and conveniently availability of capital, caused by the increased money supply, the local economic activities pick up too in large quantities, while the interest rate also goes up due to the fact that the central bank, in order to control the inflation, revises the interest rate and attempts to limit the money supply.

At this very moment of financially prosperity, assume that a huge amount of foreign investments suddenly leave in the $(t + j)$-th period, where $j > i$ because of the much higher prices in assets, in capital investments, etc., for them to take profit in order to move their capitals to other regions to capture new economic opportunities. So, in the $(t + j)$-th period, when a huge amount of foreign investment leaves, most of the economic activities that got started because of the foreign investments become stalled and/or negatively impacted. Therefore, a good portion of the local investments is forced to be retained with the interrupted economic activities. That is, the investors of the local investments can no longer receive their expected dividends, and at the same time a large amount of the investments evaporates totally if many of the stalled activities can no longer be continued until their expected completion, or are indefinitely delayed. That of course costs additional local capital to do the cleanup of what is left behind, unfinished, and unusable ruins. In particular, in equs. (11.1) and (11.2), the dividends D_{t+j} of the $(t + i)$-th period decreases drastically and the remaining future dividends, if they still come fortunately as expected, would have to be used to bail out (to finish up) some of the other potentially possibly profitable projects.

That is, right before the large amount of foreign investments leaves suddenly and strategically, the local economy is more active than ever before. Hence, the stochastic gross discount rate R^q_{t+i} of the $(t + i)$-th period would be much lower than R^q_{t+j} of the $(t + j)$-th period, because much higher returns on earlier investments are optimistically expected. So the present ratio:

$$\frac{(1 - \delta)^i D_{t+i}}{R^q_{t+1} \cdots R^q_{t+i} \cdots R^q_{t+j}}$$

would in reality be very close to zero.

Summarizing what is just analyzed above, one can see that when a large amount of foreign investments gathers in one place over either a long period or a short period of time and then leaves suddenly and massively, that local economy has to suffer through a positive bubble, caused by the increased money supply as a consequence of the foreign investments, and then a following negative, disastrous bubble, caused by the sudden dry-out of the money supply. And due to a large number of economic activities that are either unexpectedly delayed or totally impossible to complete, the local investors are actually unable to continue to collect their originally expected dividends for many periods to come. That is, foreign investments can be employed as a weapon of mass destruction, if they leave strategically and suddenly, no matter whether they come quickly in a short period of time or slowly over a relatively longer period of time.

11.3.2 A strategy of self-protection

When combining the previous theoretical analysis with the recent cases of speculative attacks in the arena of international finance, we surely see the following predicament. When a nation tries to develop economically, due to its loosening economic and monetary policies, large amounts of foreign investments would be welcomed; and at the same time, a lot of such foreign investments would strategically rush into the nation in order to ride along with the forthcoming economic boom. Now what we have shown earlier is that if a large amount of foreign investments leave suddenly, then the nation would most likely suffer from a burst of the economic bubble with a large percentage of economic activities interrupted either temporarily or indefinitely. So, a natural question at this junction is: How could we design a measure to counter such sudden leaves of foreign investments in order to avoid the undesirable disastrous consequences?

We will address this problem in this section.

11.3.2.1 A model for categorized purchasing power

Let us look at the following model that relates the purchasing power of money with the demand and supply of the money of a national economy:

$$\frac{dP}{dt} = k(D - S) \tag{11.9}$$

where D stands for the demand for money, S the money supply, P the purchasing power of money, $k > 0$ is a constant, and t represents time.

What this model says is that the rate of change in purchasing power is directly proportional to the difference between the demand and supply of money. In particular, the model says that with all other variables staying constant, if the money supply S increases by a large amount and satisfies $S > D$, then the purchasing power of the money P decreases so that more money is needed to buy essentials of living and inflation will increase due to the increase in the money supply S.

Now, let us divide the overall national economy into three sectors E_1, E_2, and E_3 as follows: E_1 stands for the goods, services, and relevant production of these goods and services that are needed for maintaining the basic living standard, E_2 the goods, services, and relevant productions that are used to acquire desired living conditions, and E_3 the goods, services, and relevant productions that are used for the enjoyment of luxurious living conditions. Accordingly, let us divide the overall demand D of money into three corresponding categories as follows:

D_1 = the demand of money for meeting the minimum requirement to
    maintain the basic living standard;
D_2 = the demand of money for acquiring desired living conditions; and
D_3 = the demand of money for enjoying luxurious living conditions.

Assume that in a stable economy, we have the following allocation of the money demand:

$$D = D_1 + D_2 + D_3 = \alpha_1 D + \alpha_2 D + \alpha_3 D \tag{11.10}$$

where the weights α_i, $i = 1, 2, 3$, stands for the average allocation of the citizens of the economy over the three categories as described above, satisfying $\alpha_1 + \alpha_2 + \alpha_3 = 1$, and $D_i = \alpha_i D$, $i = 1, 2, 3$. For instance, in the stable economy, an average family allocates half of its monthly income on necessities of living, such as food, utilities, etc., 4 tenth of the income on acquiring the desired quality of life, and one tenth of the income on luxurious items, then $\alpha_1 = 0.5$, $\alpha_2 = 0.4$, and $\alpha_3 = 0.1$.

If the money supply S increases drastically, along with the decreasing purchasing power of money, all goods will cost more. If somehow the goods in the category of living necessities rise more rapidly, then a re-allocation of household income will appear. For instance, due to rumors about potential interruptions in the supply of food and clean water accompanying a substantial increase in the money supply, the average family has to reallocate its income as follows: $\alpha_1 = 0.625$, $\alpha_2 = 0.3$, and $\alpha_3 = 0.075$.

When such a reallocation of income of the average family is forced to take place, the stability of the economy would very likely be in trouble. So, to stabilize the economy, the purchasing power of money in category D_1 should stay relatively constant, while in D_2 increases somehow slightly, and in D_3 it should be allowed to increase in order to attract and trap the additional money supply away from category D_1. So, let us assume

P_1 = the purchasing power of money in category D_1;
P_2 = the purchasing power of money in category D_2; and
P_3 = the purchasing power of money in category D_3.

Similarly, let us define

S_1 = the money supply that goes into category D_1;
S_2 = the money supply that goes into category D_2; and
S_3 = the money supply that goes into category D_3.

So, equ. (11.9) would look as follows:

$$
\begin{cases}
\dfrac{dp_1}{dt} = k_{11}(D_1 - S_1) + k_{12}(D_2 - S_2) + k_{13}(D_3 - S_3) + \displaystyle\sum_{j=1}^{n} q_{1j}x_j \\[3mm]
\dfrac{dp_2}{dt} = k_{21}(D_1 - S_1) + k_{22}(D_2 - S_2) + k_{23}(D_3 - S_3) + \displaystyle\sum_{j=1}^{n} q_{2j}x_j \\[3mm]
\dfrac{dp_1}{dt} = k_{31}(D_1 - S_1) + k_{32}(D_2 - S_2) + k_{33}(D_3 - S_3) + \displaystyle\sum_{j=1}^{n} q_{3j}x_j
\end{cases}
\tag{11.11}
$$

where k_{ij}, and q_{ij} are constants, and x_j stands for monetary policies, $i = 1, 2, 3$, and $j = 1, 2, \ldots, n$. By using matrix notations, equ. (11.11) can be rewritten as follows:

$$
\dot{P} = Kz + Qx
\tag{11.12}
$$

where $P = [P_1 \; P_2 \; P_3]^T$, $\dot{P}$ is Newton's original notation for derivatives such that $\dot{P} = \left[\dfrac{dP_1}{dt} \; \dfrac{dP_2}{dt} \; \dfrac{dP_3}{dt} \right]^T$, $K = [k_{ij}]_{3\times 3}$ the coefficient matrix of the variables $(D_i - S_i)$, $i = 1, 2, 3$, $Q = [q_{ij}]_{3\times n}$ the coefficient matrix of the variables $x_j, j = 1, 2, \ldots, n$.

In terms of systems research, the mode in equ. (11.10) can be seen as a feedback system as depicted in Figure 11.1, where S represents the initial state of the economy. After the monetary policies $x_1, x_2, \ldots, x_n$ are introduced, the participants of the economy introduce either consciously or unconsciously a feedback component system S_f so that the overall system with the added feedback produces the desired output P_1, P_2, and P_3.

What is shown in Figure 11.1 is the fact that each and every market economy, where the participants are allowed to design their own methods (without violating the established laws) to achieve their individually defined financial successes, then the economy constitutes a "rotational" field. Here, the word of rotation means that as soon as a monetary policy is introduced, the market participants will find ways to take advantage of the policy so that the policy and the individually designed methods jointly produce the individually desired outputs.

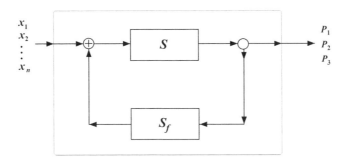

Figure 11.1 The feedback loop between monetary policies and categorized purchasing powers.

11.3.2.2   Functional relationship between P and (D − S)

In this section, we investigate the relationship between the purchasing power vector $[P_1\ P_2\ P_3]^T$ of money and the difference vector $[D_1 - S_1\ D_2 - S_2\ D_3 - S_3]^T$ of demand and supply of money. In particular, we discuss why the trends found in purchasing power of money could righteously be described by using linear models, although the purchasing power is clearly not linear in terms of the difference of the demand and the supply of money.

To this end, let us introduce the following three models, the first one is linear, the second one quadratic, and the third one cubic, to demonstrate the effects of supply and demand of money on purchasing power of money in each case:

$$P(t) = a(D(t) - S(t)) + \varepsilon \tag{11.13}$$

$$P(t) = a(D(t) - S(t))^2 + b(D(t) - S(t)) + \varepsilon \tag{11.14}$$

and

$$P(t) = a(D(t) - S(t))^3 + b(D(t) - S(t))^2 + c(D(t) - S(t)) + \varepsilon \tag{11.15}$$

where $P(t)$ stands for the purchasing power of money, $D(t)$ the demand of money, $S(t)$ the supply of money, ε a random variable with mean $C \neq 0$, and a, b, and c are constant. More specifically, the random variable ε is the error term in the sense that it compensates for any unpredicted event or factor that could impact the purchasing power and that is not taken into account in the model.

When the nonlinearity in the trend of purchasing power is considered, such as in the case of Japanese yen (Figure 11.2), the quadratic and cubic models in equs. (11.14) and

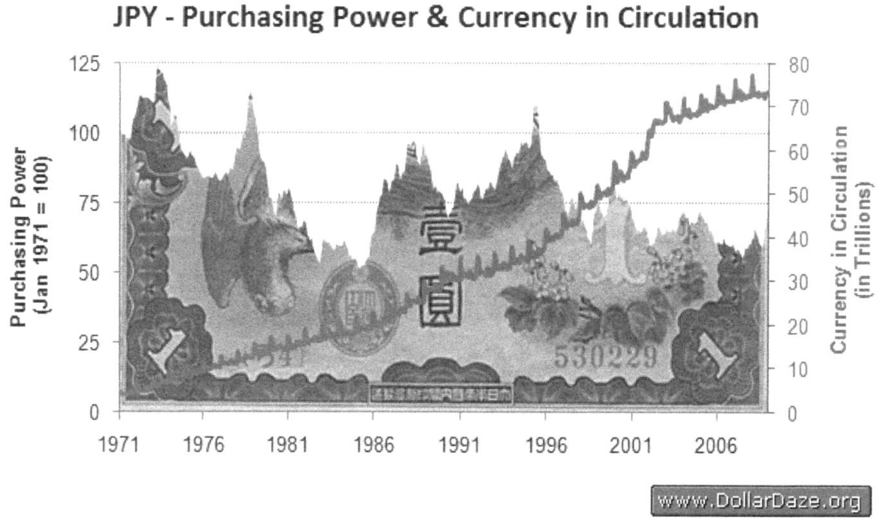

Figure 11.2 Purchasing power and currency in circulation, from www.dollardaze.com as accessed on April 4, 2012.

(11.15) seem to model adequate. In particular, the non-linear graph of the purchasing power of Japanese yen has two distinguishable patterns: one is parabolic and the other cubic. These two types of general patterns are respectively depicted by the nonlinear models in equs. (11.14) and (11.15).

Figure 11.3 depicts both the purchasing power and the amount of money in circulation of the U.S. dollars over time. Here a clear inverse relationship between the currency in circulation, supply of money, and the purchasing power of money can be seen. That is to say, as the amount of money in circulation increases the purchasing power of the US dollars decreases. The graph of purchasing power is fairly linear, especially if broken up into two segments: from 1971 to approximately 1981 and from 1981 until the present day. This trend in the purchasing power of U.S. dollars suggests that the linear model in equ. (11.13) seems to adequately reflect what happened to the purchasing power over time.

In particular, if we employ the linear model in equ. (11.13) to describe the relationship between the purchasing power of the U.S. dollars and the difference between the demand and supply of the money with respect to time, then this model well explains how the U.S. dollar declines in purchasing power. For instance, the initial high purchasing power was due to the fact that the currency in circulation was fairly low and the value of the U.S. dollar was fixed at $35 an ounce of the gold. However, starting in May 1971 the U.S. dollar suffered from a major crisis, and consequently began to depreciate against the gold from the initial $35 an ounce to $38 an ounce on August 15, 1971, then to $42.22 an ounce at the start of 1973, and then to as high as $96 an ounce in March of the same year (Wang and Hu, 2005, p. 12). In 1976 a new international agreement was reached in the capital city Kingston of Jamaica; with several rounds of modifications the international financial system entered the Jamaica system in 1978. Within this new system, gold is no longer considered as a form of money.

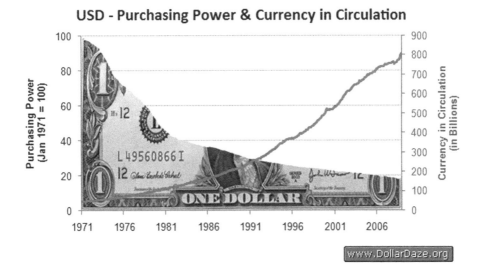

Figure 11.3 Purchasing power and currency in circulation of the US dollars, from www.dollardaze.com as accessed on April 4, 2012.

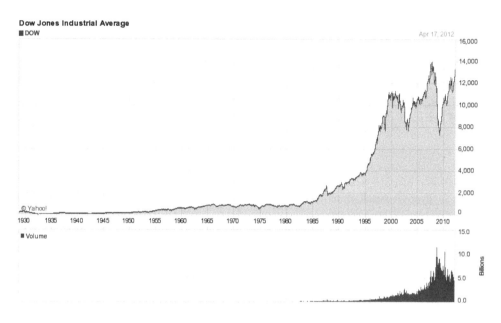

Figure 11.4 Historical data of DJIA (from Yahoo finance, accessed on April 17, 2012).

That is, when the demand for the U.S. dollars is brought into the equation, it can justify why there are more severe or gradual drops in purchasing power as the amount of money in circulation increases. For example, if the change in the supply of money is equal to the increase in demand of money, then the purchasing power of the money should remain constant. To cause the severe drop in purchasing power as seen from 1971 until approximately 1981, there was an increase in the money supply accompanied by a decrease in the demand for money. This decrease in the demand for money was caused by the transition from the U.S. dollar being backed by gold to that not having gold backing (dollardaze.org). This also makes sense intuitively. If we already know that a rise in the money supply decreases the purchasing power of money, then people wanting the money less would further exacerbate that decrease in purchasing power. From that point on, the demand of money must have increased but still not greater than the supply of money because the decline in the purchasing power flattens out instead of becoming more severely negative. As the graph of the stock prices of the Dow Jones Industrial Average indicates, Figure 11.4, the stock market began to rise in the early 1980s and had a sharp incline until the start of the 21st century. This rise is an indicator that supports the claim that the demand for money increased during this time period and, in conjunction with the supply of money, influenced the purchasing power to decrease less severely than during the time period from 1971 to 1981.

Looking at the same graph created for the Japanese economy, Figure 11.2, we see a similar but different story. By breaking the graph up into more intervals, Figure 11.5, we are able to form various linear patterns. From 1971 until around 1986 there was a general decline in the purchasing power of Japanese yen. Then once it came to a peak again in 1996 there again was an overall decline in the purchasing power. These are

JPY - Purchasing Power & Currency in Circulation

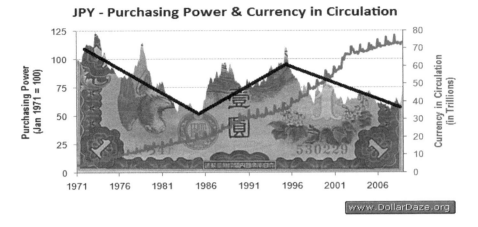

Figure 11.5 The fluctuation in the purchasing power of Japanese yen.

clearly not linear graphs; but looking at the general motion of the purchasing power in the graph they are fairly linear in two segments, like the situation with the U.S. dollar. Therefore, it is also reasonable for us to use the linear model in equ. (11.13) to explain the evolutionary trends in the purchasing power of Japanese yen.

Under the assumption that the supply and purchasing power of money are inversely related when all other factors remain constant, we can see that the purchasing power is a function of the demand and supply with respect to time. By using the linear model in equ. (11.13) and by breaking up the graph in Figure 11.5 into two intervals, we see a similar relationship that was discussed with the supply and demand of money in the U.S. dollar. If both the supply and demand increase at the same rate, then the purchasing power will remain constant. If the supply increases with the demand increasing at a faster rate than before but still less than the increase in the supply, then a less severe decline in the purchasing power would be observed. However, if the demand increases at a rate greater than that of the supply of money, then this would lead to an increase in the purchasing power. Furthermore, an increase in the supply matched with a decreasing rate in demand would result in a more severe decrease in the purchasing power. These explanations help explain varying linear slopes that are reflected in the purchasing power of money in the Japanese economy. For example, notice that for Japan 1986 was the first year of the asset pricing bubble. In the graph of the money supply and the Nikkei 225 stock market, Figure 11.6, for details, see (Okina, Shirakawa and Shiratsuka, 2001), during this time there was a small increase in the money supply while the Nikkei 225 increased drastically. With the stock market increasing during this time, the demand for money also increased because people wanted to take advantage of the market rise. As we mentioned earlier, a relatively low increase in the money supply accompanied by a large increase in the demand for money would result in an increase in the purchasing power. This explanation corresponds with the depiction of the purchasing power of Japanese yen until 1990 when it burst reflecting the sharp decline in the purchasing power as reflected in the graph and by our model.

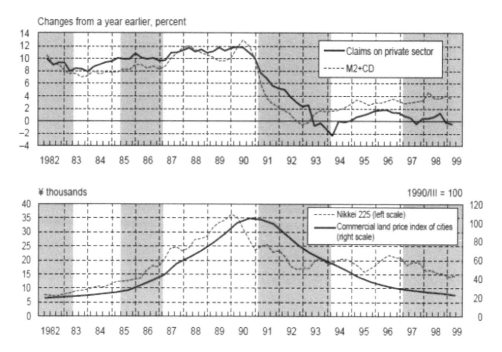

Figure 11.6 The money supply and the Nikkei 225 stock market (Okina, Shirakawa and Shiratsuka, 2001).

The sharp rise in the purchasing power that appeared at around the year 1973, Figure 11.3, could have been due to the oil crisis which caused a shift in Japanese economy toward huge investments in the electronic industries. From (Hutchison, 1986), we can see that Japan's money supply was fairly constant in the year 1973 but the demand of money from the oil crisis increased drastically. That made the purchasing power of the money to increase drastically during the early 1970s; this conclusion is further validated by the discussions of (Okina, Shirakawa and Shiratsuka, 2001). Figure 11.7 is a graph that reflects the Japanese inflation rate over the time period from April 1971 until April 2012, where we can notice the continuing patterns and similarities with those of the purchasing power.

As a matter of fact, the similarities between the graphic patterns of the purchasing power of money and the inflation rate, as described above, are also seen in the current movements of oil prices. On the Ed Show of the MSNBC at 11:00 p.m., Barny Frank referenced certain people who bought crude oil to drive up the prices only to sell it at a later time for a windfall of profits. The market system unconsciously allows the oil prices to rise and to consequently manipulate the gasoline prices, Figure 11.8. That situation works in conjunction with the demand and supply of money.

Summarizing what is discussed above we conclude that it is theoretically reasonable for us to analyze the relationship between the purchasing power of money and the difference of the demand and supply of money by using the linear model in equ. (11.13), where the random variable ε accounts for all the unexpected factors that are not included in the model.

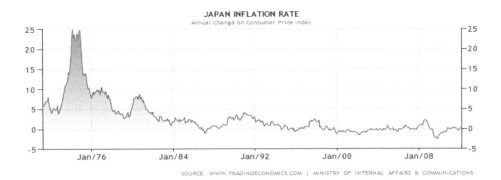

Figure 11.7 Japanese inflation rate from April 1971 to April 2012. The original source was accessed on April 4, 2012.

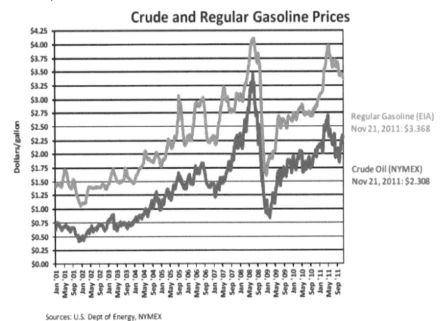

Figure 11.8 The prices of crude and gasoline move in concert, the original source was accessed on April 17, 2012.

By combining what is obtained in Section 1.1 with the linear model in equ. (11.13), we have the following relationship between categorized purchasing power $[P_1 \ P_2 \ P_3]^T$ of money and categorized difference $[D_1 - S_1 \ D_2 - S_2 \ D_3 - S_3]^T$ of demand and supply of money:

$$
\begin{bmatrix} P_1 \\ P_2 \\ P_3 \end{bmatrix} = R_{3\times3} \begin{bmatrix} D_1(t) - S_1(t) \\ D_2(t) - S_2(t) \\ D_3(t) - S_3(t) \end{bmatrix} + \begin{bmatrix} \varepsilon_1 \\ \varepsilon_2 \\ \varepsilon_3 \end{bmatrix}
\tag{11.16}
$$

where $R_{3\times3}$ is a constant square matrix with real entries, and $[\varepsilon_1 \ \varepsilon_2 \ \varepsilon_3]^T$ a random vector with a none zero mean.

11.3.2.3 Separating economic categories by using feedback

If we consider the mathematical expectations of the variables in equ. (11.16), we have

$$
\begin{bmatrix} P_1 \\ P_2 \\ P_3 \end{bmatrix} = A_{3\times3} \begin{bmatrix} D_1(t) - S_1(t) \\ D_2(t) - S_2(t) \\ D_3(t) - S_3(t) \end{bmatrix} + \begin{bmatrix} c_1 \\ c_2 \\ c_3 \end{bmatrix} \tag{11.17}
$$

where $E(\varepsilon_i) = c_i \neq 0$, $i = 1, 2, 3$. By substituting equ. (11.17) into equ. (11.12), we have

$$
R_{3\times3}\dot{z} = Kz + Qx \tag{11.18}
$$

Without loss of generality, we assume that $R_{3\times3}$ is invertible. That is, we assume in general the categorized purchasing power of money is completely determined by the categorized differences of demand and supply of money. Then, equ. (11.18) can be rewritten as follows:

$$
\dot{z} = Az + Bx \tag{11.19}
$$

where $A = R^{-1}K$ and $B = R^{-1}Q$. To make the model in equ. (11.19) technically manageable, we assume without loss of generality that B is a 3×3 matrix, meaning that the monetary policies $x_1, x_2, \ldots, x_n$ are accordingly categorized into three groups:

$X_1 = $ the set of all those monetary policies that deal with the population meeting the minimum need to maintain the basic living standard;

$X_2 = $ the set of all those monetary policies that deal with the population's need for acquiring desired living conditions; and

$X_3 = $ the set of all those monetary policies that deal with the population's need for enjoying luxurious living conditions.

Without loss of generality, we will still use the same symbol x to represent the vector $[X_1 \ X_2 \ X_3]^T$ of categorized monetary policies.

Similar to the concept of consumer price index (CPI), let us introduce an economic index vector $y = [y_1 \ y_2 \ y_3]^T$ such that y_i measures the state of the economic sector i, $I = 1, 2, 3$. Then from equ. (11.19) we can establish the following model for the national economy of our concern:

$$
\begin{cases} \dot{z} = Az + Bx \\ y = Cz + Dx \\ z(0) = 0 \end{cases} \tag{11.20}
$$

where z is the 3×1 matrix $[D_1 - S_1 \ D_2 - S_2 \ D_3 - S_3]^T$ of the categorized difference of demand and supply of money, referred to as the state of the economic system, A, B, C, and D are respectively constant 3×3 matrices, such that D is non-singular (meaning that each introduction of monetary policies does have direct, either positive or negative, effect on the performance of the economy), and the input space X and output space Y are following:

$$X = Y = \{r : [0, +\infty) \rightarrow R^3 : r \text{ is a piecewise continuous function}\} \qquad (11.21)$$

where R stands for the set of all real numbers and R^3 the nth dimensional Euclidean space.

What is described by equ. (11.20) is that the state of the national economy is representable through the use of the state variable z that helps the economy to absorb the positive and negative effects of the monetary policies X_1, X_2, and X_3. Then, both the internal mechanism z of the economy and the monetary policies x jointly have a direct effect on the overall performance y of the economy. The condition, as imposed on the input space X and the output space Y means that monetary policies, which form the input space X, are introduced based on the effects of the previously implemented policies, while the overall performance, indexes of which constitutes the output space Y, of the economy evolves from previous states mostly continuous. Such assumptions historically speaking are not always true, while failures occur rarely in terms of frequencies. For instance, in modern China, the so-called Cultural Revolution occurred abruptly (MacFarquhar and Schoenhals, 2008); and several times in the history of Russia, the Tsars had introduced rushed social reforms creating a torn country (Huntington, 1996).

Geometrically, the systemic model in equ. (11.20) of the national economy is depicted in Figure 11.9. The monetary policy input X splits into two portions $F_1 F_2 F_3 F_4$ and $F_5 F_6 F_7 F_8$. The portion labeled $F_1 F_2 F_3 F_4$ directly affects the performance Y of the economy. The other portion, labeled $F_5 F_6 F_7 F_8$, is fed into the initial state of the economy S, leading to the introduction of the feedback component system S_f, which stands for the market reactions to the introduced monetary policies, and the formation of a feedback loop within the economy. This feedback loop in fact constitutes the main body of the economy, while the overall performance vector Y of indexes is merely an artificially designed measure. What needs to be noted is that within each of our three economic sectors E_1, E_2, and E_3, the market reactions, which constitute parts of the feedback component system S_f, to monetary policies in general are unique and

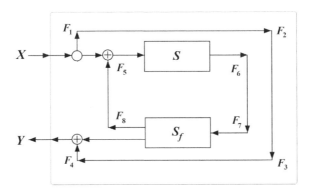

Figure 11.9 The geometry of the systemic model of the national economy.

economic sector specific. That is, the market reactions in one economic sector are different from those of another sector. The inner most loop $F_1F_2F_3F_4$ is the second stage of the economy after the introduction of new monetary policies. They by the joint effect of the market reactions (S_f) and the non-reactionary aspects $(F_5F_6F_7F_8)$ of the monetary policies, the final numerical readings Y of the economy are produced.

To see how monetary policy input X could have aspects, one is reactionary and the other non-reactionary, let us assume that X stands for such a monetary policy that allows the inflation to inch higher. Therefore, the price of crude oil is expected to rise accordingly. Now, the reactionary aspects of the policy X make the gradual and calculated increase in the price of crude oil more or less random. On the other hand, the expected outcome of the policy is the theoretical, non-reactionary aspect of the policy. When the actual outcome deviates from the expectation, the difference is caused by the market reactions to the policy.

According to (Lin, 1994), for the convenience of the reader, the main results of this reference are provided in the appendix of this chapter, the 3-dimensional system in equ. (11.20), meaning that both the input x and the output y are elements from R^3, can be decoupled into three independent systems of the same kind with one-dimensional input and out. Specifically, if we let

$$S = \{(x, y) \in X \times Y : \exists z \in Z \text{ such that } x, y, z \text{ satisfy equ. (11.20)}\}$$
$$= \text{ the system of all the ordered pairs } (x, y) \text{ satisfying equ. (11.20)}$$

where the state space $Z = \{r : [0, +\infty) \to R^3 : r \text{ is a piecewise continuous function}\}$, and for each $i = 1, 2, 3$, define a system S_i as follows:

$$\begin{cases} \dot{z} = Az + B_i x_i \\ y_i = C_i z + D_i x_i \\ z(0) = 0 \end{cases} \tag{11.22}$$

where B_i is the ith column of B, C_i the ith raw of C, D_i a non-zero constant, and the input space X_i and the output space Y_i are given as follows:

$$X_i = Y_i = \{r : [0, +\infty) \to R : r \text{ is a piecewise continuous function}\}.$$

In particular, the system S can be decoupled through feedback into the factor systems $S_i, i = 1, 2, 3$, as follows. Let

$$\alpha = \begin{bmatrix} A & 0 & 0 \\ 0 & A & 0 \\ 0 & 0 & A \end{bmatrix}, \quad \beta = \begin{bmatrix} B_1 & 0 & 0 \\ 0 & B_2 & 0 \\ 0 & 0 & B_3 \end{bmatrix}, \quad \gamma = \begin{bmatrix} C_1 & 0 & 0 \\ 0 & C_2 & 0 \\ 0 & 0 & C_3 \end{bmatrix}, \quad \delta = \begin{bmatrix} D_1 & 0 & 0 \\ 0 & D_2 & 0 \\ 0 & 0 & D_3 \end{bmatrix}.$$

Then, the Cartesian product system $S_d = S_1 \times S_2 \times S_3$ is represented by the set of all ordered pairs (x, y) satisfying

$$\begin{cases} \dot{z} = \alpha z + \beta x \\ y = \gamma z + \delta x \\ z(0) = 0 \end{cases} \tag{11.23}$$

Because δ is non-singular, the inverse system S_d^{-1} is obtained as follows:

$$S_d^{-1} = \{(y,x) : (y,x) \text{ satisfies equ. (11.24)}\}$$

where equ. (11.25) is given as follows:

$$\begin{cases} \dot{z} = (\alpha - \beta\delta^{-1}\gamma)z + \beta\delta^{-1}y \\ x = -\delta^{-1}\gamma z + \delta^{-1}y \\ z(0) = 0 \end{cases} \tag{11.24}$$

The particular feedback component system $S_f : Y \to X$ used in this decoupling is given as follows:

$$\begin{cases} \begin{bmatrix} \dot{z} \\ \dot{z}' \end{bmatrix} = \begin{bmatrix} A - BD^{-1}C & 0 \\ 0 & \alpha - \beta\delta^{-1}\gamma \end{bmatrix} \begin{bmatrix} z \\ z' \end{bmatrix} + \begin{bmatrix} BD^{-1} \\ \beta\delta^{-1} \end{bmatrix} y \\ x = \begin{bmatrix} -D^{-1}C & \delta^{-1}\gamma \end{bmatrix} \begin{bmatrix} z \\ z' \end{bmatrix} + (D^{-1} - \delta^{-1})y \\ \begin{bmatrix} z \\ z' \end{bmatrix}(0) = 0 \end{cases} \tag{11.25}$$

In terms of economics, what the concept of decoupling the 3-dimensional system S into component systems S_i, $i = 1, 2, 3$, as discussed above, implies is that when monetary policies are established individually and respectively for each of the three economic sectors E_1, E_2, and E_3, although these economic-sector specific policies will most definitely have joint effects on the economy, there is at least one way to design a feedback component system S_f so that the overall feedback system $F(S, S_f)$, which represents the whole system as depicted in Figure 11.9, can be controlled through adjusting individually each of the economic sectors E_1, E_2, and E_3.

11.3.2.4 A strategy for national defense

The significance of the previous discussion is that we can now propose based on the sound analytical reasoning presented above a strategy for national defense against currency warfare in case that initial 'friendly' foreign investments turn out to be an aggressive act of war by suddenly withdrawing all or significant amount of the investments.

In particular, let us assume that at time period t, the total money supply of the specific nation of our concern is $1,000, $100 of which is from foreign investments. That is, the foreign investments amount to 10% of the total domestic money supply. With the additional money supply (due to the foreign investments) the speed of money circulation increases, signaled by increased spending and elevated levels of economic activities. If these foreign investments leave the nation suddenly, then it is reasonable to expect that more than 10% of the economic activities from around the nation will be more or less affected adversely due to the sudden exhaustion of the money flow, because accompanying the foreign investments there tends to be domestic investments attached too, making the total investments on the related economic activities more than 10% of the national economy.

Now, if before the foreign investments depart while leaving behind disastrous aftermath, the national government has been keeping the exchange rate (to this end not all nations from around the world are currently able to do this successfully) the same while gradually and strategically increased its money supply, say, to the level of $10,000, then in the entire monetary circulation around the nation the proportion of the foreign investments shrinks to about 1% from the original 10%. And if the money supply had been increased to the level of $1,000,000, then the proportion of the foreign investments would have shrunk from the original 10% to about 0.01%, which is nearly zero. So, if at this moment of time the foreign investments are suddenly withdrawn as an aggressive act of war, only around 1% or 0.01% of the consumption and economic activities of the nation will be affected adversely and the overall economic health of the nation will be relatively stable.

On the other hand, along with the drastically increased money supply, all prices in the nation will most certainly go through the roof, placing a large portion of the national population in financial crises due to the run-away inflation. For instance, a certain commodity A, which is piece of living necessity, used to cost $1.00 a unit; now it requires $10.00 or even $1,000 to purchase. Such dramatic increase in prices will surely cause hardships for a good number of citizens of the nation. To this end, the national government needs to work on how to redistribute the additional money supply. In fact, as long as the distribution of the increased money supply does not cause social upheaval, then to this nation, nothing disastrous will really happen and the potentially damaging impacts of sudden departure of the foreign investments will be under control, too.

However in reality, the distribution of the extra money supply could easily lead to major societal instabilities to the nation due to the increased unevenness in the economic scene: The rich become richer while the poor become poorer. The situation here is similar to that of the normal inflation, which has been employed to make the economic structure more uneven than before so that the yoyo structure of the economy spins with more strength. In other words, with additional money supply injected into the economy, the redistribution of the wealth that is represented by the increased money supply is surely uneven, where some people receive more than their share, some simply keep pace with the decrease of the purchasing power of their income, while others fall behind or further behind with their financial status. So, if our suggested measure is adopted to counter the damaging effects of sudden departure of foreign investments, the nation needs to develop a practical plan to distribute the extra money supply in order to:

1   Keep societal peace and national stability so that a normal and operational economy can be maintained.
2   Increase the nation's economic prosperity by taking advantage of the foreign investments even though they leave sooner or later either slowly or suddenly.

Specifically, as suggested by the systemic model in the previous section, to help protect the innocent citizens of the nation from suffering from the potential economic turmoil, the national government could purposely divide the economy into three sectors E_1, E_2, and E_3 as described earlier to meet the following goals. In Sector E_1, which consists of living necessities, the sector performance, such as the sector specific CPI, evolves as normally as possible; in Sector E_2, which consists of such goods, services, and relevant productions that are used by citizens to acquire desired living conditions, the sector performance index, say the particular CPI of E_2, could outpace that of Sector E_1 by a large amount; and critically, the national government needs to manage to trap most of the additional money supply in the economic sector E_3, which consists of such goods, services, and relevant productions that are used by the citizens for their enjoyment of luxurious living. The previous discussion on the systemic model of the national economy indicates that by managing the market reactions correctly, that is the design of the feedback component system S_f, these three economic sectors can be well separated from each other.

When the economic sector E_1 evolves normally as expected based on the history of the nation, the citizens would not need to worry about their basic living and survival. That naturally leads to the desired societal stability and peace. Although the prices in the economic sector E_2 increase drastically when compared to those in E_1, that in general should not affect the mood or the happiness of the population much, because desired living conditions change from time to time and vary from one family to another. In other words, isolated, temporary personally desires, which are not yet satisfied and which are generally inconsistent with each other and even contradictory to each other, could not amalgamate into a political force to cause turmoil. Now, the key is how to keep most of the additional money supply in the economic section E_3. To address this problem, let us pay attention to our earlier assumption that a large amount of foreign investments entered the nation. That means commercial products and services that are new to the people in the specific nation are expected to and will become available soon. Because these products and services are brand new to the region, high prices can of course be charged for these commercial goods and services. The situation is similar to the scenarios with developed nations, where the focus has been placed on innovations, design and production of unprecedented products and services. It is these never-seen products and services that generate the most profits.

At this junction, let us address the problem of how to classify a product or a service into one of the three economic sectors E_1, E_2, and E_3. First of all, goods and services that should be classified into E_1 are clear. They represent such commodities that directly relate to the survival of any human being, such as foods, utilities, and shelters of the basic quality. Now, if a family lives in a rented two-bedroom apartment and wants to own a house, then houses to that particular family will be in the economic sector E_2. If the people in a community go to work, shopping, and entertainment mostly on foot, then individually owned transportation tools, such as bicycles, motorcycles, cars, private jets, etc., will belong to the economic sector E_3. That is, as time evolves forward, what belongs to the economic sector E_3 gradually lowers itself into the economic sector E_2, and what used to be in E_2 also gradually moves to E_1. That is, the classification of one particular product or service is a function of time and space. Just by looking around shopping centers, one can see vividly that such classification of commercial goods and services have been successfully done by merchants throughout the world without much trouble.

Now, let us consider the scenario that when the nation purposefully and strategically increases its money supply in order to prevent disastrous aftermaths that could be potentially left behind by sudden withdrawals of foreign investments, additional foreign investments could continue to pour in. In this case, if the nation could not pick up its corresponding speed of economic development, then it will be taken over by the run-way high rate of inflation. In this case, if the foreign investments depart strategically, then the nation will be in real major political, societal, and economic crises. To prevent such disastrous consequences, the continued inflow of foreign money has to been managed so that they cannot depart quickly.

Another practical scenario that is different of what has been discussed above is that at a high speed develops the nation itself into a super economic power such that there is no longer such a single amount of foreign investments that can amount to an influencing percentage in the nation's domestic consumption and/or economic activities. In this case the nation will be safe in the face of any potentially sudden withdrawal of foreign investments.

APPENDIX 11A FEEDBACK: A GENERAL SYSTEMS APPROACH

This appendix is attached here for the purpose of completeness of our presentation in this chapter. What is presented below is based on (Lin, 1994) and contains a detailed procedure that can be employed to decouple a high dimensional multivariable linear system into a family of mutually independent one-dimensional linear systems.

11A.1 INTRODUCTION

The behavior of the following system becomes quite complicated as the dimension of the input $x(t)$ and the output $y(t)$ increases:

$$\begin{cases} \dot{z} = Az + Bx \\ y = Cz + Dx \\ z(0) = 0 \end{cases} \tag{11.26}$$

where z is an $m \times 1$ variable vector, called the state of the system, A, B, C and D constant matrices of sizes $m \times m$, $m \times n$, $n \times m$ and $n \times n$, respectively, such that D is non-singular, and the input space X and the output space Y are the following:

$$X = Y = \{r : [0, \infty) \to R^n : r \text{ is piecewise continuous}\} \tag{11.27}$$

where R is the set of all real numbers.

The main result of this appendix is to show that no matter how high the dimension n of the input $x(t)$ and the output $y(t)$ is, the system (11A.1) can always be studied as n linear systems with one-dimensional input and output. The method used is part of the general systems theory, developed by Mesarovic and Takahara (1975; 1989), Saito and Mesarovic (1985), Saito (1986), Lin and Ma (1990) and Lin (1991). This application of the general systems theory shows the interesting fact that this general theory can be applied not only to model and to study problems from various disciplines (Mesarovic and Takahara,1989), but also to its origin – the systems theory of ordinary equations.

This appendix is organized as follows. Section 11A.2 introduces the concept of general feedback transformation and its characteristics. As examples, Section 11A.3

studies some properties that are inherited by feedback systems so that it is reasonable to expect the use of feedback transformation in the investigation of the original system. An equivalent condition for feedback decoupling is given in Section 11A.4. This result is employed in Section 11A.5 to show how the system (11A.1) can be decoupled into n linear systems of the same kind with one-dimensional input and output. Then a detailed procedure of decoupling is given.

11A.2 FEEDBACK AND ITS PROPERTIES

Let A be a field, X and Y linear spaces over the field A. A non-empty subset $S \subset X \times Y$ is called an (input-output) linear system (Mesarovic and Takahara, 1975), provided S is a linear subspace of the product linear space $X \times Y$, where X and Y are known as the input and output spaces of the system S, respectively. Without loss of generality, let the range $R(S)$ be the same as the output space Y and the domain $D(S)$ the input space X.

Let

$$\mathcal{S} = \{S \subset X \times Y : S \text{ is a linear system}\} \tag{11A.3}$$

$$\mathcal{S}_f = \{S_f \subset X \times Y : S_f \text{ is a linear system}\} \tag{11A.4}$$

$$\mathcal{S}' = \{S' \subset X \times Y : S' \text{ is a subset}\} \tag{11A.5}$$

and define a function

$$F : \mathcal{S} \times \mathcal{S}_f \to \mathcal{S}' \tag{11A.6}$$

By letting $\forall \, (S, S_f) \in \mathcal{S} \times \mathcal{S}_f$, $F(S, S_f) \in \mathcal{S}'$ satisfies

$$(x, y) \in F(S, S_f) \text{ iff } (\exists z \in X)((x + z, y) \in S \text{ and } (y, z) \in S_f) \tag{11A.7}$$

This function F is known as the feedback transformation over the linear spaces X and Y. The systems S and S_f are referred to as an original system and a feedback component system (Saito, 1986).

Proposition 11A.1. The following statements hold true for the feedback transformation F:

1   Let $S \in \mathcal{S}$, $S_f \in \mathcal{S}_f$, and $S' \in \mathcal{S}'$ be a linear system such that $F(S, S_f) = S'$, then $F(S', -S_f) = S$, where $-S_f \subset Y \times X$ is the linear system defined by $(y, x) \in -S_f$ iff $(y, -x) \in S_f$, $\forall \, (y, x) \in Y \times X$.
2   For each $S \in \mathcal{S}$, there is a linear functional system $S_f : Y \to X$ such that $F(S, S_f) = S$. In other words, the system S is a feedback system of itself.
3   Assume axiom of choice. For each $S \in \mathcal{S}$ and arbitrary linear subspace $S' \in \mathcal{S}'$, there is a linear functional system $S_f : Y \to X$ such that $F(S, S_f) = S'$, if and only if $R(S') = R(S)$ and $N(S') = N(S)$, where $N(W)$ stands for the null space of the linear system $W \subset X \times Y$, defined by

$$N(W) = \{x \in X : (x, 0_y) \in W\} \tag{11A.8}$$

Proof.   We will only show statement 3, because both statements 1 and 2 are obvious.

Necessity: Suppose that there is a linear functional system $S_f: Y \rightarrow X$ such that $F(S, S_f) = S'$. Then equ. (11A.7) implies that $R(S') = R(S)$ and that

$$
\begin{aligned}
N(S') &= \{x \in X : (x, 0_y) \in S'\} \\
&= \{x \in X : (x + S_f(0_y), 0_y) \in S\} \\
&= \{x \in X : (x + 0_x, 0_y) \in S\} \\
&= N(S).
\end{aligned}
$$

Sufficiency: Suppose that the systems S and S' have the same output space and the null space: $R(S') = R(S)$ and $N(S') = N(S)$. We will define a linear functional system S_f: $Y \rightarrow X$ such that $F(S, S_f) = S'$.

For each element $y \in R(S')$, let

$$X'_y = \{x' \in X : (x', y) \in S'\} \tag{11A.9}$$

$$X_y = \{x \in X : (x, y) \in S\} \tag{11A.10}$$

From the axiom of choice (Kunen, 1983), it follows that there are mappings:

$$C' : \{X'_y : y \in R(S')\} \rightarrow X \tag{11A.11}$$

$$C : \{X_y : y \in R(S)\} \rightarrow X \tag{11A.12}$$

such that $C'(X'_y) = C'(y) \in X'_y$; i.e. $(C'(y), y) \in S'$; and that $C(X_y) = C(y) \in X_y$; i.e. $(C(y), y) \in S$. Here it is easy to check that for each $y \in R(S') = R(S)$,

$$X'_y \neq \varnothing \neq X_y \tag{11A.13}$$

So, the mappings C and C' are well-defined.

Let $\{y_i : i \in I\}$ be a base in the linear space $R(S')$, where I is an either finite or infinite index set. We now define a linear functional system

$$S_f : Y \rightarrow X \tag{11A.14}$$

satisfying

$$S_f(y_i) = C(y_i) - C'(y_i), \quad \forall i \in I \tag{11A.15}$$

We then have the following lemma.

Lemma 11A.1. The three systems S, S' and S_f, as defined above, satisfy the relation $F(S, S_f) = S'$.

In fact, for each element $(x, y_i) \in S'$, for $i \in I$, we have $(x + S_f(y_i), y_i) \in S$. This is because

$$
\begin{aligned}
(x + S_f(y_i), y_i) &= (x + C(y_i) - C'(y_i), y_i) \\
&= (x - C'(y_i), 0_y) + (C(y_i), y_i) \in N(S') \times \{0_y\} + S \\
&= N(S) \times \{0_y\} + S \\
&= S
\end{aligned}
\tag{11A.16}
$$

Equ. (11A.16) implies that

$$
(x, y_i) \in S' \to (x + S_f(y_i), y_i) \in S
\tag{11A.17}
$$

Now, let $(x, y_i) \in S$. We then have $(x - S_f(y_i), y_i) \in S'$. It is because

$$
\begin{aligned}
(x - S_f(y_i), y_i) &= (x - C(y_i) + C'(y_i), y_i) \\
&= (x - C(y_i), 0_y) + (C'(y_i), y_i) \in N(S) \times \{0_y\} + S' \\
&= N(S') \times \{0_y\} + S' \\
&= S'
\end{aligned}
\tag{11A.18}
$$

So, we have from equ. (11A.18) the opposite implication of equ. (11A.17). That is, for each $i \in I$, we have

$$
(x, y_i) \in S' \leftrightarrow (x + S_f(y_i), y_i) \in S
\tag{11A.19}
$$

It now remains to show that for each pair $(x, y) \in X \times Y$, equ. (11A.7) holds true. To this end, let (x, y) be an arbitrary element in S'. Then $y = \sum_i c_i \circ y_i$, for a finite number of y_i's from the base $\{y_i : i \in I\}$, where each c_i is a non-zero element in $\mathcal{A}$. Then $(x + S_f(y), y) \in S$. In fact,

$$
\begin{aligned}
(x + S_f(y), y) &= \left(x + S_f\left(\sum_i c_i \circ y_i\right), y\right) \\
&= \left(x + \sum_i c_i \circ (C(y_i) - C'(y_i)), \sum_i c_i \circ y_i\right) \\
&= \left(x - \sum_i c_i \circ C'(y_i), 0_y\right) + \left(\sum_i c_i \circ C(y_i), \sum_i c_i \circ y_i\right) \\
&= \left((x, y) - \left(\sum_i c_i \circ C'(y_i), y_i\right)\right) + \sum_i c_i \circ (C(y_i), y_i) \in N(S') \times \{0_y\} + S \\
&= N(S) \times \{0_y\} + S \\
&= S
\end{aligned}
\tag{11A.20}
$$

This gives us the implication from the right-hand side to the left-hand side in equ. (11A.7). The opposite implication can be shown as follows.

For every pair $(x, y) \in S$, we claim that $(x - S_f(y), y) \in S'$. In fact, let $y = \sum_i c_i \circ y_i$, for some finite number of non-zero coefficients c_i, where each y_i is from the base $\{y_i : i \in I\}$. Then,

$$
\begin{aligned}
(x - S_f(y), y) &= \left(x - \sum_i c_i \circ S_f(y_i), y \right) \\
&= \left(x - \sum_i c_i \circ \left(C(y_i) - C'(y_i) \right), \sum_i c_i \circ y_i \right) \\
&= \left(x - \sum_i c_i \circ C(y_i), 0_y \right) + \left(\sum_i c_i \circ C'(y_i), \sum_i c_i \circ y_i \right) \\
&\in N(S) \times \{0_y\} + S' \\
&= N(S') \times \{0_y\} + S' \\
&= S'
\end{aligned}
\tag{11A.21}
$$

This completes the proof of Proposition 11A.1 (statement 3). QED.

11A.3 PROPERTIES THAT ARE FEEDBACK INVARIANT

A property of a linear system $S \subset X \times Y$ is referred to as feedback invariant (respectively, S_f-feedback invariant), if the property also holds true for the feedback system $F(S, S_f)$, for each $S_f \in \mathcal{S}_f$ (respectively, for some $S_f \in \mathcal{S}_f$). In this section, we will show what important properties are feedback invariant or S_f-feedback invariant.

Proposition 11A.2. Range space, null space, linearity, injectivity, surjectivity, and bijectivity of original systems are feedback invariant.

Proof. The argument for each of the properties listed is straightforward and is omitted. QED.

In the rest of this section, we will focus on the feedback invariant properties of Mesarovic and Takahara (MT) time systems. Let us first list some notations of MT time systems from (Mesarovic and Takahara, 1975). Let T be the positive half of the real number line, i.e., $T = [0, +\infty)$, and $\mathcal{A}$ and $\mathcal{B}$ be two linear spaces over the field $\mathcal{A}$. Define

$$
\mathcal{A}^T = \{x : x \text{ is a mapping } T \to \mathcal{A}\}
\tag{11A.22}
$$

$$
\mathcal{B}^T = \{y : y \text{ is a mapping } T \to \mathcal{B}\}
\tag{11A.23}
$$

Then, $\mathcal{A}^T$ and $\mathcal{B}^T$ can be made linear spaces over the field $\mathcal{A}$ as follows: $\forall f, g \in \mathcal{A}^T$ (respectively, $\in \mathcal{B}^T$) and $\forall \alpha \in \mathcal{A}$,

$$
(f + g)(t) = f(t) + g(t), \quad \forall t \in T
\tag{11A.24}
$$

$$(\alpha f)(t) = \alpha \circ f(t), \quad \forall t \in T \tag{11A.25}$$

Assume that the input and output spaces X and Y in the previous discussion are linear subspaces of $\mathcal{A}^T$ and $\mathcal{B}^T$, respectively. Each input-output system $S \subset \mathcal{A}^T \times \mathcal{B}^T$ is termed as an MT time system. A subset $S \subset X \times Y \subset \mathcal{A}^T \times \mathcal{B}^T$ is referred to as a linear time system (Mesarovic and Takahara, 1975), provided that S is a linear subspace of $X \times Y$ and the input space $D(S)$ satisfies the following conditions:

$$\forall x, x' \in X \quad \forall t \in T(x, x' \in D(S) \rightarrow x^t \circ x'_t \in D(S)) \tag{11A.26}$$

where $x^t = x|[0,t), x'_t = x|[t, +\infty)$ and $x^t \circ x'_t \in \mathcal{A}^T$ is called the concatenation of x^t and x'_t defined by

$$x^t \circ x'_t(s) = \begin{cases} x(s), & \text{if } s < t \\ x'(s), & \text{if } s \geq t \end{cases} \tag{11A.27}$$

Without loss of generality, we will in the following always assume that $D(S) = X$ and $R(S) = Y$. Let σ^τ be the shift operator defined as follows: For each $x \in X$,

$$\sigma^\tau(x)(\xi) = x(\xi - \tau), \quad \forall \xi \in [\tau, +\infty) \tag{11A.28}$$

where τ can be any real value if $\sigma^\tau(x)$ is meaningful.

Definition 11A.1. (Mesarovic and Takahara, 1975; Saito and Mesarovic, 1985). Let $S \subset X \times Y$ be a linear time system.

1   S is strongly stationary, provided

$$(\forall t \in T) \left(\sigma^{-t}(S|[t, +\infty)) = S \right) \tag{11A.29}$$

2   S is precausal, provided

$$(\forall t \in T)(\forall x \in X)(x|[0,t] = 0|[0,t] \rightarrow S(x)|[0,t] = S(0)|[0,t]) \tag{11A.30}$$

3   If $S: X \rightarrow Y$ is functional, then S is time invariantly realizable, provided

$$(\forall t \in T)(\forall x \in X)\left(\lambda^t S(0^t \circ \sigma^t(x)) = S(x)\right) \tag{11A.31}$$

where $\lambda^t(\circ) = \sigma^{-t}(\circ|(t, +\infty))$.

Theorem 11A.1. The following statements hold true:

1   Let $S \subset X \times Y$ and $S_f: Y \rightarrow X$ be a linear time system and a strongly stationary linear time system, respectively. The feedback system $F(S, S_f)$ is strongly stationary, if and only if the original system S is strongly stationary.

2   Let $S: X \rightarrow Y$ and $S_f: Y \rightarrow X$ be linear functional precausal time systems. The feedback system $F(S, S_f)$ is precausal, if and only if the composition time system $S \circ S_f \circ F(S, S_f): X \rightarrow Y$ is precausal.

3   Suppose that $S: X \to Y$ and $S_f: Y \to X$ are linear functional time systems such that S and S_f are time invariantly realizable, then the feedback system $F(S, S_f)$ is time invariantly realizable, if and only if the system $S \circ S_f \circ F(S, S_f)$ is time invariantly realizable.

Proof.   The arguments are very technical and are omitted here. QED.

Definition 11A.2. (Zhu and Wu, 1987). Let $S \subset X \times X$ be a binary system and $D \subset X$ a subset.

1   If $D^2 \cap S = \emptyset$, then the subset D is known as a chaos of the input-output system S.
2   If for each $x \in X - D$, $S(x) \cap D \neq \emptyset$, then the subset D is known as an attractor of the system S.
3   If D is both a chaos and an attractor of the system S, then the subset D is known as a strange attractor of S.

Proposition 11A.3. For each linear system $S \subset X \times X$, there is a feedback component system $S_f: X \to X$ such that a subset $D \subset X$ is a chaos, attractor, or strange attractor, respectively, of the system S, if and only if the subset D is a chaos, attractor, or a strange attractor of the feedback system $F(S, S_f)$, respectively.

Proof.   The arguments follow from the fact that each linear system $S \subset X \times X$ is a feedback system of itself. QED.

11A.4   A CHARACTERIZATION OF DECOUPLING

Let $S \subset X \times Y$ and the input and output spaces be

$$X = \prod \{X_i : i \in I\} \quad \text{and} \quad Y = \prod \{Y_i : i \in I\} \tag{11A.32}$$

For each $i \in I$, define the projection mapping

$$p_i = (p_{ix}, p_{iy}): X \times Y \to X_i \times Y_i \tag{11A.33}$$

by letting $p_{ix}: X \to X_i$ and $p_{iy}: Y \to Y_i$ satisfying $p_{ix}(x) = x_i$ and $p_{iy}(y) = y_i$, for each $x \in X$ and $y \in Y$.

If we write $S_i = p_i(S)$, for each $i \in I$, the system S is decomposed into a family of factor systems $\bar{S} = \{S_i : i \in I\}$. In general, if we identify each element $((x_i)_{i \in I}, (y_i)_{i \in I}) \in S$ with $((x_i, y_i)_{i \in I}) \in \prod \{S_i : i \in I\}$, it is obvious that $S \subset \prod \{S_i : i \in I\}$. If $S = \prod \{S_i : i \in I\}$, the system S is referred to as non-interacted (Saito and Mesarovic, 1985). The following theorem lists the properties that are transferrable both from the 'whole' system S to the factor systems S_i and from the S_i's to S.

Theorem 11A.2. (Saito and Mesarovic, 1985). Let $S \subset X \times Y$ be an input-output system that has been decomposed into a family $\bar{S} = \{S_i : i \in I\}$ of factor systems such that $S = \{S_i : i \in I\}$. Then the following hold true:

1   S is a linear system, if and only if each S_i is a linear system.
2   S is a linear time system, if and only if each S_i is a linear time system.
3   S is strongly stationary, if and only if each S_i is strongly stationary.

4   S is precausal, if and only if each S_i is precausal.
5   Functional system $S: X \to Y$ is time invariantly realizable, if and only if each S_i is time invariantly realizable. QED.

The significance of this result is that the 'whole' system S and the factor systems S_i's behave the same. Because of this important feature of the non-interacted systems, it is interesting and meaningful to study the following question: Since in the general case we do not have the property $S = \prod\{S_i : i \in I\}$, can we find a feedback component system $S_f \subset Y \times X$ such that $F(S, S_f)$ is non-interacted? Equivalently, the question can be recast as follows: Under what conditions is there a feedback system $S_f \subset Y \times X$ such that the behavior of the system S can be studied through the behavior of the factor systems of $F(S, S_f)$? By virtue of Section 11A.3, the properties of the original system S can be inherited by the feedback system $F(S, S_f)$.

Definition 11A.3. (Saito and Mesarovic, 1985). Let $S \subset X \times Y$ be a linear system. S is said to be decoupled by feedback, if there is an $S_f \in \mathcal{S}_f$ such that

$$F(S, S_f) = \prod \{p_i(F(S, S_f)) : i \in I\} \tag{11A.34}$$

Decoupling by feedback means to take the following actions. The original system S may not be non-interacted, there are then properties that cannot be lifted up from the factor systems S_i's to the 'whole' system S. If the system S can be decoupled by feedback, after applying appropriate feedback transformation, the system S is transferred into a system that is non-interacted. We now can use this non-interacted system representation to analyze and to control the original system S.

Theorem 11A.3. Assume the axiom of choice. Suppose that $S \subset X \times Y$ is a linear system such that S is decomposed into a family of factor systems $\bar{S} = \{S_i : i \in I\}$. The system S can be decoupled by feedback, if and only if $R(S) = \prod\{R(S_i) : i \in I\}$ and $N(S) = \prod\{N(S_i) : i \in I\}$.

Proof.   Necessity: Suppose that S can be decoupled by feedback. By the definition of decoupling by feedback, there is a functional linear system $S_f \in \mathcal{S}_f$ such that

$$F(S, S_f) = \prod \{p_i(F(S, S_f)) : i \in I\} \tag{11A.35}$$

Therefore, by Proposition 11A.1 (statement 3) we have the following:

$$\begin{aligned}
R(S) &= R(F(S, S_f)) \\
&= \prod \{R(p_i(F(S, S_f))) : i \in I\} \\
&= \prod \{R(F(S_i, S_{fi})) : i \in I\} \\
&= \prod \{R(S_i) : i \in I\}
\end{aligned} \tag{11A.36}$$

and

$$N(S) = N(F(S, S_f))$$

$$= \prod \{N(p_i(F(S, S_f))) : i \in I\}$$

$$= \prod \{N(F(S_i, S_{fi})) : i \in I\}$$

$$= \prod \{N(S_i) : i \in I\} \tag{11A.37}$$

Sufficiency. Let $S' \subset X \times Y$ be the system defined by

$$S' = \prod \{S_i : i \in I\} \tag{11A.38}$$

Because $R(S) = \prod \{R(S_i) : i \in I\}$ and $N(S) = \prod \{N(S_i) : i \in I\}$ due to our assumption, it can be seen that $R(S) = R(S')$ and $N(S) = N(S')$. By applying Proposition 11A.1 (statement 3), there ia functional linear system $S_f \in \mathcal{S}_f$ such that

$$F(S, S_f) = S' = \prod \{S_i : i \in I\} = \prod \{p_i(F(S, S_f)) : i \in I\} \tag{11A.39}$$

That is, the system S can be decoupled by feedback. QED.

11A.5 THE MAIN THEOREM

We are now ready to show how the multivariable linear system (11A.1) can be decoupled into n linear systems of the same kind with one-dimensional input and output. To this end, let S be the system defined as follows:

$$S = \{(x, y) \in X \times Y : \exists z \in Z(x, y, z \text{ satisfy equ. (11A.1)})\} \tag{11A.40}$$

where $X = Y = \{r : [0, +\infty) \to R^n : r \text{ is piecewise continuous}\}$ and $Z = \{r : [0, +\infty) \to R^m : r \text{ is piecewise continuous}\}$.

For each $i = 1, 2, \ldots, n$, define a system S_i as follows:

$$\begin{cases} \dot{z} = Az + B_i x_i \\ y_i = C_i z + D_i x_i \\ z(0) = 0 \end{cases} \tag{11A.41}$$

where B_i is an $m \times 1$ constant matrix such that $B = [B_1 \ B_2 \ldots B_n]$, C_i a $1 \times m$ constant matrix satisfying

$$C = [C_1 C_2 \ldots C_n]^T \tag{11A.42}$$

D_i is a non-zero constant, and the input space X_i and the output space Y_i are given by

$$X_i = \{x_i : [0, +\infty) \to R : x_i \text{ is piecewise continuous}\} \tag{11A.43}$$

$$Y_i = \{y_i : [0, +\infty) \to R : y_i \text{ is piecewise continuous}\} \tag{11A.44}$$

Proposition 11A.4. The systems S and S_i, $i = 1, 2, \ldots, n$, satisfy that

$$R(S) = \prod \{R(S_i) : i = 1, 2, \ldots, n\} \tag{11A.45}$$

$$N(S) = \prod \{N(S_i) : i = 1, 2, \ldots, n\} \tag{11A.46}$$

Proof. For each $x \in X$, the solution of the differential equation in equ. (11A.1) is given by

$$z = \phi(t, 0, x) = \int_0^t e^{(t-s)A} Bx(s)ds \tag{11A.47}$$

Therefore, the corresponding output y can be computed by using

$$y = C\phi(t, 0, x) + Dx \tag{11A.48}$$

That is, it is shown that $D(S) = X$. At the same time, because the matrix D is non-singular, the inverse system S^{-1} of S is given as follows:

$$S^{-1} = \{(y, x) \in Y \times X : \exists z \in Z(x, y, z \text{ satisfy equ. (11A.50)})\} \tag{11A.49}$$

where equ. (11A.50) is given below:

$$\begin{cases} \dot{z} = (A - BD^{-1}C) z + BD^{-1}y \\ x = -D^{-1}Cz + D^{-1}y \\ z(0) = 0 \end{cases} \tag{11A.50}$$

Therefore, a similar argument as above shows that $R(S) = D(S^{-1}) = Y$. For the same reason, it can be shown that for each $i = 1, 2, \ldots, n$, $R(S_i) = Y_i$. That is,

$$R(S) = Y = \prod \{Y_i : i = 1, 2, \ldots, n\} = \prod \{R(S_i) : i = 1, 2, \ldots, n\} \tag{11A.51}$$

For the null spaces of the systems S and S_i, $i = 1, 2, \ldots, n$, we have

$$N(S) = \{-D^{-1}Cz : \dot{z} = (A - BD^{-1}C)z\} \tag{11A.52}$$

and

$$N(S_i) = \{-D_i^{-1}C_i z : \dot{z} = (A - B_i D_i^{-1} C_i)z\} \tag{11A.53}$$

Therefore, $N(S) = \prod \{N(S_i) : i = 1, 2, \ldots, n\}$. QED.

Theorem 11A.4. The system S can be decoupled by feedback into factor systems S_i, $i = 1, 2, \ldots, n$.

Proof. The argument is based on Theorem 11A.3 and Proposition 11A.4. All the details are omitted. QED.

To conclude this appendix, we will describe a detailed procedure on how to decouple the system S by feedback into the factor systems S_i, $i = 1, 2, \ldots, n$. To this end, let us define

$$
\alpha = \begin{bmatrix} A & & & 0 \\ & A & & \\ & & \ddots & \\ 0 & & & A \end{bmatrix}, \quad \beta = \begin{bmatrix} B_1 & & & 0 \\ & B_2 & & \\ & & \ddots & \\ 0 & & & B_n \end{bmatrix},
$$

$$
\gamma = \begin{bmatrix} C_1 & & & 0 \\ & C_2 & & \\ & & \ddots & \\ 0 & & & C_n \end{bmatrix}, \quad \delta = \begin{bmatrix} D_1 & & & 0 \\ & D_2 & & \\ & & \ddots & \\ 0 & & & D_n \end{bmatrix}.
$$

Then the system $S_d = S_1 \times S_2 \times \cdots \times S_n$ is represented by the set of all ordered pairs (x, y) satisfying

$$
\begin{cases} \dot{z} = \alpha z + \beta x \\ y = \gamma z + \delta x \\ z(0) = 0 \end{cases} \tag{11A.54}
$$

Because δ is non-singular, S_d^{-1} is obtained as follows: $= \{(y, x): (y, x)$ satisfies equ. (11A.55)$\}$, where

$$
\begin{cases} \dot{z} = \left(\alpha - \beta \delta^{-1} \gamma \right) z + \beta \delta^{-1} y \\ x = -\delta^{-1} \gamma z + \delta^{-1} y \\ z(0) = 0 \end{cases} \tag{11A.55}
$$

We now define a functional linear system $S_f \colon Y \to X$ as follows:

$$
\begin{cases} \begin{bmatrix} \dot{z} \\ \dot{z}' \end{bmatrix} = \begin{bmatrix} A - BD^{-1}C & 0 \\ 0 & \alpha - \beta \delta^{-1} \gamma \end{bmatrix} \begin{bmatrix} z \\ z' \end{bmatrix} + \begin{bmatrix} BD^{-1} \\ \beta \delta^{-1} \end{bmatrix} y \\ x = \begin{bmatrix} -D^{-1}C & \delta^{-1} \gamma \end{bmatrix} \begin{bmatrix} z \\ z' \end{bmatrix} + (D^{-1} - \delta^{-1}) y \\ \begin{bmatrix} \dot{z} \\ \dot{z}' \end{bmatrix}(0) = 0 \end{cases} \tag{11A.56}
$$

Proposition 11A.5. The original system S can be transformed into S_d by the feedback component system S_f.

Proof.   It suffices to show that for each $(x, y) \in X \times Y$, $(x, y) \in S_d$, if and only if the input x and the output y satisfy the system representation equ. (11A.55), if and only if $(x + S_f(y), y) \in S$. The second "if and only if" condition is shown as follows.

First, suppose that the input x and the output y satisfy the system representation (11A.55). Let z' be the solution of the differential equation in the representation and z the solution of the differential equation in the systems representation (11A.50).

Then $[z \ z']^T$ is the solution of the differential equation in the systems representation (11A.56). From the definition of S_f, it follows that

$$S_f(y) = -D^{-1}Cz + \delta^{-1}\gamma z' + D^{-1}y - \delta^{-1}y$$
$$= \left(-D^{-1}Cz + D^{-1}y\right) - \left(-\delta^{-1}\gamma z' + \delta^{-1}y\right) \tag{11A.57}$$

Therefore, we have that

$$\left(x + S_f(y), y\right) = \left(-D^{-1}Cz + D^{-1}y, y\right) \in S \tag{11A.58}$$

where the last membership relation comes from the systems representation (11A.49) of the inverse system S^{-1}.

Secondly, suppose that a pair $(x, y) \in X \times Y$ satisfies that $(x + S_f(y), y) \in S$. From the systems representations (11A.49) and (11A.56), we have that

$$x + S_f(y) = -D^{-1}Cz + D^{-1}y \tag{11A.59}$$

That is,

$$x = \left(-D^{-1}Cz + D^{-1}y\right) - S_f(y)$$
$$= \left(-D^{-1}Cz + D^{-1}y\right) - \left(-D^{-1}Cz + \delta^{-1}\gamma z' + D^{-1}y - \delta^{-1}y\right)$$
$$= -\delta^{-1}\gamma z' + \delta^{-1}y \tag{11A.60}$$

By combining equ. (11A.60) with the systems representation equ. (11A.55), we have proved that the pair (x, y) satisfies the systems representation equ. (11A.55). QED.

Chapter 12

Modern China: A quick glance

As a new, heavy weight player in the international arena, it is very important for other members of this arena to understand China with reasonable expectations in order to bring about mutually beneficial economic consequences. To achieve this goal, this chapter is devoted to the discussion of modern China by looking at its current state of economic and financial operations in particular and by looking at China at the height of civilization in general. By doing so, it is expected that all the players of the international economy and the international financial system could understand how decisions are made in China and what could be expected from Chinese leadership.

This chapter is organized as follows. In Section 12.1, after providing a brief development history of the central banks of China, it introduces the current hierarchy of Chinese central bank, the business coverage of the People's Bank of China, and how money is supplied in China. Section 12.2 is devoted to the study of Chinese financial system by looking at Chinese banks, different kinds of financial organizations, and the practice of lending and investing. Section 12.3 focuses on the financial reform China has embarked on. In Section 12.4 the international reserve of China is the topic of discussion. It looks at how fast the foreign reserves in China have been increasing, how these foreign reserves are managed, and how the risk associated with the large foreign reserves could potentially be avoided. Section 12.5 attempts to understand China at the height of civilization in order to understand how China might behave differently from the norm of the Western nations in the international economy and the international financial system. The concluding Section 12.6 looks at the question: Will economic prosperity visit China soon?

The presentation of this chapter has made good use of (Wang, 2005; Wu, 2006; Jiang, 2008; Lin and Forrest, 2011) and references listed therein.

12.1 APPEARANCE AND DEVELOPMENT OF CHINESE CENTRAL BANK

The goal of this section is to familiarize the reader with the short history of Chinese central banks, and then with the current hierarchy of Chinese central bank, the business coverage of the People's Bank of China, and how money is supplied in China.

12.1.1 A brief history

In 1905, Qing government established its first central bank of China, named the Bank of Ministry of Finance. In 1908, this bank was renamed as Bank of Great Qing. During the time periods of Xinhai Revolution, which was the domestic revolution of capitalists that broke out in 1911 and ended in 1912, and Beiyang (or Warlords) Government (1912–1928), in 1908, Bank of Transportation was established, and was later evolved from a commercial bank into the central bank of Beiyang Government; and in 1911, Bank of Great Qing was reorganized into Bank of China.

In 1924, under the leadership of Sun Yat-sen, a new Chinese central bank was established in Guangzhou; and in 1926, a different central bank was established in Hankou.

During the time of Guomindang, a central bank was established and opened to business on November 1, 1928. Together with Bank of China, Bank of Transportation, and Chinese Farmer's Bank, this central bank occupied the position of financial center that was monopolized by the government of Guomindang.

In the revolutionary bases of the communist party, a central bank, named the Soviet National Bank, was established in February 1932. In November 1935, the northwestern branch of the Soviet National Bank was established. And in 1948, the People's Bank of China was established.

With the birth of the communist modern China, during 1948–1978, a greatly unified composite central bank system that fulfills two duties at the same time was established. During 1979–1983, the greatly unified system was improved so that various specialized banks were either revived or reestablished one by one. During 1984–1998, the People's Bank of China started to only focus on the responsibilities of a central bank. After 1998, nine inter-provincial first-tier branch banks were introduced (October 1998), China Banking Regulatory Commission became independent from the central bank on April 28, 2003, and the People's Bank of China began its sole functionality of issuing currency, making and carrying out monetary policies.

12.1.2 Hierarchy of Chinese central bank

Currently, the organization of Chinese central bank consists of four levels: The headquarter bank, branch banks of large districts, city center banks, and county branch banks. The nine large district branch banks include Tianjin Branch Bank, Shenyang Branch Bank, Xi'an Branch Bank, Jinan Branch Bank, Shanghai Branch Bank, Nanjing Branch Bank, Guangzhou Branch Bank, Chengdu Branch Bank, and Wuhan Branch Bank. Additionally, in both Beijing and Chongqing two business management departments are established. Each and every branch organization of the People's Bank of China is a field office of the headquarters, its main responsibilities are authorized directly by the headquarter office and are carried out within its specialized areas, representing the central bank.

There are three channels for Chinese central bank, representing the asset business of the bank, to issue currency: (1) provide loans to commercial banks and other financial institutions; (2) rediscount commercial papers; and (3) purchase gold and foreign currencies. The promises behind the issuance of Chinese currency include the fully backed system, fixed guarantee system, maximum issuance system, proportional issuance system, etc. The issuance of Renmingbi (RMB) is not required by law to have any issuance guarantee. However, it has been practically done on the principle of credit issuance, meaning that the issuance of RMB is guaranteed by the credit worthiness of the nation and that of the central bank.

The current state of payment and settlement system in China consists of same-city clearance, nationally computerized banking system, and the system of bank cards. Since the 1980s, Chinese financial monitoring system has gradually evolved from the original, single doing-all entity into a multi-layered hierarchical organization, where different units are specialized in different sectorial monitoring responsibilities. Starting in 1995, the system of sectorial monitoring system, consisting of the central bank, Security and Futures Commission, China Insurance Regulatory Commission, and China Banking Regulatory Commission has been basically formed. Since 1998, the central bank has been authorized to monitor all commercial banks, trusts and investment companies, credit unions, accounting firms, and financial leasing companies, where commercial banks constitute the focus of monitoring.

12.1.3 Business of the people's bank of China

As clearly spelled out in the banking law of China, the business of the People's Bank of China include the following activities:

1   In order to carry out monetary policies, the people's Bank of China can make use of the following monetary policy tools: (i) Demand financial institutions to hand over their required deposit reserves according to the regulation; (ii) Determine the base interest rate of the central bank; (iii) Provide rediscount services to all financial institutions that have accounts with the People's Bank of China; (iv) Provide loans to commercial banks; (v) Purchase and sell national debts, governmental bonds, and foreign currencies on the open market; (iv) Apply any other monetary policy tools as authorized by the Department of State from time to time.

2   According to the law and administrative regulations, the People's Bank of China manages the national treasury.

3   On the behalf of the accounting department of the Department of State, the People's Bank of China can issue and repay national debts and other governmental bonds to other financial institutions.

4   Based on need, the People's Bank of China can establish accounts for financial institutions. However, no financial institutions can overdraw against their accounts.

5   The People's Bank of China organizes or helps to organize the clearance system for financial institutions, coordinates the clearance business between financial institutions, and provides clearance services.

6   Based on the need that arises when attempting to carry out monetary policies, the People's Bank of China can determine the amounts, terms, interest rates, and forms of loans given to commercial banks. However, the terms of loans cannot be longer than one year.

7   The People's Bank of China cannot overdraw against the government finance, cannot directly subscribe or underwrite national debts and other governmental bonds.

8   The People's Bank of China cannot provide loans to local governments and agencies of governments of different levels, and cannot give loans to non-banking financial institutions, other organizations, and individuals. But, it can provide

loans to certain non-banking financial institutions as specified from time to time by the Department of State

9   The People's Bank of China cannot provide guarantee to any company or individual.

12.1.4 Money supply of China

Other than cash, the media of trades and payments in Chinese economic activities also include settlements by transfers between individual organizations and firms. The base of settlement by transfers is the demand deposits of these organizations and firms in their banks. Because of this reason, many scholars recognize that the money supply in China should be defined as follows:

$$
\begin{aligned}
M_1 &= \text{cash} + \text{demand deposits that can be transferred} \\
&= M_0 + \text{demand deposits of commercial firms} \\
&\quad + \text{demand deposits of government entities} \\
&\quad + \text{demand deposits of basic constructions} \qquad (12.1)
\end{aligned}
$$

The demand and time deposits of individual Chinese residents cannot be included in the computation of M_l. Instead, they should be included in M_2, because in China residents cannot write checks against their demand deposits in banks. The liquidity of time deposits of commercial and governmental agencies is very low. So, these deposits naturally belong to M_2. Fiscal deposits can only materialize their purchasing power after the funds are allocated to business units and institutions. So, these deposits should also belong to the range of M_2. Therefore, the definition of general money M_2 for China should be

$$
\begin{aligned}
M_2 &= M_1 + \text{savings deposits of individual residents} \\
&\quad + \text{time deposits of business units and institutions} \\
&\quad + \text{fiscal deposits} \qquad (12.2)
\end{aligned}
$$

There are also some scholars who classify demand deposits of individual residents into the calculation of M_1, believing that because these deposits can be withdrawn at any time the purchasing power of these funds can be materialized instantly with little costs. That makes using these funds roughly the same as writing checks against these funds. So, they should be considered in the calculation of M_1 instead of M_2.

In calculating the total quantity of money, each category occupies a specific proportion in the overall computation. These specific proportions are jointly known as the composite of the total quantity of money. Through analyzing the composite of the total quantity, one can see which monetary asset, whether it is goods or labor, that plays a greater role in the payment activities. Within Chinese monetary composite of M_1 in 1987 cash (currency) only accounted for about one fourth of the total. It means that cash transactions amounted for only a very small part of the overall landscape of the economic activities. The similar situation also holds true for the United States. For example, in 1993 cash only occupied about 35% in the composite of American M_1.

12.2 CHINESE FINANCIAL SYSTEM

Reform the recent reform and opening up of China, Chinese economic system has gone through a complete reorganization with its financial system being reformulated and adjusted. The reformulation of Chinese financial system consists roughly of two stages. The first stage took place during the time period from 1979 to 1989. In that period of time, the central bank and specialized banks are separated; various different kinds of financial institutions were initially introduced. The second stage represents the further reform since the year 1990 when the People's Bank of China started to function as the central bank and to perform various macroscopic financial adjustments to the economy, and the original specialized banks began to be transformed to nationally owned commercial banks. Also in this stage of development, commercial banks evolved further, and different kinds of financial institutions started to appear, leading to a new financial landscape that consists of many kinds of financial organizations with major weights placed on the nationally owned commercial banks that center around the People's Bank of China as the core.

12.2.1 Chinese banks

In July 1982, Department of State authorized the People's Bank of China to function as the nation's central bank. However, due to various reasons, it has not truly played its role as the central bank. It was until January 1986 that Department of State issued a document, entitled "Provisional Regulations of the People's Republic of China". This document clearly stated that the People's Bank of China is an entity of the national government directly under the authority of Department of State and that the bank manages the sector of national finance, and is the central bank of China. It was then that the position for the People's Bank of China to function as the central bank is lawfully determined. After entering into the 1990s, the People's Bank of China started to truly perform its various duties of macroscopically monitoring and adjusting the national finance.

Currently, the commercial banking system of China consists of two parts. One part is made up of the nationally owned commercial banks that were converted earlier from the original nationally owned specialized banks. The other part contains the general commercial banks newly established during the process of reform and opening up of China. In particular, the Bank of China has been transformed into a nationally owned commercial bank, headquartered in Beijing, with its governor responsible for daily operations under the auspices of the board of directors. Although its business focus is the management of foreign currencies and clearance of foreign trades, its range of business is no longer limited by specialties. It can participate in all business activities commercial banks are allowed to do with emphasis placed on developing domestic RMB deposit business and providing comprehensive financial services. Currently, Bank of China has developed into such a nationally owned commercial bank with the amount of assets only behind that of Industrial and Commercial Bank of China. It has been listed as one of the major banks from around the world.

Industrial and Commercial Bank of China is the last nationally owned commercial bank with the largest amount of assets. It was initially established in January 1984 as approved by Department of State. It was originally a nationally owned specialty bank, focusing only on industrial and commercial loans, savings, and settlement services, which were earlier handled by the People's Bank of China. Although its main business is to provide domestic financial services, it also involves itself in international transactions. Currently, Industrial and Commercial Bank of China has been transformed into a nationally owned commercial bank with services in such areas as attracting deposits, giving off loans of various kinds, clearing financial transactions, playing as an agent to issue bonds and to trade foreign currencies, and other services that international commercial banks involve in. And its range of business activities is still constantly expanding.

Starting with the reform and opening up of China, in order to quickly modernize the very obsolete agriculture the central government proposed in February 1979 to revive Agricultural Bank of China and to develop the credit business in the countryside. Since then, Agricultural Bank of China has played an important role in the development of financial business in the countryside and in the management of the assets of its branches, which in turn has helped to make the bank itself evolve greatly. Currently, Agricultural Bank of China has been evolved into a comprehensive nationally owned commercial bank, providing all kinds of services an international commercial bank provides, including

Savings deposits for the countryside,
Various rural credits,
The management of various government funds allocated to the development of agriculture,
The settlements by transfers and management of cash for the countryside,
The organization and management of credit unions in the countryside,
Trusts and credits of the countryside,
Engagement in nationally own farms, and
Lending to and cash management for supply and marketing cooperatives.

It has also been involved in the clearance of foreign transactions and international trades, making it a nationally designated bank for foreign exchange settlements, and has started to explore business opportunities abroad by opening branch offices in foreign countries.

Before the year 1985, Bank of Construction had always been an institution that managed various investments under the auspices of Treasury Department. It oversaw the appropriations allocated for constructions of infrastructures within the national budget and managed all the construction loans. After the year 1985, the credit business of the Bank of Construction was included into the modern management system of the People's Bank of China; and Bank of Construction is required to maintain its deposit reserves at the People's Bank of China. That has made Bank of Construction a nationally owned specialty bank focusing on the business of investment management of and loans against real estate properties. After the year 1994, Bank of Construction has been reformed into a comprehensive nationally owned commercial bank. Its business has been expanded beyond the limitation of investing and giving out loans for real estate assets into the general business areas of commercial banks, making its balance sheet completely different from before.

As for other commercial banks, they exist in the form of shareholdings, where shareholders can be private citizens, business entities, in additional to the country. Instead of being limited by any specialty, they generally offer various financial products and provide comprehensive financial services. Some of these banks are national, while others could also be either regional or local. For example,

(i)   The reconstructed Bank of Transportation is a comprehensive commercial bank held by shareholders based mainly on national assets. Its headquarters was moved from Beijing to Shanghai on April 1, 1987. Its service is not limited to any specialty. The development and growth of this bank have provided successful lessons for other Chinese commercial banks.

(ii) CITIC (China International Trust and Investment Corporation) Industrial Bank is a comprehensive commercial bank that was launched in February 1987 on the basis of the banking department of China Trust and Investment Corporation with the approval of the State Department. Its business operation, which was started officially in April of the same year, is not limited by either any specialty or specific geographic region. Its organization takes the form of governor responsibility system under the supervision of a board of directors.

(iii) China Everbright Bank was established with approval from the State Department on February 9, 1992. It was permitted by the People's Bank of China to open business on April 25, 1992. It is a nationally owned financial company with its initial funds completely provided by China Everbright Corporation. This bank employs independent accounting and is completely self-managed and self-financed, and is under the centralized leadership of the People's Bank of China. It mainly engages in the business of issuing jumbo equipment loans for such industries as machineries and electricity, energy, transportation, etc., and other categories of business, such as leasing aircrafts.

(iv) Huaxia Bank is the first national comprehensive commercial bank organized by industrial firms and approved by the State Department. It opened its door for business on December 22, 1992. It is a financial institution wholly owned by shareholders and wholly financed by Capital Iron and Steel Corporation. Administratively, this bank is under the supervision of Capital Iron and Steel Corporation; and in terms of business, it is under the monitoring of the People's Bank of China. In terms of business, it employs independent accounting and is self-managed and financed.

(v)   Along with the deepening of the reform of the economic system and the development of the economic market, other than the afore-mentioned national commercial banks, there have also appeared a group of regional and local commercial banks, such as the Merchants Bank in Shengzheng special economic zone, the Development Bank of Guangdong, the Industrial Bank of Fujiang, the Development Bank of Pudong, the Development Bank of Hainan, (local) Savings Banks of Housing, etc. The rapid development of commercial banks means that the monopoly of nationally owned commercial banks has been broken, and that Chinese financial industry has stepped forward toward multitude in the direction of competition.

Since the year 1994, in order to adapt to the transition of nationally owned specialty banks to commercial banks and to separate policy-related businesses from commercial businesses, three nationally owned policy-related banks that would be under the direct auspices of the State Department were officially organized: National Development Bank, Import-Export Bank of China, and Agricultural Development Bank of China, where the main task of the National Development Bank is to establish stable sources of capital for major projects of national construction by attracting and guiding societal flows of money. The source of funds of the National Development Bank is mainly from the fiscal appropriation of the government and issuance of financial bonds. The main task of the Import-Export Bank of China is to provide policy-related financial supports for the imports and exports of such capital-heavy goods as mechanical and electrical products, complete sets of equipment, etc., according to the national industrial policies and foreign trade policies. Its capital is financed by the national treasury, and its source of capital is mainly from national treasury bonds sold both inside and outside China. The main task of the Agricultural Development Bank of China is to provide policy-related supports to basic constructions and major engineering projects of agriculture and the production and circulation of agricultural products. Its capital is from the national treasury allocation, and its source of capital is mainly from the issuance of financial bonds both inside and outside China.

12.2.2 Other financial organizations

Currently, non-banking financial institutions are also developed greatly in China, constituting an important aspect of Chinese financial system. These non-banking financial institutions include: trust and investment institutions, insurance companies, cooperative financial organizations, savings institutions of the postal service, financing firms, leasing firms, accounting firms, and security firms.

Specifically, at the present time, there are more than 300 trust and investment companies in China, totaling more than 4000 billion Yuan in assets. The development in this area has shown the tendency of multitude. There are trust and investment companies that are organized by the national government, trust and investment companies that are organized by various business entities, and attached entities funded by banks. The businesses of some of these companies cover the entire nation, while the businesses of others are only regional.

As for the present insurance industry in China, there are three major domestic insurance companies. People's Insurance Company of China is a nationally owned insurance company that offers various domestic and international insurance and reinsurance products. It is the largest insurance company in China. Its business covers all areas of the insurance industry. China Ping-An Insurance Company is a shareholding company that is initially financed jointly by the Bureau of Merchants, the Industrial and Commercial Bank of China, Ocean Shipping Company of China and the Finance Bureau of Shenzhen City. Its business covers all areas of commercial and property insurance. Pacific Insurance Company of China is initially financed by the Bank of Transportation. It provides all forms of commercial insurance. This company does its business mainly in the coastal cities of open economic policies.

Cooperative financial organizations mainly include credit cooperative institutions in China, including the credit unions located in either the countryside or the urban areas. They do not belong to the banking industry. However, their business is quite similar to those of commercial banks that offer deposit services. For example, the savings institution of the postal service is established by Chinese Postal Service. It targets and attracts deposits of individual residents. In developed countries, such savings institutions have enjoyed a history of hundreds of years. In China, it started to appear in 1986. All of the savings deposits of the Postal Service are handed over to the People's Bank for its coordinated use. When the People's Bank has a limited need for cash, it offers the additional funds to other financial institutions that are in need of cash.

Financing and financial leasing companies are those financial institutions that specialize in financing and leasing. For example, accounting firms are such financial firms that specialize in helping circulate the funds between different units of a business organization. Starting at the onset of reform and opening up in China, most of the Chinese financial leasing companies are financed by Chinese financial trusts solely or jointly with foreign legal entities. Finance companies are organized by banks under the auspices of the People's Bank. Accounting firms are developed on the basis of the original clearance centers internal to individual business organizations.

Chinese financial reform has provided the golden opportunity for these three non-banking financial institutions to develop rapidly. Chinese security companies started to appear on top of the rapid development of financial markets and continuous expansion of the securities sector. As of the present day, there have formed national and regional security companies. Among these firms, some are financed by trusts, while some others are financed by financial institutions. Some of these companies are wholly owned, while some others are shareholding. All of their business conducts are under the supervision of the national securities commissions.

12.2.3 Business of lending and investing

Chinese law on commercial banks divides the loan business into two categories. One category is about commercial banks issuing financial bonds or borrowing money from outside China; and the other category is about interbank lending. For the former category of money borrowings, commercial banks must first obtain approval according to the relevant law and administrative regulations before they can issue bonds and/or take loans from abroad. As for interbank lending, banks must comply with the terms as specified by the People's Bank of China. Generally, the terms of lending cannot be beyond four months.

Industrial and commercial loans constitute the largest portion of the lending business of commercial banks. They belong to whole-sale lending. In China, industrial and commercial loans represent the most important form of loans given out by commercial banks. These loans can be grouped into three classes: 1) short-term loans of liquid capital, also known as loans of seasonal liquid capital. They are short-term loans, mainly used to cover temporary or seasonal needs of the general liquid capital of industrial companies and commercial firms; 2) long-term loans of liquid capital. They are mid-term loans, mainly utilized to meet the circulative needs of long-term liquid capital of industrial companies and commercial firms; and 3) project loans. They are large long-term loans, generally involving constructions projects of great risk and high costs.

Consumer loans banks provide to individual persons. They are mainly used for purchasing durable goods and/or for paying toward the loans taken out for other purposes. They belong to the category of retail loans, also known as consumption credit loans. This business of lending is commonly seen in industrialized nations, where people of the working class take out bank loans to purchase houses or automobiles with periodic repayment plans.

In China, the law of commercial banks stipulates that commercial banks are not allowed to involve in businesses that are related to stock trading. So, when commercial banks are said to be investing in securities, it means that they invest in bonds.

Bank Credit cards stand for a method for banks to give out consumer loans. By issuing credit cards to individuals, banks literally provide consumers with the convenience of consumption first and payment second. And in general, consumers are allowed to go beyond their credit limits slightly. There are two largest credit card organizations in the world. One is the Visa Group, consisting of Bank of America and other banks from 30 plus different countries, that issues the visa cards; and the other is the American multinational financial services corporation, which issues the master cards. At the present, China has already joined these two organizations.

The capital of Chinese commercial banks include four classes: funds that are actually collected, capital funds, surplus reserves, and profits that are yet to be distributed. By the funds that are actually collected, it means the various types of assets that investors actually invested in banks' products. They stand for the original inputs of the banks from the investors. Based on the differences of the ownerships of the assets, the funds that are actually collected can be grouped into national investments, other units' investments, individual residents' investments, and foreign investments. The funds that are actually collected can be in different forms depending on in which form the investors invested in the bank assets, including monetary investments, physical objects investments, and investments of assets that have no shape. In particular, by monetary investments, it means those investments in the forms of RMB, bank deposits, or foreign currency, etc. By physical objects investments, it means those investments that have tangible shapes, such as residential houses, buildings, machines, etc. By assets that have no shape, it includes such shapeless assets as patents, proprietary technology, use rights of certain spaces, commercial reputations, etc. Capital funds represent the increased value of assets incurred through non-business operations of the banks, such as capital premium experienced when banks are raising capital, gifts of cash or physical objects received by banks, increased re-evaluation values of banks' assets, etc. Surplus reserves denote the banks' shares of the funds collected from the after-tax profits based on the relevant agreements. They can be immediately used either to pay off the losses of the banks or to increase the levels of the banks' assets. By profits that are yet to be distributed, it means the profits that remain after the total profits at the end of the fiscal year are categorized into relevant forms.

The deposit service of Chinese banks can be classified into company deposits and savings deposits. By company deposits, it means those funds deposited by business units and institutions, governmental agencies, social organizations, military units, schools, etc. By savings deposits, it represents the funds deposited by individual residents that are kept either as savings from daily expenses or as ear-marked funds only to be applied for pre-determined purposes. Company deposits can be divided according to their terms and nature of funds into three classes: demand deposits, time deposits, and fiscal deposits. By demand deposits, it represents those deposited funds that are not constrained by any terms and can be withdrawn at any time. By time deposits, it means the deposited funds that can be mobilized or used by the depositors only when their terms are up. By fiscal deposits, it stands for the funds that are deposited by companies either for the purpose of tax payment to be paid to the treasury of the government later or as the distributed capital of the treasury department. Fiscal deposits include national treasury deposits, funds that are not included in local governments' fiscal budgets, deposits of various government agencies and organizations, funds of the military budgets, etc. Based on their different deposit forms, savings deposits can be divided into demand savings deposits and time savings deposits. The latter carries pre-determined mature dates; the principals can be deposited either only once at the beginning or at multiple pre-determined times; and the account balances can be withdrawn once all together or multiple times at different pre-determined time moments.

The funds borrowed by Chinese commercial banks are mainly from the People's Bank, interbank lending, rediscounting, transferred discounting, etc. The funds borrowed from the People's Bank are divided into three classes according to their terms: annual loans, seasonal loans, and daily loans. Here, the annual loans are used by commercial banks to cover their annual money shortages as caused by reasonable economic growth. Their terms are one year with the longest not more than two years. Seasonal loans are mainly used by commercial banks to safeguard temporary capital shortages as caused either by their advanced payments and delayed collections of money or by such factors as seasonal variations in their deposits. Seasonal loans carry two-month terms with the longest not over four months. Daily loans are mainly designed to help cover temporary cash shortages of commercial banks as caused by such factors as delayed arrival of payments. Their terms are about ten days with the longest not over 20 days. The interbank lending between Chinese banks can be roughly divided into two layers. Firstly, they are used for position lending in order to shelter the payment differences as caused by delayed exchanges of commercial papers; and secondly, they are used for fund positions in order to cover the credit gap. Position lending involves short-term loans borrowed by banks that do not have adequate amounts of reserves. Fund positions represent those loans that are short-terms used to adjust the fund positions in order to make up the time difference, space difference, and interbank difference as experienced in the flows of capital between financial organizations. They were introduced in January 1986 and occupy a very large weight in interbank lending.

The cash assets of Chinese commercial banks include cash on hands and bank deposits, various moneys maintained with the central bank, deposited interbank payments, and cash assets of other forms. Banks deposits represent the funds of the banks deposited by the banks' administrations in the business departments of their own banks to be used for their daily expenses. Both cash and bank deposits together reflect the situation of cash on hands of Chinese commercial banks. The funds deposited at the central bank include the deposits and reserves of the banks maintained at the central bank. The central bank, as the issuing bank and the last lender, provides such services to commercial banks as cash withdrawals, remit business, and loans. As the national center of clearance, the central bank organizes the settlement to make same-city exchanges of commercial papers for commercial banks. It helps to adjust the situation when a commercial bank's branch office sits on too much excess cash while another commercial bank's branch office is in bad need of cash as that might be and are often caused by their respective daily business operations. All of these services are provided by the central bank through the assets deposited by the commercial banks at the central bank. The deposit of a commercial bank at the central bank consists of two parts: policy-related deposits and general deposit. All policy-related deposits absorbed by the commercial bank need to be handed over to the central bank as the source of capital of the central bank. These policy-related deposits the commercial bank cannot use and include the deposits of budgets, deposits of funds not in the budgets, deposits of the government investment for basic constructions, etc., of different levels. Commercial banks need to handover statutory proportions of the general purpose deposits they attract to the central bank to be used coordinately by the central bank as the credit funds.

The classification and characteristics of the lending business of Chinese commercial banks are roughly the same as those of the Western commercial banks. However, in terms of the business coverage of lending and the flexibility in the forms of lending, Chinese commercial banks are very different from their Western counterparts. Depending on how borrowed many is used, the loans provided by Chinese commercial banks are grouped mainly into two classes: loans of liquid capital and loans of fixed capital. So-called loans of liquid capital stand for those loans the borrowing companies need to use to cover their needs that arise in their reasonable capital flows of business. Loans of liquid capital in turn are grouped into two types depending on their lengths of terms. One type participates jointly with the liquid capitals of the borrowing companies in the production and circulation of the companies' goods. Because these borrowed funds will be occupied over long periods of time, they are known as circulative loans of liquid capital. And circulative loans are again divided into two parts. One part is used to pay for the shortage in the statutory minimum working capital of commercial firms, and the other part is used to cover the reasonable physical inventories that are needed to meet the annual production targets and that are beyond the amount of the statutory minimum working capital. Another form of the loans of liquid capital is used to meet businesses' temporary monetary need of reasonable capital reserves that are beyond the limits of circulative loans of liquid capital, as caused by seasonal and temporary reasons. Because these funds are generally occupied no more than 6 months, these loans are known as temporary loans of liquid capital. Loans of fixed capital stand for the business loans borrowed from banks according to relevant Chinese regulations for such purposes as maintaining companies' real estate assets, updating obsolete equipment, exploring business potentials, renovating the existing business infrastructures, constructing new facilities, and expanding the existing facilities, etc.

Investments represent an important component in banks' assct business. The investment business of Chinese commercial banks consists of two classes of investments: short term and long term. For short-term investments, banks mainly purchase various priced securities that can be converted into cash at any time and hold these securities for less than one year. For long-term investments, it means such investments that banks do not plan to convert back to cash within one year, including stocks, bonds, and other types of assets. As for other types of investments, banks invest in other companies in the forms of cash, tangible objects, and intangible assets. Speaking generally, by banks' investment business, it mainly means banks' investments in securities, that is, the business activities banks participate in in terms of buying and selling of priced securities on the financial markets. The objects of banks' investments contain mainly the government debts, business companies' bonds, stocks, and other priced securities. The banks' investment business in securities is different of their lending business. Firstly, lending is initiated by the applications of the borrowers, where banks only react passively, while investment in securities is a proactive behavior of the banks. Based on their financial abilities and the needs of asset management, banks proactively select and purchase relevant securities based on the liquidities and terms of the securities on the financial markets. Secondly, lending involves bilateral relationships between banks and the borrowers. It has the characteristics of individualism, while investing in securities does not involve any individual; instead, it is a societalized and standardized market behavior. It is exactly because lending is of the essence of individualism that in order to reduce risk, banks often require the borrowers to provide some kind of guarantee or collateral. Because investing in securities is a market behavior supported by law and legal procedures, as investors banks do not need to face the problem of providing guarantee or collateral. Thirdly, in the transactions of lending, banks in general are in the position of the sole creditor, while in the investment of securities, banks are only one of the many creditors. Fourthly, the liquidities involved are different. Loans are mostly unlikely transferrable, while investments in securities can be freely traded or transferred on the securities markets with high degrees of liquidity.

12.2.4 Other banking businesses

Other than lending and investing, as mentioned above, other bank businesses include clearance business, trust business, agency business, and leasing business.

Specifically, clearance is an abbreviation of currency settlement. It clears the monetary relationship between payments and receipts as caused by such economic activities as the trading of goods, supply of labor, movement of money, etc. The papers banks use in their clearance of transactions include checks, bills, and promissory notes. Starting on April 1, 1989, China has stopped using five of its domestic clearance tools, such as letter of credit, escrow, collection without commitment, check of guaranteed payments, and limit settlement within province. The clearance tool, known as remote collection commitment, was also eliminated on August 1, 1989. However, due to various objective reasons, it has been called back and been used once again since April 1, 1990. So, currently, Chinese commercial banks use a total of seven different tools for their basic clearance: bank drafts, commercial drafts, promissory notes of banks, checks, exchanges, entrusted collections, and collection commitments.

The clearance of using bank's promissory notes means that after the sender of funds hands over his funds to a local bank, the bank provides a draft for the sender to complete his transfer of funds or to receive his funds in cash at a different geographic location. The clearance of using commercial drafts means that the draft is signed and issued by either the fund receiver or the fund payer and authorizes the bank where both parties have their active accounts to pay the receiver of funds on specified dates. The clearance of using banks' promissory notes means that the applicant hands over his funds to a bank that in turn issues back to the applicant a promissory note for him to transfer funds or take the funds in cash. The clearance of using checks means either that the owner of a bank's deposit account writes a signed check in an amount within the account's limit to clear his payment with the fund receiver, or that he authorizes the bank to pay directly to the fund receiver out of his bank account. The clearance of using exchange means that the sender of funds authorizes his bank to send his funds to the receiver located at a different geographic location. Each exchange can be one of two kinds: telegraphic transfer and postal transfer, where telegraphic transfer is done in the form of either telegraph or wire transfer through the sending bank based on the instruction of the fund sender, informing the receiving bank to make the payment to the fund receiver. Postal transfer is done in the form of using a document sent through the postal service for the receiving bank to make payment to the fund receiver. The only difference between these two kinds fund transfer and payment is that the former is faster and more expensive. Banks' promissory notes, banks' drafts, and commercial drafts are jointly known as ticket exchange. The clearance of using collection means that the fund receiver fills out an authorization for his bank to receive funds on his behalf with his bank along with a document proving his ownership of the funds, all of which authorizes the bank to collect money from the fund payer. Collection commitments have been one of the traditional methods of clearance involving fund payers and receivers located in different geographic areas for Chinese banks. Because many constraints exist with this method, including long periods of time are involved, it had been taken out of use on August 1, 1989. However, when the method of entrusted collection is employed, payers do not have to make the payments, while banks are not obligated to check the reasons why the payers refused to make the payments and are not authorized to collect the funds so that there appeared a large number of companies that refused to make or delayed their payments. Because of this reason, the People's Bank of China has revived the use of collection commitments on April 1, 1990. By collection commitment, it means that when the selling company has sent out its products to fill an order according to the sales agreement, it authorizes its bank to collect the funds from the buying company that is located at the different geographic location. When the buying company receives the invoice for payment from the bank, after checking the invoice against the purchase order or against the goods received, it processes the paperwork for payment with its bank and then its bank transfers the funds to complete the payment.

Other than the international method of clearance using letters of credit, China has stopped on April 1, 1989 using its domestic clearance using letters of credit, although it represents one of the important methods of clearance involving different locations used by Western commercial banks. The so-called clearance using letters of credit entails that the payer deposits his money in his bank in advance as his guarantee for payment. After then he requests the bank to issue him a letter of credit and to let the bank of the fund receiver at a different location to inform the payee that if he sends off his goods according to the conditions in the sales agreement and the letter of credit, then the payer's bank will make the payment on behalf of the payer according to the conditions listed in the letter of credit. This method of settlement is most appropriate for situations when the parties of trade do not know the credit worthiness of each other. When a commercial bank is involved in the dealings of letters of credit, other than collecting handling and service fees, it can also make partial use of the clients' funds.

The trust business of modern China was initially introduced from Great Britain, the United States, and other Western countries at the beginning of the 20th century. In 1917, Shanghai Commercial Savings Bank first established the storage unit, managing the rentals of safety boxes. In 1921 a group of specialty trust institutions was created. In 1935 the trust bureau of the central government was officially established. As soon as the government of the communist China was established in 1949, Shanghai City branch office of the People's Bank became on November 1 its storage unit, which had been essentially idling since 1952. Along with the recent reform of the economic system in China, business companies started to have financial strengths and powers that they can freely control; after allocating their appropriated funds government agencies and organizations also started to have remnant funds to be used for un-budgeted items; and the financing needs of various economic entities have become increasingly resilient. Hence, starting in the year 1980, Chinese banks began to be involved in trust business. According to the law, the trust business of Chinese banks includes mainly trust deposits, trust loans, trust investments, entrusted deposits and loans, entrusted investments, etc. By trust deposits, it means that trustor gives his savings to the trust department of his bank to manage and to invest no matter how. He will only collect the income according to the promised interest rate. By trust loan, it means the form of bank lending that the trust units of banks select borrowers of their own choosing based on the banks' criteria to lend their funds that are either the assets of the banks or collected by the banks, while the banks are fully responsible for the potential risks involved. By trust investment, it means that the trust units directly participate, as individual investors, in the investment, management, and distribution of the returns of jointly managed business activities. By entrusted deposits and loans, it means the kind of banking business that the banks' trust units accept, apply, and manage the entrusted loan funds from individual trustors according to the agreed upon objects and purposes. By entrusted investments, it means that the banks' trust units accept the entrust of the trustors to invest in specified companies, while monitoring the companies' operations, management, profit distributions, etc.

Currently, the often seen agency business of Chinese banks, which are similar to those of the banks in the Western World, include:

1   Issuing securities for clients. That is, to raise funds, business firms authorize banks to issue on the firms' behalf stocks, bonds, etc.
2   Delegated trading. That is, banks accept clients' authorization to buy and sell priced securities, precious metals, and foreign currencies.
3   Delegated receiving and making of payments. It is a service that banks provide to their clients, where the banks either receive the incomes for specified funds on behalf of individual clients and business firms or make payments within limits of the clients' available funds. The range of this service is very wide, including the delegated receipts and payments of wages, capitals, stock dividends, interests, school tuitions, welfare fees, telephone bills, etc.
4   Insurance agency business. That is a bank service, where the banks play the role of brokers of insurance companies, including purchasing the necessary insurances for business firms, providing insurance protections for business firms on behalf of insurance companies, etc. Through such services, the banks charge a certain amount of handling fees.
5   Custody business, where banks keep valuable items, such as gold bullions, silver bars, important documents, contracts, stock certificates, etc., for their individual and/or company clients

Leasing business is one of the economic behaviors that have been practiced since antiquity. However, the modern leasing, as a bank service, was only initially introduced in the 1950s in the United States and became to be widely used since the 1960s in Western Europe and Japan. Since 1981, all specialty banks in China have also been providing various kinds of leasing services. The advantage for banks to be involved in the business of leasing is that leasing does not involve the transfer of ownership and can guarantee the safety of the assets. Through collecting rents, the entire investment plus interest can be recovered and additional added values can be materialized. In nations where speedy depreciations are provided and the taxes on investment income are waived, leasing can help to acquire preferential tax treatments. In terms of the operation of leasing, additional returns can be generated through providing service and maintenance.

12.3 FINANCIAL REFORMS OF CHINA

Since entering the 1990s, the economic reform in China has been continuously deepened; finance has become increasingly important in the development of Chinese economy; and monetary policies have been playing the role of lever that directly affects the development speed of Chinese economy. However, there have been financial crises in China, where dangerous take-offs and painful landings have more than once brought Chinese economy to the brink of collapse. It has been totally possible that Chinese leaders had expected that hidden financial crises could very well endanger the economic foundation of China. That would explain why at the national level a vice prime minister has also taken the governor's position of the People's Bank of China, which in itself is unprecedented in Chinese history. However, that in fact signals exactly that the era of planned economy has been evolved into the era of market economy, where China will utilize the financial lever to control its economy.

12.3.1 Need for reform

The financial system is the Great Wall of Chinese economy. Although the current financial system behaved like a levee that blocked the Asian financial storm of the 1990s from entering China, it still contains many imperfections. By reflecting on the Asian financial storm of the 1990s, it is not hard for anyone to realize that enormous amounts of foreign capital and abundant foreign debts can provide the necessary engine for Chinese national economy to take off within a short period of time and within a certain scope. However, after the take-off, how high, how far, and how long the economy can sustain its flight are determined by whether or not China can grasp the advantageous opportunity when the nation is still braced by foreign capitals and foreign debts to establish and to perfect its internal structure and financial system. If the advantageous opportunity were lost mistakenly, the long-term plan as envisioned for Chinese national economy would be handicapped half way through its materialization on one of those future days when the foreign capital were leaving China.

In order for Chinese economy to grow sustainably and healthily while avoiding the breakout and invasion of financial crises, China has to improve its economic structure, strengthen its industrial system, create employment opportunities, perfect its banking system, and create a stable and peaceful social environment and a sound financial order. Maintaining its social and economic stability is a great contribution China can make to the whole world. What China had shown during the Asian financial crisis of the 1990s has very well proven this end.

12.3.2   Key aspects of financial reform

From the Asian financial crisis of the 1990s, the following point has been widely recognized: The wide-ranging reform, including the reform in the financial industry, in China needs to be continued without any interruption. In order to make better use of the capital of the international market, China must speed up the reform of its financial system and formulate a better mechanism. To this end, the current focus of reform of Chinese financial system has been about the following aspects:

1   Deepening its banking system reform so that Chinese financial order can be normalized and maintained based on laws;
2   Raising the quality of bank assets;
3   Establishing a modern financial system;
4   Perfecting the financial monitoring and control system; and
5   Strengthening the management of foreign currencies.

More specifically, under the current conditions, the financial crises that are similar to the Southeast Asian financial crisis of the 1990s will not occur in China. However, due to its ongoing transition from the originally totally planned economic system to a semi-market economic system, there are hidden risks for financial crises to happen within the Chinese financial industry. These hidden risks appear to be mainly in the following areas:

a)   The proportions of bad debts on the balance sheets of the nationally owned commercial banks are quite high;
b)   There are many challenges facing the non-banking financial institutions, some of which could no longer afford to cover maturing debts on a timely basis;
c)   There have appeared a good number of illegal financial institutions that have severely interrupted the working environment of Chinese financial industry;
d)   There exist various kinds of illegal behaviors in the stock and futures markets.

The resolution of these and other problems facing Chinese financial system will closely related to the continued and deepening reform of the banking system, strengthened monitoring of the general financial market in general and banking system, stock and futures markets in particular. The reinforcement of the relevant laws and regulations has to be firstly carried out and secondly fortified. As soon as a problem surfaces, it has to be dealt with seriously; and severely affected financial organizations will have to be reorganized or closed through the means of the legal system or the bankruptcy court depending the problems involved.

It has been a problem for many years that Chinese companies do not have a strong sense of credit worthiness. For example, Chinese firms have large amounts of savings deposited in their bank accounts. However, for some reason these firms just do not like to pay off their debts, including bank loans, even though they might have to pay penalties for overdue debts. On the other hand, the terms of bank loans tend to be two short, the average of which is only about five to six months long. That surely does not agree with the industrial and commercial production cycles of around 200 days. Along with the accumulation of these and other relevant problems for many years, the quality of assets on the balance sheets of Chinese nationally owned commercial banks have been quite low, including huge amounts of delinquent loans, sluggish loans, which have been overdue for over two years, and doubtful loans, which represent loans that can no longer be collected due to reasons like bankruptcy, dead debtors, and others. Such a problem of bed debts should be resolved as soon as possible by separating financial assets, a method commonly used in the international arena. Only by increasing the ratings of Chinese commercial banks and by improving the management capability of assets, the hidden danger of bank panics and bank runs can be avoided.

In order to lower financial risks, China also needs to establish and perfect its modern financial system that can operate smoothly along with its specific economic market. The government's direct, tight control over the nationally owned commercial banks should be loosened up and make it an indirect control. In particular, these nationally owned commercial banks should be managed through adjusting liquidity, asset composition, debt quality, and sufficiency of capital, The limitation on the magnitude of loans Chinese nationally owned commercial banks are allowed to give out should be removed so that these commercial banks could become truly commercial and are able to compete with other commercial banks, be they domestic or international, at the global level. After removing the limitation on the ceiling of loans, the Chinese mechanism of financial control would be quite similar to that used in the United States. Also, interest rate should no longer be determined by the People's Bank of China. Instead, the interest rate on deposits should be adjusted by commercial banks based on the supply and demand of the market place and their individual management strategies. At the same time when commercial banks need to pay close attention to the management of credit risk and interest rate risk, the nationally owned commercial banks should become truly commercialized without much interference from the national government. Only after all these works of reform have been accomplished, the marketization of Chinese financial system will be basically complete, where a modern and internationalized financial system will appear in China.

Currently, the development of Chinese financial institutions is very unbalanced. The internal controls of banks and non-banking financial institutions have very different origins and are of different quality. In order to deepen the financial reform in China, there is a strong need for the central bank to strengthen its monitoring of financial institutions so that the business operations and internal managements of banks and non-banking financial institutions can be carried to a much higher level.

Externally, the monitor, management and control of the flows of foreign currencies in and out of China need to be continued and strengthened so that financial attacks from international speculators could be either avoided or their aftermath damages could be minimized. Only by constantly improving the quality and ability of managing Chinese financial system, adverse impacts of international financial crises can be minimized. In terms of materializing the goal of making RMB completely floating with the market demand and supply, much caution is needed.

In short, after having experienced the Asian financial storm and witnessed other speculative financial attacks from around the world, it can be clearly see that no matter whether it is forced by the outside world or self-motivated from within Chinese system, a complete reform of Chinese financial system is a must choice for China to survive and to grow in the current ever-increasingly globalized world. The feeling of luck cannot be dependent, China has to proactively eliminate any and all potential dangers in order to improve and strengthen its ability to withstand financial storms that might occur either externally or internally to China.

12.4   INTERNATIONAL RESERVES OF CHINA

During the time period from 1949 to 1979, China has carried out a highly central-ized, planned economic system, and employed a strict control on foreign currencies. All foreign currencies were solely managed and used centrally by the People's Bank of China. Because during that time period China adopted a policy of having closed itself from the outside world in terms of foreign affairs, there were very limited eco-nomic exchanges with the rest of the world. Because of this reason, the development of foreign trades was slow and the scale of international reserves was extremely small so that the management of international reserves was not emphasized. After the third plenary session of the eleventh central committee of the Communist Party of China, China started to open up itself to the rest of the world. Economic exchanges with the external world had been increasingly expanded. Consequently, the amounts of foreign trades and foreign currencies mobilized increased unprecedentedly, making interna-tional reserves significantly more important in Chinese economic activities than ever before. The management of international reserves had gradually become a part of the macroeconomic management of China. In the year 1980, after receiving its seats in both International Monetary Fund and the World Bank, China started to fully enter the international market. Same as other member countries of these organizations, China also enjoyed its ordinary drawing right and special drawing right of International Monetary Fund. Since then, the international reserves of China have been consisted of four parts: gold reserve, foreign reserve, ordinary drawing right, and special drawing right.

In terms of gold, China had applied the policy of stable gold reserve. During the three years from 1978 to 1980, the gold reserve of China was kept at the level of 12,800,000 ounces. During the time period from 1981 to 2000, Chinese gold reserve dropped slightly and was maintained at the level of 12,670,000 ounces. In 2001 and 2004, the gold reserve rose respectively to 16,080,000 ounces and 19,290,000 ounces.

Because during the time period from 1949 to the mid-1980s, China implemented a highly planned economic system, its gold reserve was managed in the fashion of house-keeping. Other than focusing on the physical safety of the gold, the operational aspects of the gold reserve were totally ignored. Along with the development of the market economic system, the management of Chinese gold reserve has gradually evolved from its original simple housekeeping mode to the current operational mode. Currently, there are two main methods for China to apply its gold reserve:

1   By participating in the international gold market, a good amount of return has been made through the usage of spot contracts, future contracts, options of future contract, etc.
2   Through issuing and selling various kinds of gold coins, the value of the gold in storage has increased. As time goes on, along with the reform of the gold management system, the management quality of the gold reserve is expected to improve further.

In terms of the ordinary drawing right and the special drawing right, consid-ering the magnitude of foreign reserves China currently holds, they occupy a small proportion in Chinese international reserves.

In terms of foreign reserves, they constitute the most important part of Chinese international reserves. Since 1949, Chinese foreign reserves have gone from not standardized management to standardized management through the process from slowing growth to fact growth. Because of the important position of foreign reserves in Chinese international reserves, in the following subsection, let us look at the foreign reserves of China in for details.

12.4.1   Fast increase in foreign reserves

During the time period from 1949 to 1978, what was implemented in China was a planned economic system with little economic exchange with the outside world. In the area of foreign currencies, all expenses were measured by how much income there was; income was used to determine how much to spend, while trying to maintain a small amount of balance. Because of this practice, the amount of foreign reserve was kept at an annual average of about US$500,000,000 without any foreign debt. Correspondingly, the management of foreign reserves did not seem that important.

Starting with the beginning of reform in 1979, along with the increasingly expanding scale of international exchanges, the magnitude of Chinese foreign reserves rose continuously. During the time period from 1979 to 1992, Chinese foreign reserves consisted of two parts. One was the foreign currencies in the hands of the national government, which was the total of the differences between the annual incomes and annual payments of all the past international trades and non-trade business transactions. The other was the foreign currency balance on the balance sheet of the Bank of China, which consisted of the total of foreign currencies owned by the bank plus foreign currency deposits attracted and foreign loans the bank took from abroad minus the bank's lending and investments in foreign currencies. So, it can be seen that the foreign currencies in the hands of the national government in fact were the external debts Chinese monetary authorities held, and could be mobilized at any time without any restriction. On the other hand, the accumulation of foreign currencies of the Bank of China was in fact the bank's foreign debts, and could not be used unconditionally by the national government. So, speaking strictly, this accumulation of foreign currencies could not considered as the international reserves of China. If this part of capital were treated as foreign currency reserve, it would exaggerate the true scale of Chinese international reserves. Hence, in order to employ the same concept and method of statistics as commonly used internationally regarding foreign reserves, since October 1992, China has been using the foreign currencies held in the hands of the national government as the sole component of Chinese foreign reserves. From that day on, the official statistics on Chinese foreign reserves have no longer included the accumulation of foreign currencies of the Bank of China.

During the time period from 1979 to 1984, Chinese foreign reserves grew steadily. By introducing and implementing a whole series of policies that encouraged exports while restricting imports, Chinese foreign reserves were gradually increased, reaching US$8,200,000 at the end of 1984. During the time period from 1985 to 1989, the scale of Chinese foreign reserves fluctuated up and down drastically. Starting in the year 1985, Chinese national economy had been developing rapidly, leading to drastically expanded demand for imports. That caused Chinese lost their control over their foreign reserves so that the scale of the foreign reserves dropped significantly to the level of US$2,600,000,000 at the end of 1985. After then, although the scale of foreign reserves increased somehow, it only reached US$5,500,000,000 at the end of 1989. During the time period from 1990 to 1993, Chinese foreign reserves grew quickly so that, starting in 1990, China strengthened its macroeconomic control, adopted the corresponding monetary policies, and implemented procedures to encourage exports. As a consequence, the foreign reserves for the year 1990 jumped to US$11,000,000,000. After that year, except 1992, Chinese foreign reserves have been continuously grown year after year, reaching the level of US$21,100,000,000 in 1993.

In the year 1994, China implemented a reform in the management system of foreign reserves. That made Chinese foreign reserves grow at an extremely high speed since 1994. Nineteen ninety four was the first year when exchange rates were unified and when an enforced system of collecting and selling of foreign currencies was implemented for domestic enterprises. The foreign currencies that had been accumulated and managed by individual companies that had been involved in foreign trades in all the past years were centralized and replaced with RMB within a short period of time. Another characteristic of the growth in Chinese foreign reserves in this period is where the foreign currencies were originated. During the time period from 1994 to 1996, the main source of growth in Chinese foreign reserves was from the surplus between capitals and financial accounts. In 1997 after the Southeastern Asian financial crisis broke out, Chinese government adopted a series of preferential policies in order to secure more ways to expand the national exports so that the surplus of the current account has been growing enormously since 1997. That has been the main reason behind the growth of Chinese foreign reserves. In 2001, China became a member of the World Trade Organization, which in turn helped to greatly boost the growth of Chinese capital and surplus of financial accounts, while at the same time, Chinese current account maintained a relatively large surplus.

Since 1994 when China started the reform of its foreign reserve system, China's foreign reserves have been increasing tremendously grown. In particular, China's foreign reserves grew from the level of US$21,200,000,000 at the end of 1993 to that of

US$51.620 billion in 1994, then to
US$73.597 billion in 1995, then to
US$105.000 billion in 1996, then to
US$154.675 billion in 1999, then to
US$165.575 billion in 2000, then to
US$212.165 billion in 2001, then to
US$286.407 billion in 2002, then to
US$403.251 billion in 2003, then to
US$609.932 billion in 2004, then to
US$818.872 billion in 2005, then to
US$1066.344 billion in 2006, then to
US$1528.249 billion in 2007, then to
US$1946.030 billion in 2008, then to
US$2399.152 billion in 2009, then to
US$2847.380 billion in 2010, and then to
US$3181.148 billion in 2011.

The swelling growth in the scale of its foreign reserves has made China's ability to pay foreign debts increasingly stronger and China's comprehensive national strength lifted unmistakably. However, the excessive growth in the foreign reserves has also brought along with great pressures on the domestic inflation and on raising the exchange rate of RMB. In the following, let us look at both the positive and negative effects of the rapidly increasing foreign reserves in China.

In terms of positive effects, the rapid growth in foreign reserves has strengthened China with its comprehensive national power and its financial worthiness in the international arena. The scale of foreign reserves is an important indicator of a nation's comprehensive power. The scale of China's foreign reserves has in the recent past jumped to the second place in the world. The greatly improved comprehensive national power has provided China with sufficient capability to select how it could develop itself based on the actual situations both within and outside China. At the same time, the 784,902 billion foreign reserves have also provides a reliable guarantee for China to borrow foreign debts and to repay all of its debts along with all the interests. That greatly improves the international trustworthy of China as a nation. Other than the improved ability to repay foreign debts, the rapidly growing scale of foreign reserves also strengthens China's ability to amend its magnitude of international transactions so that Chinese government could make proactive adjustments to its policies regarding international trades while creating a reasonable growth rate for its domestic economy. Additionally, the sufficient foreign reserves provide a necessary condition to eventually realize the floating exchange system for RMB, which has been one of the top priorities for the reform of China's foreign reserve system. After China realized the convertibility of RMB current account at the end of 1996, its next goal has been to remove the foreign currency control over its current and capital projects so that normal activities of international currency exchanges and capital flows will no longer be constrained. That will lead to the actualization of market exchangeability of RMB. Sufficient foreign reserves provide the capability for Chinese central bank to effectively monitor and adjust the market and to maintain the basic balance between international incomes and payments. That helps to make the maintenance of a relatively stable exchange rate possible in the process of materializing the free exchange of RMB.

In terms of the negative effects, the rapid growth in China's foreign reserves has pushed the value of RMB to go up continuously. That has weakened the competitiveness of Chinese products in the world market. Speaking generally, when a nation's totality of foreign reserves grows, the nation's foreign exchange rate will rise accordingly. From what actually happens to China, it can be seen that the exchange rate of RMB has been climbing higher and higher along with China's growing foreign reserves. That has surely created adverse impacts on China's exports. Additionally, the rapid growth in foreign reserves has placed an increasing pressure on inflation. Essentially, the funds foreign exchange occupies belong to the base currency. So, any growth in the funds foreign exchange occupies directly increases the amount of the base currency, which, through the effect of monetary multipliers, causes the amount of money supply to increase drastically. That of course places a great deal of pressure on inflation. The rapid growth in China's foreign reserves inevitably leads to ineffective use and even waste of capital resources, which affects the speed of development of Chinese national economy.

Foreign reserves are in fact a symbol of actual resources. What they hold is opportunity cost, which in turn is equal to the domestic productivity of capital minus the return from holding foreign reserves. When seeing from this angle, for China to hold a massive amount of foreign reserves and to take a large amount of foreign debts, it is equivalent to provide Chinese capital for foreigners to use at low prices, and at the same time China borrows foreign capital at high prices. The potential loss is pretty high. Additionally, holding foreign reserves also means temporarily giving up the use of a certain amount of actual resources. That is to say, the economic growth and rise in the level of incomes, as might be caused by these resources, have been lost.

The rapid increase in China's foreign reserves makes China lose its preferred loans from the international monetary fund (IMF). According to the rules of the IMF, when a member nation experiences a deficit in its balance of foreign payments, it can take a loan at a low interest rate from the monetary fund in such an amount that is equivalent to its paid share. If a member nation experiences a structural problem in the areas of production and trade, it can also obtain mid- to long-term loans totaling to such an amount that is equal to 160% of its share at also a preferred interest rate. Conversely, not only can nations with sufficient foreign reserves not enjoy these preferred low interest rate loans, but also have to provide loans to other nations that experience difficulties in their balance of international payments when necessary.

12.4.2 Management of foreign reserves

As discussed above, a nation should maintain an appropriate amount of foreign reserves, which should be neither excessive nor insufficient. When determining the appropriate scale of foreign reserves, China should have considered the following several factors: the circumstances of foreign trades, the ability of financing, the exchange rate system and method of adjustment, and the situation of foreign debts.

In particular, in terms of the circumstances of foreign trades, among the products exported from China the proportion of labor intensive products is too high. Such products in general do not have the needed adequate flexibility to adjust to price fluctuations of the international market. Among the products imported into China, because many products of China and prices of labor had been higher than those of the international market, after joining the WTO, along with the lowering and elimination of tariffs, the demand for imports has been steadily increasing. The international trade of China has led to increasing instability in the balance of China's current account. Because of this reason, China has held a large amount of foreign reserves. In terms of the ability to finance, since the start of its reform, China's political and economic status in the international arena has been steadily improving. According to the evaluations of internationally credible organizations, China has enjoyed a relatively high credit rating. That helps China to have a relatively strong ability to finance so that the magnitude of foreign capitals employed in China has been rising year after year. On the other hand, a nation's ability to finance is closely related to its scale of foreign reserves. So, in order to guarantee a relatively strong ability to finance, China has to maintain a relatively sufficient level of foreign reserves. In terms of the exchange rate system and method of adjustment, what is implemented in China currently is a managed floating exchange rate system based on the market supply and demand. Under this system, Chinese authorities of finance must maintain a sufficient amount of foreign reserves so that it

Table 12.1 The situation of China's foreign debts in recent past (unit: 1 billion US$).

Year	1994	1995	1996	1997	1998	1999	2000	2001	2002
Foreign debt balance	92.81	106.59	116.28	130.96	146.04	151.83	145.73	203.30	202.63

Year	2003	2004	2005	2006	2007	2008	2009	2010	2011
Foreign debt balance	219.36	262.99	296.54	338.59	389.22	390.16	428.65	548.94	695.00

Source of data: Website of China's state administration of foreign exchange.

can interfere the foreign currency market when necessary. So, due to this reason, in the foreseeable future, China will continue to maintain a high level of foreign reserves. Of course, along with the evolution of the reform on its exchange rate system and the management system of foreign currencies, after RMB can be freely exchanged, the effect of exchange rate on the adjustment of the international payments will be more obvious. So, looking at long term, the scale of China's foreign reserves will be lowered.

In terms of the situation of foreign debts, the greater the total amount of foreign debts a nation has, the more foreign reserves the nations needs to maintain. Especially for short-term debts, because the repayments will need to be made within relatively short periods of time, the guarantee of sufficient foreign reserves will be more necessary. From the current situation in China, it can be seen that the scale of Chinese foreign reserves has reached the level of 15 times of that of short-term debts. If this indicator is employed, then China's foreign reserves are quite sufficient. For China's situation of foreign debt balance, please consult Table 12.1.

Other than the qualitative factors, as mentioned above, other quantitative analysis methods widely employed internationally, such as the method that considers the proportion of reserves over imports, etc., should also be considered in determining the appropriate level of foreign reserves for China. Only by doing so, the scale of foreign reserves of China can be soundly determined so that the allocation of the resource of foreign currencies can be optimized.

As for the structural management of foreign reserves, it contains two aspects: 1) The selection of different kinds of currencies and their corresponding proportions, and 2) the selection of the asset forms of the reserved foreign currencies. Speaking generally, when selecting reserve currencies, the following factors should be considered: the structure of trades, the structure of foreign debts, the exchange rate system, and risk and return.

In terms of structure of trades, China is a large international country of trade. Since joining the WTO, China's proportion of international trades has gradually risen to the recent rank of number 1 ahead of Germany and the United States. In order to guarantee a certain level of capability of making payments, China has considered such factors as the origins of imports, the direction of exports, quantities of imports and exports, the accustomed ways of payments used by trading partners, etc., in its selection of reserve currencies. Because of this consideration, China has placed priorities in its selection of reserve currencies on U.S. dollars, Euros, Japanese yen, British pound, etc., as its major reserve currencies. In terms of the structure of foreign debts, in order to make sure the principals and interests of foreign debts are repaid on time, while possibly minimizing the trading cost and risk potentially existing in the exchange of currencies, China has also considered proportional structure of different foreign currencies of its foreign reserves. Speaking practically, because the main foreign currencies involved in the foreign debts of China are U.S. dollar, Euro, Japanese yen, British pound, etc., where U.S. dollar and Euro occupy the largest weights, most of China's foreign reserves are made up of these two currencies. In terms of the exchange rate system, as mentioned earlier, the essence of the current Chinese exchange rate system is that of pegging the U.S. dollar. Although the exchange rate of RMB, as determined by the People's Bank of China, references a basket of foreign currencies, U.S. dollar carries the most weight in this basket. That fact means that the foreign reserves of China have to include mostly U.S. dollars. In terms of the risk and return of the reserve currencies, the safety, liquidity, and profitability of foreign reserves represent the commonly employed criteria internationally. They are also the foundations on which Chinese manage and employ their foreign reserves. That has been how China adjusts the currency structure of its foreign reserves in order to spread risk while maximize the profitability and liquidity by closely following the international market exchange rates, interest rates, and the price changes within the nations which issue the reserve currencies.

After having considered the structural distribution and adjustment of reserve currencies, China also experienced with the problem of in which assets forms foreign reserves should be kept. In other words, in its foreign reserves, what proportions of Chinese debts, stocks, agency bonds, company bonds, etc., should be considered? Regarding this problem, the general belief is that safety should be considered with the number 1 importance along with appropriate consideration giving to liquidity and profitability. Currently, the main asset form of China's foreign reserves is the national debts of the United States, amounting to roughly 40% of the total foreign reserves. That proportion makes China the largest creditor of the government of the United States. The reason behind this proportion is that American national debts are relatively safer than those of other nations with a complete array of different kinds of debts, while the market of American national debts is huge, making it easy to convert American debts into cash. Additionally, it is also because of the relatively low level of skills China currently possesses in its management of foreign reserves. At the same time of guaranteeing safety, how to consider jointly profitability and liquidity has been a major challenge facing China's management of its foreign reserves.

12.4.3   How can the risk of US$3,442.65 billion reserves be resolved?

As of March 2013, China's foreign reserves have been recorded at the level of US$3,442.65 Billion. Now, a natural question is how the risk of holding such a massive amount of foreign reserves can be minimized. To this end, a lot of discussions have focused on increasing China's petroleum inventory. These discussions believe that doing so can not only partially ease the safety problem of petroleum oil supply, but also release the pressure ad risk of holding such a enormous amount of foreign reserves. Indeed, with the current scale of foreign reserves China completely possesses the capability of purchasing petroleum oil as its national strategic reserve. However, the next important question before taking the action of actual purchase is: Is it an optimal choice for China to exchange its foreign reserves for petroleum oil? To address this problem, let us first look at how the international society has traditionally dealt with large scales of foreign reserves.

Developed economic powers generally convert their foreign reserves into gold. As of January 2005, the proportions of gold in individual nations' foreign reserves were respectively as follows: 61.1% for the United States, 55.8% for Italy, 55.1% for France, 51.1% for Germany, 50.5% for Holland, while as of January 2005, the proportion of gold in China's foreign reserves was merely 1.5%. Therefore, some Chinese scholars believe that China should increase his gold reserve somehow.

Because of the unification of currencies and the appearance of Euro in Europe, some scholars suggested drastically promoting the status of Euro in China's foreign reserves so that corresponding scale of U.S. dollar reserve could be reduced accordingly.

Various nations employed their released foreign reserves to make domestic investments and to improve the living quality of their citizens. During the economic take-off period of time when Japan was very tight with foreign reserves, it still spent a great deal of money to introduce advanced technologies from abroad. After having sufficiently absorbed and digested the new knowledge, Japanese companies designed and produced more competitive products that brought back more foreign currencies than they initially spent. Along with the continued expansion in its positive difference between international incomes and payments, Japan then converted large quantities of foreign currency reserves into strategic reserves of goods. For example, after having imported large amounts of coal from China and other countries, Japan sank the coal into the ocean as its coal reserve. Additionally, from North to South along its various islands, Japan constructed 10 new artificial oil fields (national bases of petroleum reserves), while the government provided its citizens with large sums of interest subsidy for them to store petroleum oil individually. As for Russia, during the recent time when the purchasing power of U.S. dollar was continuously dropping and the petroleum oil price unceasingly rising, it made a fortune by exporting petroleum oil. To the rapid increase of its foreign currency incomes, Russia did the following: 1) It purchased a large amount of gold in order to increase the proportion of gold in Russia's reserves; 2) It greatly increased the proportions of non-U.S. dollar currencies in its reserves; 3) It repaid foreign debts ahead of time; and 4) It spent heavily on areas of high technology.

Then, what's appropriate for China to do with its extreme high level of foreign reserves? A recent report of investment banks shows that in the year 2004, the U.S. dollar asset in China's foreign reserves had been reduced by 6 percentage points from 82% in 2003 to 76%; and ever since this total proportion has been gradually reduced considering the risk hidden in the U.S. dollar with its continuously decreasing value. As for how to spread the risk faced by the large scale of China's foreign reserves, one of the many effective methods is to increase the capital injection into nationally owned commercial banks. For example, since 1998, Chinese central government has multiple times both directly and indirectly injected several thousands of billions yuan into its nationally owned commercial banks in order to reduce their rates of bad debts. However, each year the newly added non-performing loans still reached as high as 500–600 billion yuan. In other words, to save the nationally owned banks, a great deal of more capital is needed for these banks.

Additionally, scholars also suggest that the government could continue to encourage investments overseas, convert a greater proportion of the foreign reserves into material reserves, and employ many other ways of conversion. Looking around the modern society of China, another extremely urgent problem is the huge capital shortage in social welfare programs. To maintain the stability of the society, the government could consider mobilizing the excessive foreign reserves that are beyond the reasonable level to further supplement pension funds in order to release the current urgency.

Currently, China's foreign reserves have way pasted the reasonable level. For example, as of March 2013, China's foreign reserves had reached the level of US$3,442.65 billion. However, economic studies show that the ideal scale of the foreign reserves of a developing country is the amount that is enough to cover the payments of 3-month worth of imports. And the standard given by the IMF is the amount that is enough to support 4-month worth of imports. On the other hand, reports from the World Bank suggest that the foreign reserves of an emerging-market nation should be adjusted according to its capital account instead of how much foreign currencies are needed to cover the demand of imports. It is because when the nation is closer to the international capital market, the magnitude of that nation's overall capital flow will be much higher than the need for covering the demand for foreign currencies from imports. In terms of China in particular, some scholars believe that the appropriate scale of China's foreign reserves should be somewhere between US$300–400 billion, while some others think that China should maintain such a scale of foreign reserves that is enough to cover 6-month worth of imports as well as considering the need of short-term foreign debts for foreign currencies. So, if all these suggestions are combined, the ideal scale for China to maintain its foreign reserves should be at the scale of US$500 billion. As of March 2013, China's foreign reserves have reached the level of US$3,442.65 billion, which is roughly US$3,000 billion beyond the reasonable scale.

International experience and lessens indicate that one has to pay additional cost and take extraordinarily risk for holding excessive amount of foreign reserves. For most of the developing nations, especially for emerging markets, foreign reserves are not only a necessary method of making international payments, but also a steady embankment necessary for their self-protection against international financial, speculative attacks. However, it is not true that the more the foreign reserves a nation has the better. Foreign reserves represent domestic capital that is idling; and they in turn introduce foreign funds at a higher cost. An increasing scale of foreign reserves also alters the supply structure of the domestic base money, weakens the control of the central bank on the money supply, and upsurges the pressure for the domestic currency to strengthen. For China, the rapid expansion of its foreign reserves also means that the natural and labor resources of China are consumed at a fast rate. That surely affects the sustainability of China's social and economic development. What is worth noticing specifically is that a major part of China's foreign reserves are used to purchase United States' national debts, which equalizers American fiscal deficits and saves the United States from suffering from economic contractions. However, doing so has been criticized by Americans as increasing the financial risk of the United States, while at the same time China experiences the risk of devaluation of the U.S. dollar and the potential freeze of Chinese assets. Hence, when considering the long term, a series of natural questions, which need to be addressed urgently, includes how to reduce the pressure of China's large foreign reserves and how to appropriately spread the associated risk over a wider range.

12.5 UNDERSTANDING CHINA CULTURALLY

As a new, heavy weight player in the international economy and financial system, in this section, we learn how China might behave differently from the expected norm of the Western civilization. Only by knowing each other, misunderstandings could be avoided and beneficial communications could be greatly enhanced.

12.5.1 What is China as a civilization?

To address this question, it should be noted that human history is basically composed of stories of civilizations from ancient Sumerian and Egyptian to Classical and Mesoamerican to Christian and Islamic civilizations and through successive manifestations of Sinic and Hindu civilizations. On different perspectives, methodology, focus, and concepts, many distinguished historians, sociologists, and anthropologists have well studied civilizations (Kissinger, 1994).

The concept of civilization was initially developed by eighteenth century French thinkers; nineteenth century Europeans elaborated the criteria for judging non-European societies as to which were sufficiently civilized to be accepted as members of the European-dominated international system. German thinkers of the nineteenth century distinguished civilizations by mechanics, technology, and material factors, the cultures that involved values and ideals, and the higher intellectual, artistic, moral qualities of a society. Some anthropologists conceived cultures as characteristic of primitive unchanging nonurban societies and complex developed urban and dynamic societies as civilizations. In general, civilization and culture both refer to the overall way of life of a people with a civilization as a culture writ large. They both involve the values, norms, institutions, and modes of thinking to which successive generations in a given society have attached primary importance (Bozeman, 1975). By a civilization, it is meant to be a space, a cultural area, a collection of cultural characteristics, and phenomena (Braudel, 1980, pp. 177, 202), a particular concatenation of worldview, customs, structures, and culture (both material culture and high culture) which forms some kind of historical whole and which coexists with other varieties of this phenomenon (Wallerstein, 1991, pp. 215), and a kind of moral milieu encompassing a certain number of nations, each national culture being only a particular form of the whole (Durkheim and Mauss, 1971, pp. 811). For example, blood, language, religion, and way of life were what the ancient Greeks had in common and what distinguished them from the Persians and other non-Greeks. When people are divided by cultural characteristics, they are grouped into civilizations, which are comprehensive, just as the concept of general systems (Lin, 1999), none of their constituent units (or elements) can be fully understood in isolation without reference to the encompassing whole (civilization). A civilization is a maximal whole or totality in terms of its fundamental values and philosophical assumptions. Civilizations comprehend without being comprehended by others (Toynbee, 1937, pp. 455).

For example, European nations share cultural features that separate them from others such as Chinese or Hindu communities. And, Chinese, Hindus, and Westerners are not part of any broader cultural entity so that they are from different civilizations. Hence, the concept of civilizations stands for the broadest cultural grouping of people and is defined both by common objective elements, such as language, history, religion, customs, institutions, and by the subjective self-identification of people (Huntington, 1996, pp. 43). Intuitively, the civilization we belong to represents the biggest "we" within which we feel culturally at home as separated from all "thems" out there. Some civilizations may contain large numbers of people, such as Chinese civilization, and others very small numbers, such as the Anglophone Caribbean. According to the literature, civilizations do not seem to have clear-cut boundaries, precise endings or beginnings. Throughout history people have redefined their identities and the composition and shape of civilizations have gone through changes over time. Civilizations interact and overlap and are meaningful entities with ambiguous but real dividing line between them.

Civilizations have life of their own. They tend to be long lived by appropriately evolving and adapting to the environment. (For example, virtually all major civilizations of the contemporary world either have existed for a millennium, or as with Latin America, are the immediate offspring of another long-lived civilization.) They represent some of the most enduring forms of human associations and possess long historical continuity, representing the longest story of all (Braudel, 1980, pp. 209–210). (Bozeman, 1992, pp. 26) concludes that "international history rightly documents the thesis that political systems are transient expedients on the surface of civilizations, and that the destiny of each linguistically and morally unified community depends ultimately upon the survival of certain primary structuring ideas around which successive generations have coalesced and which thus symbolize the society's continuity."

When comparing peoples in terms of their values, social relations, customs, and overall outlooks on life, one can readily see significant differences due to their different underlying philosophical assumptions. And the evolution of all peoples from across the world reflects the varied approaches employed for political and economic development. For example, recent economic successes and difficulties in achieving stable democracies in Eastern Asian have their sources in relevant cultures. The difficulty of emerging democracy in most of the Muslim world can be explained through the Islamic culture. The paths of development taken by the post-communist societies in east Europe and the former Soviet Union were determined by their cultural identities. In terms of the hierarchy of social structures, the Western, Japanese, Chinese, Hindu, Muslim, and African civilizations share little common ground in terms of religion, social structure, institutions, and prevailing values. National interests are pervasively defined by domestic values, cultures, and institutions and shaped by international norms and institutions so that states with similar cultures and institutions will see common interest (Huntington, 1996).

In order to understand why different civilizations have different underlying assumptions and values of philosophy, let us first look at Bjerknes' Circulation Theorem (1898) (Hess, 1959). It shows that nonlinearity mathematically stands (mostly) for singularities, and in terms of physics it represents eddy motions. Such motions are a problem of structural evolutions, a natural consequence of uneven evolutions of materials. In particular, at the end of the 19th century, V. Bjerknes discovered the eddy effects due to changes in the density of the media in the movements of the atmosphere and ocean. By a circulation, it is meant to be a closed contour in a fluid. Mathematically, each circulation Γ is defined as the line integral about the contour of the component of the velocity vector locally tangent to the contour. In symbols, if $\vec{V}$ stands for the speed of a moving fluid, S an arbitrary closed curve, $\delta \vec{r}$ the vector difference of

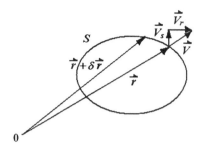

Figure 12.1   The definition of a closed circulation.

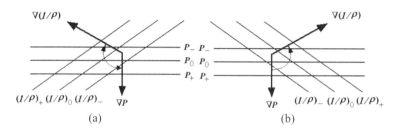

Figure 12.2   A diagram for solenoid circulations.

two neighboring points of the curve S (Figure 12.1), then a circulation Γ is defined as follows:

$$\Gamma = \oint_S \vec{V} \delta \vec{r}. \tag{12.3}$$

Through some ingenious manipulations (Wu and Lin, 2002), the following well-known Bjerknes' Circulation Theorem is obtained:

$$\frac{d\vec{V}}{dt} = \iint_\sigma \nabla\left(\frac{1}{\rho}\right) \times (-\nabla p) \cdot \delta\sigma - 2\Omega\frac{d\sigma}{dt}, \tag{12.4}$$

where σ is the projection area on the equator plane of the area enclosed by the closed curve S, p the atmospheric pressure, ρ the density of the atmosphere, and Ω the earth's rotational angular speed.

The left-hand side of equ. (12.3) represents the acceleration of the moving fluid, which according to Newton's second law of mechanics is equivalent to the force acting on the fluid. On the right-hand side, the first term is called a solenoid term in meteorology. It is originated from the interaction of the p- and ρ-planes due to uneven density ρ so that a twisting force is created. Consequently, materials' movements must be rotations with the rotating direction determined by the equal p- and ρ-plane distributions (Figure 12.2). The second term in equ. (12.4) comes from the rotation of the earth.

This theorem reveals the commonly existing and practically significant eddy effects of fluid motions and implies that uneven eddy motions are the most common form of movements observed in the universe. Because uneven densities create twisting forces, fields of spinning currents are naturally created. Such fields do not have uniformity in terms of types of currents. Clockwise and counter clockwise eddies always co-exist, leading to destructions of the initial smooth, if any, fields of currents. What's important is that the concept of uneven eddy evolutions reveals that forces exist in the structures of evolving objects, and do not exist independently outside of the objects.

Now we are ready to look at why different civilizations have different underlying assumptions and values of philosophy. In particular, at the start of time when still living in the primitive conditions, due to the existing natural conditions and available resources within the environment, people formed their elementary beliefs, basic values, and fundamental philosophical assumptions. Since the population density was low, tools available for production, conquering surroundings, etc. were extremely limited and inefficient, minor obstacles of nature in today's standard easily divided people on the same land into small tribes. Since these individual and separated tribes in really lived in the same natural environment, they naturally held an identical value system and an identical set of philosophical assumptions, on which they reasoned in order to explain whatever inexplicable, developed approaches to overcome hardships, and established methods to administrate members in their individual tribes. As time went on, better tools for production and transportation and better practices of administration were designed and employed in various individual tribes. The natural desire for better living conditions paved the way for the inventions of new tools, discovery of new methods of reasoning, and the introduction of more efficient ways of administration to pass around the land through word of mouth. So, a circulation of people with special skills started to form. Along with these talented people, information, knowledge, and commercial goods were also parts of the circulation. As a circulation started to appear, Bjerknes's Circulation Theorem guarantees the appearance of abstract eddy motions over the land consisting of migration of people, spread of knowledge and information, and transportation of goods. That explains how civilizations are initially formed and why different civilizations have different underlying assumptions and values of philosophy.

At this junction, there is a natural need for us to justify the validity for us to employ the Bjerknes' Circulation Theorem as in the previous paragraph, because in theory this theorem holds true only for fluids. Firstly, in Chapter 1, when the systemic yoyo model is introduced, we have given a relevant explanation for how and why each economic entity is a spinning pool of fluid, consisting of flows of such fluids as energy, information, materials, etc., that circulate within the inside of, go into, and are given off from the entity. Secondly, at the end of the chapter 2, it is concluded that the universe is a huge ocean of eddies, which changes and evolves constantly. That is, the totality of the physically existing world can be studied as fluids. Thirdly, as described in the previous paragraph, people in the land helped to circulate beliefs, basic values, fundamental philosophical assumptions, knowledge, commercial goods, etc., all of which are studied using continuous or differentiable functions in social sciences in general and economics in particular. When these aspects of a civilization are modelled by such functions, they are generally seen in physics and mathematics as flows of fluids and are widely known as flow functions. Specifically, in the formation of a civilization, it is these commonly shared aspects (or fluids) that make the land to have a living culture, where individual persons are simply local "impurities" of the fluids; and each of the "impurities" carries some concentrated amount of "energy", "information", etc.

The previous discussion also indicates how civilizations redefine their identities, composition and shapes throughout history: It was all accomplished through the greater desire of controlling more natural resources that would make one's own yoyo structure more powerful and fighting off different beliefs and value systems that might very well destroy one's own yoyo structure. That is, when each civilization is seen as a spinning yoyo, differences in the natural environments make the abstract but very realistic civilizational yoyos spinning in different ways, such as different spinning speeds, angles, and directions. When two civilizations met either in cooperation or in conflict, in order to sustain themselves viably in the contact, each of them absorbed some elements of the other to benefit themselves individually. To accomplish this without destroying the existing set of values and philosophical assumptions, the receiving civilization had to reformulate the needed elements from the other civilization in its own terms before adding these external elements into their basic system of values and philosophical assumptions. Even though such activities had been carried out throughout the history by various cultures, nations, and civilizations unconsciously, the underlying theoretical guarantee for success is given in the well-known Gödel's theorem below:

Theorem 12.1. (Gödel's Incompleteness Theorem (Hewitt 2008)) For any consistent formal, recursively enumerable theory that proves basic arithmetic truths, an arithmetical statement that is true, but not provable in the theory, can be constructed. That is, any effectively generated theory capable of expressing elementary arithmetic cannot be both consistent and complete.

Here, the word "theory" refers to an infinite set of statements, some of which are taken as true without proof, which are called axioms, and others (the theorems) that are taken as true because they are implied by the axioms. The phrase "provable in the theory" means derivable from the axioms and primitive notions of the theory using standard first-order logic. A theory is consistent if it never proves a contradiction. The phrase "can be constructed" means that some mechanical procedure exists that can construct the statement, given the axioms, primitive notions, and first-order logic. The elementary arithmetic consists merely of additions and multiplication over the natural numbers. The resulting true but not provable statement is often referred to as the Gödel sentence for the theory, although there are infinitely many other statements in the theory that share with the Gödel sentence the property of being true but not provable from the theory.

This yoyo model analysis and discussion lead naturally to explanations on how civilizations are separated from each other and how civilizations remain and survive political, social, economic, even ideological upheavals, although empires rise and fall, and governments come and go. In particular, a good indicator for telling different civilizations apart from each other is the natural environment and geographic conditions in which people live. For example, let us think of some islanders who are cut off from other varieties of environmental conditions by large bodies of water. If throughout history their closest neighboring lands are always occupied by well-formulated civilizations (vigorously spinning yoyos), then the islanders (constituting a small and barely spinning yoyo) have no alternative other than forming their own civilization while periodically absorbing useful and beneficial elements from the neighboring civilizations. And when the neighboring civilizations (as seen as spinning fluids in the dishpan experiment) experience internal chaos (that is when the spinning fluids contain traveling eddy leaves), the small civilizational yoyo of the islanders might have the opportunity to expand temporarily onto the land occupied by the organizationally chaotic people.

As for political, social, economic, even ideological upheavals, they only appear on the surface of civilizations without fundamentally altering the underlying systems of values and philosophical assumptions. They represent some of the eddy leaves periodically appearing in the "whirlpools" of civilizations. So, they are natural phases of the evolution of civilizations. As for the rise and fall of empires, it is like the situation of an ocean of spinning yoyos, the appearance of which is similar to what is shown on the charts of airstreams of the earth as seen from above the North Pole, where one spinning pool, helped by some invisible joint forces from other whirlpools, attempts to conquer many other pools of various sizes by forcing each of them to "spin" in exactly the same way as itself. When the forces that combine internal and external strengths become exhausted, the most likely consequence is that the conquering whirlpool structure is greatly weakened, while the other pools either recover back to their original spinning motion or spins more energetically than before, because their basic sets of values and philosophical assumptions have been just renewed and/or expanded by adopting some of the useful elements from the failed conquering whirlpool. Similar to the situation of political, social, economic, and ideological upheavals, governments are just temporary centers of cultural yoyo structures. As cultures evolve, their centers also move from one "location" to another, similar to the case of the magnetic poles of the earthly yoyo structure.

With all the previous discussions in place, we are now ready to address the question in the title of this subsection: What is China As a civilization? To this end, let us treat each culture and every social organization as an abstract, high dimensional spinning yoyo field. The geographic location of Europe, where the land is mainly surrounded by oceans and has rivers well distributed throughout the land, indicates that the Europeans should be naturally accustomed to open thinking and freedom of movements (or individualism) without much need for difficult negotiation and compromise with anyone. In comparison, China is enclosed by deserts or extreme natural conditions in the north, cold, mountainous, and dry yellow-earth plateau in the west, the Himalayas in the southwest, lush jungles in the south, and open seas in the east. And throughout much of the land, people depend on one of the two river systems, the Yellow River and the Yangtze River, to survive. So, the geographic environment and condition of China more resembles that of the spinning dish in Fultz's dishpan experiment with only the east open to large bodies of water. However, in ancient times without much capability to travel overseas, China was completely situated in a "dish with a solid periphery." So, if we imagine the Chinese civilization as a realization of a high dimensional spinning yoyo (of fluids), then Chinese history is pretty much a social version of the periodic pattern changes observed in the dishpan experiment, where as a nation, it has gone through divisions and unifications alternately. In terms of how China would behave politically and economically in the international arena, this civilizational characteristic – collectivism and China-centered – should be considered very series. That actually

explains why China and Chinese have been known respectively as the Central Kingdom and the people of the Central kingdom, the historical names of that vast land and
the people who live on that land.

12.5.2   How will China revive through adopting
beneficial elements

To revive itself culturally, economically and politically, China has been in recent years
adopting elements of other prosperous economies in order to reconstruct its own civilizational structure. To make this process work, Chinese economic structure has to
be reconstructed with the objects m out of scenarios of Figures 12.3(b) and (c) and
Figures 12.4(a) and (d). For scenarios in Figures 12.3(a) and (d) and Figures 12.4(b)
and (c) no viable sub-eddies can be fruitfully produced. If a new economy, such as
the reviving Chinese economy, modeled as a sub-eddy in these figures, is indeed created in cracks between existing spinning fields, elements of these economies useful
to China might not be naturally adopted by China. They could also be forced on to

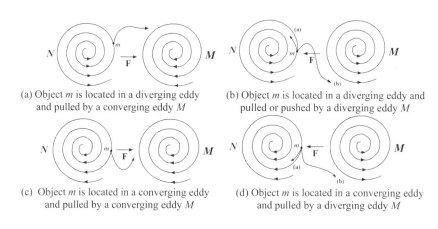

(a) Object m is located in a diverging eddy
and pulled by a converging eddy M

(b) Object m is located in a diverging eddy and
pulled or pushed by a diverging eddy M

(c) Object m is located in a converging eddy
and pulled by a converging eddy M

(d) Object m is located in a converging eddy
and pulled by a diverging eddy M

Figure 12.3 Acting and reacting models with yoyo structures of harmonic spinning patterns.

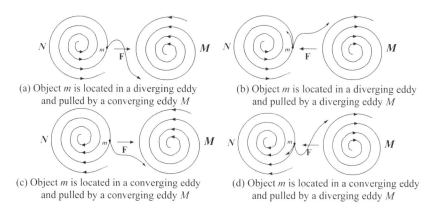

(a) Object m is located in a diverging eddy
and pulled by a converging eddy M

(b) Object m is located in a diverging eddy
and pulled by a diverging eddy M

(c) Object m is located in a converging eddy
and pulled by a converging eddy M

(d) Object m is located in a converging eddy
and pulled by a diverging eddy M

Figure 12.4 Acting and reacting models with yoyo structures of inharmonic spinning patterns.

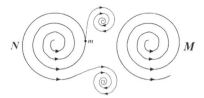

Figure 12.5 Object *m* might be thrown into a sub-eddy created by the spin fields of *N* and *M* jointly.

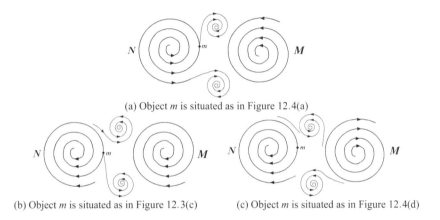

(a) Object *m* is situated as in Figure 12.4(a)

(b) Object *m* is situated as in Figure 12.3(c)   (c) Object *m* is situated as in Figure 12.4(d)

Figure 12.6 Additional ways for new civilizations to be created by the spin fields of *N* and *M* jointly.

China. For example, Japan's forcible opening to the West by Commodore Perry in 1854 vividly shows that after tasting the greatness of the Western civilization by external force or pressure, Japan voluntarily made its dramatic decision to learn from the West following the Meiji Restoration in 1868. Other than the potential scenario for China to revive as shown in Figure 12.5, other possibilities are given in Figure 12.6.

12.5.3 What kind of democracy would China potentially embrace?

Based on the precious discussion, one could expect how and with what levels of transparency decisions are made in China. In other words, will China eventually become as democratic as the nations of the Western civilization? The following discussion shows that the answer to this question is: No, China will not become as democratic as the Western wants.

In particular, to arrive at the previous conclusion, one only needs to ask in terms of scientific belief that the fittest survive why the Western democracy was not originated from Eastern Asian considering its long-lasting civilizational identity?

Other than what was concluded earlier that the political and economic evolution of China closely resembles that of Fultz's dishpan experiment and Chinese history is pretty much a social version of the periodic pattern changes observed in the dishpan experiment, where as a nation, it has gone through divisions and unifications alternately, in terms of social organizations, seen as spinning pools of fluids, China, as any other nation, goes through stages of expansion and contraction alternately and at each transition from expansion to contraction or from contraction to expansion, the social entity goes through a blown-up. That represents a weakest link in the national history. If China could not successfully transform from the current stage of contraction to that of expansion, then China, as an abstract spinning yoyo field, will no longer exist as a viable nation. To support this theory, let us now look at how we can prove mathematically that each two-dimensional eddy current can constitute an alternating transformation between converging and diverging flows.

By introducing a flow function, the equations for two-dimensional fluid dynamics can be written as follows (As it has been argued in different occasions so far in this book, modelling and investigating social organizations as pools of fluids are a valid and legitimate approach. For more details, see the related paragraphs about the systemic yoyo model, and the conclusions of Chapter 2):

$$\begin{cases} (\Delta \psi)_t = J(\Delta \psi, \psi) \\ \Delta \psi(x,t)|_{t=0} = \Delta \psi_0 \end{cases}. \tag{12.5}$$

If in the first equation in equ. (12.5) we add $J(f, \psi)$, $f = 2\Omega \sin \phi$, where ϕ stands for the earthly latitude, then we obtain the general equation of eddy motion for a spinning fluid, such as the atmosphere, where $J(A, B) = \frac{\partial A}{\partial x} \frac{\partial B}{\partial y} - \frac{\partial A}{\partial y} \frac{\partial B}{\partial x}$ is the Jacobian operator. No doubt, $\psi(x, y, t)$ is a two dimensional flow function, $u = -\frac{\partial \psi}{\partial y}, v = \frac{\partial \psi}{\partial x}$; $\Delta \psi = \nabla^2 \psi = \varsigma$ the vorticity in the vertical direction. Equ. (12.5) represents the Cauchy initial value problem of a two-dimensional spinning fluid. It in form is equivalent to the two dimensional equation of fluid mechanics in Euler language. Since this equation comes from the fluid mechanics of continuous media, we have to introduce the assumption that $\psi_0(x, y)$ is continuously differentiable on $-\infty < x, y < +\infty$ and

$$|\psi_0(x, y)| \leq k, \quad k = \text{const.} \tag{12.6}$$

Now, assume that $\psi(x, y, t)$, for $t \in [0, T] (T < \infty)$; $x, y \in (0, L)$ is uniformly continuous. And for the convenience of discussion, let us take

$$\begin{cases} x, y = 0, \Delta \psi(x, 0, t) = \Delta \psi(x, 0, t) = 1 \\ x, y = L, \Delta \psi(L, y, t) = \Delta \psi(x, L, t) = \Delta \psi_L \end{cases} \tag{12.7}$$

and, without loss of generality, let us introduce the separate variables as follows:

$$\begin{aligned} \psi(x, y, t) &= A(t)\Psi(x, y), \\ \psi_0(x, y, 0) &= A(0)\Psi(x, y) \end{aligned} \tag{12.8}$$

where assume that $A(t)$ is positive for $t \in [0, T] (T > 0)$. Substituting equ. (12.8) into equ. (12.5) and applying some simplification produce

$$\Delta \Psi \frac{dA}{dt} = A^2 [\Psi_y (\Delta \Psi)_x - \Psi_x (\Delta \Psi)_y]. \tag{12.9}$$

Evidently, the two terms within [ ] in equ. (12.9) are essentially the same. So, let us take the first term for our discussion. By taking $\frac{dA}{dt} = \dot{A}$ we have

$$\Delta\Psi\dot{A} = A^2\Psi_y(\Delta\Psi)_x. \tag{12.10}$$

By using condition equ. (12.6) and factoring the not expanded variables, let us take

$$\frac{\dot{A}}{A^2} = -\lambda,$$

$$(\Delta\psi)_x + \frac{\lambda}{\Psi_y}\Delta\Psi = 0 \tag{12.11}$$

From equs. (12.11) and (12.8), it follows that

$$A = \frac{A_0}{1 + \lambda A_0 t}. \tag{12.12}$$

When $1 + \lambda A_0 t = 0$ or $t = t_b$, we have

$$\lambda = -\frac{1}{A_0 t_b} \quad \text{or} \quad t_b = -\frac{1}{\lambda A_0}. \tag{12.13}$$

Substituting this expression into equ. (12.12) and then point-multiplying the result by $\Delta\Psi$ produce

$$\Delta\psi = A(t)\Delta\Psi(x,y) = \frac{A_0\Delta\Psi}{1 + \lambda A_0 t} = \frac{\Delta\psi_0}{1 - t/t_b} = \frac{\zeta_0}{1 - t/t_b}, \tag{12.14}$$

where ζ_0 represents the initial vorticity. So, equ. (12.14) uncovers a transformation of the eddy current. When $t \geq t_b$, a blown-up of transitional changes occur. What deserves our attention is that this change is reversible.

If the initial vorticity satisfies $\zeta_0 > 0$, then when $t < t_b$, the movement will be a continuation of the positive vorticity. When $t = t_b$, the movement experience a blown-up. When $t > t_b$, since $(1 - t/t_b) < 0$, the movement changes from the initial $\zeta_0 > 0$ to the vorticity of $\zeta_0 < 0$. If the initial value is a negative vorticity $\zeta_0 < 0$, then the changes will be opposite of what is described above. So, for two dimensional eddy currents, positive and negative rotations can be transformed into each other.

Summarizing the previous discussions, it can be concluded that if China were successful in its current attempt to revive from the ongoing stage of contraction, it would emerge as an economically prosperous, psychologically confident, and militarily powerful nation. It is because other than absorbing beneficial elements from other economies and civilizations, as it has been done currently, Chinese people still live in the particular land that geographically resembles the rotational dish used in Fultz's dishpan experiment. And because due to the modern technology of communication and transportation the land does not have an as solid periphery as the dish used in the experiment, as soon as the economy of the land starts to transform from its current stage of divergence into a convergent field, it will soon become an economic "black hole" that attracts and absorbs human talents, natural resources, among others from the rest of the world.

12.5.4 External pressures China currently experiences

As what is shown in the previous subsection for two-dimensional eddy currents, positive and negative (that is, converging and diverging) rotations are transformed into each other alternately by going through transitional changes (blow-ups). So, in a world that is filled with many co-existing, rotational economic yoyo fields, the rise and fall of a particular economy are simply parts of the natural cycle of the specific region's evolution with the time periods of transitional changes (blown-ups) as the weakest links of the whole evolution, as indicated by the Theorem of Never-Perfect Value Systems (Lin and Forrest, 2010a; Lin and Forrest, 2008b). Also, interactions between economies exist universally (since the spinning fields and meridian fields of the yoyo structures extend infinitely into the space) so that elements of one economy are constantly imposed on other economies and each economy relentlessly experiences novel ideas from other economies that are not provable within their own systems of values and philosophical assumptions.

In particular, when two economies N and M coexist side by side (Figure 12.7), their infinitely expanding eddy fields have to interference with each other so that their action and reaction pressure their would-be circular motions into elliptical motions. So, the practical operations of the basic values and fundamental philosophical assumptions of one economy are greatly affected by those of the other economy as shown in Figure 12.8, where we took out one loop from the spin field of N in Figure 12.7 and

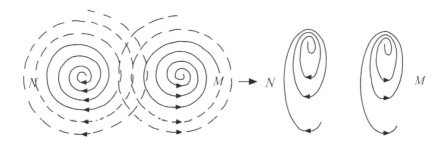

Figure 12.7 Imaginary action and reaction spin fields of our solar system M and a neighboring system N.

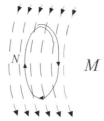

Figure 12.8 The effect of the spin field of M on one loop of N.

take a closer look. The downward motion along the loop of N on the right-hand side is helped by the spin field of M (some novel elements of M that are not provable in N). At the same time, the upward traveling on the left-hand side of the loop of N has to overcome the encountering effects of M's spin field (those M's novel but not provable elements in N). So for the economy N to evolve smoothly (not necessarily for its ultimate survival), it would naturally absorb some of those novel elements from other economies. Similarly, the same also holds true for the economy M.

This discussion also helps us understand why an economy might choose to isolate itself from other economies such as China and Japan in history. In particular, if the spin field N in Figure 12.7 stands for an economy that is naturally separated from all other similar size economies except some minor connections (controllable from within by N) with some of the similar size economies, then to a degree, N would become content with how circular its spin field is (this happens only when this economy is at its peak strength and prosperity), since in this case, the behaviors and the system of values and philosophical assumptions of N would not be placed under any pressure from or scrutiny of other economies as analyzed above. Combining this conclusion with China's and Japan's natural surroundings, when outside forces, which try to act upon these countries, were relatively weak, the rulers of these countries would like to cut off the cultural and economic connections from within the countries with the outside world. By doing so, these rulers could potentially strengthen their control of the respective countries. However, as a consequence of doing so, as suggested by the Theorem of Never-Perfect Value Systems (Lin and Forrest, 2008b), the great and prosperous economy will soon face its dismay by going through a transitional change (blown-up). That will be when the society has to open up itself to accept useful and beneficial elements from other powerful and prospering economies.

12.5.5 Will China be peaceful with neighboring economies?

After seven centuries of serving as the principal arena of great power, conflict, and cooperation, Western Europe is currently peaceful and war is unthinkable. In comparison, Asia is a cauldron of economies developed on very distinct civilizational bases. For example, East Asia alone contains economies of six civilizations: Sinic, Japanese, Orthodox, Buddhist, Muslim, and Western, and the south Asia additionally includes Hinduism. The South Asia adds India. And, Indonesia is a rising Muslim power. And, Asia involves such major nations as China, Japan, Russia, and the United States. So, as China revives itself economically, politically, and militarily, one has to wonder: Will China be peaceful with its neighboring economies, considering Asia's current economic dynamism, territory disputes, resurrected rivalries, and political uncertainties? In other words, will the past of Europe become the future of Asia, as suggested by (Friedberg, 1993/94, pp. 7)?

Based on our yoyo model analysis, the answer to this question is: No; the bloody past of Europe will not become the future of Asia. In particular, due to the natural balance in resources necessary for human survival, Europe is made up of independent nation states that have played the political game of balancing power in the past and will continue to do so in the foreseeable future. It is because the European states are quite even in strength if seen as spinning yoyo fields. On the other hand, the land-scape of East Asia has produced many yoyo structures of different scales, which is evidenced by the good number of different civilizations in Asia, with China sitting right in the biggest "dishpan with a solid periphery" of the world. This end explains why throughout history Chinese international policies have been more about influencing others than conquering as what the West did in the recent history. It is because as soon as the rotational yoyo field that exists in the land of China starts to pick up its spinning strength, all the surrounding much smaller yoyo fields will be naturally sucked inwardly toward it. As for the question of whether or not the current economic dynamism, territory disputes, resurrected rivalries, and political uncertainties will create large scale, bloody conflicts in Asia, the answer drawn on our analysis is NO. It is because as long as the natural yoyo field, which sits right in China, picks up its vigor and strength in a timely fashion, all other smaller civilizations in the region will experience irresistible attraction from this biggest spin field of the world and in history so that instead of fighting against such irresistible pull, the Asian societies will learn from each other to see how they can all benefit from the natural spinning structure they are in.

12.6 Will economic prosperity visit China soon?

During the 8th and the 9th centuries, European Christendom started to emerge as a distinct civilization and lagged behind many other civilizations, such as China, which was respectively in Tang, Song, and Ming dynasties, the Islamic world, and the Byzantium, in its level of civilization and in the following several hundred years in terms of wealth, territory, military power, and artistic, literary, and scientific achievements (Toynbee, vol. VIII, pp. 347–348). During the 11th and 13th centuries, European culture began to develop by absorbing suitable elements from the higher civilizations of Islam and Byzantium together with adapting the inheritance to the special conditions and interests of the West. During the same time period, Western Christianity successfully converted Hungry, Poland, Scandinavia, and the Baltic coast; and Roman law and other aspects of Western civilization were established soon after that time period. During the 12th and 13th centuries, Westerners struggled to expand their control of Spain and did establish effective dominance of the Mediterranean; however, the rise of Turkish power brought about the collapse of Western Europe's first overseas empire (McNeill, 1992, p. 547). By the 1500s, the renaissance of European culture was well underway and social pluralism, expanding commerce, and technological achievements provided the basis for a new era in global politics. Multidirectional encounters among civilizations gave way to the sustained impact of the West on all other civilizations in the following hundred years, during which it generated a significant political ideology without generating a major religion.

Now, natural questions arise as follows: What brings prosperity, such as the renaissance of European culture, to a specific geographic region? Will economic prosperity visit China next? To this end, the First Law on State of Motion indicates that any newly found prosperity of a civilization has to come from interactions with other civilizations. Internal appearances of eddy leaves, as seen in dishpan experiment (Hide, 1953; Fultz, et al., 1959), only mean that periodically local prosperities could be expected within the civilization. To address these two questions successfully, let us look at the following:

The Third Law on State of Motion   (Lin 2007): *When two yoyos N and M act on each other, their interaction falls in one of the six scenarios as shown in Figure 12.9(a)–(c) and Figure 12.10(a)–(c). And, the following are true:*

1   *For the cases in (a) of Figures 12.9–12.10, if both N and M are relatively stable temporarily, then their action and reaction are roughly equal but in opposite directions during the temporary stability. In terms of the whole evolution involved, the divergent spin field (N) exerts more action on the convergent field (M) than M's reaction peacefully in the case of Figure 12.9(a) and violently in the case of Figure 12.10(a).*
2   *For the cases (b) in Figures 12.9–12.10, there are permanent equal, but opposite, actions and reactions with the interaction more violent in the case of Figure 12.9(b) than in the case of Figure 12.10(b).*
3   *For the cases in (c) of Figures 12.9–12.10, there is a permanent mutual attraction. However, for the former case, the violent attraction may pull the two spin fields together and have the tendency to become one spin field. For the later case, the peaceful attraction is balanced off by their opposite spinning directions. And, the spin fields will coexist permanently.*

In all the scenarios discussed in the Third Law on State of Motion, only when a civilization's yoyo structure contains a convergent spinning field, the civilization has the potential to bring about prosperity, because each divergent field constantly loses its assets. In comparison, the convergent civilization M in Figure 12.9(a) has a better chance than the civilization M in Figure 12.10(a) to obtain wealth, because due to opposite spinning directions the spinning yoyo M in Figure 12.10(a) receives assets from N with delay and struggle, while the harmonic spinning pattern in Figure 12.9(a) makes transfers from N to M very easy. The convergent N in Figure 12.10(b) is in a similar situation as the convergent M in Figure 12.9(a). For the scenario in Figure 12.10(c), both civilizations N and M can be prosperous at the same time. However,

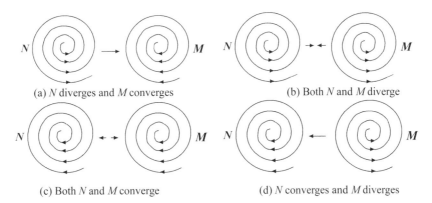

(a) *N diverges and M converges*   (b) *Both N and M diverge*

(c) *Both N and M converge*   (d) *N converges and M diverges*

Figure 12.9 Same scale acting and reacting spinning yoyos of the harmonic pattern.

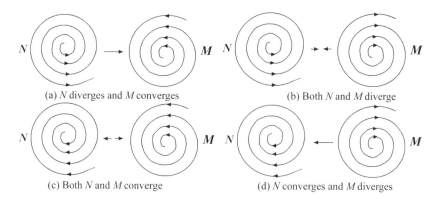

(a) N diverges and M converges   (b) Both N and M diverge

(c) Both N and M converge   (d) N converges and M diverges

Figure 12.10 Same scale acting and reacting spinning yoyos of inharmonic patterns.

due to their opposite spinning directions, they will not be on friendly terms in dealing with global affairs for at least 50% of time. What is interesting is the scenario in Figure 12.9(c), where both of the convergent civilizations N and M have the tendency to combine into a much greater spinning yoyo structure. To this end, one natural problem of practical significance is that when faced with an objective social organization, how can we tell if it is convergent or divergent? There are many ways to accomplish this end. For example, if one organization has a rigorous mechanism in place to prevent its members and/or properties from leaving, then the underlying yoyo structure of the organization is divergent. If one organization has a tighter control than another organization, then the former organization is more divergent than the later.

Now, the story of the West is different from all the scenarios discussed in the previous paragraph, because in history the Western civilization is a new species when compared to other age-old civilizations. To this end, let us look at the following:

The Fourth Law on State of Motion (Lin 2007): *When the spin field M acts on an object m, rotating in the spin field N, the object m experiences equal, but opposite, action and reaction, if it is either thrown out of the spin field N and not accepted by that of M (Figure 12.11(a), (d), Figure 12.12(b) and (c)) or trapped in a sub-eddy motion created jointly by the spin fields of N and M (Figure 12.11(b), (c), Figure 12.12(a) and (d)). In all other possibilities, the object m does not experience equal and opposite action and reaction from N and M.*

Since the Western civilization is not closely attached to any other civilization other than receiving beneficial elements from others, it means that it is an offspring of these lending civilizations, each of which contains a divergent spinning field (Figure 12.13), while the newborn is convergent. So, the answer to our questions is that riding on the strengths of all the surrounding, divergent civilizations, the independent area, existing between these civilizations, that is blessed with rich natural resources, will experience cultural and economic prosperity. It will be more so, if the surrounding civilizations at the same time suffer from internal chaos (as during the time period of having eddy

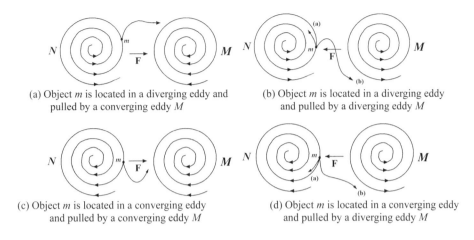

(a) Object m is located in a diverging eddy and pulled by a converging eddy M

(b) Object m is located in a diverging eddy and pulled by a diverging eddy M

(c) Object m is located in a converging eddy and pulled by a converging eddy M

(d) Object m is located in a converging eddy and pulled by a diverging eddy M

Figure 12.11 Acting and reacting models with yoyo structures of harmonic spinning patterns.

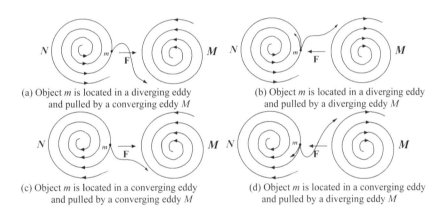

(a) Object m is located in a diverging eddy and pulled by a converging eddy M

(b) Object m is located in a diverging eddy and pulled by a diverging eddy M

(c) Object m is located in a converging eddy and pulled by a converging eddy M

(d) Object m is located in a converging eddy and pulled by a diverging eddy M

Figure 12.12 Acting and reacting models with yoyo structures of inharmonic spinning patterns.

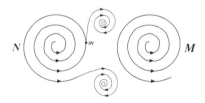

Figure 12.13 Object m might be thrown into a sub-eddy created by the spin fields of N and M jointly.

leaves in the dishpan experiment (Hide, 1953; Fultz, et al., 1959). In particular, in order for economic prosperity to visit China in the near future, China has to transform its currently diverging yoyo field into a convergent field with vigor and strength. To this end, China seems to be doing currently and effectively.

Bibliography

Ando, C. A., and Modigliani, F. (1965). The relative stability of monetary velocity and the investment multiplier. *American Economic Review*, vol. 55, pp. 673–728.

Armson, R. (2011). *Growing Wings On The Way: Systems Thinking for Messy Situations*. Devon, UK: Triarchy Press.

Beechey, M., Gruen, D., and Vickrey, J. (2000). The efficient markets hypothesis: A survey. Reserve Bank of Australia in its series *RBA Research Discussion Papers* numbered rdp2000-01.

Berlinski, D. (1976). *On Systems Analysis*. Cambridge, MA: MIT Press.

Bernanke, B., and Gertler, M. (1999). Monetary policy and asset price volatility. *Economic Review, Federal Reserve Bank of Kansas City*, fourth quarter, pp. 17–51.

Blauberg, I. V., Sadovsky, V. N., and Yudin, E. G. (1977). *Systems Theory, Philosophy and Methodological Problems*. Moscow: Progress Publishers.

Bozeman, A. B. (1975). "Civilizations under stress." *Virginia Quarterly Review* 51 (winter): 1–18.

Bozeman, A. B. (1992). *Strategic Intelligence and Statecraft*. Washington: Brassey's (US).

Brandelik, R. et al. (1979). Evidence for Planar Events in e^+e^- Annihilation at High Energies. *Phys. Lett. B* **86**: 243–249.

Braudel, F. (1980). *On History*. Chicago: University of Chicago Press.

Brunner, K., and Meltzer, A. H. (1968). Liquidity traps for money, bank credit, and interest rates. *Journal of Political Economy*, University of Chicago Press, vol. 76, pp. 1–37.

Campbell, J. Y., Lo, A. W., and McKinley, A. C. (1997). *The Econometrics of Financial Markets*. Princeton, New Jersey: Princeton University Press.

Campbell, J. Y., and Shiller, R. J. (1987). Cointegration and tests of present value models. *Journal of Political Economy*, vol. 95, no. 5, 1062–1088.

Campbell, J. Y., and Shiller, R. J. (1991). Yield spreads and interest rate movement: A bird's eye view. *Review of Economic Studies*, vol. 58, no. 3, pp. 495–514.

Chen, G. R. (2007). *The Original State of the World: The Theory of Ether Whirltrons*. Tianma Books Limited, Hong Kong.

Chenery, H. B., and Brune, (1962). Development alternatives in an open economy: The case of Israel. *Economic Journal*, March, pp. 79–103.

Corsetti, G. (1998). Interpreting the Asian financial crisis: Open issues in theory and policy. *Asian Development Review*, vol. 16, no. 2, pp. 1–47.

Diamond, D. W., and Dybvig, P. H. (1983). Bank runs, deposit insurance, and liquidity. *Journal of Political Economy*, vol. 91, no. 3, pp. 401–419.

Durkheim, E. and M. Mauss (1971). "Note on the notion of civilization." *Social Research* 38: 808–813.

Einstein, A. (1983). *Complete Collection of Albert Einstein*. Trans. by L. Y. Xu. Beijing: Commercial Press.

Einstein, A. (1997). *Collected Papers of Albert Einstein*. Princeton: Princeton University Press.

Eiteman, D. Stonehill, A. I., and Moffett, M. (2001). *Multinational Business Finance* (9th edition). Addison Wesley.

English, J. and G. F. Feng (1972). *Tao De Ching*. New York: Vintage Books.

Fama, E. (1984). The information in the term structure. *Journal of Financial Economics*, vol. 13, pp. 509–528.

Fama, E., and Bliss, R. (1987). The information in long-maturity forward rates. *American Economic Review*, vol. 77, no. 4, pp. 680–692.

Fisher, I., and Brown, H. G. (2010). *The Purchasing Power of Money: Its Determination and Relation to Credit, Interest and Crises*. Charleston, SC: Nabu Press.

Friedberg, A. (1993/94). "Ripe for rivalry: Prospects for peace in multi-polar Asia." *International Security* 18 (Winter): 5–33.

Friedman, M. (1956). The quantity theory of money: A restatement. In: M. Friedman (editor), *Studies in the Quantity Theory of Money, Chicago*: University of Chicago Press.

Friedman, M. (1969). Factors affecting the level of interest rates. In: *Proceeding of the 1968 Conference on Saving and Residential Financing*, Chicago: United States Saving and Loan League, 1969, 11–27. Reprinted in: Thomas M. Havrilesky and John T. Boorman (eds.), *Current Issues in Monetary Theory and Policy*, Arlington Heights, IL: AHM Publishing Corp., 1976, 362–78.

Friedman, M., and Friedman, R. D. (1998). *Two Lucky People*. Chicago, Illinois: University of Chicago Press.

Friedman, M., and Meiselman, D. (1963). The relative stability of monetary velocity and the investment multiplier in the United States, 1898–1958. In: Commission on Money and Credit, *Stabilization Policies*, Englewood Cliffs, NJ: Prentice-Hall.

Friedman, M., and Schwartz, A. J. (1963). Money and business cycles. *Review of Economics and Statistics*, vol. 45, suppl., pp. 32–64.

Friedman, M., and Schwartz, A. J. (1971). *A Monetary History of the United States, 1867–1960*. Princeton, NJ.: Princeton University Press.

Friedman, M., and Schwartz, A. J. (1993). *A Monetary History of the United States, 1867–1960*. Princeton: Princeton University Press.

Fultz, D., R. R. Long, G. V. Owens, W. Bohan, R. Kaylor and J. Weil (1959). Studies of thermal convection in a rotating cylinder with some implications for large-scale atmospheric motion. *Meteorol. Monographs* (American Meteorological Society) vol. 21, no. 4.

Gillan, S. 2006. Recent developments in corporate governance: An overview. *Journal of Corporate Finance*, 12, 381–402.

Gleick, J. (1987). *Chaos: Making a New Science*. New York: Viking.

Griffiths, D. J. (2004). *Introduction to Quantum Mechanics* (2nd edition). Prentice Hall, New Jersey.

Hall, A. D. and Fagen, R. E. (1956). Definitions of systems. *General Systems*, vol. 1, pp. 18–28.

Havel, V. (1995). "Civilization's thin veneer." *Harvard Magazine* 97 (July–August): 32.

Hayek, F. A., (1966). *Monetary Theory and the Trade Cycle*. New York: Augustus M. Kelley Publications.

Hendrix, H. (2001). *Getting the Love You Want: A Guide for Couples*. New York: Owl Books.

Hess, S. L. (1959). *Introduction to Theoretical Meteorology*. New York: Holt, Rinehart and Winston.

Hewitt, C. (2008). Large-scale Organizational Computing requires Unstratified Reflection and Strong Paraconsistency. In: *Coordination, Organizations, Institutions, and Norms in Agent Systems III*, edited by J. Sichman, P. Noriega, J. Padget and S. Ossowski, Berlin: Springer-Verlag.

Hicks, J. R. (1937). Mr. Keynes and the classics – A suggested interpretation. *Econometrica*, vol. 5 (April), pp. 147–159.

Hide, R. (1953). Some experiments on thermal convection in a rotating liquid. *Quart. J. Roy. Meteorol. Soc.* 79: 161.

Huang, C. F., and Litzenberger, R. H. (1988). *Foundations for Financial Economics*. Amsterdam: North-Holland.

Huntington, S. P. (1996). *The Clash of Civilizations and the Remaking of World Order*. New York: Simon & Schuster.

Hutchison, M. M. (1986). Japan's "money focused" monetary policy. *Economic Review* (Federal Reserve Bank of San Francisco), Summer, no. 3, pp. 33–46.

Jiang, B. K. (2008). *International Finance* (4th edition). Shanghai: Fudan University Press.

Johnson, B., and Echeverria, C. (1999). Sequencing capital account liberation: lessons from the experiences in Chili, Indonesia, Korea, and Thailand. *IMF Working Paper*.

Johnson, R., Darbar, S., and Echeverria, C. (1997). Sequencing capital account liberalization: lessons from experiences in Chile, Indonesia, Korea, and Thailand. Working Paper, WP/97/157, *International Monetary Fund*, Washington, D.C.

Kaminsky, G., Lizondo, S., and Reinhart, C. M. (1998). Leading indicators of currency crises. *IMF Staff Papers, Palgrave Macmillan*, vol. 45, no. 1, pp. 1–48, March.

Keynes, J. M. (1936). *The General Theory of Employment, Interest and Money*. Hampshire, England: Palgrave Macmillan.

Kissinger, H. A. (1994). *Diplomacy*. New York: Simon & Schuster.

Kline, M. (1972). *Mathematical Thought from Ancient to Modern Times*. Oxford: Oxford University Press.

Klir, G. (1985). *Architecture of Systems Problem Solving*. New York, NY: Plenum Press.

Klir, G. (2001). *Facets of Systems Science*. New York: Springer.

Krugman, P. (1979). A model of balance-of-payments crises. *Journal of Money, Credit, and Banking*, vol. 11, pp. 311–325.

Krugman, P. (1979a). A model of balance-of-payments crises. *Journal of Money, Credit, and Banking*, vol. 11, pp. 311–25.

Krugman, P. (1997). *Pop Internationalism*. Cambridge: MA: The MIT Press.

Krugman, P. R., and Obstfeld, M. (1997). *International Economics: Theory and Policy*. Addison Wesley.

Krugman, P. (1999). Balance sheets, the transfer problem, and financial crises. International Tax and Public Finance, vol. 6, pp. 459–472.

Kunen, K. (1983). *Set Theory: An Introduction to Independence proofs*. Amsterdam: Elsevier Science.

Li, Z. Y. (2005). *Western Economics*. Beijing: Press of Higher Education.

Lilienfeld, D. (1978). *The Rise of Systems Theory*. New York: Wiley.

Lin, Y. (1987). A model of general systems. *Mathematical Modeling: An International Journal* 9: 95–104.

Lin, Y. (1988). Can the World be Studied in the Viewpoint of Systems? *Mathl. Comput. Modeling*, Vol. 11, pp. 738–742.

Lin, Y. (1988b). An application of systems analysis in sociology. *Cybernetics and Systems: An International Journal*, vol. 19, pp. 267–278.

Lin, Y. (1989). Multi-level systems. *International Journal of Systems Science* 20: 1875–1889.

Lin, Y. (1990). The concept of fuzzy systems. *Kybernetes: The International Journal of Systems and Cybernetics* 19: 45–51.

Lin, Y. (1991). More results on decomposition and decoupling of single relation systems. *Cybernetics and Systems: An International Journal*, vol. 22, pp. 265–281.

Lin, Y. (1994). Feedback transformation and its application. *Journal of Systems Engineering*, vol. 1, pp. 32–38.

Lin, Y. (1995). Developing a theoretical foundation for the laws of conservation. *Kybernetes: The International Journal of Systems and Cybernetics* 24: 52–60.

Lin, Y. (1998). Discontiunity: a weakness of calculus and beginning of a new era. *Kybernetes: The International Journal of Systems and Cybernetics* 27: 614–618.

Lin, Yi (guest editor) (1998). Mystery of Nonlinearity and Lorenz's Chaos. A special double issue, *Kybernetes: The International Journal of Cybernetics, Systems and Management Science*, vol. 27, nos. 6–7, pp. 605–854.

Lin, Y. (1999). *General Systems Theory: A Mathematical Approach*. New York: Plenum and Kluwer Academic Publishers.

Lin, Y. (2007). Systemic yoyo model and applications in Newton's, Kepler's laws, etc. *Kybernetes: The International Journal of Cybernetics, Systems and Management Science*, vol. 36, no. 3–4, pp. 484–516.

Lin, Y. (2008). *Systemic Yoyos: Some Impacts of the Second Dimension*. New York: Auerbach Publications, am imprint of Taylor and Francis.

Lin, Y. (guest editor) (2008). Systematic Studies: The Infinity Problem in Modern Mathematics. *Kybernetes: The International Journal of Systems, Cybernetics, and Management Science*, vol. 37, no. 3–4, pp. 387–578.

Lin, Y. (2009). *Systemic Yoyos: Some Impacts of the Second Dimension*. An Auerbach Book, New York: CRC Press.

Lin, Y. (guest editor) (2010). Research Studies: Systemic Yoyos and Their Applications. *Kybernetes: The International Journal of Systems, Cybernetics, and Management Science*, vol. 39, no. 2, pp. 174–378.

Lin, Y., and Ma, Y. H. (1990). General feedback systems. *International Journal of General Systems*, vol. 18, no. 2, pp. 143–154.

Lin, Y., and Forrest, B. (2010a). The state of a civilization. *Kybernetes: The International Journal of Cybernetics, Systems and Management Sciences*, vol. 39, no. 2, pp. 343–356.

Lin, Y., and Forrest, B. (2010b). The life form of civilizations. *Kybernetes: The International Journal of Cybernetics, Systems and Management Science*, vol. 39, no. 2, pp. 357–366.

Lin, Y., and Forrest, B. (2010c). Interaction between civilizations. *Kybernetes: The International Journal of Cybernetics, Systems and Management Science*, vol. 39, no. 2, pp. 367–378.

Lin, Y., and Forrest, B. (2011). *Systemic Structure behind Human Organizations: From Civilizations to Individuals*. New York: Springer.

Lin, Y., and Forrest, D. (2008). Economic yoyos and Becker's Rotten Kid Theorem. *Kybernetes: The International Journal of Systems and Cybernetics*, vol. 37, no. 2, pp. 297–314.

Lin, Yi and Forrest, D. (2008b). Economic yoyos and never-perfect value system. *Kybernetes: The International Journal of Cybernetics and Systems*, vol. 37, no. 1, pp. 149–165.

Lin, Y. and T. H. Fan (1997). The fundamental structure of general systems and its relation to knowability of the physical world. *Kybernetes: The International Journal of Systems and Cybernetics* 26: 275–285.

Lin, Y., and Hopkins, Z. (to appear). One potential way currency wars can be launched and one possible method of self-defense. Submitted for publication.

Lin, Y., Ma, Y., and Port, R. (1990). Several epistemological problems related to the concept of systems. *Math. Comput. Modeling*, vol. 14, pp. 52–57.

Lin, Y., and OuYang, S. C. (2010). *Irregularities and Prediction of Major Disasters*. New York: CRC Press, an imprint of Taylor and Francis.

Lorenz, E. N. (1993). *The Essence of Chaos*. Seattle: Washington University Press.

Lucas, R. (1972). Expectations and the neutrality of money. *Journal of Economic Theory*, vol. 4, no. 2, pp. 103–124.

MacDougall, D. G. A. (1960). The benefits and costs of private investment from abroad: A theoretical approach. *Economic Records*, vol. 36, pp. 13–35.

MacFarquhar, R., and Schoenhals, M. (2008). *Mao's Last Revolution*. Cambridge, MA: Harvard University Press.

Mankiw, N. G. (2007). *Macroeconomics* (6th ed.). New York: Worth Publishers.

Marx, K. (2008). *Das Kapital*. Boston: Mobile Reference.

Mathematical sciences: A unifying and dynamic resource. 1985. *Notices Am. Math. Soc.,* vol. 33, pp. 463–479.

Mckinnon, R. I. (1989). Macroeconomic instability and moral hazard in banking in a liberalizing economy. In: P. Brock, M. Connolly, and C. Gonzalez-Vega (eds.), *Latin American Debt and Adjustment*, New York: Praeger, 1989.

Mckinnon, R. I., and Pill, H. (1999). Exchange rate regimes for emerging markets: Moral hazard and international overborrowing. *The Oxford Review of Economic Policy*, vol. 15, no. 3, (Autumn), pp. 19–38.

McNeill, W. H. (1992). *Rise of the West: A History of Human Community*. University of Chicago Press, Chicago.

Mesarovic, M. D., and Takahara, Y. (1975). *General Systems Theory: Mathematical Foundations*. New York: Academic Press.

Mesarovic, M. D., and Takahara, Y. (1989). *Abstract Systems Theory*. Berlin: Springer-Verlag.

Mishkin, F. S. (1995). *The Economics of Money, Banking and Financial Markets* (4th edition). New York: Happer-Collins College Publishers.

Mishkin, F. S. (2007). *The Economics of Money, Banking, and Financial Markets (Alternate Edition)*. Boston: Addison Wesley.

Modigliani, F., and Brumberg, R. H. (1954). Utility analysis and the consumption function: an interpretation of cross-section data. In: Kenneth K. Kurihara, ed., *Post-Keynesian Economics*, New Brunswick, NJ.: Rutgers University Press, pp. 388–436.

Obstfeld, M. (1986). Rational and self-fulfilling balance-of-payments crises. *American Economic Review*, vol. 76, no. 1, pp. 72–81.

Obstfeld, M., and Rogoff, K. S. (1996). *Foundations of International Macroeconomics*. Boston: MIT Press.

O'Hara, M. (1998). *Market Microstructure Theory*. New York: Wiley.

Okina, K., Shirakawa, M., and Shiratsuka, S. (2001). The asset price bubble and monetary policy: Japan's experience in the late 1980s and the lessons. *Monetary and Economic Studies* (Special Edition), 2001, February, pp. 395–450.

OuYang, S. C. (1994). *Break-Offs of Moving Fluids and Several Problems of Weather Forecasting*. Chengdu: Press of Chengdu University of Science and Technology.

OuYang, S. C. (1998). *Weather evolutions and Structural Forecasting* (in Chinese). Beijing: Meteorological Press.

OuYang, S. C., and Chen, G. Y. (2006). End of stochastics and quantitative comparability. *Scientific Research Monthly* 14: 141–143.

OuYang, S. C., Chen, Y. G., and Lin, Y. (2009). *Digitization of Information and Prediction*. China Meteorological press, Beijing.

OuYang, S. C., McNeil, D. H., and Lin, Y. (2002). *Entering the Era of Irregularity* (in Chinese). Beijing: Meteorological Press.

OuYang, S. C., Miao, J. H., Y. Wu, Y. Lin, T. Y. Peng and T. G. Xiao (2000). Second stir and incompleteness of quantitative analysis. *Kybernetes: The International Journal of Cybernetics, Systems and Management Science*, vol. 29, pp. 53–70.

Phillips, A. W. (1958). The relationship between unemployment and the rate of change of money wages in the United Kingdom 1861–1957. *Economica*, vol. 25, no. 100, pp. 283–299.

Pigou, A. C. (1917). The value of money. *Quarterly Journal of Economics*, vol. 32, no. 1, pp. 38–65.

Poole, W., and Kornblith, E. B. F, (1973). The Friedman-Meiselman CMC paper: New evidence on an old controversy. *American Economic Review*, vol. 63, no. 5, pp. 908–917.

Radelet, S., and Sachs, J. (1998). The East Asian financial crisis: Diagnosis, remedies, prospects. *Brookings Papers on Economic Activity*, I, pp. 1–90.

Ren, Z. Q. (1996). The resolution of the difficulties in the research of earth science. In: F. H. Wu and X. J. He (eds.), *Earth Science and Development*, Beijing: Earthquake Press, pp. 298–304.

Ren, Z. Q., Lin, Y., and OuYang, S. C. (1998). Conjecture on law of conservation of informational infrastructures. *Kybernetes Int. J. Syst. Cybernetics*, vol. 27, pp. 543–52.

Ren, Z. Q. and Nio, T. (1994). Discussions on several problems of atmospheric vertical motion equations. *Plateau Meteorology* 13: 102–105.

Saito, T. (1986). Feedback transformation of linear systems. *International Journal of Systems Science*, vol. 17, no. 9, pp. 1305–1315.

Saito, T., and Mesarovic, M. D. (1985). A meaning of the decoupling by feedback of linear functional time system. *International Journal of General Systems*, vol. 11, no. 1, pp. 47–61.

Salvatore, D. (1998). Capital flows, current account deficits, and financial crises in emerging market economies. *International Trade Journal*, Spring, pp. 5–22.

Salvatore, D. (1999). Could the financial crisis in Asia have been predicted? *Journal of Policy Modeling*, May, pp. 341–348.

Sargent, T. J. (1971). A note on the accelerationist controversy. *Journal of Money, Credit and Banking* (Blackwell Publishing), vol. 3, no. 3, pp. 721–25.

Smith, V. L., Suchanek, G. L., and Williams, A. W. (1988). Bubbles, crashes, and endogenous expectations in experimental spot asset markets. *Econometrica*, vol. 56, no. 5, pp. 1119–1151.

Tobin, J. (1956). The interest-elasticity of transactions demand for cash. *Review of Economics and Statistics*, vol. 38, no. 3, pp. 241–247.

Tobin, J. (1969). A general equilibrium approach to monetary theory. *Journal of Money Credit and Banking*, vol. 1, no. 1, pp. 15–29.

Toynbee, A. (1934–1961). *A Study of History*. 12 volumes, Oxford University Press, Oxford.

Triffin, R. (1966). *Gold and the Dollar Crisis: the Future of Convertibility* (Revised edition). New Haven, CT: Yale University Press.

Triffin, R. (1972). The International Monetary System of the Year 2000. In: *Economics and World Order, from the 1970's to the 1990's*, edited by Jakdish N. Bhagwati, London, Macmillan, pp. 183–197.

von Bertalanffy, L. (1924). Einfuhrung in Spengler's werk. *Literaturblatt Kolnische Zeitung*, May.

Wallerstein, I. M. (1991). *Geopolitics and Geoculture: Essays on the Changing World-System*. Cambridge University Press, Cambridge.

Wang, R. X., and Hu, G. H. (2005). *International Finance*. Wuhan, Hubei: Wuhan University of Science and Technology.

Wilhalm, R. and C. Baynes (1967). *The I Ching or Book of Changes* (3rd edition). Princeton University Press, Princeton, NJ.

Williamson, J. H. (1996). The Crawling Band as an Exchange Rate Regime: Lessons from Chile, Colombia, and Israel. *Washington: Institute for International Economics*.

Wu, K. P. (2006). *Money and Banking* (2nd edition). Beijing: Tsinghua University Press.

Wu, X. M. (1990). *The Pansystems View of the World*. People's University of China Press, Beijing.

Wu, Y., and Lin, Y. (2002). *Beyond Nonstructural Quantitative Analysis: Blown-ups, Spinning Currents and Modern Science*. River Edge NJ: World Scientific.

Ye, Z. X., and Lin J. Z. (1998). *Mathematical Finance*. Beijing: Science Press.

Zhu, Y. Z. (1985). *Albert Einstein: The Great Explorer*. Beijing: Beijing People's Press.

Zhu, X. D., and Wu, X. M. (1987). An approach on fixed pansystems theorems: Panchaos and strange panattractors. *Applied Mathematics and Mechanics*, vol. 8, no. 4, pp. 339–344.

Subject index

Communications in Cybernetics, Systems Science and Engineering

Book Series Editor: Jeffrey 'Yi-Lin' Forrest

ISSN: 2164-9693

Publisher: CRC Press/Balkema, Taylor & Francis Group

T - #0119 - 191219 - C0 - 246/174/31 - PB - 9780367378790